T0200570

AIRCRAFT STRUCTURES

David J. Peery

Dover Publications
Garden City, New York

Bibliographical Note

This Dover edition, first published in 2011, is an unabridged republication of the work originally published in 1950 by the McGraw-Hill Book Company, Inc., New York.

Library of Congress Cataloging-in-Publication Data

Peery, David J.
    Aircraft structures / David J. Peery.
        p. cm.
    Includes bibliographical references and index.
    ISBN-13: 978-0-486-48580-5 (pbk.)
    ISBN-10: 0-486-48580-3 (pbk.)
        1. Airframes. I. Title.

TL671.6.P4 2011
629.134'31—dc23

2011019335

Printed in the United States of America
48580313
www.doverpublications.com

PREFACE

The subject of aircraft structures includes a wide range of topics. Most of the classical methods of analysis for heavy structures must be considered as well as methods which have particular application to lightweight structures. It is obviously impossible to treat all phases of the analysis and design of aircraft structures in a single book. In this book, which is written as an undergraduate college text, an attempt is made to emphasize basic structural theory which will not change as new materials and new construction methods are developed. Most of the theory is applicable for any design requirements and for any materials. The design engineer may then supplement this theory with the detail design specifications and the material properties which are applicable to his particular airplane.

It is believed that most of the serious mistakes made by college students and by practicing engineers result from errors in applying the simple equations of statics. Heavy emphasis is therefore placed on the application of the elementary principles of mechanics to the analysis of aircraft structures. The problems of deflections and statically indeterminate structures are treated in later chapters. The topics of airload distribution and flight loading conditions require a knowledge of aerodynamics. These topics are therefore placed in the latter part of the book because students usually study the prerequisite aerodynamics topics in concurrent courses.

The author appreciates the suggestions of Dr. Joseph Marin, Dr. Alexander Klemin, Prof. Raymond Schneyer, and George Cohen, who read and criticized the first part of the manuscript; Dr. P. E. Hemke, Dr. A. A. Brielmaier, and C. E. Duke also made valuable suggestions and criticisms.

DAVID J. PEERY

STATE COLLEGE, PA.
August, 1949

CONTENTS

CHAPTER 1

EQUILIBRIUM OF FORCES

1.1. Equations of Equilibrium. One of the first steps in the design of a machine or structure is the determination of the loads acting on each member. The loads acting on an airplane may occur in various landing or flight conditions. The loads may be produced by ground reactions on the wheels, by aerodynamic forces on the wings and other surfaces, or by forces exerted on the propeller. The loads are resisted by the weight or inertia of the various parts of the airplane. Several loading conditions must be considered, and each member must be designed for the combination of conditions which produces the highest stress in the member. For practically all members of the airplane structure the maximum loads occur when the airplane is in an accelerated flight or landing condition and the external loads are not in equilibrium. If, however, the inertia loads are also considered, they will form a system of forces which are in equilibrium with the external loads. In the design of any member it is necessary to find all the forces acting on the member, including inertia forces. Where these forces are in the same plane, as is often the case, the following equations of static equilibrium apply to any isolated portion of the structure:

$$\left. \begin{array}{l} \Sigma F_x = 0 \\ \Sigma F_y = 0 \\ \Sigma M = 0 \end{array} \right\} \tag{1.1}$$

The terms ΣF_x and ΣF_y represent the summations of the components of forces along x and y axes, which may be taken in any two arbitrary directions. The term ΣM represents the sum of the moments of all forces about any arbitrarily chosen point in the plane. Each of these equations may be set up in an infinite number of ways for any problem, since the directions of the axes and the center of moments may be chosen arbitrarily. Only three independent equations exist for any free body, however, and only three unknown forces may be found from the equations. If, for example, an attempt is made to find four unknown forces by using the two force equations and moment equations about two points, the four equations cannot be solved because they are not independent, *i.e.*, one of the equations can be derived from the other three.

1

The following equations cannot be solved for the numerical values of the three unknowns because they are not independent.

$$x + y + z = 3$$
$$x + y + 2z = 4$$
$$2x + 2y + 3z = 7$$

The third equation may be obtained by adding the first two equations, and consequently does not represent an independent condition.

In the analysis of a structure containing several members it is necessary to draw a free-body diagram for each member, showing all the forces acting on that member. It is not possible to show these forces on a composite sketch of the entire structure, since equal and opposite forces act at all joints and an attempt to designate the correct direction of the force on each member will be confusing. In applying the equations of statics it is desirable to choose the axes and centers of moments so that only one unknown appears in each equation.

Many structural joints are made with a single bolt or pin. Such joints are assumed to have no resistance to rotation. The force at such a joint must pass through the center of the pin, as shown in Fig. 1.1, since the moment about the center of the pin must be zero. The force at the pin joint has two unknown quantities, the magnitude F and the direction θ. It is usually more convenient to find the two unknown components, F_x and F_y, from which F and θ can be found by the equations:

$$F = \sqrt{F_x^2 + F_y^2} \tag{1.2}$$

$$\tan \theta = \frac{F_y}{F_x} \tag{1.3}$$

The statics problem is considered as solved when the components F_x and F_y at each joint are obtained.

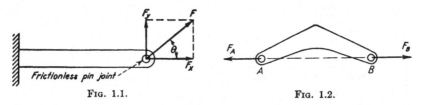

FIG. 1.1.   FIG. 1.2.

1.2. Two-force Members. When a structural member has forces acting at only two points, these forces must be equal and opposite, as shown in Fig. 1.2. Since moments about point A must be zero, the force F_B must pass through point A. Similarly the force F_A must pass through point B for moments about point B to be zero. From a summation of forces, the forces F_B and F_A must have equal magnitudes but opposite directions. Two-force members are frequently used in aircraft and other structures, since simple tension or compression members are usually the lightest members for transmitting forces. Where possible, two-force members are straight, rather than curved as shown in Fig. 1.2. Structures made up entirely of two-force members are called trusses and are frequently used in fuselages, engine mounts, and other aircraft structures, as well as in bridge and building structures. Trusses represent an important special type of structure and will be treated in detail in the following articles.

Many structures contain some two-force members, as well as some members which resist more than two forces. These structures must first be examined carefully in order to determine which members are two-force members. Students frequently make the serious mistake of assuming that forces act in the direction of a member, when the member resists forces at three or more points. In the case of curved two-force members such as shown in Fig. 1.2, it is important to assume that the forces act along the line between pins, rather than along the axis of the members.

While Eqs. 1.1 are simple and well known, it is very important for a student to acquire proficiency in the application of these equations to various types of structures. A typical structure will be analyzed as an example problem.

Example. Find the forces acting at all joints of the structure shown in Fig. 1.3.

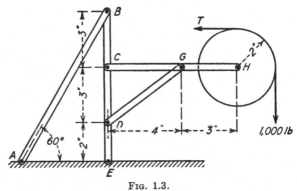

Fig. 1.3.

Solution. First draw free-body diagrams of all members, as shown in Fig. 1.4. Since AB and GD are two-force members, the forces in these members are along the line joining the pin joints of these members, and free-body diagrams of these members are not shown. The directions of forces are assumed, but care must be taken to show the forces at any joint in opposite directions on the two members, that is, C_z is assumed to act to the right on the horizontal member, and therefore must act to the left on the vertical member. If forces are assumed in the wrong direction, the calculated magnitudes will be negative.

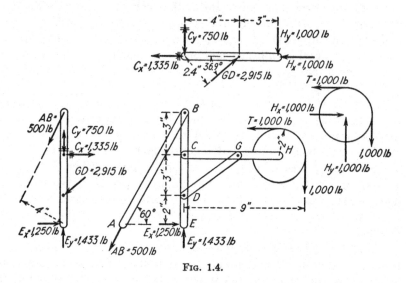

Fig. 1.4.

For the pulley,

$$\Sigma M_H = 2 \times 1,000 - 2T = 0$$
$$T = 1,000 \text{ lb}$$
$$\Sigma F_y = H_y - 1,000 = 0$$
$$H_y = 1,000 \text{ lb}$$
$$\Sigma F_x = H_x - 1,000 = 0$$
$$H_x = 1,000 \text{ lb}$$

Once the numerical values of these forces are obtained, they are shown on the free-body diagram. Subsequent equations contain the known numerical values rather than the algebraic symbols for the forces.

For member CGH,

$$\Sigma M_C = 1,000 \times 7 - 2.4GD = 0$$
$$GD = 2,915 \text{ lb}$$
$$\Sigma F_x = C_x + 2,915 \cos 36.9° - 1,000 = 0$$
$$C_x = -1,335 \text{ lb}$$
$$\Sigma F_y = C_y + 2,915 \sin 36.9° - 1,000 = 0$$
$$C_y = -750 \text{ lb}$$

Since C_x and C_y are negative, the directions of the vectors on the free-body diagrams are changed. Such changes are made by crossing out the original arrows rather than by erasing, in order that the analysis may be checked conveniently by the original designer or by others. Extreme care must be observed in using the proper direction of known forces.

For member $BCDE$,

$$\Sigma M_E = 1{,}335 \times 5 - 2{,}915 \times 2 \cos 36.9° - 4.0AB = 0$$
$$AB = 500 \text{ lb}$$
$$\Sigma F_x = E_x - 2{,}915 \cos 36.9° + 1{,}335 - 500 \cos 60° = 0$$
$$E_x = 1{,}250 \text{ lb}$$
$$\Sigma F_y = E_y - 2{,}915 \sin 36.9° + 750 - 500 \sin 60° = 0$$
$$E_y = 1{,}433 \text{ lb}$$

All forces have now been obtained without the use of the entire structure as a free body. The solution will be checked by using all three equations of equilibrium for the entire structure.

Check using entire structure as free body,

$$\Sigma F_x = 1{,}250 - 1{,}000 - 500 \cos 60° = 0$$
$$\Sigma F_y = 1{,}433 - 1{,}000 - 500 \sin 60° = 0$$
$$\Sigma M_E = 1{,}000 \times 9 - 1{,}000 \times 7 - 500 \times 4 = 0$$

This check should be made wherever possible in order to detect errors in computing moment arms or forces.

PROBLEMS

1.1. A 5,000-lb airplane is in a steady glide with the flight path at an angle θ below the horizontal. The drag force in the direction of the flight path is 750 lb. Find the lift force L normal to the flight path and the angle θ.

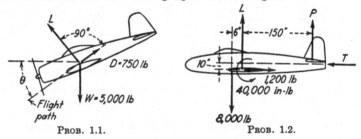

PROB. 1.1.   PROB. 1.2.

1.2. A jet-propelled airplane in steady flight has forces acting as shown. Find the jet thrust T, lift L, and tail load P.

1.3. A wind-tunnel model of an airplane wing is suspended as shown. Find the loads in members B, C, and E if the forces at A are $L = 43.8$ lb, $D = 3.42$ lb, and $M - -20.6$ in-lb.

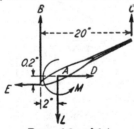

PROB. 1.3 and 1.4.

1.4. For the model of Prob. 1.3 find the forces L, D, and M at a point A, if the measured forces are $B = 40.2$ lb, $C = 4.16$ lb, and $E = 3.74$ lb.

1.5. Find the horizontal and vertical components of the forces at all joints. The reaction at point *B* is vertical.   Check results by using the three remaining equations of statics.

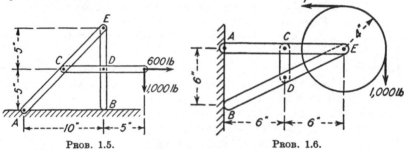

PROB. 1.5.   PROB. 1.6.

1.6. Find the horizontal and vertical components of the forces at all joints. The reaction at point *B* is horizontal.   Check results by three equations.

1.7. Find the horizontal and vertical components of the forces at all joints. The reaction at point *B* is vertical.   Check results by three equations.

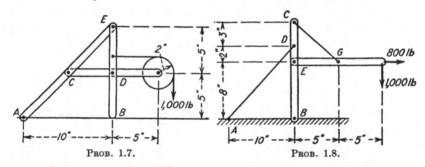

PROB. 1.7.   PROB. 1.8.

1.8. Find the forces at all joints of the structure.   Check results by three equations.

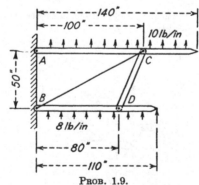

PROB. 1.9.

1.9. Find the forces on all members of the biplane structure shown.   Check results by considering equilibrium of entire structure as a free body.

1.10. Find the forces at points A and B of the landing gear shown.

1.11. Find the forces at points A, B, and C of the structure of the braced-wing monoplane shown.

1.12. Find the forces V and M at the cut cross section of the beam.

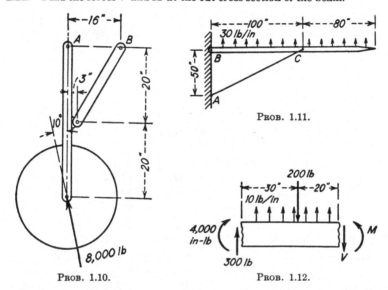

PROB. 1.11.

PROB. 1.10.

PROB. 1.12.

1.3. Truss Structures.

A truss has been defined as a structure which is composed entirely of two-force members. In some cases the members have a single bolt or pin connection at each end, and the external loads are applied only at the pin joints. In other cases the members are welded or riveted at the ends, but are assumed to be pin-connected in the analysis because it has been found that such an analysis yields approximately the correct values for the forces in the members. The trusses considered in this chapter are assumed to be coplanar; the loads resisted by the truss and the axes of all of the truss members lie in the same plane.

Trusses may be classified as statically determinate and statically indeterminate. The forces in all of the members of a statically determinate truss may be obtained from the equations of statics. In a statically indeterminate truss, there are more unknown forces than the number of independent equations of statics, and the forces cannot be determined from the equations of statics. If a rigid structure is supported in such a manner that three nonparallel, nonconcurrent reaction components are developed, the three reaction forces may be obtained from the three equations of statics for the entire structure as a free body. If more than three support reactions are developed, the structure is statically indeterminate externally.

Trusses which have only three reaction force components, but which contain more members than required, are statically indeterminate internally. Trusses are normally formed of a series of triangular frames. The first triangle contains three members and three joints. Additional triangles are each formed by adding two members and one joint. The number of members m has the following relationship to the number of joints j.

$$m - 3 = 2(j - 3)$$

or

$$m = 2j - 3 \tag{1.4}$$

If a truss has one less member than the number specified by Eq. 1.4, it becomes a linkage or mechanism, with one degree of freedom. A linkage is not capable of resisting loads, and is classified as an unstable structure. If a truss has one more member than the number specified by Eq. 1.4, it is statically indeterminate internally.

If each pin joint of a truss is considered as a free body, the two statics equations $\Sigma F_x = 0$ and $\Sigma F_y = 0$ may be applied. The equation $\Sigma M = 0$ does not apply, since all forces act through the pin, and the moments about the pin will be zero regardless of the magnitudes of the forces. Thus, for a truss with j joints, there are $2j$ independent equations of statics. The equations for the equilibrium of the entire structure are not independent of the equations for the joints, since they can be derived from the equations of equilibrium for the joints. For example, the equation $\Sigma F_x = 0$ for the entire structure may be obtained by adding all the equations $\Sigma F_x = 0$ for the individual joints. The equations $\Sigma F_y = 0$ and $\Sigma M = 0$ for the entire truss may similarly be obtained from the equations for the joints. Equation 1.4 may therefore be derived in another manner by equating the number of unknown forces for m members and three reactions to the number of independent equations $2j$, or $m + 3 = 2j$.

It is necessary to apply Eq. 1.4 with care. The equation is applicable for the normal truss which contains a series of triangular frames and has three external reactions, such as the truss shown in Fig. 1.5(a). For other trusses it is necessary to determine by inspection that all parts of the structure are stable. The truss shown in Fig. 1.5(b) satisfies Eq. 1.4; yet the left panel is unstable, while the right panel has one more diagonal than is necessary. The truss shown in Fig. 1.5(c) is stable and statically determinate, even though it is not constructed entirely of triangular frames.

Some trusses may have more than three external reactions, and fewer members than are specified by Eq. 1.4, and be stable and statically determinate. The number of reactions r may be substituted for the three in Eq. 1.4.

$$m = 2j - r \qquad (1.4a)$$

The number of independent equations, $2j$, is therefore sufficient to obtain the $m + r$ unknown forces for the members and the reactions. An example of a stable and statically determinate truss which has four reactions may be obtained from the truss of Fig. 1.5(a) by adding a horizontal reaction at the upper left-hand corner and removing the right-hand diagonal member.

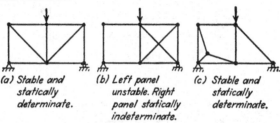

(a) Stable and
statically
determinate.

(b) Left panel
unstable. Right
panel statically
indeterminate.

(c) Stable and
statically
determinate.

All trusses have 9 members and 6 joints, or m = 2j -3.

FIG. 1.5.

1.4. Truss Analysis by Method of Joints. In the analysis of a truss by the method of joints, the two equations of static equilibrium, $\Sigma F_x = 0$ and $\Sigma F_y = 0$, are applied for each joint as a free body. Two unknown forces may be obtained for each joint. Since each member is a two-force member, it exerts equal and opposite forces on the joints at its ends. The joints of a truss must be analyzed in sequence by starting at a joint which has only two members with unknown forces. After finding the forces in these two members, an adjacent joint at the end of one of these members will have only two unknown forces. The joints are then analyzed in the proper sequence until all joints have been considered.

In most structures it is necessary to determine the three external reactions from the equations of equilibrium for the entire structure, in order to have only two unknown forces at each joint. These three equations are used in addition to the $2j$ equations at the joints. Since there are only $2j$ unknown forces, three of the equations are not necessary for finding the unknowns, but should always be used for checking the numerical work. The analysis of a truss by the method of joints will be illustrated by a numerical example.

Example. Find the loads in all the members of the truss shown in Fig. 1.6.

Solution. Draw a free-body diagram for the entire structure and for each joint, as shown in Fig. 1.7. Since all loads in the two-force members act along the members, it is possible to show all forces on a sketch of the truss, as shown in Fig. 1.7, if the forces are specified as acting on the joints. Care must be used in

the directions of the vectors, since at every point there is always a force acting
on the member which is equal and opposite to the force acting on the joint. If
a structure contains any members which are not two-force members, it is always
necessary to make separate free-body sketches of these members, as shown in
Fig. 1.4.

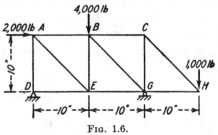

Fig. 1.6.

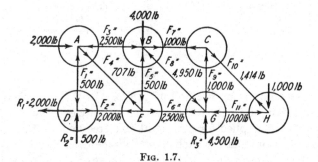

Fig. 1.7.

Considering the entire structure as a free body,

$$\Sigma M_D = 2{,}000 \times 10 + 4{,}000 \times 10 + 1{,}000 \times 30 - 20R_3 = 0$$
$$R_3 = 4{,}500 \text{ lb}$$
$$\Sigma F_y = R_2 - 4{,}000 - 1{,}000 + 4{,}500 = 0$$
$$R_2 = 500 \text{ lb}$$
$$\Sigma F_x = 2{,}000 - R_1 = 0$$
$$R_1^* = 2{,}000 \text{ lb}$$

The directions of unknown forces are assumed, as in the previous example,
and vectors changed on the sketch when they are found to be negative. Some
engineers prefer to assume all members in tension, in which case negative signs
designate compression. Joints must be selected in the proper order, so that
there are only two unknowns at each joint.

Joint D:

$$\Sigma F_x = F_2 - 2{,}000 = 0$$
$$F_2 = 2{,}000 \text{ lb}$$
$$\Sigma F_y = 500 - F_1 = 0$$
$$F_1 = 500 \text{ lb}$$

Joint A:

$$\Sigma F_y = 500 - F_4 \sin 45° = 0$$
$$F_4 = 707 \text{ lb}$$
$$\Sigma F_x = 2{,}000 + 707 \cos 45° - F_3 = 0$$
$$F_3 = 2{,}500 \text{ lb}$$

Joint E:

$$\Sigma F_x = F_6 - 2{,}000 - 707 \cos 45° = 0$$
$$F_6 = 2{,}500 \text{ lb}$$
$$\Sigma F_y = 707 \sin 45° - F_5 = 0$$
$$F_5 = 500 \text{ lb}$$

Joint B:

$$\Sigma F_y = 500 - 4{,}000 + F_8 \sin 45° = 0$$
$$F_8 = 4{,}950 \text{ lb}$$
$$\Sigma F_x = 2{,}500 - 4{,}950 \cos 45° + F_7 = 0$$
$$F_7 = 1{,}000 \text{ lb}$$

Joint C:

$$\Sigma F_x = F_{10} \cos 45° - 1{,}000 = 0$$
$$F_{10} = 1{,}414 \text{ lb}$$
$$\Sigma F_y = F_9 - 1{,}414 \sin 45° = 0$$
$$F_9 = 1{,}000 \text{ lb}$$

Joint G:

$$\Sigma F_x = 4{,}950 \cos 45° - 2{,}500 - F_{11} = 0$$
$$F_{11} = 1{,}000 \text{ lb}$$

Check: $\Sigma F_y = 4{,}500 - 4{,}950 \cos 45° - 1{,}000 = 0$

Joint H:

Check: $\Sigma F_y = 1{,}414 \sin 45° - 1{,}000 = 0$

Check: $\Sigma F_x = 1{,}000 - 1{,}414 \cos 45° = 0$

Arrows acting toward a joint show that a member is in compression, and arrows acting away from a joint indicate tension.

1.5. Truss Analysis by Method of Sections. It is often desirable to find the forces in some of the members of a truss without analyzing the entire truss. The method of joints is usually cumbersome in this case, since the forces in all members to the left of any member must be obtained before finding the force in that particular member. An analysis by the method of sections will yield the force in any member by a single operation, without the necessity of finding the forces in the other members. Instead of considering the joints as free bodies, a cross section is taken through the truss, and the part of the truss on one side of the cross section is considered as a free body. The cross section is chosen so that it cuts the members for which the forces are desired and so that it preferably cuts only three members.

If the forces in members BC, BG, and EG of the truss of Fig. 1.7 are desired, the free body will be as shown in Fig. 1.8. The three unknowns may be found from the three equations for static equilibrium.

$$\Sigma M_G = 10F_7 - 10 \times 4,000 + 2,000 \times 10 + 500 \times 20 = 0$$
$$F_7 = 1,000 \text{ lb}$$
$$\Sigma F_y = F_8 \sin 45° - 4,000 + 500 = 0$$
$$F_8 = 4,950 \text{ lb}$$
$$\Sigma F_x = F_6 + 1,000 - 4,950 \cos 45° + 2,000 - 2,000 = 0$$
$$F_6 = 2,500 \text{ lb}$$

These values check those obtained in the analysis by the method of joints.

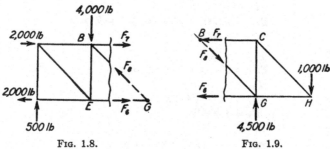

FIG. 1.8.   FIG. 1.9.

The portion of the truss to the right of the section through the members might have been taken as the free body, as shown in Fig. 1.9. The equations of equilibrium would be as follows:

$$\Sigma M_G = 1,000 \times 10 - 10F_7 = 0$$
$$F_7 = 1,000 \text{ lb}$$
$$\Sigma F_y = 4,500 - 1,000 - F_8 \sin 45° = 0$$
$$F_8 = 4,950 \text{ lb}$$
$$\Sigma F_x = 4,950 \cos 45° - 1,000 - F_6 = 0$$
$$F_6 = 2,500 \text{ lb}$$

It would also be possible to find the force F_6 by taking moments about point B, thus eliminating the necessity of first finding the forces F_7 and F_8.

1.6. Truss Analysis—Graphic Method. In the analysis of trusses by the method of joints, two unknown forces were obtained from the equations of equilibrium. It is also possible to find two unknown forces at each joint graphically by the use of the force polygon for the joint. The joints must be analyzed in the same sequence as used in the method of joints. In the graphic truss analysis, it is convenient to use Bow's notation, in which each space is designated by a letter and forces are designated by the two letters corresponding to the spaces on each side of the force. The truss of Fig. 1.6 will be analyzed graphically, and the notation used will be as shown in Fig. 1.10. The capital letters designating the joints are usually omitted but are included here only for reference during the discussion.

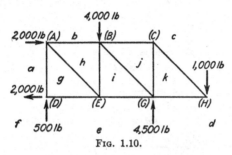

Fig. 1.10.

The external reactions can be determined graphically, but since it is usually more convenient to use algebraic methods the graphic solution will not be considered here. Using the external reactions found in Art. 1.4, a force polygon for the entire structure as a free body is shown in Fig. 1.11(a). While *ab* and *fa* are on the same horizontal line, and *bc* and *de* are on the same vertical line and will be shown that way in future work, they are shown displaced slightly for purposes of explanation. The notation shown is such that when the letters are read clockwise around the structure of Fig. 1.10, *a, b, c, d, e,* and *f,* the directions of the forces in the force polygon will be *a* to *b, b* to *c, c* to *d, d* to *e,* and *e* to *f,* with the polygon closing by the force *fa.*

Joint *D* is first considered as a free body and the known forces *ef* and *fa* drawn to scale. The unknown forces *ag* and *ge* are then drawn in the proper directions from *a* and *e,* and the magnitudes are determined from the intersection *g,* as shown in Fig. 1.11(*b*). Joint *A* is next analyzed by drawing known forces *ga* and *ab* and finding *bh* and *hg* by the intersection at *h* in Fig. 1.11(*c*). Similarly, joints *E, B, C,* and *G* are analyzed as shown in Figs. 1.11(*d*) to (*g*). All forces have now been determined without considering joint *H.* The force polygon for joint *H,* shown in Fig. 1.11(*h*), is used for checking results, as in the algebraic solution.

A study of the force polygons shows that each force appears in two polygons. Force *ge* is a force to the right on joint *D,* and *eg* is a force to the left on joint *E,* but points *e* and *g* have the same relative position in both diagrams. All the polygons can be combined into one stress diagram as shown in Fig. 1.11(*i*). Arrows are omitted from the stress diagram, since there would always be forces in both directions and the arrows would have no significance. To determine the direction of the forces at any joint, the letters are read clockwise around the joint. Thus, at joint D in Fig. 1.10, the letters ge are read, which are seen in Fig. 1.11(i) to represent a force to the right on joint D.   Proceeding

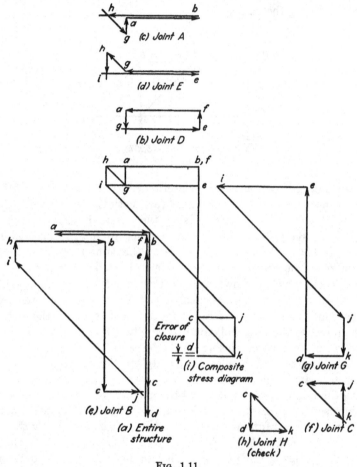

Fɪɢ. 1.11.

clockwise around joint E, the letters eg are read, which represent a force to the left in the stress diagram.   The member is therefore a tension member exerting a force to the left on joint E.

Example. Construct a stress diagram for the steel-tube fuselage truss shown in Fig. 1.12.   The structure is stable and statically determinate although space c is not triangular.   A lighter structure would be obtained if a single diagonal were used in place of members ce, cd, and de, but this would not permi⁺

enough space for a side door in the fuselage. The type of framing shown is often used to permit openings in the truss for access purposes.

Solution. The reactions R_1 and R_2 are first obtained algebraically. The stress diagram is then constructed as shown in Fig. 1.12(b). The forces in all members are then obtained by scaling the lengths from the stress diagram. The

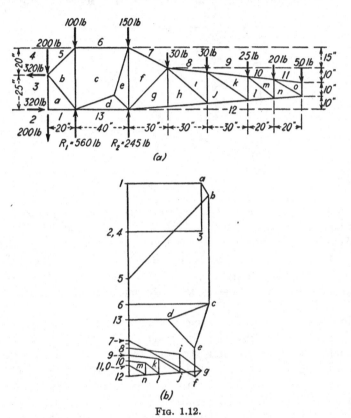

Fig. 1.12.

directions of the forces are obtained from the stress diagram in the same manner as in the previous example. For member 8–i, the line in the stress diagram is from left to right, indicating a tension force acting to the right on the joint to the left of the member. Reading clockwise around the joint to the right of the member, the force is i–8, which in the stress diagram is a force to the left. This checks the conclusion that the member is in tension.

An algebraic solution of problems such as the fuselage truss may become rather tedious, since each of the inclined members is at a different angle. The graphic solution has many advantages for a problem of this type, because it is much easier to draw the truss to scale and to project lines parallel to the members than to calculate the angles and forces for the members.

1.7. Trusses Containing Members in Bending. Many structures
are made up largely of two-force members but contain some members
which are loaded laterally, as shown in Fig. 1.13. These structures are
usually classed as trusses, since the analysis is similar to that used for
trusses. The horizontal members of the truss shown in Fig. 1.13 are

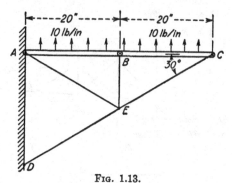

Fig. 1.13.

not two-force members, and separate free-body diagrams for these
members, as shown in Figs. 1.14(a) and (b), are required. Since each
of these members has four unknown reactions, the equations of statics
are not sufficient for finding all four forces. It is possible to find the
vertical forces $A_y = B_{y1} = B_{y2} = C_y = 100$ lb and to obtain the relations $A_x = B_{x1}$ and $B_{x2} = C_x$ from the equilibrium equations for the
horizontal members.

When the forces obtained from the horizontal members are applied to
the remaining structure as a free body, as shown in Fig. 1.14(c), it is
apparent that the remaining structure may be analyzed by the same
methods that were used in the previous truss problems. The loads
obtained by such an analysis are shown in Fig. 1.14(d). All members except the horizontal members may now be designed as simple
tension or compression members. The horizontal members must be
designed for bending moments combined with the compression load of
173.2 lb.

In the trusses previously analyzed, the members themselves have been
assumed to be weightless. The effects of the weight of the members may
be considered by the method used in the preceding example. It will be
noticed that the correct axial loads in the truss members may be obtained
if half the weight of the member is applied at each of the panel points at
the ends of the member. The bending stresses in the member resulting
from the weight of the member must be computed separately and combined with the axial stresses in the member.

Many trusses used in aircraft and other structures do not have frictionless pins at each end. Aircraft trusses are usually made of steel tubes with welded ends. While the truss members are not free to rotate at the ends in the same manner as frictionless pin joints, the members are flexible in bending when compared with the flexibility of the entire

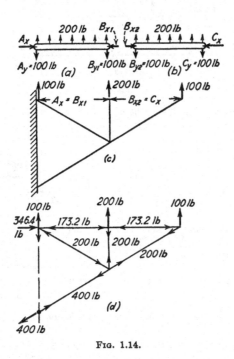

Fig. 1.14.

truss. It is customary and reasonably accurate to assume such trusses as pin-ended for analysis. This same assumption is also used for heavy bridge and building truss members, although bending stresses resulting from truss deflections are occasionally calculated for some bridge members. It can be shown that the ultimate strength of tension members can be predicted more accurately when these secondary bending stresses are neglected, since the bending stresses are relieved when the material yields slightly. Compression members are much stronger when the ends are rigidly welded than when the ends are pinned. The centroidal axes of all truss members should meet in one point at a joint. When clearance or manufacturing requirements do not permit members to meet at a point, the members must be designed for bending stresses produced by the eccentric loading.

PROBLEMS

1.13. Find the loads in all members of the truss by each of the following methods: (a) algebraically by method of joints; (b) algebraically by method of sections; (c) graphically.

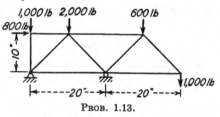

1,000 lb 2,000 lb 600 lb
800 lb
10"
1,000 lb
|------20"------|------20"------|

PROB. 1.13.

1.14. Find the loads in all members of the truss shown, by each of the three methods of analysis.

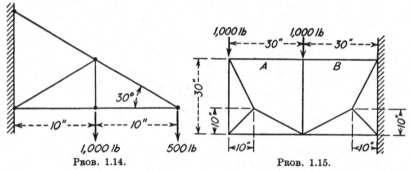

30°
|----10"----|----10"----|
1,000 lb   500 lb
PROB. 1.14.

1,000 lb   1,000 lb
|----30"----|----30"----|
A   B
30"
10"   10"
|10"| |10"|
PROB. 1.15.

1.15. Find the loads in all members graphically and algebraically by the method of joints.

Hint: First obtain the loads in members A and B by the method of sections. It is possible to solve this truss graphically by the use of "phantom" members, but it is usually less confusing to use the method of sections for members A and B.

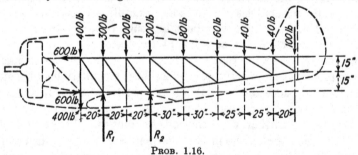

400 lb 300 lb 200 lb 300 lb 80 lb 60 lb 40 lb 40 lb 100 lb
600 lb
15"
15"
600 lb
400 lb |20"|20"|20"|--30"--|--30"--|25"|25"|20"|
R₁   R₂
PROB. 1.16.

1.16. Find the loads in all members by graphic methods. The reactions R_1 and R_2 are obtained algebraically. Show numerical values of all forces on a diagram of the structure, indicating the directions by arrows.

1.17. The structure shown represents an engine mount for a V-type engine. Find the reactions on member *AB* and the forces in other members of the structure.

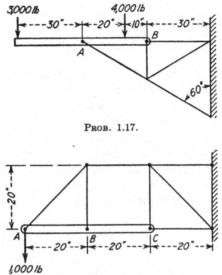

Prob. 1.17.

Prob. 1.18.

1.18. All members of the structure shown are two-force members, except member *ABC*. Find the reactions on member *ABC* and the loads in other members of the structure.

CHAPTER 2

SPACE STRUCTURES

2.1. Equations of Equilibrium. Most structures must be designed to resist loads acting in more than one plane. Consequently, the structures are actually space structures, although in many cases the loads in each plane may be considered independently and the structures analyzed by the methods of analysis for coplanar structures. When it is necessary to consider simultaneously the forces acting in more than one plane, the methods of analysis are no more difficult, but it is often more difficult to visualize the space geometry of the structure. In an analysis of space structures it is desirable to draw several views of the structure, with the forces shown in all views.

The equilibrium of any free body in space is defined by six equations:

$$\left.\begin{array}{ll} \Sigma F_x = 0 & \Sigma M_x = 0 \\ \Sigma F_y = 0 & \Sigma M_y = 0 \\ \Sigma F_z = 0 & \Sigma M_z = 0 \end{array}\right\} \tag{2.1}$$

The first three equations represent the summation of force components along three nonparallel axes, which may be chosen arbitrarily. The second three equations represent the summation of moments about three nonparallel axes. For a free body in space, it is possible to find six unknown forces from the equations of statics. Six reaction components are necessary for the stability of a space structure.

The components of a force R in space along three mutually perpendicular axes, x, y, and z, may be obtained from the following equations:

$$\left.\begin{array}{l} F_x = R \cos \alpha \\ F_y = R \cos \beta \\ F_z = R \cos \gamma \end{array}\right\} \tag{2.2}$$

where α, β, and γ are the angles between the force and the x, y, and z axes, respectively, as shown in Fig. 2.1. When the three components are known, the resultant may be obtained from the following equation:

$$R = \sqrt{F_x^2 + F_y^2 + F_z^2} \tag{2.3}$$

Two-force members are frequently used in space structures as well as in coplanar structures. Theoretically, such members would require end

20

conditions similar to ball and socket joints in order to eliminate bending in any direction, but the rigidity of the usual types of trusses is such that bending of the members may be neglected, as in coplanar structures. A two-force member in space resists only tension or compression, and the force in such a member represents only one unknown quantity to be

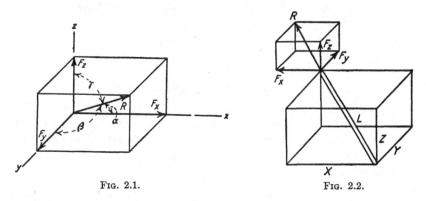

FIG. 2.1.　　　　　　　　　　　　　　FIG. 2.2.

obtained from the equations of statics. If the resultant force or one of its components is found, the remaining components may be obtained from the geometric relationships shown in Fig. 2.2:

$$\frac{R}{L} = \frac{F_x}{X} = \frac{F_y}{Y} = \frac{F_z}{Z}\qquad(2.4)$$

where X, Y, and Z are the components of the length L along the mutually perpendicular reference axes.

2.2. Moments and Couples. The moment of a force about a line is obtained by projecting the force to a plane perpendicular to the line and finding the moment of the component of the force in that plane. The force P, in Fig. 2.3, has components P_2, parallel to the axis of moments, and P_1, in a plane perpendicular to the axis of moments. The moment of the force about the line OO is $P_1 d$, since the component P_2 has no moment about

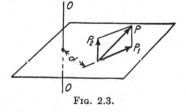

FIG. 2.3.

the line. It will be noticed that a force has no moment about any line that is in the same plane as the force.

A couple consists of two equal parallel forces acting in opposite directions. It is found by taking moments about any arbitrary point in the plane of a couple that the moment of the couple is the same about all points in the plane. Thus the effect of the couple is not changed by moving the couple in the plane. The couples shown in Fig. 2.4 are equivalent to each other, since they all have a clockwise moment Pd about any point in the plane. One quantity is therefore sufficient to define a couple in a plane.

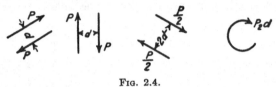

Fig. 2.4.

A couple in space may tend to produce rotation about all three coordinate axes. The method of obtaining components of a couple is similar to that for obtaining components of forces. The two parallel forces P, in Fig. 2.5(a), form a couple of magnitude Pd. The forces may

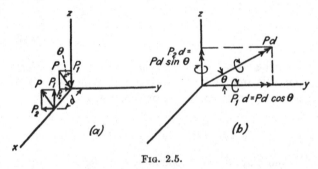

Fig. 2.5.

be resolved into components P_2 and P_1 parallel to the y and z axes. The components form couples P_1d and P_2d about the y and z axes which have magnitudes $Pd \cos \theta$ and $Pd \sin \theta$, respectively. These components are

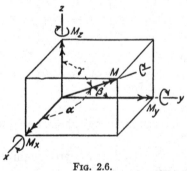

Fig. 2.6.

seen from Fig. 2.5(b) to be the same as would be obtained by representing the couple Pd by a single vector perpendicular to the plane of the couple and projecting this vector on axes perpendicular to the planes of the desired component couples. Couples will be represented by double-arrow vectors in order to distinguish them from force vectors, and the direction of a couple vector will be obtained by the left-hand rule; the arrows will be in the direction of the left thumb when the fingers are curved in the direction of rotation. While the couple shown has no moment about the x axis, the method of resolving a couple vector into three components is identical with the method of resolving a force vector into three components. Considering Fig. 2.6,

$$M_x = M \cos \alpha \left.\vphantom{\begin{matrix}a\\b\\c\end{matrix}}\right\}$$
$$M_y = M \cos \beta \;\;\;(2.5)$$
$$M_z = M \cos \gamma$$

Since a couple can be moved to any point in a plane, the couple vector, unlike a force vector, may be moved laterally to any point. A couple vector, like a force vector, may be moved along its line of action, since a couple may be moved from one plane to any parallel plane.

2.3. Analyses of Typical Space Structures

Example 1. Find the loads in the two-force members *OA*, *OB*, and *OC* of the structure shown in Fig. 2.7.

Solution. Since all forces in the structure act through point *O*, the moments about any axis through point *O* will be zero regardless of the magnitudes of the forces. Consequently only the three equations of statics for the summation of forces along the axes are used to find the three unknowns. In Fig. 2.7, the

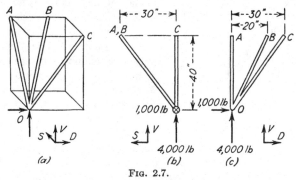

Fig. 2.7.

three mutually perpendicular axes are designated as the *V*, *D*, and *S* axes. This notation is customary in the analysis of airplane landing gears to represent the vertical, drag, and side components of forces. The direction cosines of a member are obtained as *V/L*, *D/L*, and *S/L*, where *V*, *D*, and *S* represent the projected length of the member along the reference axes and *L* represents the true length of the member. The direction cosines are obtained in Table 2.1.

TABLE 2.1

Member	V	D	S	$L = \sqrt{V^2 + D^2 + S^2}$	$\dfrac{V}{L}$	$\dfrac{D}{L}$	$\dfrac{S}{L}$
A	40	0	30	50	0.8	0	0.6
B	40	20	30	53.9	0.743	0.371	0.557
C	40	30	0	50	0.8	0.6	0

All members are assumed to be in tension, and summations of forces along the axes are as follows:

$$\Sigma F_v = 0.8A + 0.743B + 0.8C + 4{,}000 = 0$$
$$\Sigma F_d = 0.371B + 0.6C + 1{,}000 = 0$$
$$\Sigma F_s = 0.6A + 0.557B = 0$$

Solving these equations simultaneously, the forces are obtained as follows:

$$A = -5{,}000 \text{ lb} \qquad \text{compression}$$
$$B = +5{,}390 \text{ lb} \qquad \text{tension}$$
$$C = -5{,}000 \text{ lb} \qquad \text{compression}$$

Alternate Solution. The above solution requires that three simultaneous equations be solved for the three unknowns. In coplanar structures it is usually convenient to set up the equations of statics so that each equation contains only one unknown. In space structures this

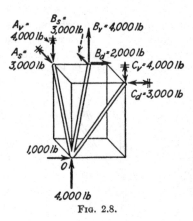

$A_v = 4{,}000\ lb$ · $B_s = 3{,}000\ lb$ · $B_v = 4{,}000\ lb$

$A_s = 3{,}000\ lb$ · $B_d = 2{,}000\ lb$ · $C_v = 4{,}000\ lb$

$C_d = 3{,}000\ lb$

$1{,}000\ lb$

O

$4{,}000\ lb$

Fig. 2.8.

is not always convenient, because of the more complicated geometry. If a summation of forces perpendicular to plane AOB is taken, only the unknown C will appear in the equation, but it is more difficult to solve for the necessary angles than to solve the simultaneous equations. A better method of obtaining equations which have only one unknown in each equation is to use the moment equations. Since the forces are concurrent, only three independent equations of statics exist, but three moment equations may be used, or any combination of moment and force equations. The unknowns will be taken as

shown in Fig. 2.8. When one component of a force is found from the equations of statics, the other components are obtained by Eqs. 2.4. First taking moments about axis AB,

$$\Sigma M_{AB} = 4{,}000 \times 30 + 30C_v = 0$$
$$C_v = -4{,}000 \text{ lb}$$

From Eqs. 2.4,

$$\frac{-4{,}000}{40} = \frac{C_d}{30} = \frac{C}{50}$$

$$C_d = -3{,}000 \text{ lb} \qquad C = -5{,}000 \text{ lb}$$

These forces are now shown in the correct direction on Fig. 2.8. Force B may now be found from a summation of forces along the D axis:

$$\Sigma F_d = B_d - 3{,}000 + 1{,}000 = 0$$
$$B_d = 2{,}000 \text{ lb}$$

From Eqs. 2.4,

$$\frac{2{,}000}{20} = \frac{B_v}{40} = \frac{B_s}{30} = \frac{B}{53.9}$$

$$B_v = 4{,}000 \text{ lb} \qquad B_s = 3{,}000 \text{ lb} \qquad B = 5{,}390 \text{ lb}$$

The unknown force at A may now be found from any of the four remaining equations of statics:

$$\Sigma F_s = A_s + 3,000 = 0$$
$$A_s = -3,000 \text{ lb}$$

From Eqs. 2.4,

$$\frac{-3,000}{30} = \frac{A_v}{40} = \frac{A}{50}$$

$$A_v = -4,000 \text{ lb} \qquad A = -5,000 \text{ lb}$$

Check: $\quad \Sigma F_v = 4,000 - 4,000 + 4,000 - 4,000 = 0$

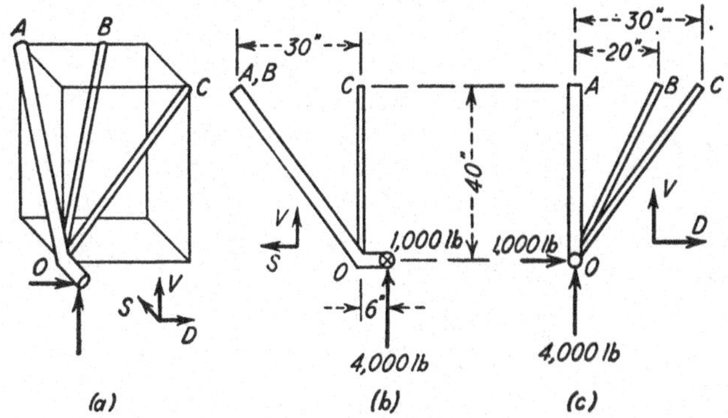

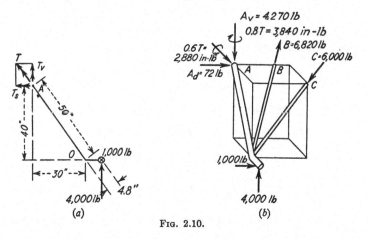

Fig. 2.9.

Example 2. Find the forces at points A, B, and C for the landing gear of Fig. 2.9. Members OB and OC are two-force members. Member OA resists bending and torsion, but point A is hinged by a universal joint so that the member can carry torsion but no bending in any direction at this point.

Fig. 2.10.

Solution. First consider the components of the torsional couple at point A. The resultant couple vector T, shown in Fig. 2.10(a), must be along the member, and it has components T_v about a vertical axis and T_s about a side axis. By proportion,

$$\frac{T_v}{40} = \frac{T_s}{30} = \frac{T}{50}$$

or

$$T_v = 0.8T$$

and

$$T_s = 0.6T$$

In the free-body diagram for the entire structure, shown in Fig. 2.10(b), there are six unknown forces. The forces at B and C act along the member; therefore there is only one unknown at each point. The force at A is unknown in direction and must be considered as three unknown force components, or as an unknown force and two unknown direction angles. Usually it is more convenient to find the components, after which the resultant force may be obtained from Eq. 2.3. The couple T is also resolved into components about the S and V axes. The direction cosines for members OB and OC are the same as those obtained in Example 1 and will be used in the following equations. Taking moments about an axis through points A and B,

$$\Sigma M_{AB} = 4{,}000 \times 36 - 0.8C \times 30 = 0$$
$$C = 6{,}000 \text{ lb}$$

All the unknown forces and the 4,000-lb load act through member OA. The torsional couple T may be found by taking moments about line OA. The 1,000-lb drag load has a moment arm of 4.8 in., as shown in Fig. 2.10(a).

$$\Sigma M_{AO} = 1{,}000 \times 4.8 - T = 0$$
$$T = 4{,}800 \text{ in-lb}$$
$$0.6T = 2{,}880 \text{ in-lb}$$
$$0.8T = 3{,}840 \text{ in-lb}$$

The other forces are obtained from the following equations, which are chosen so that only one unknown appears in each equation.

$$\Sigma M_{OS} = 2{,}880 - 40A_d = 0$$
$$A_d = 72 \text{ lb}$$

The subscripts OS designate an axis through point O in the side direction.

$$\Sigma F_d = 1{,}000 + 72 - 6{,}000 \times 0.6 + 0.371B = 0$$
$$B = 6{,}820 \text{ lb}$$
$$\Sigma F_s = A_s - 6{,}820 \times 0.557 = 0$$
$$A_s = 3{,}800 \text{ lb}$$
$$\Sigma F_v = 4{,}000 + 6{,}820 \times 0.743 - 6{,}000 \times 0.8 - A_v = 0$$
$$A_v = 4{,}270 \text{ lb}$$

Check: $\Sigma M_{AV} = -1{,}000 \times 36 + 6{,}000 \times 0.6 \times 30 - 6{,}820 \times 0.557 \times 20 + 3{,}840 = 0$

Example 3. Find the forces acting on all members of the landing gear shown in Fig. 2.11.

Solution. For convenience, the reference axes, V, D, and S, will be taken as shown in Fig. 2.11, with the V axis parallel to the oleo strut. Free-body dia-

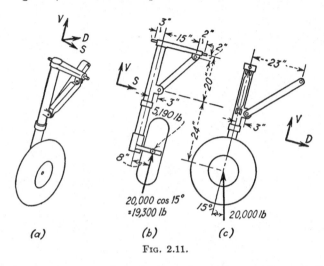

(a) (b) (c)

Fig. 2.11.

grams for the oleo strut and the horizontal member are shown in Fig. 2.12. Forces perpendicular to the plane of the paper are shown by a circled dot (⊙) for forces toward the observer and a circled cross (⊗) for forces away from the observer. The V component of the 20,000-lb force is

$$20,000 \cos 15° = 19,300 \text{ lb}$$

The D component is

$$20,000 \sin 15° = 5,190 \text{ lb}$$

The angle of the side brace member CG with the V axis is

$$\tan^{-1} 12\frac{2}{18} = 33.7°$$

The V and S components of the force in member CG are

$$CG \cos 33.7° = 0.832CG$$
$$CG \sin 33.7° = 0.555CG$$

The drag-brace member BH is at an angle of 45° with the V axis, and the components of the force in this member along the V and D axes are

$$BH \cos 45° = 0.707BH$$
$$BH \sin 45° = 0.707BH$$

The six unknown forces acting on the oleo strut are now obtained from the following equations:

$$\Sigma M_{EV} = 5,190 \times 8 - T_e = 0$$
$$T_e = 41,720 \text{ in-lb}$$
$$\Sigma M_{ES} = 5,190 \times 44 - 0.707BH \times 20 - 0.707BH \times 3 = 0$$
$$BH = 14,050 \text{ lb}$$

$$0.707BH = 9,930 \text{ lb}$$
$$\Sigma M_{ED} = 0.555CG \times 20 + 0.832CG \times 3 - 19,300 \times 8 = 0$$
$$CG = 11,350 \text{ lb}$$
$$0.555CG = 6,300 \text{ lb}$$
$$0.832CG = 9,440 \text{ lb}$$
$$\Sigma F_v = 19,300 + 9,930 - 9,440 - E_v = 0$$
$$E_v = 19,790 \text{ lb}$$
$$\Sigma F_s = E_s - 6,300 = 0$$
$$E_s = 6,300 \text{ lb}$$
$$\Sigma F_d = -5,190 + 9,930 - E_d = 0$$
$$E_d = 4,740 \text{ lb}$$

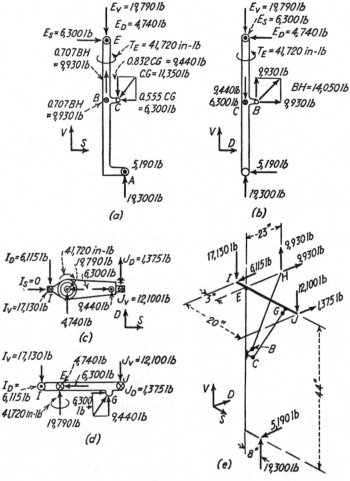

Fig. 2.12.

The horizontal member IJ will now be considered as a free body. The forces obtained above are applied to this member as shown in Figs. 2.12(c) and (d), and the five unknown reactions are obtained as follows:

$$\Sigma F_s = I_s = 0$$
$$\Sigma M_{ID} = 19,790 \times 3 + 9,440 \times 18 + 6,300 \times 2 - 20J_v = 0$$
$$J_v = 12,100 \text{ lb}$$
$$\Sigma F_v = 19,790 + 9,440 - 12,100 - I_v = 0$$
$$I_v = 17,130 \text{ lb}$$
$$\Sigma M_{IV} = 41,720 - 4,740 \times 3 + 20J_d = 0$$
$$J_d = -1,375 \text{ lb}$$
$$\Sigma F_d = 4,740 + 1,375 - I_d = 0$$
$$I_d = 6,115 \text{ lb}$$

The reactions are now checked by considering the entire structure as a free body, as shown in Fig. 2.12(e).

$$\Sigma F_v = 19,300 - 17,130 - 12,100 + 9,930 = 0$$
$$\Sigma F_d = -5,190 + 1,375 - 6,115 + 9,930 = 0$$
$$\Sigma F_s = 0$$
$$\Sigma M_{IV} = 5,190 \times 11 - 1,375 \times 20 - 9,930 \times 3 = 0$$
$$\Sigma M_{ID} = 19,300 \times 11 - 12,100 \times 20 + 9,930 \times 3 = 0$$
$$\Sigma M_{IJ} = 5,190 \times 44 - 9,930 \times 23 = 0$$

Example 4. The conventional landing-gear shock absorber contains two telescoping tubes, as shown in Fig. 2.13(a). As the shock absorber is compressed, oil is forced through an orifice into an airtight chamber, and the energy

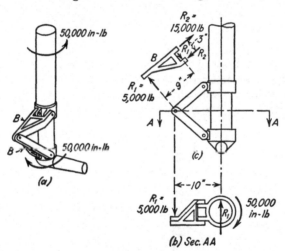

Fig. 2.13.

from the landing impact is absorbed by the oil and air. This shock-absorber mechanism, called an *oleo strut*, transmits bending moment through the two telescoped tubes. The tubes, however, are free to rotate with respect to each other, and additional structural members, called *torque links*, are required to resist torsion. The loads on the torque links B of the landing gear shown in Fig. 2.13 are desired.

Solution. Considering the lower tube and lower torque link as a free body, the forces are as shown in Fig. 2.13(b). Taking moments about the tube center line,
$$\Sigma M = 50,000 - 10R_1 = 0$$
$$R_1 = 5,000 \text{ lb}$$

Now considering one of the torque links as a free body, as shown in Fig. 2.13(c), and taking moments about an axis perpendicular to the plane of the torque link,
$$\Sigma M = 3R_2 - 5,000 \times 9 = 0$$
$$R_2 = 15,000 \text{ lb}$$

PROBLEMS

2.1. Find the forces in the two-force members AO, BO, and CO of the structure shown.

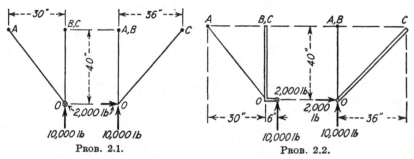

PROB. 2.1.   PROB. 2.2.

2.2. Members AO and BO of this landing-gear structure are two-force members. Member CO is attached at C by a joint which transmits torsion but no bending. Find the forces acting at all joints.

2.3. The bending moments about x and z axes in a plane perpendicular to the spanwise axis of a wing are 400,000 in-lb and 100,000 in-lb as shown. Find the

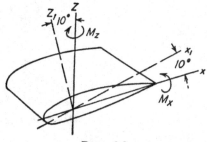

PROB. 2.3.

bending moments about x_1 and z_1 axes which are in the same plane but rotated 10° counterclockwise.

2.4. The main beam of the wing shown has a sweepback angle of 30°. The moments of 300,000 in-lb and 180,000 in-lb are first computed about axes x and y

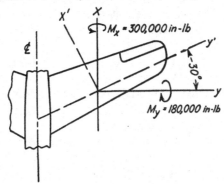

PROB. 2.4.

which are parallel and perpendicular to the center line of the airplane. Find the moments about axes x' and y'.

2.5. Find the forces in all members of the structure shown.

2.6. Omit the front diagonal member of the structure in Prob. 2.5, and add a diagonal member in the bottom plane. Find the forces in all members.

2.7. Find the forces acting on all members of the nose-wheel structure shown. Assume the V axis parallel to the oleo strut.

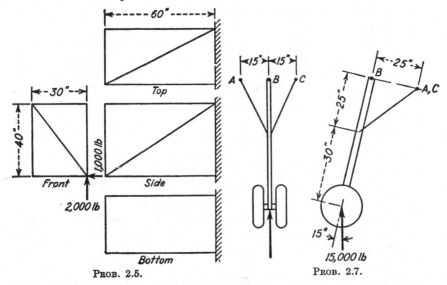

PROB. 2.5.   PROB. 2.7.

2.8. Analyze the landing-gear structure of Example 3, Art. 2.3, for a 15,000-lb load up parallel to the V axis and a 5,000-lb load aft parallel to the D axis. The loads are applied at the same point of the axle as the load in the example problem.

2.4. Torsion of Space Frameworks. Space structures composed of two-force members may be loaded in various ways. For a structure such as a truss-type airplane fuselage it is often convenient to consider separately conditions such as vertical bending, side bending, and torsion, and then superimpose the results obtained from each analysis. In the analysis for a symmetrical loading condition such as vertical bending it is usually possible to consider each side of the structure as a coplanar truss and to determine the loads in all members from the equilibrium of coplanar forces. It is only in the analysis for torsional loads that the methods for space structures need be used.

A proportionality between the components of length and the components of force in a two-force member was shown in Eqs. 2.4. A tension coefficient μ is often used and is defined as follows:

$$\mu = \frac{R}{L} = \frac{F_x}{X} = \frac{F_y}{Y} = \frac{F_z}{Z} \tag{2.6}$$

The terms are identical with those in Eqs. 2.4. The tension coefficient was proposed by Prof. R. V. Southwell [1] and was applied in the torsional analysis of space frameworks by Prof. H. Wagner.[2] The tension coefficients are considered as the unknowns when writing the equations of static equilibrium for a structure, instead of considering the force components as the unknowns. After the tension coefficient is determined for any member, the force components can be found from Eqs. 2.6.

The space frameworks which are commonly used in aircraft structures have bulkheads which are in parallel planes. The bulkheads resist loads in their planes but are too flexible to resist loads normal to the planes of the bulkheads. For the structure shown in Fig. 2.14(a), the bulkheads $BCDE$ and $B'C'D'E'$ are in parallel planes and are assumed to be loaded with equal and opposite torsional couples T in the planes of the bulkheads. From a summation of forces along the x axis at joints D and B', the forces in members BB' and DD' are found to be zero, since the bulkheads cannot resist forces normal to their planes. The remaining members, EE', $E'B$, BC', $C'C$, CD', and $D'E$, called the envelope members, have equal components of length X. The envelope members are also found to have equal force components F_x from a summation of forces along the x axis at joints E, E', B, C', C, and D'. The tension coefficients, $\mu = F_x/X$, are therefore equal for all the envelope members.

The value of the tension coefficient μ for the envelope members is now obtained from the equilibrium of torsional moments about an axis normal to the bulkheads. The forces in the envelope members must be projected to the plane of the bulkhead in order to take moments about an axis normal to the bulkheads. In Fig. 2.14(*c*), the lengths of the envelope members are projected to the plane of bulkhead *BCDE*, and the shaded area is enclosed by these projected lengths. An arbitrary point, *O*, is chosen as the center of moments. Figure 2.14(*d*) represents the

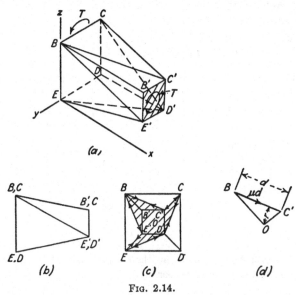

Fɪɢ. 2.14.

same projected view as Fig. 2.14(*c*), but the member *BC′* is isolated. The member *BC′* has a projected length *d* and a force component of *μd* in the plane of the bulkhead. The force component in this member has a moment *rμd*, where *r* is the moment arm shown. Since the area of the triangle *OBC′* shown in Figs. 2.14(*c*) or (*d*) is *rd*/2, the moment of the force in member *BC′* about the axis through *O* may be written as follows.

$$\Delta T = 2\mu \times (\text{area } OBC')$$

The moments of the forces in all the envelope members may be obtained in a similar manner. The sum of all the triangular areas will be equal to the shaded area shown in Fig. 2.14(*c*), and the sum of all the torsional increments will be equal to the external torque, *T*.

$$T = 2\mu \times (\text{area } EE'BC'CD')$$

This area, which is always equal to the area enclosed by the projection of the envelope members on the plane of a bulkhead, will be designated as *A*. The equation for the tension coefficient may now be written in the following form:

$$\mu = \frac{T}{2A} \tag{2.7}$$

The point O may be chosen arbitrarily, since the total area A will be the same for any position of O. The directions of the forces may be obtained by considering the forces on one end bulkhead, such as $B'C'D'E'$. The external load is a clockwise couple, and the forces exerted by the envelope members on the bulkhead must produce a counterclockwise couple. The forces on the bulkhead at any point must have opposite signs. At point E', for example, the member EE' must be in tension if member BE' is in compression.

Example 1. Find the forces in all members of the structure shown in Fig. 2.15.

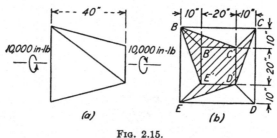

Fig. 2.15.

Solution. From Fig. 2.15(b), the shaded area is 800 sq in. The tension coefficient for each envelope member is

$$\mu = \frac{T}{2A} = \frac{10{,}000}{2 \times 800} = 6.25 \text{ lb/in.}$$

The loads are now obtained as the product of the length of the member and the tension coefficient and are shown in the following table.

The directions of the forces are determined by inspection, and are designated by positive signs for tension members and negative signs for compression members.

TABLE 2.2

Member	Length L	Force μL
EE'	42.5	+266
BE'	51.0	−319
BC'	51.0	+319
CC'	42.5	−266
CD'	51.0	+319
ED'	51.0	−319

Example 2. Find the forces in members EE', BE', BC', CC', CD', and ED' of the structure shown in Fig. 2.16. Assume rigid bulkheads in planes $BCDE$

and $B'C'D'E'$. Assume also that the vertical load of 1,000 lb is equivalent to vertical loads of 500 lb on each side truss plus a torsional moment of 10,000 in-lb, as shown in Fig. 2.16.

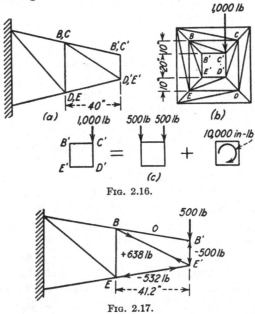

FIG. 2.16.

FIG. 2.17.

Solution. The side trusses are first analyzed as coplanar trusses carrying loads of 500 lb. The forces obtained are shown in Fig. 2.17. These forces are then combined with the forces obtained in Example 1 for the torsional loading.

TABLE 2.3

Member	Forces resulting from 500 lb per side	Forces resulting from 10,000 in-lb, torsion	Total forces
EE'	−532	⎸266	−266
BE'	+638	−319	+319
BB'	0	0	0
BC'	0	+319	+319
CC'	0	−266	−266
CD'	+638	+319	+957
DD'	−532	0	−532
ED'	0	−319	−319

This structure is in reality statically indeterminate, and the torsional analysis gives only approximate values of the forces in the members. An exact analysis could be made by the methods used for statically indeterminate structures, which will be treated in a later chapter. Since a statically indeterminate analysis depends on the deflections of various members, it is first necessary to know the areas of all members before making the analysis.

The approximation involved in the torsional analysis of Example 2 is the assumption that there are no resultant forces perpendicular to the plane of bulkhead $BCDE$. While this assumption was correct for the structure of Example 1, it is slightly in error here because the structure to the left of the bulkhead supplies additional restraints. The force components along the x axis may not be the same for all envelope members. Another distribution of forces would exist if the members in plane $BB'EE'$ were very flexible and those in plane $CC'DD'$ extremely rigid. The side truss in plane $CC'DD'$ would then transfer all the 1,000-lb load into the wall at the left of the structure, with the members carrying forces twice as large as those shown in Fig. 2.17, and other members of the structure would carry no appreciable load. If the structure were supported rigidly at points B, C, D, and E, the torsional analysis would be considerably in error, since the loaded truss $CC'DD'$ would carry more of the load directly to the supports.

In most aircraft structures, the rigidity is such that torsional forces may be computed separately and superimposed on the forces resulting from symmetrical loads. Rigid supports, such as those shown at the left of Fig. 2.16(a), seldom exist, and the structure would normally extend further to the left. Most bulkheads would be free to warp from their initial plane, and would have no restraining forces normal to the plane of the bulkhead. A statically indeterminate analysis, while comparatively simple for the structure with rigid supports, would be more difficult for the actual aircraft structure, and it is seldom used. Where bulkheads are not free to warp, as in the case of wing bulkheads at the center line of the airplane, local corrections are applied to the analysis made by superimposing bending and torsional forces.

2.5. Wing Structures. Most of the early types of airplanes were biplanes, because it was possible to design an efficient, lightweight wing structure with external bracing. The wing loading, which is the ratio of the gross weight of the airplane to the total wing area, was quite low for the early biplanes in order to permit slow landing speeds and to permit cruising at a low engine-power output. The wing weight had to be small for each square foot of wing area, because of the low wing loading. The wings were usually constructed with two wooden spanwise beams, or spars, which were braced externally to form a space truss. The spars, which supported light chordwise former ribs, were braced horizontally by occasional compression ribs with wire x bracing between them.

The structural advantages of the biplane were far outweighed by the aerodynamic disadvantages. The drag of the numerous struts and braces, and the aerodynamic interference of the wings, external bracing, and fuselage, had a more adverse effect on performance than did the additional structural weight required for monoplane construction. The early monoplane wings were also externally braced, and the construction was similar to that used for biplane wings. For light private airplanes, in which a slow landing speed is often more important than other performance features, the wing loading must be kept quite low, and the wing must consequently have a very light weight per square foot. It is possible that externally braced, fabric-covered wing construction will continue to be used for some airplanes of this type.

In military and commercial airplanes, where high speed performance and efficient cruising are more important than very slow landing speeds, only full-cantilever, internally braced wings can be used. The added structural weight is preferable to the aerodynamic drag of external brace members. With better materials and methods of construction to reduce weight, and better landing fields, visibility, and landing-gear and wing-flap design to permit safer landing with higher wing loadings, it is possible that external wing bracing will become obsolete even for the light private airplane.

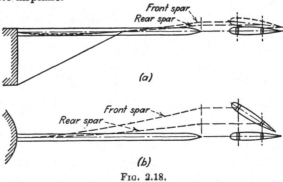

Fig. 2.18.

The air pressure on a wing changes for different flight conditions. The resultant pressure is nearer the leading edge at high angles of attack than at low angles of attack and also shifts when the ailerons or flaps are deflected. The wing must be rigid torsionally, so that it cannot twist enough to affect its aerodynamic properties when the load shifts. The simple two-spar, fabric-covered wing with a single drag truss would twist excessively if used for a full-cantilever wing. In Fig. 2.18(a), the deflections (exaggerated) of a braced wing in which the front spar deflects more than the rear spar are compared with the deflections of a full-cantilever wing in Fig. 2.18(*b*) in which the front spar deflects more than the rear spar. It is seen that the angle of twist of the braced wing is negligible compared with the angle of twist of the full-cantilever wing.

In full-cantilever wings it is necessary to provide a structure which is rigid torsionally, so that both spars must deflect approximately the same distance. This can be done in a fabric-covered wing by providing drag trusses at both the upper and lower surfaces of the wing, as shown in Fig. 2.19, rather than the single drag truss shown in Figs. 2.18 and 2.20.

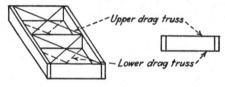

FIG. 2.19.

It is desirable to have the area of the wing cross section enclosed by the spars and the drag trusses as large as possible, since this is the area *A* to be used in Eq. 2.7 for a torsional analysis.

High-performance airplanes have much higher wing loadings than light private airplanes. The wing weight per square foot must be much greater in order to provide adequate strength. The total wing weight may be about the same percentage of the airplane gross weight as it is for the light airplane. Wings for high-performance airplanes are of the semimonocoque, all-metal type. The metal covering on the top and bottom surfaces of the wing acts in the same manner as the drag trusses of the two-spar, fabric-covered wing. A wing of this type is very rigid torsionally, and also has a much better aerodynamic surface than a fabric-covered wing. For very high-speed airplanes, the slight wrinkling of even the metal skin is objectionable, and thick metal skin, or plastic-coated metal, is used to prevent wrinkling and to provide a smooth aerodynamic surface. The structural analysis of semimonocoque wings is treated in detail in later chapters.

Example 1. Find the loads on the lift and drag-truss members of the externally braced monoplane wing shown in Fig. 2.20. The air load is assumed to be uniformly distributed along the span of the wing. The diagonal drag-truss members are wires, with the tension diagonal effective and the other diagonal carrying no load.

Solution. The vertical load of 20 lb/in. is distributed to the spars in inverse proportion to the distance of the center of pressure from the spars. The load on the front spar is therefore 16 lb/in., and that on the rear spar 4 lb/in. If the front spar is considered as a free body, as shown in Fig. 2.21(*a*), the vertical forces at *A* and *G* may be obtained.

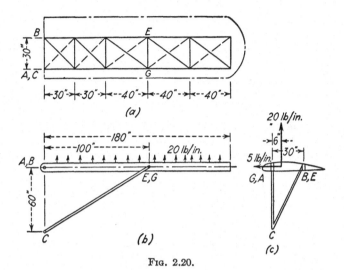

Fig. 2.20.

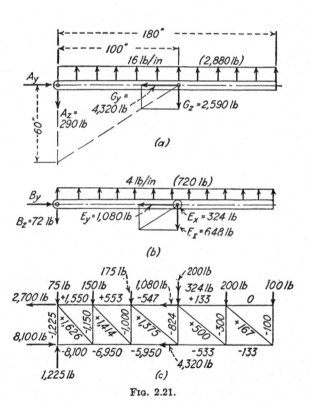

Fig. 2.21.

$$\Sigma M_A = -16 \times 180 \times 90 + 100G_z = 0$$
$$G_z = 2,590 \text{ lb}$$
$$\frac{G_y}{100} = \frac{2,590}{60}$$
$$G_y = 4,320 \text{ lb}$$
$$\Sigma F_z = 16 \times 180 - 2,590 - A_z = 0$$
$$A_z = 290 \text{ lb}$$

Force A_y cannot be found at this point in the analysis, since the drag-truss members exert forces on the front spar which are not shown in Fig. 2.21(a).

If the rear spar is considered as a free body, as shown in Fig. 2.21(b), the vertical forces at B and E may be obtained.

$$\Sigma M_B = -4 \times 180 \times 90 + 100E_z = 0$$
$$E_z = 648 \text{ lb}$$
$$\frac{E_x}{30} = \frac{E_y}{100} = \frac{648}{60}$$
$$E_x = 324 \text{ lb} \qquad E_y = 1,080 \text{ lb}$$
$$\Sigma F_z = 4 \times 180 - 648 - B_z = 0$$
$$B_z = 72 \text{ lb}$$

The loads in the plane of the drag truss may now be obtained. The forward load of 5 lb/in. is applied as concentrated loads at the panel points, as shown in Fig. 2.21(c). The components of the forces at G and E which lie in the plane of the truss must also be considered. The remaining reactions at A and B and the forces in all drag-truss members may now be obtained by the methods of analysis for coplanar trusses, shown in Fig. 2.21(c).

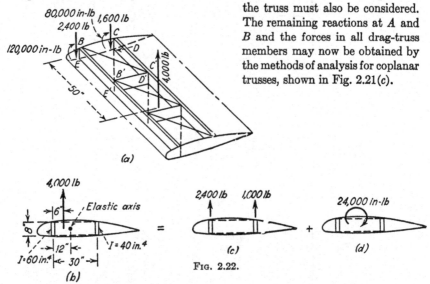

(a)

4,000 lb

Elastic axis

$I = 40 \text{ in.}^4$

$I = 60 \text{ in.}^4$

(b)

2,400 lb   1,000 lb

(c)

$=$

24,000 in-lb

(d)

Fig. 2.22.

Example 2. Find the forces acting at the cut cross section of the two-spar, full-cantilever wing shown in Fig. 2.22. The wing has two drag trusses in horizontal planes 8 in. apart, which are assumed to supply enough torsional rigidity that the two spars have equal bending deflections.

Solution. The bending deflections of the spars are proportional to their loads and inversely proportional to their moments of inertia. Since the rear spar has a moment of inertia of 40 in.[4] and the front spar a moment of inertia of 60 in.[4], the rear spar must carry 40 per cent and the front spar 60 per cent of the total load in order to have the bending deflections of the beams equal. If the 4,000-lb resultant force acted at a point 40 per cent of the distance between spars from the front spar, the former ribs would distribute the loads to the spars in the correct proportion, and there would be no torsional moment producing load in the drag-truss members. The spanwise line on which a load may be applied without producing torsion is called the elastic axis, and the distances of this axis from the spars are inversely proportional to their moments of inertia. The wing is analyzed by superimposing the forces produced by torsion about the elastic axis and those produced by loads applied at the elastic axis.

The elastic axis for this wing is 12 in. from the front spar, and the torsional moment is the product of the 4,000-lb resultant air load and its moment arm about the elastic axis.

$$T = 4,000 \times 6 = 24,000 \text{ in-lb}$$

The area obtained by projecting the envelope members on the plane of the bulkhead *BCDE* is 240 sq in. The tension coefficient is now obtained from Eq. 2.7 as follows:

$$\mu = \frac{T}{2A} = \frac{24,000}{2 \times 240} = 50$$

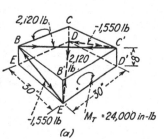
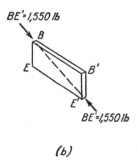

Fig. 2.23.

The forces in the envelope members of the torsion structure shown in Fig. 2.23(a) are as follows:

TABLE 2.4

Member	L, in.	Force, μL
BE'	31.0	−1,550
BC'	42.4	+2,120
DC'	31.0	−1,550
DE'	42.4	+2,120

It will be noticed that members BE' and DC' are fictitious members which represent the portions of the spars acting as torsion structure. The loads on the actual spar members are obtained, as shown in Fig. 2.23(b), by applying the loads obtained for the fictitious members to the actual spar members.

The 4,000-lb load applied at the elastic axis is distributed by the ribs as 2,400 lb to the front spar and 1,600 lb to the rear spar. The bending moment in the front spar at the cut cross section is

$$M = 2,400 \times 50 = 120,000 \text{ in-lb}$$

The bending moment in the rear spar is

$$M = 1,600 \times 50 = 80,000 \text{ in-lb}$$

These forces are shown in Fig. 2.22(a).

Example 3. The wing structure shown in Fig. 2.24 is similar to that used for some full-cantilever fabric-covered wings. One main spar, member $A_1A_5B_5B_1$, carries all the wing bending moment. The structure aft of this spar carries wing torsion and is somewhat similar to landing-gear torque links. The structure aft of the spar also serves as a truss to resist drag loads on the wing. Unlike the two-spar wing, the structure is statically determinate, since a section through the wing would cut only six members, which could be analyzed by the equations of statics.

Solution. The 4,000-lb load 9 in. aft of the spar can be replaced by a 4,000-lb load in the plane of the spar and a 36,000 in-lb couple. The forces in the spar truss members resulting from the 4,000-lb load in the plane of the spar are shown in Fig. 2.25(a) and are tabulated in column (2) of Table 2.5, as the forces in the members resulting from bending of the spar.

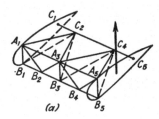

(a)

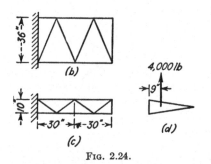

(b)

(c)

4,000 lb

$\rightarrow$|9"|$\leftarrow$

(d)

Fig. 2.24.

The envelope members of the torsion structure are shown in Fig. 2.25(b) as members A_3C_4, C_4B_3, B_3B_4, and B_4A_3. The tension coefficients for these members are

$$\mu = \frac{T}{2A} = \frac{36,000}{2 \times 180} = 100 \text{ lb/in.}$$

Members A_3C_4 and B_3C_4 have a length of 39.3 in. and forces of $39.3 \times 100 = 3,930$ lb. Member A_3B_4 has a length of 18 in. and a force of 1,800 lb. The force in member B_3B_4 is 1,500 lb, and the directions of all forces are shown in Fig. 2.25(b). These forces are tabulated in column (3) of Table 2.5 as forces

TABLE 2.5

Member (1)	Forces in members, lb		
	Bending (2)	Torsion (3)	Total (4)
A_1A_3	−18,000	0	−18,000
A_1B_2	−7,210	+1,800	−5,410
B_1B_2	+24,000	−1,500	+22,500
B_2A_3	+7,210	−1,800	+5,410
B_2B_3	+12,000	+1,500	+13,500
A_3B_3	0	0	0
A_3A_5	−6,000	0	−6,000
A_3B_4	−7,210	+1,800	−5,410
B_3B_4	+12,000	−1,500	−10,500
B_4A_5	+7,210	−1,800	+5,410
B_4B_5	0	+1,500	+1,500
A_1C_2	0	−3,930	−3,930
B_1C_2	0	+3,930	+3,930
A_3C_2	0	+3,930	+3,930
B_3C_2	0	−3,930	−3,930
A_3C_4	0	−3,930	−3,930
B_3C_4	0	+3,930	+3,930
A_5C_4	0	+3,930	+3,930
B_5C_4	0	−3,930	−3,930

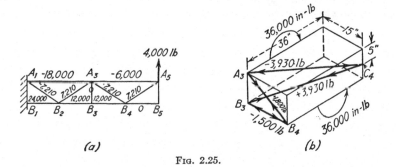

FIG. 2.25.

resulting from torsion. The resultant forces in the members from combined bending and torsion are tabulated in column (4). There are no forces in members C_1C_2, C_2C_4, and C_4C_5 for a vertical loading, but these members would resist chordwise bending from drag loads.

Care must be exercised in finding the load in member A_3B_3. In this case, a summation of vertical forces at joint B_3 gives no load in the member, but the member would be loaded if the torque varied along the span, or if there were drag loads acting.

PROBLEMS

2.9. Solve Example 1, Art. 2.3, by a summation of forces along each axis, considering the tension coefficients for the three members as the unknowns. It is not necessary to find the direction cosines, as the force components will be the product of the tension coefficients and the length components. After finding the tension coefficients the forces are found as the product of the coefficients and the lengths of the members.

2.10. Analyze the structure of Example 1, Art. 2.5, assuming the vertical load of 20 lb/in. to act midway between the spars and the drag load of 5 lb/in. to act in the aft direction.

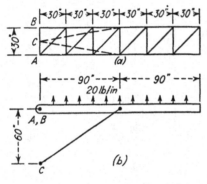

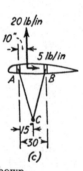

PROB. 2.11.

2.11. Analyze the braced-wing monoplane structure shown.

2.12. Find the loads in the fuselage truss structure shown by analyzing the side trusses as coplanar structures each carrying half the vertical load and by

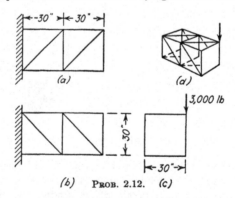

PROB. 2.12.

superimposing the forces resulting from torsion about the center line. Tabulate the forces in the manner shown in Table 2.5.

2.13. Find the torsional forces in the structure of Example 3, Art. 2.5, by the method used for landing-gear torque links in Example 4, Art. 2.3. After finding the forces on the aft structure, apply these forces to the spar, and analyze the spar as a coplanar truss. The results should check the values in column (3), Table 2.5.

REFERENCES FOR CHAPTER 2

1. SOUTHWELL, R. V.: Primary Stress Determination in Space Frames, *Engineering*, Feb. 6, 1920.
2. WAGNER, H.: The Analysis of Aircraft Structures as Space Frameworks, *NACA TM* 522, 1929.
3. NILES, A. S., and J. S. NEWELL: "Airplane Structures," Vol. I, Chap. 8. John Wiley & Sons, Inc., New York, 1943.

CHAPTER 3

INERTIA FORCES AND LOAD FACTORS

3.1. Pure Translation. The maximum load on any part of the airplane structure occurs when the airplane is being accelerated. The loads produced by landing impact or when maneuvering or encountering gusts in flight are always greater than the loads occurring when all the forces on the airplane are in equilibrium. Before any member can be designed, it is therefore necessary to determine the inertia forces acting on the structure. If the inertia forces are included, it is possible to draw a free-body diagram for any member showing the forces in equilibrium.

In many of the loading conditions, the airplane may be considered as being in pure translation, since the rotational velocities and accelerations

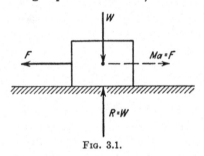

FIG. 3.1.

are small. The inertia force on any element of mass is equal to the product of the mass and the acceleration and acts in a direction opposite to the acceleration. If the applied loads and inertia forces act on an element as a free body, these forces are in equilibrium. If a force F acts on a block on a frictionless plane, as shown in Fig. 3.1, the block will be accelerated in the direction of the force. The inertia force Ma, shown by the dotted vector, acts in the opposite direction and is equal to the accelerating force.

$$F = Ma \qquad (3.1)$$

In engineering problems, the common unit of mass is the slug. The mass in slugs is obtained by dividing the weight in pounds by the acceleration of gravity g in feet per second per second.

$$M = \frac{W}{g} \frac{\text{lb-sec}^2}{\text{ft}} \qquad (3.2)$$

If g is 32.2 ft/sec², the value normally used, a body having a mass of 1 slug will have a weight of 32.2 lb. In many problems the inertia force Ma can be found as a force in pounds by the equations of static equilibrium, without first finding the mass and the acceleration.

46

The airplane shown in Fig. 3.2 is moving forward after landing, and has a braking force F applied to the right. The acceleration is to the right, and the inertia force on each element of mass is to the left, or opposite to the direction of acceleration. The sum of the inertia forces on all elements will be the product of the total mass of the airplane and the acceleration, or Ma. Since the inertia forces are distributed in proportion to the mass dM of each element, the resultant force will act at the center of gravity of the airplane, as shown in Fig. 3.2. If some part of the airplane is considered as a free body, it is necessary to obtain the inertia force acting on that part. The inertia force on any part will be the product of its mass and the acceleration and will act at the center of gravity of that part of the structure.

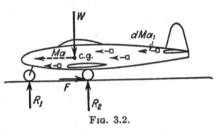

Fig. 3.2.

While the motion of the airplane is not within the scope of a book on aircraft structures, it is occasionally necessary to consider the motion in order to estimate the duration or the magnitude of loads. The velocity v is defined as the time rate of change of the displacement s.

$$v = \frac{ds}{dt} \tag{3.3}$$

The acceleration a is the time rate of change of the velocity,

$$a = \frac{dv}{dt} \tag{3.4}$$

A combination of Eqs. 3.3 and 3.4 gives other forms for the acceleration:

$$a = \frac{d^2s}{dt^2} \tag{3.5}$$

$$a = v\frac{dv}{ds} \tag{3.6}$$

For a motion of pure translation of a rigid body, all elements of the body must have the same displacement, velocity, and acceleration. If the acceleration is constant, the following equations are obtained by integrating Eqs. 3.4 to 3.6:

$$v - v_0 = at \tag{3.7}$$
$$s = v_0t + \tfrac{1}{2}at^2 \tag{3.8}$$
$$v^2 - v_0^2 = 2as \tag{3.9}$$

where s is the distance moved in time t, v_0 is the initial velocity, and v is the final velocity after t sec.

Example 1. Assume that the airplane shown in Fig. 3.3 weighs 20,000 lb. and that the braking force F is 8,000 lb.

a. Find the wheel reactions R_1 and R_2.

b. Find the landing run if the airplane lands at 100 mph (146.7 ft/sec).

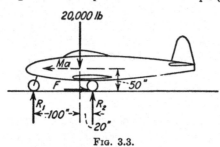

Fig. 3.3.

Solution.   *a.* $\Sigma F_x = 8{,}000 - Ma = 0$

$$Ma = 8{,}000 \text{ lb}$$
$$\Sigma M_{R_1} = 120R_1 - 8{,}000 \times 50 - 20{,}000 \times 20 = 0$$
$$R_1 = 6{,}670 \text{ lb}$$
$$\Sigma F_y = 6{,}670 - 20{,}000 + R_2 = 0$$
$$R_2 = 13{,}330 \text{ lb}$$

b. From Eqs. 3.1 and 3.2,

$$a = \frac{F}{M} = \frac{Fg}{W} = \frac{-8{,}000 \times 32.2}{20{,}000} = -12.88 \text{ ft/sec}^2$$

From Eq. 3.9,

$$v^2 - v_0^2 = 2as$$
$$0 - (146.7)^2 = 2(-12.88)s$$
$$s = 835 \text{ ft}$$

When s is measured as positive to the left, it is necessary to consider a as negative, since it is an acceleration to the right.

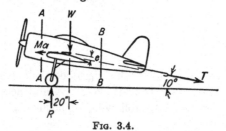

Fig. 3.4.

Example 2. When landing on a carrier, a 10,000-lb airplane is given a deceleration of $3g$ (96.6 ft/sec²) by means of a cable engaged by an arresting hook, as shown in Fig. 3.4.

a. Find the tension in the cable, the wheel reaction *R*, and the distance *e* from the center of gravity to the line of action of the cable.

b. Find the tension in the fuselage at vertical sections *AA* and *BB* if the portion of the airplane forward of section *AA* weighs 3,000 lb and the portion aft of section *BB* weighs 1,000 lb.

c. Find the landing run if the landing speed is 80 ft/sec.

Solution. *a.* First considering the entire airplane as a free body,

$$Ma = \frac{W}{g} a = \frac{10,000}{g} \times 3g = 30,000 \text{ lb}$$
$$\Sigma F_x = T \cos 10° - 30,000 = 0$$
$$T = 30,500 \text{ lb}$$
$$\Sigma F_y = R - 10,000 - 30,500 \sin 10° = 0$$
$$R = 15,300 \text{ lb}$$
$$\Sigma M_{cg} = 20 \times 15,300 - 30,500e = 0$$
$$e = 10 \text{ in.}$$

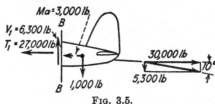

Fig. 3.5.

b. Considering the aft section of the fuselage as a free body as shown in Fig. 3.5, it is acted upon by an inertia force of

$$Ma = \frac{1,000}{g} \times 3g = 3,000 \text{ lb}$$

The tension on section *BB* is found as follows:

$$\Sigma F_x = 30,000 - 3,000 - T_1 = 0$$
$$T_1 = 27,000 \text{ lb}$$

Since there is no vertical acceleration, there is no vertical inertia force. Section *BB* has a shear force V_1 of 6,300 lb, which is equal to the sum of the weight and the vertical component of the cable force.

Considering the portion of the airplane forward of section *AA* as a free body, as shown in Fig. 3.6, the inertia force is

$$Ma = \frac{3,000}{g} \times 3g = 9,000 \text{ lb}$$
$$\Sigma F_x = T_2 - 9,000 = 0$$
$$T_2 = 9,000 \text{ lb}$$

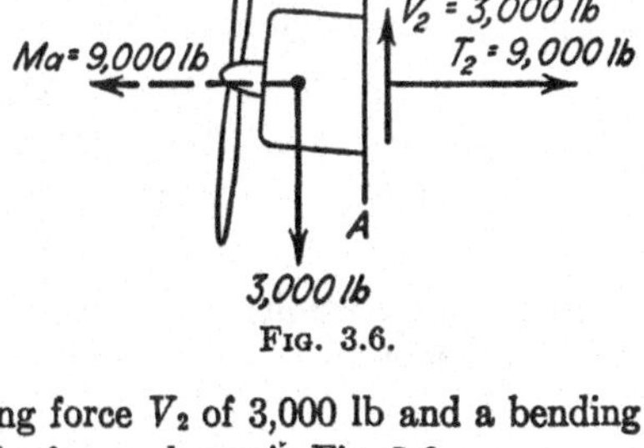

Fig. 3.6.

The section *AA* must also resist a shearing force V_2 of 3,000 lb and a bending moment obtained by taking moments of the forces shown in Fig. 3.6.

The forces T_1, T_2, V_1, and V_2 may be checked by considering the equilibrium of the center portion of the airplane, as shown in Fig. 3.7.

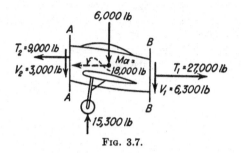

6,000 lb

A

$T_2 = 9,000\ lb$
$V_2 = 3,000\ lb$

$Ma = 18,000\ lb$

B

$T_1 = 27,000\ lb$
$V_1 = 6,300\ lb$

A

B

15,300 lb

FIG. 3.7.

$$Ma = \frac{6,000}{g} \times 3g = 18,000 \text{ lb}$$
$$\Sigma F_x = 27,000 - 18,000 - 9,000 = 0$$
$$\Sigma F_y = 15,300 - 3,000 - 6,000 - 6,300 = 0$$

c. The landing run s is obtained from Eq. 3.9.

$$v^2 - v_0^2 = 2as$$
$$0 - (80)^2 = 2(-96.6)s$$
$$s = 33 \text{ ft}$$

Example 3. A 30,000-lb airplane is shown in Fig. 3.8(a) at the time of landing impact, when the ground reaction on each main wheel is 45,000 lb.

a. If one wheel and tire weighs 500 lb, find the compression C and bending moment m in the oleo strut, if the strut is vertical and is 6 in. from the center line of the wheel, as shown in Fig. 3.8(b).

b. Find the shear and bending moment at section AA of the wing, if the wing outboard of this section weighs 1,500 lb and has its center of gravity 120 in. outboard of section AA.

c. Find the required shock strut deflection if the airplane strikes the ground with a vertical velocity of 12 ft/sec and has a constant vertical deceleration until the vertical velocity is zero. This neglects the energy absorbed by the tire deflection, which may be large in some cases.

d. Find the time required for the vertical velocity to become zero.

Solution. a. Considering the entire airplane as a free body and taking a summation of vertical forces,

$$\Sigma F_y = 45,000 + 45,000 - 30,000 - Ma = 0$$
$$Ma = 60,000 \text{ lb}$$
$$a = \frac{60,000}{M} = \frac{60,000g}{30,000} = 2g$$

Considering the landing gear as a free body, as shown in Fig. 3.8(b), the inertia force is

$$M_1 a = \frac{w_1}{g} a = \frac{500}{g} \times 2g = 1,000 \text{ lb}$$

The compression load C in the oleo strut is found from a summation of vertical forces.

$$\Sigma F_y = 45,000 - 500 - 1,000 - C = 0$$
$$C = 43,500 \text{ lb}$$

The bending moment m is found as follows:

$$m = 45,000 \times 6 - 1,000 \times 6 - 500 \times 6 = 261,000 \text{ in-lb}$$

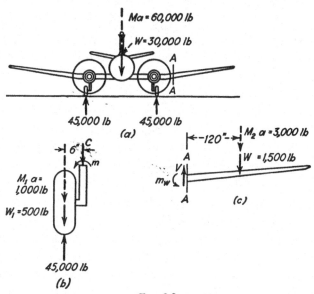

Fig. 3.8.

b. The inertia force acting on the portion of the wing shown in Fig. 3.8(c) is

$$M_2 a = \frac{w_2}{g} a = \frac{1,500}{g} \times 2g = 3,000 \text{ lb}$$

The wing shear at section AA is found from a summation of vertical forces.

$$\Sigma F_y = V \quad 3,000 - 1,500 = 0$$
$$V = 4,500 \text{ lb}$$

The wing bending moment is found by taking moments about section AA.

$$m_w = 3,000 \times 120 + 1,500 \times 120 = 540,000 \text{ in-lb}$$

c. The shock strut deflection is found by assuming a constant vertical acceleration of $-2g$, or -64.4 ft/sec², from an initial vertical velocity of 12 ft/sec to a final zero vertical velocity.

$$v^2 - v_0^2 = 2as$$
$$0 - (12)^2 = 2(-64.4)s$$
$$s = 1.12 \text{ ft}$$

d. The time required to absorb the landing shock is found from Eq. 3.7.

$$v - v_0 = at$$
$$0 - 12 = -64.4t$$
$$t = 0.186 \text{ sec}$$

Since this landing shock occurs for such a short interval of time, it may be less injurious to the structure and less disagreeable to the passengers than a sustained load would be.

Example 4. The 8,000-lb airplane shown in Fig. 3.9 is landing on soft ground with an upward acceleration, a_y, of $3.5g$ and an aft acceleration, a_x, of $1.5g$. Find the wheel reactions A and B, assuming them to be parallel.

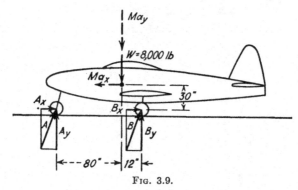

FIG. 3.9.

Solution. The vertical inertia force is found from the vertical acceleration.

$$Ma_y = \frac{8,000}{g} \; 3.5g = 28,000 \text{ lb}$$

This force acts in a direction opposite to the acceleration, or downward. The horizontal inertia force is found in a similar manner, and acts forward at the center of gravity.

$$Ma_x = \frac{8,000}{g} \times 1.5g = 12,000 \text{ lb}$$

The total vertical force at the center of gravity is the sum of the inertia force and the weight, or 36,000 lb. The horizontal force of 12,000 lb is one-third of this vertical force, so the resultant of these forces acts at an angle with the vertical which is $\tan^{-1} \frac{1}{3}$. The wheel reactions must both be parallel to this resultant, and the following relations are obtained:

$$A_x = \frac{1}{3}A_y$$
$$B_x = \frac{1}{3}B_y$$

The forces may now be found from the equations of statics:

$$\Sigma M_B = 92A_y - 12,000 \times 30 - 36,000 \times 12 = 0$$
$$A_y = 8,600 \text{ lb}$$
$$\Sigma F_y = 8,600 - 36,000 + B_y = 0$$
$$B_y = 27,400$$

$$A_x = \tfrac{1}{3}A_y = 2,870 \text{ lb}$$
$$B_x = \tfrac{1}{3}B_y = 9,130 \text{ lb}$$

Check: $\Sigma M_A = 36,000 \times 80 - 12,000 \times 30 - 27,400 \times 92 = 0$

PROBLEMS

3.1. A 20,000-lb airplane with the dimensions shown is being catapulted from a carrier with a forward acceleration of $3g$.

a. Find the tension in the launching cable T and the wheel reactions R_1 and R_2.

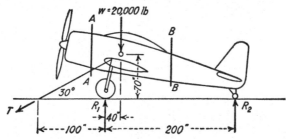

PROBS. 3.1 AND 3.2.

b. Find the horizontal tension and vertical shear in the fuselage at cross sections AA and BB if the portion of the airplane forward of section AA weighs 3,500 lb and the portion aft of section BB weighs 2,000 lb. Check the results by considering the part of the airplane between the two cross sections.

c. If the flying speed is 80 mph, what distance is required for the airplane to gain this speed? How much time is required?

3.2. If an airplane with the dimensions used in Prob. 3.1 is taxiing, what maximum braking force may be applied at the main wheels without the airplane nosing over?

3.3. An airplane weighing 5,000 lb strikes an upward gust of air which produces a wing lift of 25,000 lb. What tail load P is required to prevent a pitching accel-

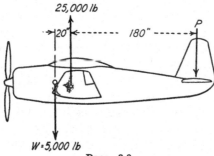

PROB. 3.3.

eration, if the dimensions are as shown? What will be the vertical acceleration of the airplane? If this lift force acts until the airplane obtains a vertical velocity of 20 ft/sec, how much time is required?

3.4. An airplane weighing 8,000 lb has an upward acceleration of $3g$ when landing.   If the dimensions are as shown, what are the wheel reactions R_1 and R_2?   What time is required to decelerate the airplane from a vertical velocity

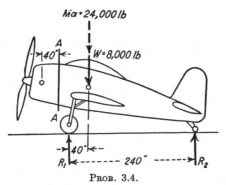

Prob. 3.4.

of 12 ft/sec?   What is the vertical compression of the landing gear during this deceleration?   What is the shear and bending moment on a vertical section AA, if the weight forward of this section is 2,000 lb and has a center of gravity 40 in. from this cross section?

3.2. Inertia Forces on Rotating Bodies.   In the preceding discussion all parts of the rigid body were moving in straight, parallel lines and had equal velocities and accelerations.   In many engineering problems it is necessary to consider the inertia forces acting on a rigid body which has other types of motion.   In many cases where the elements of a rigid body are moving in curved paths they are moving in such a way that each element moves in only one plane, and all elements move in parallel planes.   This type of motion is called plane motion, and occurs, for example, when an airplane is pitching and yet has no rolling or yawing motion.   All elements of the airplane move in planes parallel to the plane of symmetry.   Any type of plane motion can be considered as a rotation about some instantaneous axis perpendicular to the planes of motion, and the following equations for inertia forces are derived on the assumption that the rigid body is rotating about an instantaneous axis perpendicular to a plane of symmetry of the body.   The inertia forces obtained may be used for the pitching motion of an airplane, but when used for rolling or yawing motions it is necessary first to obtain the principal axes and moments of inertia of the airplane.

A rigid body in plane motion must have the same angular velocity at all points.   This angular velocity ω is defined as the time rate of change of the angle θ measured between a reference line on the body and a fixed reference axis.

$$\omega = \frac{d\theta}{dt} \quad \text{rad/sec} \tag{3.10}$$

The angle is expressed in radians, and ω is in units of radians per second. The angular acceleration α is the time rate of change of the angular velocity.

$$\alpha = \frac{d\omega}{dt} \qquad \text{rad/sec}^2 \qquad (3.11)$$

All points of a rigid body will have the same angular acceleration. For constant angular acceleration α, the following expressions similar to those in Eqs. 3.7 to 3.9 are obtained:

$$\omega - \omega_0 = \alpha t \qquad (3.12)$$
$$\theta = \omega_0 t + \tfrac{1}{2}\alpha t^2 \qquad (3.13)$$
$$\omega^2 - \omega_0^2 = 2\alpha\theta \qquad (3.14)$$

where θ is the angle of rotation in time t, ω_0 is the initial angular velocity, and ω is the angular velocity after t sec.

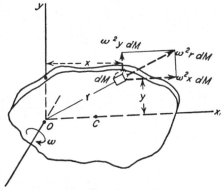

Fig. 3.10.

The rigid mass shown in Fig. 3.10 is rotating about point O with a constant angular velocity ω. The acceleration of any point a distance r from the center of rotation is $\omega^2 r$, and is directed toward the center of rotation. The inertia force acting on an element of mass dM is the product of the mass and the acceleration, or $\omega^2 r\, dM$, and is directed away from the axis of rotation. This inertia force has components $\omega^2 x\, dM$ parallel to the x axis and $\omega^2 y\, dM$ parallel to the y axis. If the x axis is chosen through the center of gravity C, the forces are simplified. The resultant inertia force in the y direction for the entire body is found as follows:

$$F_y = \int \omega^2 y\, dM = \omega^2 \int y\, dM = 0$$

The angular velocity ω is constant for all elements of the body, and the integral is zero because the x axis was chosen through the center of gravity. The inertia force in the x direction is found in the same manner.

$$F_x = \int \omega^2 x \, dM = \omega^2 \int x \, dM = \omega^2 \bar{x} M \tag{3.15}$$

The term $\bar{x}$ is the distance from the axis of rotation O to the center of gravity C, as shown in Fig. 3.10.

If the body has an angular acceleration α, the element of mass dM has an additional inertia force $\alpha r \, dM$ acting perpendicular to r and opposite to the direction of acceleration. This force has components $\alpha x \, dM$ in

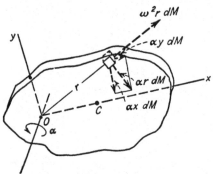

Fɪɢ. 3.11.

the y direction, and $\alpha y \, dM$ in the x direction, as shown in Fig. 3.11. The resultant inertia force on the entire body in the x direction is

$$F_x = \int \alpha y \, dM = \alpha \int y \, dM = 0$$

The resultant inertia force in the y direction is

$$F_y = \int \alpha x \, dM = \alpha \int x \, dM = \alpha \bar{x} M \tag{3.16}$$

The resultant inertia torque about the axis of rotation is found by integrating the terms representing the product of the tangential force on each element $\alpha r \, dM$ and its moment arm r.

$$T_0 = \int \alpha r^2 \, dM = \alpha \int r^2 \, dM = \alpha I_0 \tag{3.17}$$

The term I_0 represents the moment of inertia of the mass about the axis of rotation. It can be shown that this moment of inertia can be transferred to a parallel axis through the center of gravity by use of the following relationship:

$$I_0 = M \bar{x}^2 + I_c \tag{3.18}$$

where I_c is the moment of inertia of the mass about an axis through the center of gravity, obtained as the sum of the products of mass elements dM and the square of their distances r_c from the center of gravity.

$$I_c = \int r_c^2 \, dM$$

Substituting the value of I_0 from Eq. 3.18 in Eq. 3.17, the following expression for the inertia torque is obtained:

$$T_0 = M\bar{x}^2\alpha + I_c\alpha \qquad (3.19)$$

The inertia forces obtained in Eqs. 3.15, 3.16, and 3.19 may be represented as forces acting at the center of gravity and the couple $I_c\alpha$, as shown in Fig. 3.12. The force $\alpha\bar{x}M$ and the couple $I_c\alpha$ must both pro-

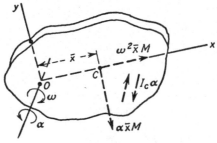

Fig. 3.12.

duce moments about point O which are opposite to the direction of α. The force $\omega^2\bar{x}M$ must act away from point O.

It is seen from Fig. 3.12 that the forces at the centroid represent the product of the mass of the body and the components of acceleration of the center of gravity. In many cases the axis of rotation is not known, but the components of acceleration of the center of gravity can be obtained. In other cases the acceleration of one point of the body and the angular velocity and angular acceleration are known. If the point O in Fig. 3.13 has an acceleration a_0, an inertia force at the center of gravity

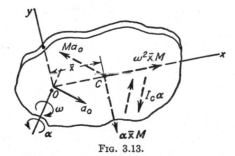

Fig. 3.13.

of Ma_0, opposite to the direction of a_0, must be considered in addition to those previously considered.

Consistent units must be used for the inertia couple $I_c\alpha$ of Eq. 3.19. If foot units are used throughout, the moment of inertia may be in units of slug-ft², or since the slug has units of lb-sec²/ft, the moment of inertia is in units of lb-sec²-ft.

$$I = \int r^2 \, dM = \int \frac{r^2}{g} \, dW = \frac{\text{ft}^2\text{-lb}}{\text{ft/sec}^2} = \text{lb-sec}^2\text{-ft}$$

The value of the acceleration of gravity g is used as 32.2 ft/sec². The angular acceleration α has units of radians per second per second, or since radians are dimensionless, has the dimension sec⁻². The couple $I\alpha$ is in units of foot-pounds.

All drawings of airplanes are dimensioned in units of inches, regardless of the size of the airplane. Most calculations for weight and balance, and for structures, are made using units of inches. The acceleration of gravity g is $32.2 \times 12 = 386$ in./sec². It is convenient to calculate the moment of inertia of an airplane by multiplying the weight of each element by the square of the distance in inches and dividing by $g = 386$ in./sec².

$$I = \int \frac{r^2 \, dW}{g} = \frac{\text{in}^2\text{-lb}}{\text{in./sec}^2} = \text{lb-sec}^2\text{-in.}$$

The inertia couple $I\alpha$ will then be in units of inch-pounds.

Example 1. A 60,000-lb airplane with a tricycle landing gear makes a hard two-wheel landing in soft ground, so that the vertical ground reaction is 270,000 lb and the horizontal ground reaction is 90,000 lb. The moment of inertia about the center of gravity is 5,000,000 lb-sec²-in., and the dimensions are shown in Fig. 3.14.

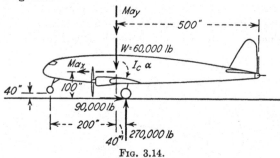

Fig. 3.14.

a. Find the inertia forces on the airplane.

b. Find the inertia forces on a 400-lb gun turret in the tail, which is 500 in. aft of the center of gravity. Neglect the moment of inertia of the turret about its own center of gravity.

c. If the nose wheel is 40 in. from the ground when the main wheels touch the ground, find the angular velocity of the airplane and the vertical velocity of the nose wheel when the nose wheel reaches the ground, assuming no appreciable change in the moment arms. The airplane center of gravity has a vertical velocity of 12 ft/sec at the moment of impact, and the ground reactions are assumed constant until the vertical velocity reaches zero, at which time the vertical ground reaction becomes 60,000 lb and the horizontal ground reaction becomes 20,000 lb.

Solution. *a.* The inertia forces on the entire airplane may be considered as horizontal and vertical forces Ma_x and Ma_y at the center of gravity and a couple $I_c\alpha$, as shown in Fig. 3.14. These correspond to the inertia forces shown on the mass of Fig. 3.12, since the forces at the center of gravity represent the product of the mass and the acceleration components of the center of gravity.

$$\Sigma F_x = 90,000 - Ma_x = 0$$
$$Ma_x = 90,000 \text{ lb}$$
$$\Sigma F_y = 270,000 - 60,000 - Ma_y = 0$$
$$Ma_y = 210,000 \text{ lb}$$
$$\Sigma M_{cz} = -270,000 \times 40 - 90,000 \times 100 + I_c\alpha = 0$$
$$I_c\alpha = 19,800,000 \text{ in-lb}$$
$$a_x = \frac{90,000}{M} = \frac{90,000}{60,000}g = 1.5g$$
$$a_y = \frac{210,000}{M} = \frac{210,000}{60,000}g = 3.5g$$
$$\alpha = \frac{I_c\alpha}{I_c} = \frac{19,800,000}{5,000,000} = 3.96 \text{ rad/sec}^2$$

b. The acceleration of the center of gravity of the airplane is now known, and the acceleration and inertia forces for the turret can be obtained by the method shown in Fig. 3.13, where the center of gravity of the airplane corresponds to

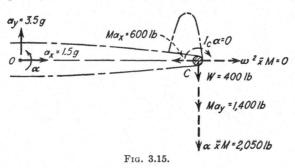

Fig. 3.15.

point O of Fig. 3.13 and the center of gravity of the turret corresponds to point C. These forces are shown in Fig. 3.15, and have the following values:

$$Ma_x = \frac{400}{g} \times 1.5g = 600 \text{ lb}$$
$$Ma_y = \frac{400}{g} \times 3.5g = 1,400 \text{ lb}$$
$$\alpha\bar{x}M = 3.96 \times 500 \times \frac{400}{386} = 2,050 \text{ lb}$$

In calculating the term $\alpha \bar{x} M$, $\bar{x}$ is in inches, and g is used as 386 in./sec². If $\bar{x}$ is in feet, g will be 32.2 ft/sec².

The total force on the turret is 600 lb forward and 3,850 lb down. This total force is seen to be almost ten times the weight of the turret.

c. The center of gravity of the airplane is decelerated vertically at 3.5g, or 112.7 ft/sec². The time of deceleration from an initial velocity of 12 ft/sec to a zero vertical velocity is found from Eq. 3.7.

$$v - v_0 = at$$
$$0 - 12 = -112.7t$$
$$t = 0.106 \text{ sec}$$

During this time the center of gravity moves through a distance found from Eq. 3.8.

$$s = v_0 t + \tfrac{1}{2}at^2$$
$$= 12 \times 0.106 - \tfrac{1}{2} \times 112.7 \times (0.106)^2$$
$$= 0.636 \text{ ft, or } 7.64 \text{ in.}$$

The angular velocity of the airplane at the end of 0.106 sec after the landing is found from Eq. 3.12.

$$\omega - \omega_0 = \alpha t$$
$$\omega - 0 = 3.96 \times 0.106$$
$$\omega = 0.42 \text{ rad/sec}$$

The angle of rotation during this time is found from Eq. 3.13.

$$\theta_1 = \omega_0 t + \tfrac{1}{2}\alpha t^2$$
$$\theta_1 = 0 + \tfrac{1}{2}(3.96)(0.106)^2 = 0.0222 \text{ rad}$$

The vertical motion of the nose wheel resulting from this rotation, shown in Fig. 3.16, is

$$s_1 = \theta_1 x = 0.0222 \times 200 = 4.44 \text{ in.}$$

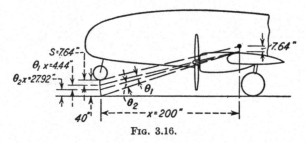

FIG. 3.16.

The distance of the nose wheel from the ground, after the vertical velocity of the center of gravity of the airplane has become zero, is

$$s_2 = 40 - 7.64 - 4.44 = 27.92 \text{ in.}$$

The remaining angle of rotation θ_2, shown in Fig. 3.16, is

$$\theta_2 = \frac{s_2}{x} = \frac{27.92}{200} = 0.1396 \text{ rad}$$

Since the ground reactions decrease by the ratio of $\dfrac{60,000}{270,000}$ after the vertical acceleration of the airplane becomes zero, the angular acceleration decreases in the same proportion, as found by equating moments about the center of gravity.

$$\alpha_2 = \frac{60,000}{270,000} \times 3.96 = 0.88 \text{ rad/sec}^2$$

The angular velocity of the airplane at the time the nose wheel strikes the ground is found from Eq. 3.14.

$$\omega^2 - \omega_0^2 = 2\alpha_2\theta_2$$
$$\omega^2 - (0.42)^2 = 2 \times 0.88 \times 0.1396$$
$$\omega = 0.65 \text{ rad/sec}$$

Since at this time the motion is rotation, with no vertical motion of the center of gravity, the vertical velocity of the nose wheel is found as follows:

$$v = \omega x$$
$$v = 0.65 \times \tfrac{200}{12} = 10.8 \text{ ft/sec}$$

This velocity is smaller than the initial sinking velocity of the airplane. The nose wheel consequently would strike the ground with a higher velocity in a three-wheel level landing.

It is of interest to find the centrifugal force on the turret, $\omega^2\bar{x}M$, at the time the nose wheel strikes the ground. This force was zero at the time the main wheels hit, because the angular velocity ω was zero. For the final value of ω, the following value is obtained

$$\omega^2\bar{x}M = (0.65)^2 \times 500 \times \tfrac{400}{386} = 219 \text{ lb}$$

This force is much smaller than other forces acting on the turret, and is usually neglected.

In part (c) of this problem, certain simplifying assumptions were made which do not quite correspond with actual landing conditions. Aerodynamic forces were neglected, and the ground reactions on the landing gear were assumed constant while the landing gear was being compressed. The actual compression of the landing gear is a combination of the tire deflection, in which the load is approximately proportional to the deformation, and the oleo-strut deflection, in which the load is almost constant during the entire deformation, as assumed in the problem. The tire deflection may be as much as one-third to one-half the total deflection. The aerodynamic forces, which have been neglected, would probably reduce the maximum angular velocity of the airplane, since the horizontal tail moves upward as the airplane pitches, and the combination of upward and forward motion would give a downward aerodynamic force on the tail, tending to reduce the pitching acceleration.

The aerodynamic effects of the lift on the wing and tail surfaces are not shown in Fig. 3.14, but they will not affect the pitching acceleration appreciably if the ground reactions remain the same. Just before the airplane strikes the ground, the lift forces on the wing and tail are in equilibrium with the gravity force of 60,000 lb. Since the horizontal velocity of the airplane and the angle of attack are not appreciably changed ($\theta_1 = 0.0222 \text{ rad} = 1.27°$), the lift forces continue to balance the weight of the airplane when the center of gravity is being decelerated. Instead of the weight of 60,000 lb shown in Fig. 3.14, there should be an additional inertia force of 60,000 lb down at the center of gravity. The moments about the center of gravity and the pitching acceleration are not changed, but the vertical deceleration a_y is increased. At the end of the deceleration of the center of gravity the ground reactions are almost zero, since most of the airplane weight is carried by the lift on the wings. The airplane then pitches forward through the angle θ_2, which appreciably changes the angle of attack ($\theta_2 = 0.1396$ rad $= 8°$). The wing lift is then decreased, and most of the weight is supported by the ground reactions on the wheels. For the structural design of the airplane, only the loads during the initial impact are usually significant.

Example 2. An airplane weighing 8,000 lb is flying at 250 mph in a

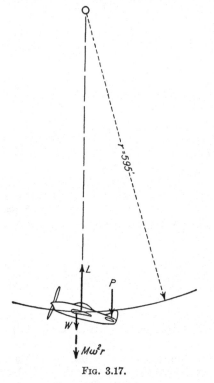

FIG. 3.17.

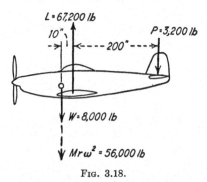

FIG. 3.18.

vertical circle with a radius of 595 ft, as shown in Fig. 3.17. Find the lift L on the wing and the load P on the tail at the time the airplane is moving horizontally, if the dimensions are as shown in Fig. 3.18.

Solution. If the airplane is moving at a constant velocity along a circular path, the angular acceleration is zero, and the only inertia force is the centrifugal force which is found as follows:

$$250 \text{ mph} = 366 \text{ ft/sec}$$

$$\omega = \frac{v}{r} = \frac{366}{595} = 0.615 \text{ rad/sec}$$

$$M\omega^2 r = \frac{8,000}{32.2} \times (0.615)^2 \times 595 = 56,000 \text{ lb}$$

This inertia force acts down at the center of gravity of the airplane and must be added to the weight of the airplane. The propeller thrust is assumed to be equal and opposite to the airplane drag, so that only the vertical forces need be considered. The forces L and P are obtained from the equations of statics.

$$\Sigma M_L = 200P - 8,000 \times 10 - 56,000 \times 10 = 0$$
$$P = 3,200 \text{ lb}$$
$$\Sigma F_y = L - 56,000 - 8,000 - 3,200 = 0$$
$$L = 67,200 \text{ lb}$$

Example 3. The airplane in Example 2 is maneuvered by giving the control stick an abrupt forward displacement so that the airplane is given a pitching acceleration of 6 rad/sec². Find the tail load and the inertia forces if the wing lift does not change while the control stick is being moved. The moment of inertia of the airplane about a pitching axis through the center of gravity is 180,000 lb-sec²-in. Find the forces acting on the engine which weighs 1,000 lb and is 50 in. forward of the center of gravity. Find the time required for the airplane to pitch through an angle of 3°, if the pitching acceleration is constant.

Solution. The inertia couple is found as follows:

$$I\alpha = 180,000 \times 6 = 1,080,000 \text{ in-lb}$$

The tail load P_1 and the inertia force at the center of gravity are shown in Fig. 3.19 and are found as follows:

$$\Sigma M_{cg} = 1,080,000 - 67,200 \times 10 - 210P_1 = 0$$
$$P_1 = 1,940 \text{ lb}$$
$$\Sigma F_y = 67,200 + 1,940 - 8,000 - Ma_y = 0$$
$$Ma_y = 61,140 \text{ lb}$$
$$a_y = \frac{61,140}{M} = \frac{61,140}{8,000} g = 7.65g$$

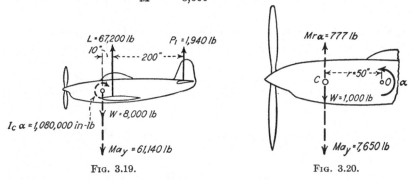

FIG. 3.19.   FIG. 3.20.

The forces acting on the engine, shown in Fig. 3.20, consist of the weight of 1,000 lb down, the inertia force Ma_y down, and the inertia force $Mr\alpha$ up.

$$Ma_y = \frac{1,000}{g} \times 7.65g = 7,650 \text{ lb}$$
$$Mr\alpha = \frac{1,000}{386} \times 50 \times 6 = 777 \text{ lb}$$

The resultant force down is

$$1,000 + 7,750 - 777 = 7,973 \text{ lb}$$

The corresponding force acting on the engine for the conditions of Example 2 would be 8,000 lb. The pitching acceleration therefore reduces this load.

The time required for the airplane to pitch through an angle of 3° is found from Eq. 3.13.

$$\theta = \frac{3}{57.3} = 0.0523 \text{ rad}$$
$$\theta = \omega_0 t + \tfrac{1}{2}\alpha t^2$$
$$0.0523 = 0 + \tfrac{1}{2} \times 6t^2$$
$$t = 0.132 \text{ sec}$$

A maneuver such as that considered here would represent an improbable condition. Even for a military airplane, which might be maneuvered violently, tests have shown that the minimum possible time required for a full displacement of the controls is about 0.1 sec. The time required for the angle of attack to decrease 3° was only 0.132 sec, and this decrease in angle of attack would decrease the wing lift at least 20 per cent. By the time the control surfaces had been displaced far enough to produce the specified tail load, the airplane would have pitched enough to reduce the wing load, and consequently the inertia load would be decreased rather than increased. It is probable that the wing lift would always decrease more than the upload on the tail would increase. A commercial or private airplane would never be intentionally maneuvered in this manner, but it is doubtful that any structural damage would occur as a result of the rotational acceleration, if the translational acceleration did not exceed the allowable limit for the airplane.

Example 4. Helicopter rotor blades are hinged at the hub so that the blades are free to flap up and down. The blades move on the surface of a cone, and are prevented from excessive coning by the centrifugal forces. The moment of the centrifugal forces and the weight of the blade about the horizontal hinge axis must balance the moment of the aerodynamic forces. The blade shown in Fig. 3.21 is rotating at 300 rpm. The length of the blade is 200 in., and the blade weight of 50 lb is uniformly distributed along the length of the blade. If the resultant aerodynamic lift force is 1,000 lb and has a moment arm of 150 in., find the coning angle β.

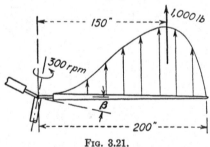

Fig. 3.21.

Solution. The resultant inertia forces previously obtained do not apply to this problem, since the rotating mass is not symmetrical about a plane perpendicular to the axis of rotation. The resultant inertia force on a blade of length L, with a weight w per unit length, is found by integrating the forces shown in Fig. 3.22.

$$F = \int \omega^2 x \, dM = \omega^2 \cos \beta \frac{w}{g} \int_0^L s \, ds = \omega^2 \cos \beta \frac{w}{g} \frac{L^2}{2}$$

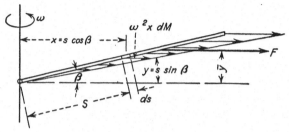

Fig. 3.22.

The moment of this force about the flapping hinge is

$$F\bar{y} = \int \omega^2 xy \, dM = \omega^2 \cos \beta \sin \beta \frac{w}{g} \int_0^L s^2 \, ds = \omega^2 \cos \beta \sin \beta \frac{w}{g} \frac{L^3}{3}$$

or

$$\bar{y} = \frac{F\bar{y}}{F} = \frac{2L}{3} \sin \beta$$

The numerical value of the centrifugal force is found as follows:

$$\omega = \tfrac{300}{60} 2\pi = 31.4 \text{ rad/sec}$$
$$F = (31.4)^2 \cos \beta \tfrac{50}{386} \tfrac{200}{2} = 12,800 \cos \beta$$

The moment arm of this force is

$$\bar{y} = \frac{2 \times 200}{3} \sin \beta = 133 \sin \beta$$

The moment of the centrifugal force is

$$F\bar{y} = 1,700,000 \sin \beta \cos \beta$$

Equating this moment to the moments due to the lift and the weight of the blade, the following equation is obtained:

$$1,700,000 \sin \beta \cos \beta = 1,000 \times 150 - 50 \times 100 \cos \beta$$

Solving this equation by trial,

$$\beta = 4.9°$$

3.3. Load Factors for Translational Acceleration. For flight or landing conditions in which the airplane has only translational acceleration, every part of the airplane is acted upon by parallel inertia forces which are proportional to the weight of the part. For purposes of analysis it is convenient to combine these inertia forces with the forces of gravity by multiplying the weight of each part by a load factor n and thus consider the combined weight and inertia forces. When the airplane is being accelerated upward, the weight and inertia forces add directly. The weight w of any part and the inertia force wa/g have a sum nw.

$$nw = w + w\frac{a}{g}$$

or

$$n = 1 + \frac{a}{g} \tag{3.20}$$

The combined inertia and gravity forces are considered in the analysis in the same manner as weights which are multiplied by the load factor n.

In the case of an airplane in flight with no horizontal acceleration, as shown in Fig. 3.23, the propeller thrust is equal to the airplane drag, and

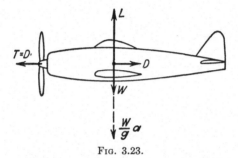

F$_{\text{IG}}$. 3.23.

the horizontal components of the inertia and gravity forces are zero. The weight and the inertia force on the airplane act down and will be equal to the lift. The airplane lift L is the resultant of the wing and tail lift forces. The load factor is defined as follows:

$$\text{Load factor} = \frac{\text{lift}}{\text{weight}}$$

or

$$n = \frac{L}{W} \tag{3.21}$$

This value for the load factor can be shown to be the same as that given by Eq. 3.20 by equating the lift nW to the sum of the weight and inertia forces.

$$L = nW = W + W\frac{a}{g}$$

or

$$n = 1 + \frac{a}{g}$$

which corresponds to Eq. 3.20.

An airplane frequently has horizontal acceleration as well as vertical acceleration. The airplane shown in Fig. 3.24 is being accelerated forward, since the propeller thrust T is greater than the airplane drag D. Every element of mass in the airplane is thus under the action of a

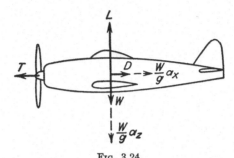

Fig. 3.24.

horizontal inertia force equal to the product of its mass and the horizontal acceleration. It is also convenient to consider the horizontal inertia loads as equal to the product of a load factor n_x and the weights. This horizontal load factor, often called the thrust load factor, is obtained from the equilibrium of the horizontal forces shown in Fig. 3.24.

$$n_x W = \frac{a_x}{g} W = T - D$$

or

$$n_x = \frac{T - D}{W} \tag{3.22}$$

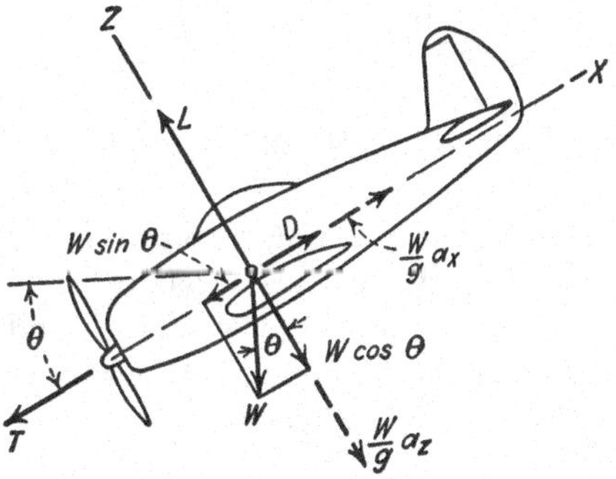

Fig. 3.25.

A more general case of translational acceleration is that shown in Fig. 3.25, in which the airplane thrust line is not horizontal. It is usually convenient to obtain components of forces along x and z axes which are parallel and perpendicular to the airplane thrust line. The combined weight and inertia load on any element has a component along the z axis of the following magnitude:

$$nw = w \cos \theta + w \frac{a_z}{g}$$

or

$$n = \cos \theta + \frac{a_z}{g} \tag{3.23}$$

From a summation of all forces along the z axis,

$$L = W\left(\cos \theta + \frac{a_z}{g}\right) \tag{3.24}$$

By combining Eqs. 3.23 and 3.24,

$$L = Wn$$

or

$$n = \frac{L}{W}$$

which corresponds with the value used in Eq. 3.21 for a level attitude of the airplane.

The thrust load factor for the condition shown in Fig. 3.25 is also similar to that obtained for the airplane in level attitude. Since the thrust and drag forces must be in equilibrium with the components of weight and inertia forces along the x axis, the thrust load factor is obtained as follows:

$$n_x W = \frac{W}{g} a_x - W \sin \theta = T - D$$

or

$$n_x = \frac{T - D}{W}$$

This value is the same as that obtained in Eq. 3.22 for a level attitude of the airplane.

In the case of the airplane landing as shown in Fig. 3.26, the landing load factor is defined as the vertical ground reaction divided by the airplane weight. The load factor in the horizontal direction is similarly defined as the horizontal ground reaction divided by the airplane weight.

$$n = \frac{R_z}{W} \tag{3.25}$$

$$n_x = \frac{R_x}{W} \tag{3.26}$$

In the airplane analysis it is necessary to obtain the components of the load factor along axes parallel and perpendicular to the propeller thrust line. Aerodynamic forces, however, are usually first obtained as lift and drag forces perpendicular and parallel to the direction of flight. If load factors are first obtained along lift and drag axes, they may be resolved into components along other axes, in the same manner that forces are resolved into components.

The force acting on any weight w is wn, and the component of this force along any axis at an angle θ to the force is $wn \cos \theta$. The component of the load factor is then $n \cos \theta$.

As a general definition, the load factor n_i along any axis i is such that the product of the load factor and the weight of an element is equal to the sum of the components of the weight and inertia forces along that axis.

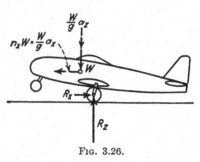

FIG. 3.26.

The weight and inertia forces are always in equilibrium with the external forces acting on the airplane, and the sum of the components of the weight and inertia forces along any axis must be equal and opposite to the sum of the components of the external forces along the axis ΣF_i. The load factor is therefore defined as follows:

$$n_i = \frac{-\Sigma F_i}{W} \tag{3.27}$$

where ΣF_i includes all forces except weight and inertia forces.

In many airplane loading conditions, only translational acceleration is considered even though the airplane has some rotational motion. In Example 2, Art. 3.2, for example, the airplane has a rotational velocity during the pull-out from a dive. The inertia forces all act through the center of curvature of the flight path, although they are assumed to be parallel for various elements of the airplane. It is obvious that the inertia force for a mass near the nose of the airplane has a forward component, and the inertia force for a mass near the tail has an aft component. Since the radius of the flight path circle is large compared with the length of the airplane, these components are negligible.

The maximum loads which an airplane may be expected to encounter at any time in service are designated as *limit* loads or *applied* loads. The load factors associated with these loads are known as *limit load factors*, or *applied load factors*. For loads which are under the control of the pilot, flight restrictions are used so that the limit load factor is never exceeded. All parts of the airplane are designed so that they are not stressed beyond the yield point at the limit load factor. It is also necessary to use a safety factor in the structural design of airplanes.

In most cases a safety factor of 1.5 based on the ultimate strength of the members is used, although higher factors are used for castings, fittings, and joints. The load factor obtained by multiplying the applied or limit load factor by the factor of safety is known as the *design load factor*, or *ultimate load factor*. The airplane structure must carry the ultimate or design loads without collapsing, even though the members may acquire permanent deformation under these loads.

The load factors for which airplanes must be designed are discussed in detail in a later chapter. The load factors are usually specified by the purchaser, or by the government licensing agency. The load factors depend on the purpose for which the airplane is intended and on the aerodynamic characteristics of the airplane. Military airplanes such as fighters or dive bombers are designed so that the airplane is strong enough to withstand any load factor that the pilot can withstand. Experience has shown that some pilots can withstand load factors of 7 or 8 before losing consciousness or "blacking out"; therefore a fighter or dive-bomber airplane would be designed for a limit or applied load factor of about 8, or a design or ultimate load factor of about 12. Large transport airplanes are never intentionally maneuvered to exceed a load factor much greater than 1. The greatest flight loads encountered by such airplanes occur as a result of air "gusts," or ascending air currents. The gust load factors depend on the gust velocity, the airplane velocity, and the airplane wing loading. Other types of airplanes are designed for load factors between those for the large transports and those for the fighters.

Landing load factors also depend on the function of the airplane. Large transport airplanes are expected to be landed on paved runways by experienced pilots, and need not be designed for large landing load factors. Training airplanes which may be operated from rough fields by inexperienced pilots must be designed for rather large landing load factors. The required load factor for any airplane is specified on the basis of the type of airplane, the airplane weight, and the wing loading.

3.4. Load Factors for Angular Acceleration. In the design of private or commercial airplanes it is seldom necessary to make an extensive analysis of the inertia forces resulting from rotational accelerations, since the loads in these conditions will determine the design for only a very few of the structural members, and conservative approximations do not result in excessive structural weight. In most of the loading conditions for military airplanes the angular accelerations are also neglected. In the design of Navy airplanes, which are often subjected to unusual loading conditions such as catapulting, arrested landings, barrier crashes, or seaplane landings, it is necessary to calculate the moment of inertia of the airplane with reasonable accuracy and to consider various conditions of angular acceleration.

In the two-wheel landing analyzed in Example 1, Art. 3.2, the landing load factor for the airplane was 4.5; yet the total load down on the 400-lb turret was 3,850 lb. Of this load, 1,800 lb results from the weight and the translational acceleration, and 2,050 lb results from the angular acceleration of the airplane. It is obviously very important that the angular acceleration be considered when designing the supporting structure for the turret, but it is doubtful that much of the remaining airplane structure would be designed from this loading condition. The aerodynamic loads on the tail surfaces during various flight conditions will be much larger than the inertia loads during landing, and consequently most of the empennage and fuselage structure will be designed for the flight condition. For airplanes with the tail-wheel type of landing gear, the angular acceleration during landing is much smaller than that computed for the airplane with the nose-wheel type of landing gear.

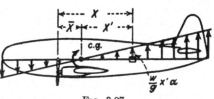

Fig. 3.27.

The vertical force F acting on an element of weight w when the airplane has a pitching acceleration α consists of one force equal to the airplane load factor times the weight and of another force proportional to the distance of the element from the center of gravity, as shown in Fig. 3.27.

$$F = nw - \frac{w}{g} x'\alpha \qquad (3.28)$$

The distance x' is measured from the center of gravity and must be multiplied by the weight. In the weight and balance calculations for the airplane, the terms wx are calculated for all items of weight, where x is measured from some reference plane a distance $\bar{x}$ forward of the center of gravity. From Fig. 3.27,

$$x' = x - \bar{x}$$

substituting this value of x' in Eq. 3.28,

$$\left.\begin{array}{l} F = nw - \dfrac{\alpha}{g} w(x - \bar{x}) \\[2mm] F = n'w - \dfrac{\alpha}{g} wx \end{array}\right\} \qquad (3.29)$$

where

$$n' = n + \frac{\alpha\bar{x}}{g}$$

The last term in Eq. 3.29 is easily evaluated by multiplying the tabulated values of wx by the ratio α/g. The load factor n' represents the load factor for an element of weight located at the reference plane.

PROBLEMS

3.5. Repeat Example 1, Art. 3.2, assuming that the airplane lands on a hard runway and that there is no horizontal ground reaction on the wheels. Find the landing load factor for a point at the center of gravity and also for a point a distance x from the center of gravity.

3.6. Assume the airplane of Example 1, Art. 3.2, to land in such a way that the nose wheel strikes the runway when the main wheels are 12 in. above the runway. The reaction on the nose wheel is vertical and has a constant value of 60,000 lb until the main wheels touch. Find the load factor for a point at the center of gravity, a point 200 in. forward of the center of gravity, and a point 500 in. aft of the center of gravity. Find the distance the nose-wheel shock strut is compressed at the time the main wheels touch if the vertical velocity of descent is 12 ft/sec before landing.

3.7. An airplane is making a horizontal turn with a radius of 1,000 ft, and with no change in altitude. Find the angle of bank and the load factor for a speed of (a) 200 mph, (b) 300 mph, and (c) 400 mph. Find the loads on the wing and tail if the dimensions of the airplane are shown in Fig. 3.18.

3.8. The airplane shown is making an arrested landing on a carrier deck. Find the load factors n and n_x, perpendicular and parallel to the deck, for a point at the center of gravity, a point 200 in. aft of the center of gravity, and a point

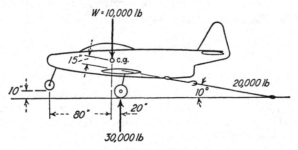

PROB. 3.8.

100 in. forward of the center of gravity. Find the relative vertical velocity with which the nose wheel strikes the deck if the vertical velocity of the center of gravity is 12 ft/sec and the angular velocity is 0.5 rad/sec counterclockwise for the position shown. The radius of gyration for the mass of the airplane about the center of gravity is 60 in. Assume no change in the dimensions or loads shown.

CHAPTER 4

MOMENTS OF INERTIA, MOHR'S CIRCLE

4.1. Centroids. The force of gravity acting on any body is the resultant of a group of parallel forces acting on all elements of the body.
The magnitude of the resultant of several parallel forces is equal to the algebraic sum of the forces, and the position of the resultant is such that it has a moment about any axis equal to the sum of the moments of the component forces. The resultant gravity force on a body is its weight W, which must be equal to the sum of the weights w_i of all elements of the body. If the forces of gravity act parallel to the z axis,

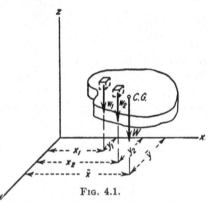

FIG. 4.1.

as shown in Fig. 4.1, the moments of all forces about the x and y axes must be equal to the moment of the resultant.

$$W\bar{x} = x_1w_1 + x_2w_2 + \cdots = \Sigma xw \quad \text{or} \quad \int x\,dW \qquad (4.1)$$

$$W\bar{y} = y_1w_1 + y_2w_2 + \cdots = \Sigma yw \quad \text{or} \quad \int y\,dW \qquad (4.2)$$

If the body and the axes are rotated so that the forces are parallel to one of the other axes, a third moment equation can be used.

$$W\bar{z} = z_1w_1 + z_2w_2 + \cdots = \Sigma zw \quad \text{or} \quad \int z\,dW \qquad (4.3)$$

The three coordinates $\bar{x}$, $\bar{y}$, and $\bar{z}$ of the center of gravity (c.g.) may be obtained from Eqs. 4.1 to 4.3.

$$\bar{x} = \frac{\Sigma xw}{W} \quad \text{or} \quad \frac{\int x\,dW}{W} \qquad (4.4)$$

$$\bar{y} = \frac{\Sigma yw}{W} \quad \text{or} \quad \frac{\int y\,dW}{W} \qquad (4.5)$$

$$\bar{z} = \frac{\Sigma zw}{W} \quad \text{or} \quad \frac{\int z\,dW}{W} \qquad (4.6)$$

73

The summations or integrals for Eqs. 4.4 to 4.6 must include all elements of the body. In many engineering problems the weights and coordinates of the various items are known, and the center of gravity is obtained by a summation procedure, rather than by an integration procedure.

In the case of a plate of uniform thickness and density which lies in the xy plane, as shown in Fig. 4.2, the coordinates of the center of gravity are

$$\bar{x} = \frac{\int x \, dW}{W} = \frac{w \int x \, dA}{wA} = \frac{\int x \, dA}{A} \tag{4.7}$$

$$\bar{y} = \frac{\int y \, dW}{W} = \frac{w \int y \, dA}{wA} = \frac{\int y \, dA}{A} \tag{4.8}$$

where A is the area of the plate and w is the weight per unit area. It is seen that the coordinates $\bar{x}$ and $\bar{y}$ will be the same regardless of the thickness or weight of the plate. In many engineering problems the properties of areas are important, and the point in the area having coordinates $\bar{x}$ and $\bar{y}$ as defined by Eqs. 4.7 and 4.8 is called the centroid of the area.

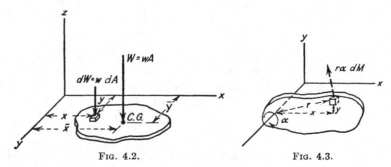

Fig. 4.2.   Fig. 4.3.

4.2. Moment of Inertia. In considering inertia forces on rotating masses, it was found that the inertia forces on the elements of mass had a moment about the axis of rotation of

$$L_z = \alpha \int r^2 \, dM$$

as shown in Fig. 4.3. The term under the integral sign is defined as the moment of inertia of the mass about the z axis.

$$I_z = \int r^2 \, dM \tag{4.9}$$

Since the x and y coordinates of the elements are easier to tabulate than the radius, it is frequently convenient to use the relation

$$r^2 = x^2 + y^2$$

or

$$I_z = \int x^2 \, dM + \int y^2 \, dM \tag{4.10}$$

An area has no mass, and consequently no inertia, but it is customary to designate the following properties of an area as the moments of inertia of the area since they are similar to the moments of inertia of masses.

$$I_z = \int y^2 \, dA \tag{4.11}$$

$$I_y = \int x^2 \, dA \tag{4.12}$$

The coordinates are shown in Fig. 4.4. The polar moment of inertia of an area is defined as follows:

$$I_p = \int r^2 \, dA \tag{4.13}$$

From the relationship used in Eq. 4.10,

$$\left. \begin{aligned} r^2 &= x^2 + y^2 \\ I_p &= \int x^2 \, dA + \int y^2 \, dA = I_y + I_z \end{aligned} \right\} \tag{4.14}$$

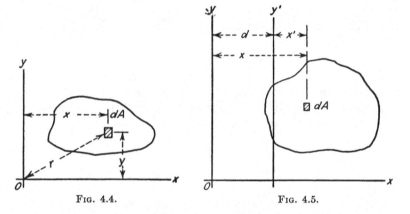

FIG. 4.4.   FIG. 4.5.

It is frequently necessary to find the moment of inertia of an area about an axis when the moment of inertia about a parallel axis is known. The moment of inertia about the y axis shown in Fig. 4.5 is defined as follows:

$$I_y = \int x^2 \, dA \tag{4.15}$$

substituting the relation $x = d + x'$ in Eq. 4.15,

$$I_y = \int (d + x')^2 \, dA$$

$$= d^2 \int dA + 2d \int x' \, dA + \int x'^2 \, dA$$

or

$$I_y = Ad^2 + 2\bar{x}'Ad + I_y' \tag{4.16}$$

where $\bar{x}'$ represents the distance of the centroid of the area from the y' axis, as defined in Eq. 4.7, I_y' represents the moment of inertia of the area about the y' axis, and A represents the total area. Equation 4.16 is simplified when the y' axis is through the centroid of the area, as shown in Fig. 4.6.

$$I_y = Ad^2 + I_c \tag{4.17}$$

The term I_c represents the moment of inertia of the area about a centroidal axis.

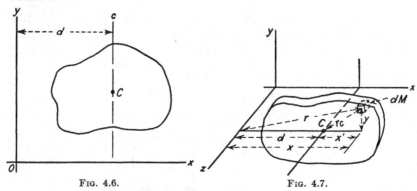

FIG. 4.6.   FIG. 4.7.

The moment of inertia of a mass may be transferred to a parallel axis by a similar procedure to that used for the moment of inertia of an area. For the mass shown in Fig. 4.7, the following relations apply where the centroidal axis C lies in the xz plane.

$$\begin{aligned}
I_z &= \int r^2 \, dM = \int (x^2 + y^2) \, dM \\
&= \int [(d + x')^2 + y^2] \, dM \\
&= \int [d^2 + 2dx' + x'^2 + y^2] \, dM
\end{aligned}$$

substituting $r_c^2 = x'^2 + y^2$,

$$I_z = d^2 \int dM + 2d \int x' \, dM + \int r_c^2 \, dM$$

Since x' is measured from the centroidal axis, the second integral is zero. The last integral represents the moment of inertia about the centroidal axis.

$$I_z = Md^2 + I_c \tag{4.18}$$

Equations 4.17 and 4.18 can be used either to find the moment of inertia about any axis when the moment of inertia about a parallel axis through the centroid is known or to find the moment of inertia about the centroidal axis when the moment of inertia about any other parallel axis is known. In transferring moments of inertia between two axes, neither of which is through the centroid, it is necessary first to find the moment of inertia about the centroidal axis, then to transfer this to the desired axis, by using Eq. 4.17 or Eq. 4.18 two times. It is seen that the moment of inertia is always a positive quantity, and that the moment of inertia about a centroidal axis is always smaller than that about any other parallel axis. If Eq. 4.16 is used, it is necessary to use the proper sign for the term $\bar{x}'$. All terms in Eqs. 4.17 and 4.18 are always positive.

The radius of gyration ρ of a body is the distance from the inertia axis that the entire mass would be concentrated in order to give the same moment of inertia. Equating the moment of inertia of the concentrated mass to that for the body,

$$\rho^2 M = I$$

or

$$\rho = \sqrt{\frac{I}{M}} \tag{4.19}$$

It is seen that the point where the mass is assumed to be concentrated is not the same as the center of gravity, except for the case where I_c in Eq. 4.18 is zero. The point at which the mass is assumed to be concentrated is also different for each inertia axis chosen.

The radius of gyration of an area is defined as the distance from the inertia axis to the point where the area would be concentrated in order to produce the same moment of inertia.

$$\rho^2 A = I$$

or

$$\rho = \sqrt{\frac{I}{A}} \tag{4.20}$$

The moment of inertia of an area is obtained as the product of an area and the square of a distance and is usually expressed in units of inches to the fourth power. The moments of inertia for the common areas shown

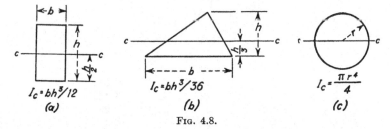

$$I_c = bh^3/12$$
(a)

$$I_c = bh^3/36$$
(b)

$$I_c = \frac{\pi r^4}{4}$$
(c)

Fig. 4.8.

in Fig. 4.8 should be memorized, since they are frequently used. Moments of inertia for other areas may be found by integration or from engineering handbooks.

Example 1. Find the center of gravity of the airplane shown in Fig. 4.9(a). The various items of weight and the coordinates of their individual centers of gravity are shown in Table 4.1. It is customary to take reference axes in the directions shown in Fig. 4.9(b) with the x axis parallel to the thrust line and the z axis vertical. While the z axis is at the wing leading edge in this problem, it may be taken through the propeller or through some other convenient reference point.

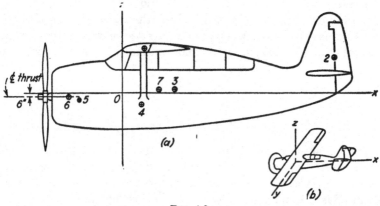

FIG. 4.9.

Solution. The y coordinate of the center of gravity is in the plane of symmetry of the airplane. The coordinates $\bar{x}$ and $\bar{z}$ are obtained from Eqs. 4.4 and 4.6. The terms W, Σxw, and Σzw are obtained by totaling columns (3), (5), and (7) of Table 4.1.

$$\bar{x} = \frac{\Sigma wx}{W} = \frac{50{,}723}{4{,}243} = 12.16$$

$$\bar{z} = \frac{\Sigma wz}{W} = \frac{26{,}109}{4{,}243} = 6.2$$

TABLE 4.1

No. (1)	Item (2)	Weight w (3)	x (4)	wx (5)	z (6)	wz (7)
1	Wing group.............	697	22.6	+15,781	40.9	+28,574
2	Tail group..............	156	198.0	30,904	33.1	5,171
3	Fuselage group..........	792	49.8	39,430	3.9	3,092
4	Landing gear (up).......	380	19.2	7,297	−11.7	−4,429
5	Engine section group.....	160	−38.6	−6,179	−7.1	−1,138
6	Power plant.............	1,302	−48.8	−63,674	−6.0	−7,782
7	Fixed equipment........	756	35.9	27,164	3.5	2,621
	Total weight empty....	4,243	.....	50,723	.....	26,109

Example 2. Find the centroid and the moment of inertia about a horizontal axis through the centroid of the area shown in Fig. 4.10. Find the radii of gyration about axis xx and about axis cc.

Solution. The area is divided into rectangles and triangles, as shown. The areas of the individual parts are tabulated in column (2) of Table 4.2. The y coordinates of the centroids of the elements are tabulated in column (3), and the moments of the areas Ay tabulated in column (4). The centroid of the total area is now obtained by dividing the summation of the terms in column (4) by the summation of the terms in column (2).

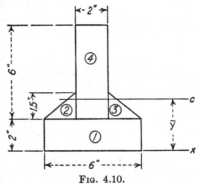

FIG. 4.10.

$$\bar{y} = \frac{\Sigma Ay}{A} = \frac{79.5}{27} = 2.94 \text{ in.}$$

TABLE 4.2

Element (1)	A (2)	y (3)	Ay (4)	Ay^2 (5)	I_0 (6)
1	12	1	12	12	4.0
2	1.5	2.5	3.75	9.4	0.2
3	1.5	2.5	3.75	9.4	0.2
4	12	5	60	300	36.0
Total...	27.0	...	79.5	330.8	40.4

The moment of inertia of the total area about the x axis will be obtained as the sum of the moments of inertia of the elements about this axis. In finding the moment of inertia of any element about the x axis, Eq. 4.17 may be written as follows:

$$I_x = Ay^2 + I_0$$

where I_x is the moment of inertia of the element of area A about the x axis, y is the distance from the centroid of the element to the x axis, and I_0 is the moment of inertia of the element about its own centroid. The terms Ay^2 for all the elements are obtained in column (5) as the product of terms in columns (3) and (4). The values of I_0 are obtained from the equations shown in Fig. 4.8. The moment of inertia of the entire area about the x axis is equal to the sum of all terms in columns (5) and (6).

$$I_x = 330.8 + 40.4 = 371.2 \text{ in.}^4$$

This moment of inertia may be transferred to the centroid of the entire area by using Eq. 4.17 as follows:

$$I_c = I_x - (\Sigma A)\bar{y}^2$$

where ΣA represents the total area and $\bar{y}$ represents the distance from the x axis to the centroid of the total area.

$$I_c = 371.2 - 27.0(2.94)^2 = 138.0 \text{ in.}^4$$

The radii of gyration may now be obtained as defined in Eq. 4.20.

$$\rho_x = \sqrt{\frac{I_x}{A}} = \sqrt{\frac{371.2}{27.0}} = 3.71 \text{ in.}$$

$$\rho_c = \sqrt{\frac{I_c}{A}} = \sqrt{\frac{138.0}{27.0}} = 2.26 \text{ in.}$$

Example 3. In a metal stressed-skin airplane wing, the sheet-metal covering acts with the supporting spanwise spars and stringers to form a beam which resists the wing bending. Figure 4.11(a) shows a cross section of a typical wing which has only one spar. The spar has a vertical web and extruded angle sections riveted to the spar web and to the skin. The stringers are extruded Z sections which are riveted to the skin. The upper surface of the wing is in compression, and the sheet-metal skin buckles between the stringers and is ineffective in carrying load. The skin is riveted to the stringers at frequent intervals, and a narrow strip of skin adjacent to each stringer is prevented from buckling and acts with the stringer in carrying compressive load. The effective width of skin acting with each stringer is usually about thirty times the skin thickness, but will be computed by more accurate equations in a subse-

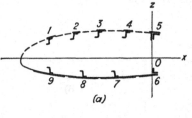

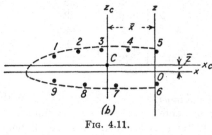

(a)

(b)

Fig. 4.11.

quent chapter. On the under side of the wing the entire width of the skin is effective in resisting tension. It is usually sufficiently accurate to assume the area of each stringer and its effective skin to be concentrated at the centroid of its area in computing the moment of inertia of the area. The wing cross section would then be represented by the nine elements of area shown in Fig. 4.11(b). The moment of inertia of each element about its own centroid is neglected. In this particular wing, the skin and stringers to the right of the spar are very light and are assumed to be nonstructural.

The moment of inertia of the area shown in Fig. 4.11(b) will be obtained about horizontal and vertical axes through the centroid of the total area. The areas and coordinates of the elements are given in columns (2), (3), and (6) of Table 4.3.

Solution. This problem is solved by the method used for Example 2, except that the column for I_0 is omitted. Table 4.3 shows the calculations for the moments of inertia about both the horizontal and vertical axes.

TABLE 4.3

Element (1)	A (2)	x (3)	Ax (4)	Ax² (5)	z (6)	Az (7)	Az² (8)
1	0.358	−34.5	−12.34	426	+8.6	3.08	26.5
2	0.204	−28.1	−5.73	161	+9.6	1.96	18.8
3	0.395	−19.9	−7.85	156	+10.0	3.95	39.5
4	0.204	−10.1	−2.06	21	+9.6	1.96	18.8
5	1.615	+0.5	+.81	0	8.8	14.21	125.2
6	1.931	+0.5	+.97	1	−5.7	−11.02	62.8
7	0.752	−10.1	−7.60	77	−5.2	−3.91	20.4
8	0.784	−22.4	−17.65	394	−4.3	−3.37	14.5
9	0.892	−34.7	−30.92	1,074	−2.4	−2.14	5.1
Total.....	7.135	.....	−82.40	2,310	.....	4.72	331.6

$$\bar{x} = \frac{-82.4}{7.135} = -11.56$$
$$I_{zc} = 2,310 - 7.135(11.56)^2 = 1,358 \text{ in.}^4$$
$$\bar{z} = \frac{4.72}{7.135} = 0.66$$
$$I_{xc} = 331.6 - 7.135(0.66)^2 = 328 \text{ in.}^4$$

PROBLEMS

4.1. Determine the moment of inertia of the area shown about a horizontal axis through the centroid.

4.2. Determine the moments of inertia and radii of gyration about horizontal and vertical axes through the centroid of the area shown.

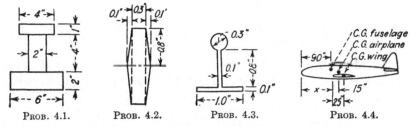

PROB. 4.1.   PROB. 4.2.   PROB. 4.3.   PROB. 4.4.

4.3. Find the moments of inertia and radii of gyration about horizontal and vertical axes through the centroid of the area shown.

4.4. For proper balance of the 10,000-lb airplane shown, the wing must be located so that the airplane center of gravity is 15 in. aft of the wing leading edge. The center of gravity of the 1,500-lb wing is 25 in. aft of the leading edge. The remaining 8,500 lb of weight is considered to be in the fuselage and has its center of gravity 90 in. aft of the nose, as shown. Find the distance x from the nose of the fuselage to the wing leading edge.

4.5. Find the moments of inertia of the area shown about the two centroidal axes. The area is symmetrical about the x axis.

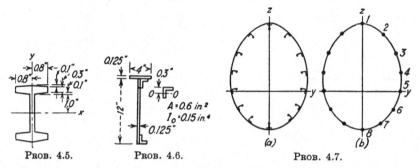

PROB. 4.5.   PROB. 4.6.   PROB. 4.7.

4.6. Find the moment of inertia of the area shown about a horizontal axis through the centroid. The top and bottom angles are similar and have areas and moments of inertia about their own centroids as shown.

4.7. An airplane fuselage is made up of extruded bulb angle stringers covered with sheet metal as shown in (*a*). The combined areas of the extrusions and the effective sheet are assumed to be concentrated at the points shown in (*b*). The areas and their coordinates are tabulated below, with one-half the areas of elements 1 and 8 assumed to act with each side. Find the moment of inertia about a horizontal axis through the centroid of the total cross section. The vertical axis is an axis of symmetry.

TABLE 4.4

Element	A	z	Element	A	z
1	$0.2 \times \frac{1}{2} = 0.1$	18.0	5	0.3	0
2	0.2	16.8	6	0.3	-7.0
3	0.2	14.2	7	0.3	-13.8
4	0.2	7.3	8	$0.3 \times \frac{1}{2} = 0.15$	-15.0

4.3. Moments of Inertia about Inclined Axes. The moment of inertia of an area about any inclined axis may be obtained from the properties of the area with respect to the horizontal and vertical axes. For the area shown in Fig. 4.12, a relationship between the moments of inertia about the inclined axes x' and y' and the axes x and y may be obtained. The moment of inertia about the x' axis is

$$I_{x'} = \int y'^2 \, dA \qquad (4.21)$$

the coordinate y' of any point is

$$y' = y \cos \phi - x \sin \phi \qquad (4.22)$$

If this value of y' is substituted in Eq. 4.21, the following value is obtained:

$$I_{x'} = \cos^2 \phi \int y^2 \, dA - 2 \sin \phi \cos \phi \int xy \, dA + \sin^2 \phi \int x^2 \, dA \quad (4.23)$$

The integrals must extend over the entire area. The angle ϕ is the same regardless of the element of area considered, and is therefore a constant with respect to the integrals. The first and last integrals of Eq. 4.23 represent the moments of inertia of the area about the x and y axes. The second integral represents a term which is called the product

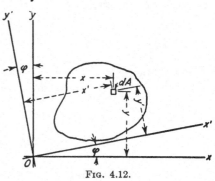

Fig. 4.12.

of inertia I_{xy} and which will be treated in detail later.

$$I_{xy} = \int xy \, dA \quad (4.24)$$

Equation 4.23 may now be written:

$$I_{x'} = I_x \cos^2 \phi - I_{xy} \sin 2\phi + I_y \sin^2 \phi \quad (4.25)$$

A similar expression may be derived for the moment of inertia about the y' axis.

$$I_{y'} = I_x \sin^2 \phi + I_{xy} \sin 2\phi + I_y \cos^2 \phi \quad (4.26)$$

If Eqs. 4.25 and 4.26 are added, the following relationship is obtained:

$$I_x + I_y = I_{x'} + I_{y'}$$

From Eq. 4.14, the sum of the moments of inertia about any two perpendicular axes is seen to be equal to the polar moment of inertia, which is the same regardless of the angle ϕ of the axes.

The product of inertia about the x' and y' axes is defined as follows:

$$I_{x'y'} = \int x'y' \, dA \quad (4.27)$$

Substituting the relations

$$x' = x \cos \phi + y \sin \phi$$

and

$$y' = y \cos \phi - x \sin \phi$$

into Eq. 4.27, the following value of $I_{x'y'}$ is obtained:

$$I_{x'y'} = \cos^2 \phi \int xy \, dA - \sin^2 \phi \int xy \, dA + \sin \phi \cos \phi \int y^2 \, dA$$
$$- \sin \phi \cos \phi \int x^2 \, dA$$

or

$$I_{x'y'} = I_{xy}(\cos^2 \phi - \sin^2 \phi) + (I_x - I_y) \sin \phi \cos \phi \quad (4.28)$$

4.4. Principal Axes. The moment of inertia of any area about an inclined axis is a function of the angle ϕ, as given in Eqs. 4.25 and 4.26. The angle ϕ at which the moment of inertia $I_{x'}$ is a maximum or minimum is obtained from the derivative of Eq. 4.25 with respect to ϕ.

$$\frac{dI_{x'}}{d\phi} = -2I_z \cos \phi \sin \phi - 2I_{xy} \cos 2\phi + 2I_y \sin \phi \cos \phi$$

This derivative is zero when $I_{x'}$ is a maximum or minimum. Equating the derivative to zero and simplifying,

$$(I_y - I_z) \sin 2\phi = 2I_{xy} \cos 2\phi$$

or

$$\tan 2\phi = \frac{2I_{xy}}{I_y - I_z} \tag{4.29}$$

Since there are two angles under 360° which have the same tangent, Eq. 4.29 defines two values of the angle 2ϕ, which will be at 180° intervals. The two corresponding values of the angle ϕ will be at 90° intervals. It can be shown that the value of $I_{x'}$ will be a maximum about one of these axes and a minimum about the other. These two perpendicular axes about which the moment of inertia is a maximum or minimum are called the *principal axes*.

The product of inertia about the inclined axes may be expressed in terms of the angle 2ϕ, by making use of the trigonometric relations

$$\sin 2\phi = 2 \sin \phi \cos \phi \tag{4.30}$$

and

$$\cos 2\phi = \cos^2 \phi - \sin^2 \phi \tag{4.31}$$

Substituting these values in Eq. 4.28, the following relation is obtained.

$$I_{x'y'} = I_{xy} \cos 2\phi + \frac{I_z - I_y}{2} \sin 2\phi \tag{4.32}$$

An important relation is obtained for the angle at which $I_{x'y'}$ is zero. Substituting $I_{x'y'} = 0$ in Eq. 4.32,

$$\tan 2\phi = \frac{2I_{xy}}{I_y - I_z}$$

This is identical to the expression defining the principal axes in Eq. 4.29. The product of inertia about the principal axes is therefore zero.

The moments of inertia about the principal axes may be obtained by substituting the value of ϕ obtained from Eq. 4.29 into Eq. 4.25.

$$I_p = \frac{I_z + I_y}{2} + \sqrt{I_{xy}^2 + \left(\frac{I_z - I_y}{2}\right)^2} \tag{4.33}$$

and

$$I_q = \frac{I_x + I_y}{2} - \sqrt{I_{xy}^2 + \left(\frac{I_x - I_y}{2}\right)^2} \tag{4.34}$$

where I_p represents the maximum value of $I_{x'}$ and I_q represents the minimum value of $I_{x'}$. These values are moments of inertia about perpendicular axes defined by Eq. 4.29.

4.5. Product of Inertia. The product of inertia of an area is evaluated by methods similar to those used in evaluating the moment of inertia. Products of inertia for various elements of the area are usually evaluated separately and then added in order to obtain the product of inertia for the entire area. When both x and y are positive or negative, the product of inertia is positive, but when one coordinate is positive and the other negative the product of inertia is negative. In the case of an area which is symmetrical with respect to the x axis, as shown in Fig. 4.13, each

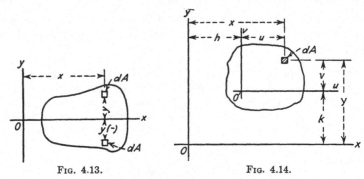

Fig. 4.13.   Fig. 4.14.

element of area dA in the first quadrant will have a corresponding area in the fourth quadrant with the same x coordinate but with the y coordinate changed in sign. The sum of the products of inertia for the two elements will be zero, and the integral of these terms for the entire area will be zero.

$$I_{xy} = \int xy \, dA = 0$$

The same relation is true if the area is symmetrical with respect to the y axis. Therefore, when either axis is an axis of symmetry, the product of inertia is zero and the axes are principal axes.

When the product of inertia of an area about one set of coordinate axes is known, it is possible to find the product of inertia about a set of parallel axes. For the area shown in Fig. 4.14, the product of inertia about the x and y axes is defined as

$$I_{xy} = \int xy \, dA$$

Substituting the values

$$x = h + u$$
$$y = k + v$$

the transfer theorem is obtained.

$$I_{xy} = \int (h + u)(k + v)\, dA$$
$$= hk\int dA + h\int v\, dA + k\int u\, dA + \int uv\, dA$$
$$I_{xy} = hkA + h\bar{v}A + k\bar{u}A + I_{uv} \tag{4.35}$$

where $\bar{u}$ and $\bar{v}$ are the coordinates of the centroid of the area and I_{uv} is the product of inertia of the area with respect to the u and v axes. If the u and v axes are through the centroid of the area, Eq. 4.35 becomes

$$I_{xy} = hkA + I_{uv} \tag{4.36}$$

If the u and v axes are also principal axes of the area, $I_{uv} = 0$, and Eq. 4.36 becomes

$$I_{xy} = hkA \tag{4.37}$$

For an area composed of several symmetrical elements, the product of inertia may be obtained as the sum of the values found by using Eq. 4.37 for each element.

Example 1. Find the product of inertia for the area shown in Fig. 4.15.

Solution. The total area is divided into the three rectangular elements A, B, and C. Rectangle A is symmetrical about both the x and y axes; hence the product of inertia is zero. Rectangle B is symmetrical about axes through its centroid; therefore the product of inertia may be found from Eq. 4.37.

$$I_{xy} = hkA = -3 \times 5 \times 8 = -120 \text{ in.}^4$$

For rectangle C,

$$I_{xy} = hkA = 3 \times -5 \times 8 = -120 \text{ in.}^4$$

For the total area,

$$I_{xy} = 0 - 120 - 120 = -240 \text{ in.}^4$$

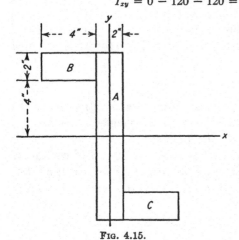

Fig. 4.15.

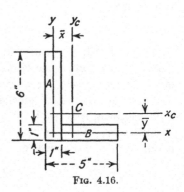

Fig. 4.16.

Example 2. Find the product of inertia about horizontal and vertical axes through the centroid of the area shown in Fig. 4.16.

Solution. The total area is divided into the two rectangular elements A and B as shown. The x and y reference axes are chosen through the centroids of these rectangles. Since rectangle A is symmetrical about the y axis and rectangle B is symmetrical about the x axis, $I_{xy} = 0$. The centroid of the area is obtained as follows:

$$\bar{x} = \frac{4 \times 2.5}{10} = 1.0$$

$$\bar{y} = \frac{6 \times 2.5}{10} = 1.5$$

The product of inertia about the centroidal axes may now be obtained from Eq. 4.36

$$I_{xy} = \bar{x}\bar{y}A + I_{x_c y_c}$$
$$0 = 1.0 \times 1.5 \times 10 + I_{x_c y_c}$$

or

$$I_{x_c y_c} = -15 \text{ in.}^4$$

Example 3. Find the product of inertia about horizontal and vertical axes through the centroid of the area shown in Fig. 4.11. The areas and coordinates of the elements are given in Table 4.3 and are repeated in columns (2), (3), and (4) of Table 4.5.

TABLE 4.5

Element (1)	A (2)	x (3)	z (4)	Axz (5)
1	0.358	-34.5	$+8.6$	-106.2
2	0.204	-28.1	$+9.6$	-55.0
3	0.395	-19.9	$+10.0$	-78.5
4	0.204	-10.1	$+9.6$	-19.8
5	1.615	$+0.5$	$+8.8$	7.1
6	1.931	$+0.5$	-5.7	-5.5
7	0.752	-10.1	-5.2	39.5
8	0.784	-22.4	-4.3	75.9
9	0.892	-34.7	-2.4	74.3
Total.....	7.135	.....	.....	-68.2

Solution. The product of inertia about the x and z axes is obtained as the summation of the terms Axz, obtained in column (5) of Table 4.5. The centroidal axes were found in Example 3, Art. 4.2, to have coordinates $\bar{x} = -11.56$, $\bar{z} = 0.66$. From Eq. 4.36,

$$I_{xz} = \bar{x}\bar{z}A + I_{x_c z_c}$$
$$-68.2 = -11.56 \times 0.66 \times 7.135 + I_{x_c z_c}$$

or

$$I_{x_c z_c} = -13.8$$

4.6. Mohr's Circle for Moments of Inertia. The equations for the moments and products of inertia about inclined axes are difficult to remember. It is often convenient to use a semigraphic solution which is easier to remember and which is an aid in visualizing the relationship between the moments of inertia about various axes. If the values of $I_{xy'}$ from Eq. 4.32 are plotted against values of $I_{x'}$ obtained from Eq. 4.25 for corresponding values of ϕ, the points would all fall on the circle shown in Fig. 4.17. The maximum and minimum values of the moments of inertia are represented by points P and Q.

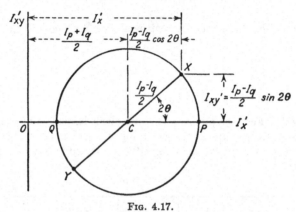

Fig. 4.17.

In order to prove that the circle shown in Fig. 4.17 represents the values given by Eqs. 4.25 and 4.32, the values of $I_{x'}$ and $I_{xy'}$ will be expressed in terms of the moments of inertia about the principal axes I_p and I_q and the angle θ from the x' axis to the principal axis. If the x and y axes are principal axes, a substitution of the values $I_{xy} = 0$, $I_x = I_p$, $I_y = I_q$, and $\phi = \theta$ into Eqs. 4.25 and 4.32 yields the following equations:

$$I_{x'} = I_p \cos^2 \theta + I_q \sin^2 \theta \qquad (4.38)$$

$$I_{xy'} = \frac{I_p - I_q}{2} \sin 2\theta \qquad (4.39)$$

The following trigonometric relations for double angles are used.

$$\sin^2 \theta = \tfrac{1}{2} - \tfrac{1}{2} \cos 2\theta$$
$$\cos^2 \theta = \tfrac{1}{2} + \tfrac{1}{2} \cos 2\theta$$

Substituting these values in Eq. 4.38,

$$I_{x'} = \frac{I_p + I_q}{2} + \frac{I_p - I_q}{2} \cos 2\theta \qquad (4.40)$$

Equations 4.39 and 4.40 correspond to the coordinates $I_{xy'}$ and $I_{x'}$ which are computed from the geometry of the circle shown in Fig. 4.17. The angle of inclination of the x' axis from the principal axis is one-half the angle 2θ measured between the corresponding points on the circle. If $I_{xy'}$ is measured as positive upward on the circle, a counterclockwise rotation of the x' axis corresponds to a counterclockwise rotation around the circle. Points at opposite ends of the diameter of the circle correspond to perpendicular inertia axes. The products of inertia about perpendicular axes are always equal numerically but opposite in sign, since rotation of the axes through 90° interchanges the numerical values of the coordinates x' and y' for any element of area and changes the sign of one coordinate.

Example 1. Find the moments of inertia about principal axes through the centroid of the area shown in Fig. 4.18. Find the moments and product of inertia about axes x_1 and y_1 and axes x_2 and y_2.

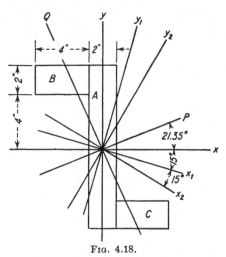

Fig. 4.18.

Solution. The moments of inertia about the x and y axes are obtained as follows:

$$I_x = \frac{2 \times 12^3}{12} + 2\left(\frac{4 \times 2^3}{12} + 8 \times 5^2\right) = 693.3 \text{ in.}^4$$

$$I_y = \frac{12 \times 2^3}{12} + 2\left(\frac{2 \times 4^3}{12} + 8 \times 3^2\right) = 173.3 \text{ in.}^4$$

From Example 1, Art. 4.5,

$$I_{xy} = -240 \text{ in.}^4$$

Mohr's circle for the moments and products of inertia about all inclined axes may now be plotted from these three values. The products of inertia $I_{xy'}$ are plotted against the moments of inertia I_x', as shown in Fig. 4.19. Point x in Fig. 4.19 has coordinates 693.3 and −240.0, as shown. If the x' axis is rotated through 90°, it will coincide with the y axis, and the coordinates of point Y will be $I_{x'} = 173.3$, $I_{xy'} = 240$. The positive sign for $I_{xy'}$ results from the fact that after rotating the x' axis through 90° from the x axis the coordinate x' is positive up and the coordinate y' is positive to the left. The value of $I_{xy'}$ is numerically equal but opposite in sign to the value of I_{xy}. Points X and Y are at opposite ends of a diameter of the circle, since a rotation of the axes of 90° corresponds to an angle of 180° on the circle.

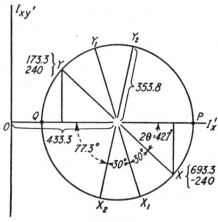

Fig. 4.19.

The center of the circle is a distance $\frac{1}{2}(693.3 + 173.3)$, or 433.3 from the origin. Point X is a distance 260 horizontally and 240 vertically from the center of the circle; therefore the following relations apply:

$$\text{Radius} = \sqrt{(240)^2 + (260)^2} = 353.8$$
$$2\theta = \arctan \tfrac{240}{260} = 42.7°$$

or

$$\theta = 21.35°$$

The principal axes are represented by points P and Q on the circle. The moments of inertia have maximum and minimum values, and the product of inertia is zero for these axes. The principal moments of inertia are equal to the distance from the origin to the center of the circle plus or minus the radius of the circle.

$$I_p = 433.3 + 353.8 = 787.1 \text{ in.}^4$$
$$I_q = 433.3 - 353.8 = 79.5 \text{ in.}^4$$

The P axis is counterclockwise from the x axis, at an angle $\theta = 21.35°$. Similarly, since point Q on the circle is counterclockwise from point Y, the Q axis is counterclockwise from the y axis.

The moments and product of inertia about the x_1 and y_1 axes are obtained from the coordinates of points on the circle. Since the x_1 and y_1 axes are 15° clockwise from the x and y axes, the points X_1 and Y_1 on the circle will be 30°

clockwise from points X and Y. The coordinates of points X_1 and Y_1 may be obtained from the geometry of the circle as follows:

$$I_{x_1} = 433.3 + 353.8 \cos 72.7° = 538 \text{ in.}^4$$
$$I_{y_1} = 433.3 - 353.8 \cos 72.7° = 328 \text{ in.}^4$$
$$I_{x_1y_1} = -353.8 \sin 72.7° = -338 \text{ in.}^4$$

The moments and product of inertia about the x_2 and y_2 axes are also obtained from the geometry of the circle.

$$I_{x_2} = 433.3 - 353.8 \cos 77.3° = 355 \text{ in.}^4$$
$$I_{y_2} = 433.3 + 353.8 \cos 77.3° = 511 \text{ in.}^4$$
$$I_{x_2y_2} = -353.8 \sin 77.3° = -345 \text{ in.}^4$$

Example 2. Find the principal axes through the centroid, and find the moments of inertia about these axes for the area shown in Fig. 4.20. The x and z axes are through the centroid, and the moments and product of inertia have the values $I_x = 320$, $I_z = 1,160$, and $I_{xz} = -120$.

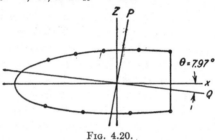

Fig. 4.20.

Solution. The coordinates of point X on the circle of Fig. 4.21 are 320 and -120. The coordinates of point Z are 1,160 and $+120$. These points are at opposite ends of the diameter and thus determine the circle. It is important to

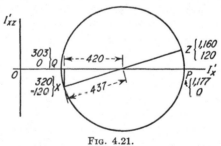

Fig. 4.21.

show I_{xz} with the correct sign at point X in order that the direction of the principal axes may be obtained correctly. The distance of the center of the circle from the origin of coordinates is $\frac{1}{2}(320 + 1,160) = 740$. From the geometry of the circle,

$$\text{Radius} = \sqrt{(420)^2 + (120)^2} = 437$$
$$\tan 2\theta = \tfrac{120}{420} \qquad 2\theta = 15.94°$$

or

$$\theta = 7.97°$$

The principal moments of inertia are represented by points P and Q on the circle, at which the moments of inertia have maximum and minimum values and the product of inertia is zero.

$$I_p = 740 + 437 = 1,177 \text{ in.}^4$$
$$I_q = 740 - 437 = 303 \text{ in.}^4$$

Since point P is 15.94° clockwise from point Z on the circle, the P axis will be half this angle, or 7.97° clockwise from the z axis, as shown in Fig. 4.20.

4.7. Mohr's Circle for Combined Stresses. The relationship between normal stresses and shearing stresses on planes at various angles of in-

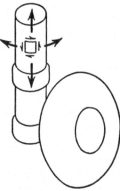

clination is similar to the relationship between moments and products of inertia about inclined axes. Most structural members are subjected simultaneously to normal and shearing stresses, and it is necessary to consider the combined effect of the stresses in order to design the members. The landing-gear strut shown in Fig. 4.22, for example, is subjected to bending stresses which produce tension in the direction of the strut, internal oil pressure which produces a circumferential tension, and torsion which produces shearing stresses on the horizontal and vertical planes. The maximum tensile stress does not occur on either the horizontal or vertical plane, but on a plane inclined at some angle to them. It can be shown that there are always two perpendicular planes on which the shearing stresses are zero.[1] These planes are called *principal planes*, and the stresses on these planes are called *principal stresses*.

Fig. 4.22.

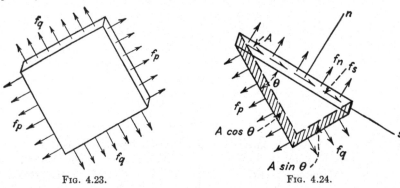

Fig. 4.23.　　　　　　　　Fig. 4.24.

Any condition of two-dimensional stresses can be represented as shown in Fig. 4.23, in which the principal stresses f_p and f_q act on the perpendicular principal planes. The orientation of these planes depends on the condition of stress and will be found from known stress conditions. The normal and shearing stresses f_n and f_s can be found on a plane at an angle θ to the principal planes, from the equations of statics. The stresses f are in pounds per square inch (psi) and must always be multiplied by the area in square inches in order to obtain the force. In Fig. 4.24, a small triangular element is shown, with principal planes forming two sides of the element and the inclined plane a third side. If the inclined plane has an area A, the sections of the principal planes have areas $A \cos \theta$ and $A \sin \theta$, as shown. From a summation of forces along the n and s axes, which are perpendicular and parallel to the inclined plane, the following equations are obtained:

$$\Sigma F_n = f_n A - f_p A \cos^2 \theta - f_q A \sin^2 \theta = 0 \tag{4.41}$$
$$\Sigma F_s = f_s A - f_p A \cos \theta \sin \theta + f_q A \sin \theta \cos \theta = 0 \tag{4.42}$$

Using the trigonometric relations for functions of double angles,

$$\cos^2 \theta = \tfrac{1}{2} + \tfrac{1}{2} \cos 2\theta$$
$$\sin^2 \theta = \tfrac{1}{2} - \tfrac{1}{2} \cos 2\theta$$
$$\sin \theta \cos \theta = \tfrac{1}{2} \sin 2\theta$$

and dividing Eqs. 4.41 and 4.42 by A, the following equations are obtained:

$$f_n = \frac{f_p + f_q}{2} + \frac{f_p - f_q}{2} \cos 2\theta \tag{4.43}$$

$$f_s = \frac{f_p - f_q}{2} \sin 2\theta \tag{4.44}$$

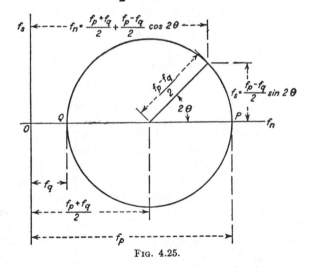

FIG. 4.25.

If values of f_s and f_n are plotted for different values of the angle θ, as shown in Fig. 4.25, all the points will lie on the circle. This construction, first used by Mohr, is similar to that used for finding moments of inertia about inclined axes.

Normal stresses are considered positive when tension and negative when compression. Compressive stresses are therefore shown to the left of the origin on Mohr's circle. In the following examples, shearing stresses will be considered positive when they tend to rotate the element clockwise and negative when they tend to rotate the element counterclockwise. Thus, on any two perpendicular planes the shearing stress on one plane would tend to rotate the element clockwise and would be measured upward on the circle, and the shearing stress on the other plane would be equal numerically but opposite in sign and would be measured downward. If this sign convention for shearing stresses is followed, a clockwise rotation of the planes of stress corresponds to a clockwise rotation on the circle. In some books the opposite sign convention for shearing stresses is used, and a clockwise rotation of the planes corresponds to a counterclockwise rotation on the circle.

Example. The small element shown in Fig. 4.26 represents the conditions of two-dimensional stress at a point in a structure. Find the normal and shearing stresses on planes inclined at an angle θ with the vertical plane, for values of θ

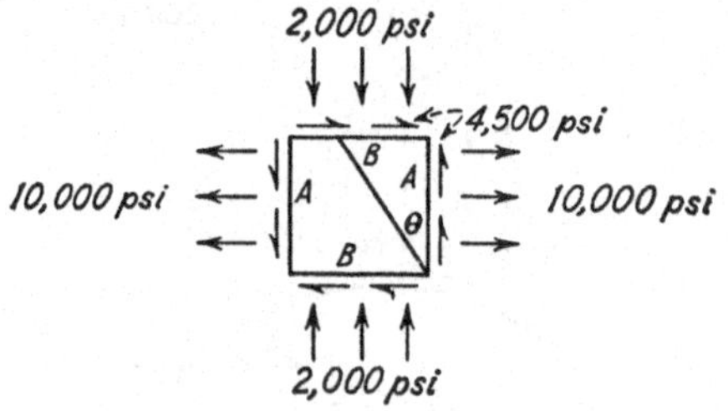

FIG. 4.26.

at 30° intervals. Find the principal planes and principal stresses. Find the planes of maximum shear and the stresses on these planes.

Solution. The values of the normal stress f_n and the shearing stress f_s on the horizontal and vertical planes are plotted as shown in Fig. 4.27. The vertical plane has a normal stress of 10,000 psi and a shearing stress of $-4,500$ psi, and these coordinates are shown for point A on Mohr's circle. The stresses on the horizontal plane are represented by point B on the circle, with coordinates of $-2,000$ and $+4,500$. The circle is now drawn with line AB as a diameter. The distance OC is $\frac{1}{2}(-2,000 + 10,000) = 4,000$. Points A and B have a horizontal distance of 6,000 and a vertical distance of 4,500 from the center of the circle.

$$\text{Radius} = \sqrt{(6,000)^2 + (4,500)^2} = 7,500$$

$$\tan 2\theta = \frac{4,500}{6,000} = 0.75$$

or

$$2\theta = 36.86° \quad \text{and} \quad \theta = 18.43°$$

The principal stresses are represented on the circle by points P and Q. The coordinates of these points are obtained by adding and subtracting the radius from distance OC.

$$f_p = 4{,}000 + 7{,}500 = 11{,}500 \text{ psi}$$
$$f_q = 4{,}000 - 7{,}500 = -3{,}500 \text{ psi}$$

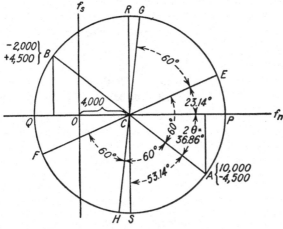

FIG. 4.27.

The point P on the circle is at an angle 2θ counterclockwise around the circle from point A. The principal plane P is therefore at the angle $\theta = 18.43°$ counterclockwise from the vertical plane A, as shown in Fig. 4.28(a). Similarly point Q is counterclockwise from the horizontal plane B, or perpendicular to plane P.

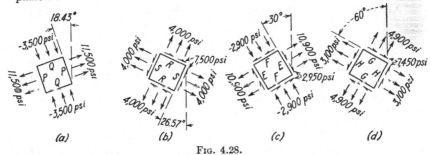

FIG. 4.28.

The planes of maximum shearing stress are always at 45° to the principal planes, regardless of the stress conditions. Points R and S at extremities of the vertical diameter of Mohr's circle represent the maximum shearing stresses. On the circle, these points are always 90° from the points at the extremities of the horizontal diameter, which represent the principal stresses. The normal stresses on the two planes of maximum shear are always equal, since they are both equal to the distance OC on the circle diagram. The maximum shearing stresses are shown on an element in Fig. 4.28(b). Plane S is 26.57° clockwise from the vertical, since point S is twice this angle clockwise from point A on the circle. Plane S is also 45° clockwise from plane P, and plane R is 45° counterclockwise from plane P. The shearing stress on plane R is positive, tending to rotate the element clockwise. The shearing stress on plane S is negative, tending to rotate the element counterclockwise.

The planes E and F are 30° counterclockwise from planes A and B. Points E and F on the circle must be 60° counterclockwise from points A and B, respectively. The stresses on plane E are obtained by calculating the coordinates of point E on the circle.

$$f_s = 7,500 \sin 23.14° = 2,950 \text{ psi}$$
$$f_n = 4,000 + 7,500 \cos 23.14° = 10,900 \text{ psi}$$

The stresses on planes E and F are shown in Fig. 4.28(c) in the correct directions.

The stresses on plane G, which is 60° counterclockwise from the vertical plane A, are obtained as follows:

$$f_s = 7,500 \sin 83.14° = 7,450 \text{ psi}$$
$$f_n = 4,000 + 7,500 \cos 83.14° = 4,900 \text{ psi}$$

The stresses on planes H, which are perpendicular to plane G, are

$$f_s = -7,500 \sin 83.14° = -7,450 \text{ psi}$$
$$f_n = 4,000 - 7,500 \cos 83.14° = 3,100 \text{ psi}$$

These stresses are shown in Fig. 4.28(d).

PROBLEMS

4.8. Find the product of inertia of the area shown about the x and y axes. The area is symmetrical by rotation about the origin.

4.9. Find the product of inertia of the area shown about the x and y axes.

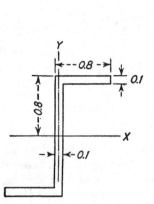

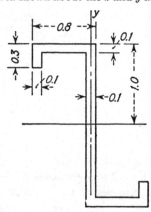

Probs. 4.8 and 4.11.   Probs. 4.9 and 4.12.

4.10. Find the product of inertia of the area shown about horizontal and vertical axes through the centroid.

4.11. Find the moments of inertia about the principal axes. Find the angles to the principal axes, and show these angles on a sketch of the area.

4.12. Find the moments of inertia about the principal axes. Show the angles to the principal axes on a sketch of the area.

4.13. Find the moments of inertia about principal axes through the centroid. Show the angles to the principal axes on a sketch of the area.

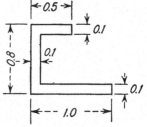

PROBS. 4.10 and 4.13.

4.14. The effective area of the wing cross section shown has the following properties about the x and z axes through the centroid: $I_x = 480$ in.[4], $I_z = 1620$ in,.[4], $I_{xz} = 180$ in.[4] Find the principal axes and the moments of inertia about the principal axes.

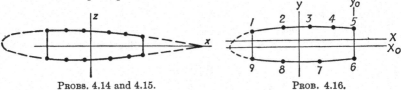

PROBS. 4.14 and 4.15.   PROB. 4.16.

4.15. Find the principal axes and the moments of inertia about these axes if the wing cross section has the following properties: $I_x = 420$ in.[4], $I_z = 1,280$ in.[4], $I_{xz} = -220$ in.[4]

4.16. The wing cross section shown is made up of elementary areas concentrated at nine points. The areas of the elements, and the coordinates of the areas with respect to reference axes x_0 and y_0 are tabulated below. Find the moments and product of inertia about the x_0 and y_0 axes. Find the location of the centroidal axes x and y and the moments and product of inertia about these axes; then find the principal axes through the centroid and the moments of inertia about the principal axes.

TABLE 4.6

Element	Area	x_0	y_0
1	0.422	−35.1	5.13
2	0.382	−28.0	5.25
3	0.382	−20.5	4.75
4	0.382	−10.5	4.25
5	0.562	0	3.55
6	0.487	0	−1.44
7	0.503	−12.4	−2.48
8	0.503	−23.2	−3.36
9	0.416	−35.1	−3.35

4.17. Find the principal stresses and principal planes for the element shown. Find stresses on inclined planes for values of θ at intervals of 15° between 0 and 180°. Show results on sketches similar to those in Fig. 4.28.

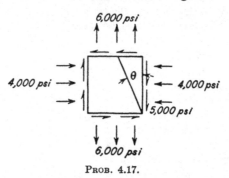

PROB. 4.17.

4.18. Find the principal stresses and principal planes for the element shown. Find the planes of maximum shear and the stresses on these planes. Find the stresses on the planes for $\theta = 30°$ and $\theta = 120°$. Show all results on sketches.

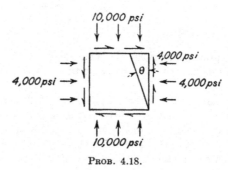

PROB. 4.18.

4.19. An element on the upper surface of a wing carries compressive stresses of 30,000 psi and shearing stresses of 10,000 psi as shown. Find the principal planes and principal stresses. Find the stresses on a plane inclined 30° as shown.

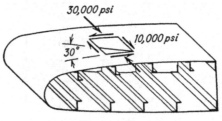

PROB. 4.19.

4.20. Find the principal stresses and principal planes for the element shown. Find the stresses on the plane inclined 20°, as shown.

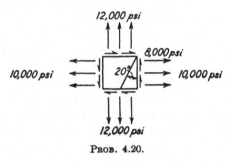

PROB. 4.20.

REFERENCE FOR CHAPTER 4

1. MARIN, J.: "Strength of Materials," Chap. VI, The Macmillan Company New York, 1948.

CHAPTER 5

SHEAR AND BENDING–MOMENT DIAGRAMS

5.1. Shears and Bending Moments. In the design of all structural members except two-force members it is necessary to obtain the shearing forces and the bending moments at various cross sections of the members. When all the loads acting on a member are known, the resultant internal forces at any cross section may readily be obtained from the static

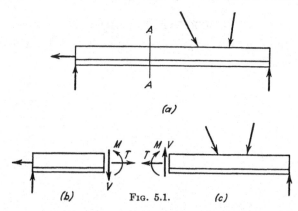

Fig. 5.1.

equilibrium of the part of the member on either side of the cross section. For members which are loaded with coplanar forces, there are three force components at any cross section, and they may be found from the three equations for static equilibrium of coplanar forces.

The member shown in Fig. 5.1(a) is loaded with coplanar forces. The internal forces at section AA may be represented by forces T and V, which are perpendicular and parallel to the cross section, and by a couple M. The internal forces at section AA which act on the right-hand part of the member must be equal and opposite to those acting on the left-hand part. The forces may be obtained from the equilibrium of either part of the member, if all the external loads are known. The internal stresses may be obtained separately for each force and superimposed. The shearing force V produces stresses parallel to the plane of the cross section, or shear stresses. The tension force T and the bending moment M produce stresses normal to the cross section. In order for the tension force to produce stresses which are uniformly distributed

100

over the cross-sectional area, it must act at the centroid of the area, as shown in Fig. 5.2, since the centroid of an area was defined as the position of the resultant of parallel forces uniformly distributed over the area. The resultant tension force is therefore considered as acting through the centroid of the cross-sectional area when computing bending moments. The bending moment M is usually computed by taking moments of the external forces to the left of the cross section about the centroid of the cross section, since the forces V and T have no moment about this point.

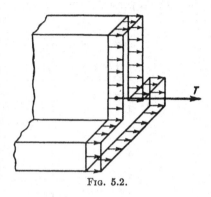

Fig. 5.2.

The values of V and M are usually plotted for all points along the length of the member, and the resulting curves are called the shear and bending-moment diagrams. The positive direction for the shear in a horizontal beam is shown in Figs. 5.1(b) and (c), in which the resultant of the external forces to the left of the cross section is up, and the resultant to the right of the cross section is down. A positive bending moment produces compressive stress on the upper side of the beam. In the case of vertical beams there is no established sign convention, and the positive direction should be designated on each diagram. A different sign convention is occasionally used for the shear in a full-cantilever airplane wing, where uploads are assumed to produce positive shear. According to the normal sign convention the uploads on the left wing, as viewed looking forward at the airplane, would produce positive shear, but uploads on the right wing would produce negative shear.

Example. Plot the shear and bending-moment diagrams for the beam shown in Fig. 5.3(a).

Solution. The reactions R_1 and R_2 are first obtained by considering the entire beam as a free body.

$$\Sigma M_{R1} = 100 \times 100 \times 10 - 20,000 \times 70 + 100R_2 = 0$$
$$R_2 = 13,000 \text{ lb}$$
$$\Sigma M_{R2} = 100 \times 100 \times 110 + 20,000 \times 30 - 100R_1 = 0$$
$$R_1 = 17,000 \text{ lb}$$

Check: $\Sigma F_y = 100 \times 100 - 17,000 + 20,000 - 13,000 = 0$

The shear and bending moment for any cross section between the left end of the beam and the left reaction are found by considering the free body shown in Fig. 5.4.

$$V = 100x$$
$$M = 50x^2$$

The equation for V represents a straight line from the point $x = 0$, $V = 0$, to the point $x = 60$, $V = 6,000$, and is plotted in Fig. 5.3(b). The equation for M represents a parabola which is concave upward, and which passes through the

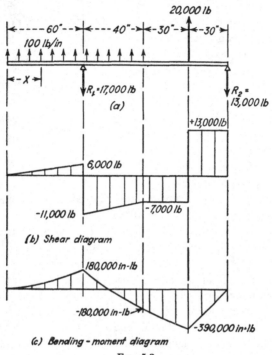

(a)

(b) Shear diagram

(c) Bending-moment diagram

Fig. 5.3.

points $x = 0$, $M = 0$, and $x = 60$, $M = 180,000$, as shown in Fig. 5.3(c).

The section of the beam under the distributed load, and to the right of the reaction, is next considered, as shown in Fig. 5.5.

$$V = 100x - 17,000$$
$$M = 50x^2 - 17,000(x - 60)$$

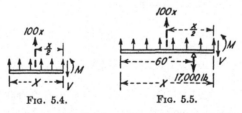

Fig. 5.4. Fig. 5.5.

The equation for V is a straight line from the point $x = 60$, $V = -11,000$, to the point $x = 100$, $V = -7,000$, as shown in Fig. 5.3(b). The equation for M is a parabola which is concave upward and which passes through the points $x = 60$, $M = 180,000$, and $x = 100$, $M = -180,000$.

The portion of the beam shown in Fig. 5.6 is now considered.

$$V = 10,000 - 17,000 = -7,000$$
$$M = 10,000(x - 50) - 17,000(x - 60)$$

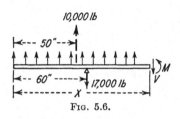

FIG. 5.6.

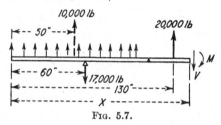

FIG. 5.7.

The equation for M represents a straight line through the points $x = 100$, $M = -180,000$, and $x = 130$, $M = -390,000$.

The portion of the beam shown in Fig. 5.7 is next considered.

$$V = 10,000 - 17,000 + 20,000 = 13,000$$
$$M = 10,000(x - 50) - 17,000(x - 60) + 20,000(x - 130)$$

The equations for V and M represent straight lines, as shown in Fig. 5.3. A numerical check is obtained from the point $x = 160$, since the shear is equal to the right reaction and the bending moment is zero. Any other point may be checked by considering the part of the beam to the right of the point.

It is not necessary to write the equations for the shear and bending-moment diagrams in most cases. The numerical values shown in Figs. 5.3(b) and (c) may be calculated directly by considering the equilibrium of the beam to the left of the points. The shape of the diagrams between these points may then be sketched accurately by making use of the relations developed in the following article.

5.2. Relations between Load, Shear, and Bending Moment. If a small element of a beam of length dx, as shown in Fig. 5.8, is considered as a free body, some useful relations may be developed. The beam carries a distributed load of w per unit length, and the load on the element is $w\,dx$, positive upward. From

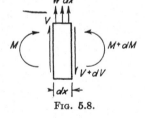

FIG. 5.8.

a summation of vertical forces on the element, the following equations are obtained:

$$V + w\,dx - (V + dV) = 0$$

or

$$\frac{dV}{dx} = w \tag{5.1}$$

Another relation is derived by a summation of moments about the right-hand cross section.

$$M + V\,dx + w\,dx\frac{dx}{2} - (M + dM) = 0$$

Simplifying and dropping the second-order differential,

$$\frac{dM}{dx} = V \tag{5.2}$$

Equation 5.1 shows that the load intensity at any point is equal to the derivative of the shear equation at that point. This relationship is seen to apply to the equations for V which were obtained in the Example problem of Art. 5.1. A differentiation of the equations for V gives values of $dV/dx = 100$ for all points under the distributed load of $w = 100$ lb/in., and $dV/dx = 0$ for points where the beam is not loaded. A study of the shear diagram shown in Fig. 5.3(b) shows that the shear increases at a rate of 100 lb/in. under the distributed load and remains constant at unloaded sections of the beam. The concentrated loads or reactions are assumed to be distributed over a zero length of the beam, and the load intensity w is infinite at these points. The shear diagram will always have discontinuities at these points.

Equation 5.2 shows that the shear is equal to the derivative of the bending-moment equation at any point. This relationship is very useful in obtaining the shape of the bending-moment diagram between plotted points. A consideration of the bending-moment diagram of Fig. 5.3(c) shows that the moment curve must have a horizontal tangent at the left end of the beam, since the shear is zero at this point. The slope of the bending-moment curve must increase gradually from the left end of the beam to the left reaction, since the shear increases in this region. Just to the left of the reaction the bending moment is increasing at the rate of 6,000 in-lb/in., since the slope must be equal to the shear at this point. The slope of the bending-moment diagram just to the right of the left reaction is $-11,000$ in-lb/in. The bending-moment diagram has a negative slope for the entire region in which the shear is negative. For the region in which the shear is constant, the bending-moment curve must be a straight line, with a slope equal to the value of the shear.

All maximum or minimum values of the bending moment must occur where the shear is zero, since the derivative of the bending-moment equation must be zero. When the maximum bending moment occurs at a point under a distributed load, it is convenient to locate the point of the maximum moment from the shear diagram.

5.3. Bending Moment as Area of Shear Diagram. Equation 5.2 may be integrated between any two points A and B of a beam.

$$\int_A^B dM = \int_A^B V \, dx$$

or

$$M_B - M_A = \int_A^B V \, dx \tag{5.3}$$

The integral of Eq. 5.3 represents the area of the shear diagram between points A and B. The increase in the bending moment from point A to point B is therefore equal to the area of the shear diagram between these points. When the shear is negative, the area must be considered as negative. This principle is quite useful in obtaining or checking values of the bending moment. This method could be applied in the Example problem of Art. 5.1. The area of the shear curve of Fig. 5.3(b) between $x = 0$, and $x = 60$, is $6{,}000 \times 60/2 = 180{,}000$ in-lb, which is the change in bending moment for this section of the beam. Since the bending moment is zero at the left end of the beam, this represents the bending moment at $x = 60$. Similarly, the area of the shear curve between points $x = 60$ and $x = 100$ is $-360{,}000$ in-lb. The bending moment at $x = 100$ is therefore $180{,}000 - 360{,}000$, or $-180{,}000$ in-lb.

For a beam such as this, where the bending moment is zero at each end, the total area of the shear diagram must be zero.

For beams carrying eccentric axial loads or couple loads, Eq. 5.3 does not apply. In deriving Eq. 5.3, the bending moment is assumed to be a continuous function between points A and B. For the beam shown in Fig. 5.9, the bending-moment curve has a discontinuity at the point where the couple is applied. The area under the shear diagram to the left of a point near the right support is obviously not equal to the bending moment for the point. The total area of the shear diagram is not zero, even though the bending moment is zero at each end of the beam. Equation 5.3 can, however, be used for parts of the beam if the discontinuity of the bending-moment curve is not between the points A and B.

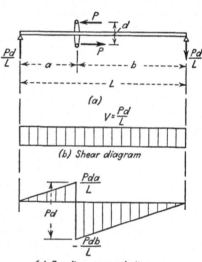

(a)

$V = \dfrac{Pd}{L}$

(b) Shear diagram

(c) Bending-moment diagram

Fig. 5.9.

For beams carrying distributed loads, the shear may be obtained as the area under the load curve. By integrating Eq. 5.1,

$$\int_A^B dV = \int_A^B w\,dx$$

or

$$V_B - V_A = \int_A^B w\,dx \qquad (5.4)$$

The integral of Eq. 5.4 represents the area under the load diagram between points A and B. This area is equal to the increase in shear between these points.

Example 1. Find the equations of the shear and bending-moment diagrams for the beam shown in Fig. 5.10(a) by the following three methods:

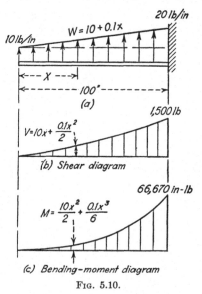

(a)

(b) Shear diagram

$V = 10x + \dfrac{0.1x^2}{2}$

1,500 lb

(c) Bending-moment diagram

$M = \dfrac{10x^2}{2} + \dfrac{0.1x^3}{6}$

66,670 in-lb

Fig. 5.10.

a. By considering forces to the left of a section, as in Art. 5.1.

b. By integration of the equations for the load and shear curves.

c. By obtaining the areas of the load and shear curves geometrically.

Solution. The load intensity increases from 10 lb/in. when $x = 0$ to 20 lb/in. when $x = 100$. The load at any point x is therefore represented by the equation

$$w = 10 + 0.1x \qquad (5.5)$$

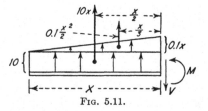

Fig. 5.11.

a. The load to the left of any cross section may be considered as shown in Fig. 5.11. One part of the load is uniformly distributed at 10 lb/in. for a length of x in., and has a resultant of $10x$ at a distance $x/2$ from the cross section. The other part of the load has a triangular distribution with a value of $0.1x$ at the cross section, and has a resultant of $0.1x^2/2$ acting at a distance $x/3$ from the cross section. From a summation of vertical forces,

$$V = 10x + 0.1\frac{x^2}{2}$$

From a summation of moments about the cross section,

$$M = 10x\frac{x}{2} + 0.1\frac{x^2}{2}\frac{x}{3}$$

or

$$M = 10\frac{x^2}{2} + 0.1\frac{x^3}{6}$$

The diagrams for V and M are plotted in Fig. 5.10.

b. The equations for V and M may be obtained by direct integration of the equation for the load w. From Eqs. 5.4 and 5.5,

$$V - 0 = \int_0^x (10 + 0.1x)\, dx$$

or
$$V = 10x + 0.1\frac{x^2}{2}$$

From Eq. 5.3,
$$M - 0 = \int_0^x V\,dx$$

$$M = \int_0^x \left(10x + 0.1\frac{x^2}{2}\right) dx$$

or
$$M = 10\frac{x^2}{2} + 0.1\frac{x^3}{6}$$

These equations for V and M correspond with those previously obtained.

c. The shear V may be obtained as the area under the load curve to the left of the point, since the shear is zero at the left end of the beam. The area under the load curve is equal to the sum of the areas of the rectangle and the triangle shown in Fig. 5.11.

$$V = 10x + 0.1\frac{x^2}{2}$$

The shear curve may be plotted in two parts as shown in Fig. 5.12. The shear resulting from the uniformly distributed load is represented by the triangular diagram having the value $10x$ at the point x. The shear resulting from the triangular load is represented by the parabola having a value of $0.1x^2/2$ at the point x. Since the bending moment at the left end of the beam is zero, the bending moment at point x is equal to the area under the shear curve to the left of the point. The area of the triangle shown in Fig. 5.12 is $10x^2/2$. The area of the parabola is one-third of the product of the base and the maximum ordinate, or $(0.1x^2/2)(x/3)$. The total bending moment is the sum of the areas,

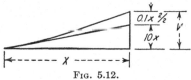

FIG. 5.12.

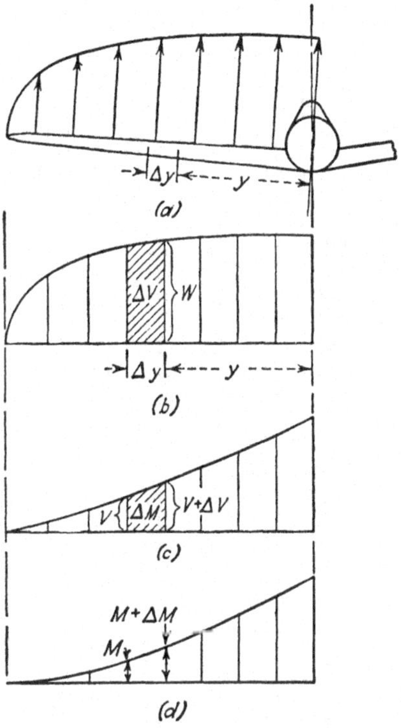

FIG. 5.13.

$$M = 10\frac{x^2}{2} + 0.1\frac{x^3}{6}$$

This checks the equation previously obtained.

Example 2. The aerodynamic loads on an airplane wing cannot be represented by a simple equation. The load per inch of span, w, for the airplane wing shown in Fig. 5.13(a) is tabulated in column (2) of Table 5.1 for various points along the span. Find the shear and bending-moment diagrams for the wing.

Solution. The values of the shear and bending moment at various points along the wing are calculated in Table 5.1. The points are called stations and

TABLE 5.1

Sta.	Load intensity	Dist. between stations	Shear increment	Shear	Moment increment	Bending moment
Dist. from center line	From air-load computations	From Col. (1)	$w_{av}\,\Delta y$	$\Sigma\,\Delta V$	$V_{av}\,\Delta y$	$\Sigma\,\Delta M$
y, in. (1)	w, lb/in. (2)	Δy, in. (3)	ΔV, lb (4)	V, lb (5)	ΔM, in-lb (6)	M, in-lb (7)
225	0			0		0
		5	90		200	
220	35			90		200
		20	930		11,100	
200	58			1,020		11,300
		20	1,290		33,300	
180	71			2,310		44,600
		20	1,510		61,300	
160	80			3,820		105,900
		20	1,690		93,300	
140	89			5,510		199,200
		20	1,870		128,900	
120	98			7,380		328,100
		20	2,030		167,900	
100	105			9,410		496,000
		20	2,160		209,800	
80	111			11,570		705,800
		20	2,270		254,100	
60	116			13,840		959,900
		20	2,360		300,400	
40	120			16,200		1,260,000
		20	2,430		348,300	
20	123			18,630		1,608,000
		20	2,480		397,400	
0	125			21,110		2,005,000

are designated by their distance y from the center line of the airplane, as shown in Fig. 5.13(a). These distances are measured along the wing rather than horizontally, since the air loads are perpendicular to the wing. The distances between the stations, Δy, are tabulated in column (3). The value of the shear at any point is obtained as the area under the load curve to the left of the point.

The load curve is assumed to be a series of straight lines between the known points, as shown in Fig. 5.13(*b*), and the area is obtained as the sum of the areas of the trapezoids. The areas of the trapezoids are obtained as the product of the average height w_{av} and the base Δy and are tabulated in column (4). The change in shear ΔV between two stations is equal to the area of the load curve between the stations, as shown in Fig. 5.13(*b*) and (*c*). The shear V is then obtained in column (5) by a summation of the terms in column (4). The change in bending moment between two stations is equal to the area under the shear curve. This area is also assumed trapezoidal and is obtained by multiplying the sum of the shears at the adjacent stations by one-half the distance between the stations. The areas of these trapezoids are tabulated in column (6). The bending moments are obtained in column (7) by a summation of the terms in column (6).

5.4. Members Resisting Noncoplanar Loads. The resultant forces on any cross section of a member resisting noncoplanar loads may be represented by six components. These are usually considered as three force components along mutually perpendicular axes and three moment components about these axes. These six unknowns may be obtained from the six equations for the static equilibrium for the portion of the structure on either side of the cross section.

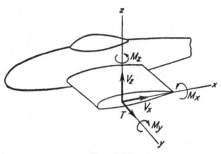

Fig. 5.14.

The forces and moments acting at a cross section of an airplane wing are shown in Fig. 5.14. The *x* and *z* axes are in the plane of the cross section, and the forces along these axes are the horizontal and vertical shearing forces on the cross section. The *y* axis should be through the centroid of the cross-sectional area if the stresses resulting from the tension force T are uniformly distributed over the area. In the case of an airplane wing, the tension force T is negligible, and the *x* and *z* axes are selected arbitrarily before calculating the position of the centroid of the area. The couples M_z and M_x represent wing bending moments resulting from vertical and horizontal loads. The couple M_y represents the wing torsion. Shear and bending-moment diagrams are obtained separately for the loads in the *yz* plane and for loads in the *xy* plane.

Example. Plot the shear and bending-moment diagrams for the landing-gear oleo strut shown in Figs. 2.11 and 2.12.

Solution. The shears and bending moments for loads in the VS plane are computed separately from the shears and bending moments for loads in the VD plane. Figure 5.15(a) shows the loads in the VS plane. The loads perpendicular to this plane do not affect the shears and bending moments in the plane, but are shown in the figure. The shear in the VS plane is produced

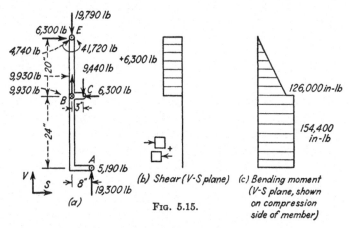

(b) Shear (V-S plane)   (c) Bending moment (V-S plane, shown on compression side of member)

(a)   Fig. 5.15.

by components of the loads along the S axis, and is zero below point B and 6,300 lb above point B, as shown in Fig. 5.15(b). The positive direction of the shear is indicated on the shear diagram since there is no established convention for positive shear in a member of this type. The bending-moment diagram for loads in the VS plane is shown in Fig. 5.15(c). The bending moment just below point B is obtained by taking moments of all forces below this point.

$$M = 19{,}300 \times 8 = 154{,}400 \text{ in-lb}$$

This bending moment is constant for the lower part of the member, since the 19,300-lb load has the same moment about any point on the center line of the member. The bending moment a small distance above point B is also obtained by taking moments of all forces below the point.

$$M = 19{,}300 \times 8 - 9{,}440 \times 3 = 126{,}000 \text{ in-lb}$$

The bending moment above point B decreases linearly to zero at the top of the member, because the moment of the horizontal 6,300-lb force at B is subtracted from the moment of 126,000 in-lb. The value at B may be checked by taking moments of the forces above the point.

$$M = 6{,}300 \times 20 = 126{,}000 \text{ in-lb}$$

The bending-moment diagram is shown on the compression side of the member, and all values on the diagram are shown as positive.

It is obvious that the bending moment at any point cannot be obtained as the area under the shear diagram below the point. The vertical loads at A and C

produce discontinuities in the bending-moment diagram, and extreme care must be used in obtaining bending moments from the shear areas in such cases. The slope of the bending-moment curve is equal to the shear, however, and the change in bending moment between two points is equal to the shear area between the points if there is no discontinuity between the points.

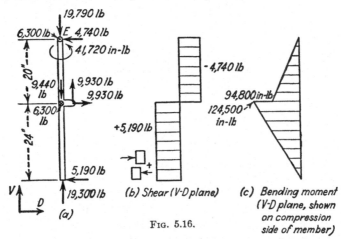

Fig. 5.16.

(b) Shear (V-D plane)

(c) Bending moment (V-D plane, shown on compression side of member)

The shear and bending-moment diagrams for loads in the VD plane are shown in Fig. 5.16. The shear diagram is obtained from the components of forces along the D axis, and is shown in Fig. 5.16(b). The bending-moment diagram shown in Fig. 5.16(c) is obtained from the moments of V and D components of forces. The bending moments are obtained by taking moments of the forces below the cross section and are checked by taking moments of the forces above the cross section, as in the previous calculations. The bending-moment diagram consists of straight lines because the shear is constant. It is necessary to compute values of the bending moment just above and just below point C and also at the ends of the member.

PROBLEMS

5.1. Plot the shear and bending-moment diagrams for the beam shown if $P_1 = 0$, $P_2 = 400$ lb, and $w = 10$ lb/in.

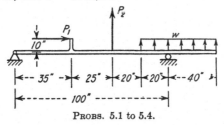

Probs. 5.1 to 5.4.

5.2. Plot the shear and bending-moment diagrams for the beam shown if $P_1 = 0$, $P_2 = 1,000$ lb, and $w = 5$ lb/in.

5.3. Plot the shear and bending-moment diagrams for the beam shown if $P_1 = 1,000$ lb, $P_2 = 400$ lb, and $w = 0$.

5.4. Plot the shear and bending-moment diagrams for the beam shown if $P_1 = 2,000$ lb, $P_2 = 100$ lb, and $w = 10$ lb/in.

5.5. Plot the shear and bending-moment diagrams for the beam shown if $P = 0$, $w_1 = 0$, and $w_2 = 10$ lb/in.

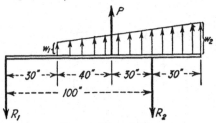

PROBS. 5.5 and 5.6.

5.6. Plot the shear and bending-moment diagrams for the beam shown if $P = 2,000$ lb, $w_1 = 10$ lb/in., and $w_2 = 20$ lb/in.

5.7. Calculate the shears and bending moments at 20-in. intervals for the beam of Fig. 5.10(a), using the tabular method of Example 2, Art. 5.3. Find the percentage error for each value of the bending moment.

5.8. Repeat Prob. 5.7, obtaining values at 10-in. intervals.

5.9. Plot the bending-moment diagrams for the landing-gear member shown in Figs. 2.12(c) and (d).

5.10. Plot the bending-moment diagrams for the landing-gear oleo strut shown in Fig. 2.11, if the loads applied to the axle are 20,000 lb in the V direction and 6,000 lb in the D direction.

CHAPTER 6

SHEAR AND BENDING STRESSES IN
SYMMETRICAL BEAMS

6.1. Flexure Formula. After finding the values of the shear and bending moment at a cross section of a beam, it is necessary to find the unit stresses on the cross section in order to design the beam. The *unit stress* is the force intensity at any point, and has units of force per unit area, or pounds per square inch (psi) in common engineering units. Where the term *stress* is used in future discussion, it will be assumed to mean *unit stress*. In this chapter, it will be assumed that the beam cross section is symmetrical about either a horizontal or a vertical axis through its centroid, so that these axes will be principal axes. The stresses resulting from axial forces, shearing forces, and bending moments will be computed separately and superimposed.

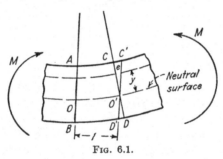

Fig. 6.1.

An initially straight beam is deflected to a circular arc when loaded in pure bending as shown in Fig. 6.1. It has been found experimentally that a plane cross section of the beam remains plane after bending. The cross sections AB and CD shown in Fig. 6.1 would both be vertical planes before bending, and would become radial planes after bending. If the beam is considered as made up of longitudinal "fibers," the fibers in the upper part of the beam will be compressed and those in the lower part will be stretched. The fibers on some intermediate surface, called the *neutral surface*, will remain the same length. The intersection of the neutral surface and any cross section is called the neutral axis for the cross section. If cross sections AB and CD are originally a unit distance apart, the unit elongation e of any fiber between these cross sections is shown in Fig. 6.1 as the distance from plane CD of the deformed beam to a plane $C'D'$ which is parallel to plane AB. The unit elongation e is proportional to the distance y of the fiber from the neutral surface, regardless of the relationship between the unit stress and the unit elongation for the material. If the stress is below the elastic limit for the material, the stress is proportional to the strain, and consequently proportional to the distance y.

$$f = Ky \tag{6.1}$$

The term K is a constant of proportionality which will be determined later, and f is the unit stress at a point a distance y from the neutral surface. If the stress exceeds the elastic limit, Eq. 6.1 no longer applies, but this condition will be considered in detail in a later chapter. The stress distribution represented by Eq. 6.1 is shown in Fig. 6.2(b).

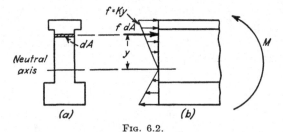

Fig. 6.2.

Since only bending stresses are being considered, the resultant horizontal component of all the stresses on any cross section must be zero in order for either part of the beam to be in equilibrium. The force acting on any element of area dA shown in Fig. 6.2(a) is equal to the unit stress times the area, or $f\,dA$. Substituting the value of f from Eq. 6.1 and summing all forces on the entire cross section, the resultant force is obtained as follows:

$$\int f\,dA = K\int y\,dA = 0 \tag{6.2}$$

Since K cannot be zero if the beam is stressed, the term $\int y\,dA$ must be zero. The centroidal axis of the area is defined as the axis about which this integral is zero, therefore Eq. 6.2 proves that the neutral axis for a beam in pure bending is the centroidal axis for the cross section.

In order for the portion of the beam shown in Fig. 6.2(b) to be in equilibrium for moments, the moment of all the forces on the cross section must be equal to the external moment M.

$$M = K\int y^2\,dA \tag{6.3}$$

The integral of Eq. 6.3 represents the moment of inertia of the area about the neutral axis and will be designated as I. From Eq. 6.3,

$$M = KI$$

or

$$K = \frac{M}{I} \tag{6.4}$$

From Eqs. 6.4 and 6.1,

$$f = \frac{My}{I} \tag{6.5}$$

Equation 6.5 is called the *flexure formula*, and is probably the most common and most useful formula in any stress analysis work. The direction of the bending stress is usually determined by inspection from the fact that a positive bending moment produces compression above the neutral axis of the beam and tension below the neutral axis. When it is desired to obtain the direction of the stress from Eq. 6.5, as in tabular computations, it is necessary to change the signs, since a positive bending moment produces a positive tensile stress for the negative values of y below the neutral axis.

6.2. Beam Shear Stresses. The shearing force V, parallel to the beam cross section, produces shear stresses f_s of varying intensity over the area of the cross section. It was shown in Chap. 4 that the shearing stresses on any two perpendicular planes are equal; hence there must be shear stresses on any horizontal plane through the beam which are equal to the shear stresses on the vertical cross section at the points of intersection of the two planes. The magnitude of the vertical shear stress at any point in the cross section will be obtained by computing the shear stress on a horizontal plane through the point.

Although the shear stresses in a beam of constant cross section are not affected by the magnitude of the bending moment, it is necessary to consider the difference between the bending moments on two cross sections. The portion of a beam between two vertical cross sections a distance a from each other is in equilibrium under the forces shown in Fig. 6.3. The distance a is assumed to be so small that there is no appreciable external load on this length of the beam. The shearing forces V on the two cross sections will be

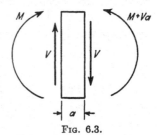

Fig. 6.3.

of equal magnitudes but opposite directions. The bending moment on the cross section to the right is equal to the sum of the bending moment M on the other cross section and the couple Va. The bending stresses on these two beam cross sections are shown in Fig. 6.4(b). At a point a distance y from the neutral axis the bending stress will be My/I on the left face and $(My/I) + (Vay/I)$ on the right face. In order to obtain the shear stress at a distance y_1 above the neutral axis, the portion of the beam above this point will be considered as a free body, as shown in

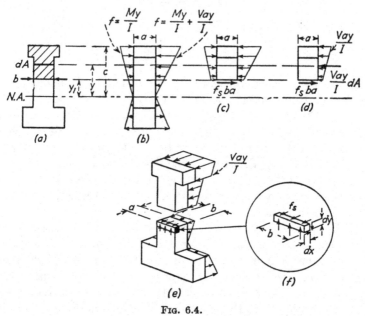

Fig. 6.4.

Fig. 6.4(c). The resultant force on the cross section to the right is larger than the resultant force on the other cross section. For equilibrium of the horizontal forces, the force produced by the shearing stress f_s on the horizontal area of width b and length a must be equal to the difference in the normal forces on the two cross sections. Since only the difference in the forces is considered, the loading shown in Fig. 6.4(d) may be used in computing the shearing stresses. Summing the horizontal forces,

$$f_s ba = \int_{y_1}^{c} \frac{Vay}{I} \, dA \qquad (6.6)$$

where the integral represents the total horizontal force on the cross section of the element shown in Fig. 6.4(d) as obtained by multiplying the unit stress on each element of area by the area dA and integrating over the area of cross section above the point where the shear is desired. Equation 6.6 may be written

$$f_s = \frac{V}{Ib} \int_{y_1}^{c} y \, dA \qquad (6.7)$$

where the integral represents the moment of the area of the cross section above the point, with the moment arms measured from the neutral axis. The cross-sectional area considered is shown by the shaded area in Fig. 6.4(a).

Since Eq. 6.7 is derived from Eq. 6.5 it is also based on the assumption that the beam cross section is symmetrical about either the horizontal or vertical axis through the centroid, or that these axes are principal axes for the area. The assumption that stress is proportional to strain was also used in Eq. 6.5; therefore Eq. 6.7 does not apply when the bending stresses are above the elastic limit for the material. The cross section of the beam is assumed to be the same at all points along the beam, and Eq. 6.7 may be considerably in error for tapered beams.

Example 1. Find the maximum bending stress in the beam shown in Fig. 6.5 and the shear stress distribution over the cross section. The beam cross section is symmetrical about both horizontal and vertical axes.

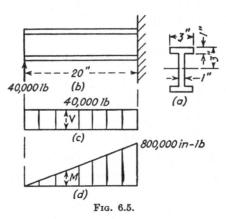

Fig. 6.5.

Solution. The shear and bending-moment diagrams for the beam are shown in Figs. 6.5(c) and (d). The maximum bending stress will occur at the point of maximum bending moment, or at the right end of the beam. Since the shearing force is constant for the entire length of the beam, the shear stress distribution will be the same on any cross section. The moment of inertia for the cross section is obtained as follows:

$$I = 2\left[\frac{3 \times 1^3}{12} + 3 \times 2.5^2\right] + \frac{1 \times 4^3}{12} = 43.3 \text{ in.}^4$$

The maximum bending stress is

$$\sigma_x = f = \frac{My}{I} = \frac{800,000 \times 3}{43.3} = 55,400 \text{ psi}$$

For a point 1 in. below the top of the beam, the integral of Eq. 6.7 is equal to the moment of the area of the upper rectangle about the neutral axis.

$$\int_{y_1}^{c} y\, dA = 2.5 \times 3 = 7.5 \text{ in.}^3$$

The average shear stress just above this point, where $b = 3$ in., will be

$$f_s = \frac{V}{Ib} \int_{y_1}^{c} y\, dA = \frac{40{,}000}{43.3 \times 3} \times 7.5 = 2{,}310 \text{ psi}$$

The average shear stress just below this point, where $b = 1$ in., will be

$$f_s = \frac{V}{Ib} \int_{y_1}^{c} y\, dA = \frac{40{,}000}{43.3 \times 1} \times 7.5 = 6{,}930 \text{ psi}$$

For a point 2 in. below the top of the beam, the integral of Eq. 6.7 will be

$$\int_{y_1}^{c} y\, dA = 2.5 \times 3 + 1.5 \times 1 = 9.0$$

The shear stress at this point will be

$$f_s = \frac{V}{Ib} \int_{y_1}^{c} y\, dA = \frac{40{,}000}{43.3 \times 1} \times 9.0 = 8{,}320 \text{ psi}$$

At a point on the neutral axis of the beam the shear stress will be

$$f_s = \frac{V}{Ib} \int_{y_1}^{c} y\, dA = \frac{40{,}000}{43.3 \times 1} (2.5 \times 3 + 1 \times 2) = 8{,}780 \text{ psi}$$

The distribution of shear stress over the cross section is shown in Fig. 6.6. The stress distribution over the lower half of the beam is similar to the distribution over the upper half because of the symmetry of the cross section about the neutral axis.

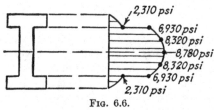

2,310 psi
6,930 psi
8,320 psi
8,780 psi
8,320 psi
6,930 psi
2,310 psi

Fig. 6.6.

Alternate Solution for Shear Stresses. In some problems it is more convenient to find shear stresses by obtaining the forces resulting from the change in bending stresses between two cross sections than it is to use Eq. 6.7. The solution of problems by this method may also help students to understand the application of Eq. 6.7 in various types of problems. Portions of the beam between two cross sections 1 in. apart are shown in Fig. 6.7. The bending moment increases by V in this 1-in. length, and the bending stresses on the right face of the beam are larger than those on the left face by an amount Vy/I. At the top of the beam, this difference in bending stress is

$$\frac{Vy}{I} = \frac{40,000 \times 3}{43.3} = 2,770 \text{ psi}$$

The expression Vy/I will be used frequently as a bending stress, and may appear to have units of pounds per cubic inch instead of pounds per square inch. It must be kept in mind that the expression is actually Vay/I and that a is a unit length which is omitted from the equation for simplicity.

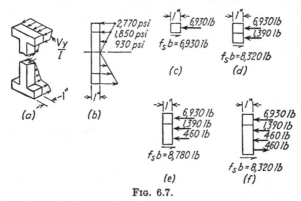

Fig. 6.7.

The differences in bending stresses for other points on the cross section are obtained by substituting other values of y and are shown in Fig. 6.7(b). The force acting on the upper rectangle is equal to the average stress of $(2,770 + 1,850)/2$, or 2,310 psi, multiplied by the area of 3 sq in. This force of 6,930 lb, shown in Fig. 6.7(c), is balanced by a shear stress on a horizontal area of 3 sq in. at a point in the upper rectangle and by a shear stress on an area of 1 sq in. at a point just below the upper rectangle. The shear stresses at these points are therefore 2,310 psi and 6,930 psi, as shown in Fig. 6.6.

The shear stress at a point 2 in. below the top of the beam must balance the force of 6,930 lb on the upper rectangle, and also the force produced by an average stress of $(1,850 + 920)/2$, or 1,390 psi, acting on the area of 1 sq in. The shear stress also acts on an area of 1 sq in., and is equal to 8,320 psi, as shown in Fig. 6.6. The shear stresses at the neutral axis and below the neutral axis are obtained in a similar manner, as shown in Figs. 6.7(o) and (f).

Example 2. In the beam cross section shown in Fig. 6.8 the webs are considered to be ineffective in resisting bending but to act in transmitting shear. Each stringer area of 0.5 sq in. is assumed to be concentrated at a point. Find the distribution of shear stress in the webs.

Solution. The moment of inertia for the area is first obtained, neglecting the moments of inertia of the webs and of the stringers about their own centroids.

$$I = 2 \times 0.5 \times 6^2 + 2 \times 0.5 \times 2^2 = 40 \text{ in.}^4$$

If two cross sections 1 in. apart are considered, the difference in bending stresses Vy/I on the two cross sections will be $8,000(6/40) = 1,200$ psi on the outside stringers and $8,000(2/40) = 400$ psi on the inside stringers. The differences in axial loads on the stringers at the two cross sections are found as the product of these stresses and the stringer areas and are shown in Fig. 6.9(a). The shear

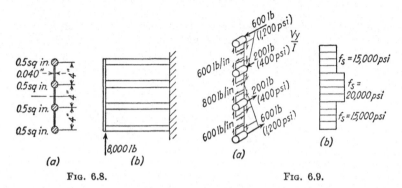

Fig. 6.8.   Fig. 6.9.

stress in the web at a point between the upper two stringers is found from the equilibrium of spanwise forces on the upper stringer.

$$f_s \times 0.040 \times 1 = 600$$

or

$$f_s = 15,000 \text{ psi}$$

If the webs resist no bending stress, the shear stress will be constant along each web, as shown in Fig. 6.9(b). If the webs resist bending stresses, the shear stress in each web will vary along the length of the web and will be greater at the end nearer the neutral axis. The shear stress in the web between the two middle stringers is found by considering spanwise forces on the two upper stringers.

$$f_s \times 0.040 \times 1 = 600 + 200$$

or

$$f_s = 20,000 \text{ psi}$$

In problems involving shear stresses in thin webs, the shear force per inch length of web is often obtained rather than the shear stress. The shear per inch or "shear flow" is equal to the product of the shear stress and the web thickness. The shear flow for each web, shown in Fig. 6.9(a), is equal to the sum of the longitudinal loads above the web.

The shear stresses may also be obtained by using Eq. 6.7. For a point between the two upper stringers,

$$f_s = \frac{V}{Ib} \int_n^c y \, dA = \frac{8,000}{40 \times 0.040} (0.5 \times 6) = 15,000 \text{ psi}$$

For a point between the two middle stringers,

$$f_s = \frac{V}{Ib} \int_n^c y \, dA = \frac{8,000}{40 \times 0.040} (0.5 \times 6 + 0.5 \times 2) = 20,000 \text{ ps}$$

PROBLEMS

6.1. Find the maximum tensile and maximum compressive stresses resulting ,rom bending of the beam shown. Find the distribution of shear stresses over the cross section at the section where the shear is a maximum, considering points in the cross section at vertical intervals of 1 in.

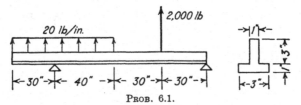

PROB. 6.1.

6.2. Find the maximum tensile and maximum compressive stresses resulting from bending of the beam shown. Find the shear stress at the neutral axis of the beam by two methods. Find the bending and shear stresses at a point 30 in. from the left end of the beam and 2 in. from the upper surface.

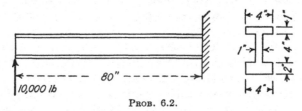

PROB. 6.2.

6.3. Find the maximum shear and bending stresses in the beam cross section shown if the shear V is 10,000 lb and the bending moment M is 400,000 in-lb. Both angles have the same cross section. Assume the web to be effective in resisting bending stresses.

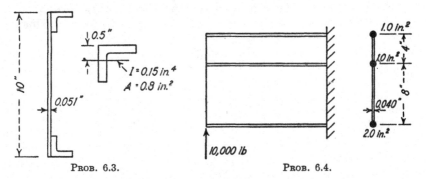

PROB. 6.3. PROB. 6.4.

6.4. Find the shear stress and the shear flow distribution over the cross section of the beam shown. Assume the web to be ineffective in resisting bending and the stringer areas to be concentrated at points.

6.5. Find the shear-flow distribution over the cross section shown for a shear V of 10,000 lb. Assume the webs ineffective in bending.

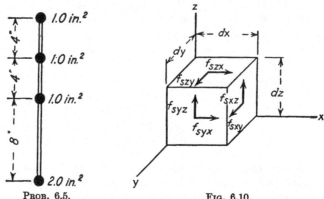

PROB. 6.5.   FIG. 6.10.

6.3. Direction of Shear Stress at a Free Surface. In a general condition of three-dimensional stress on an element, the shear stresses f_s on the faces of the element may be designated as shown in Fig. 6.10, where the second subscript indicates the plane on which the shear stress acts and the last subscript indicates the direction of the shear stress. The other three faces of the element are assumed to be under the action of shear stresses which are equal and opposite to those shown. The normal stresses on the faces of the element are not shown, but are assumed to be equal on opposite faces so that they do not affect the equilibrium of the element. The resultant forces obtained as the product of the shear stresses and the areas on which they act must be in equilibrium. For the moments about an axis through the center of the element in the z direction to be zero, the two equal and opposite forces $f_{szy}\, dy\, dz$ with a moment arm dx must form a couple equal to that formed by the forces $f_{syz}\, dx\, dz$ with a moment arm dy.

$$f_{szy}\, dy\, dz\, dx = f_{syz}\, dx\, dz\, dy$$

or

$$f_{szy} = f_{syz} \tag{6.8}$$

Similar equations are obtained by equating moments about axes through the center of the element in the x and y directions.

$$f_{syz} = f_{szy} \tag{6.9}$$

and

$$f_{szz} = f_{szz} \tag{6.10}$$

Equations 6.8 to 6.10 correspond with the relationship used in Mohr's circle analysis for two-dimensional stress conditions, in which the shear stresses on perpendicular planes were equal.

On a free unloaded surface of any structural member there can be no shear stress. Figure 6.11 represents a member in which the xy plane is a free surface on which there are no external loads. Consequently the relations $f_{szx} = f_{sxz} = 0$ and $f_{szy} = f_{syz} = 0$ must apply at the surface.

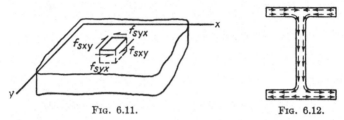

Fig. 6.11.   Fig. 6.12.

The remaining shear stress components f_{sxy} and f_{syx}, shown in Fig. 6.11, must therefore be parallel to the free surface. In applying this relationship to a beam of any cross section, the shear stress is parallel to any free surface of the beam. Since there can be no abrupt changes in stress distribution, the shear stress must be zero at any sharp corner, and may be very high at a sharp reentrant angle.

In the I beam shown in Fig. 6.12 the shear stresses must be almost horizontal near the free horizontal surfaces of the flanges. The shear stresses in the web must be vertical near the free vertical surfaces. At the juncture of the web and the flange a radius is desirable in order to permit a good distribution of shear stress. At the corners of the beam the shear stress must be zero, since there can be no shear stress on either the free vertical or the free horizontal surfaces. The shear stress distribution shown in Fig. 6.6 was obtained from the assumption of vertical shear stresses at all points in the cross section. The shear stresses shown for the flanges therefore represent only the average vertical components of the shear stresses shown in Fig. 6.12. A comparison of Figs. 6.6 and 6.12 indicates that the actual vertical shear stresses in the lower portion of the upper flange at points above the web will be approximately the same as the shear stresses in the web, while the vertical components of shear stresses nearer the sides of the flange will be zero. The actual shear stresses change gradually from point to point, rather than abruptly as shown in Fig. 6.6.

6.4. Shear Flows in Thin Webs. Since shear stresses at the free surface of a member are parallel to the surface and there can be no abrupt change in stress distribution, it will be sufficiently accurate to assume that the shear stresses in thin webs are parallel to the surfaces for the entire thickness of the web. In Fig. 6.13 a curved web representing the leading edge of a wing is shown, and the shear stresses are shown in the direction of the web at all points. Air loads normal to the surface must of course be resisted by shear stresses perpendicular to the web, but these stresses are usually negligible, and they will not be considered in this chapter. It might first appear that a thin curved web is not an efficient structure for resisting shearing stresses, but this is not the case. The diagonal tensile and compressive stresses are shown in Fig. 6.13 on principal planes at 45° from the planes of maximum shear. From a Mohr's circle construction for a condition of pure shear it is

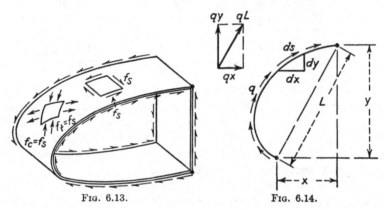

Fɪɢ. 6.13. Fɪɢ. 6.14.

found that the diagonal compressive stress f_c and the diagonal tensile stress f_t are both equal to the maximum shear stress f_s. If the diagonal compression alone were acting on the curved web, it would bend the web to an increased curvature. The diagonal tensile stress, however, tends to decrease the curvature, and the two effects counteract each other. The curved web will consequently resist a high shear stress without deforming from its original curvature. It will, in fact, resist a higher stress before buckling than a flat web of similar dimensions, although the flat web will carry an ultimate load much higher than its buckling load, whereas a curved web may collapse completely at the buckling load.

The shear flow q, which is the product of the shear stress f_s and the web thickness t, is usually more convenient to use than the shear stress. The shear flow may be obtained before the web thickness is determined, but the shear stress depends on the web thickness. It is often necessary to obtain the resultant force on a curved web in which the shear flow q is constant for the length of the web. The element of the web shown in Fig. 6.14 has a length ds, and the horizontal and vertical components of this length are dx and dy. The force on this element of length is $q\,ds$, and the components of the force are $q\,dx$ horizontally and $q\,dy$ vertically The total horizontal force is

$$F_x = \int_0^x q\,dx = qx \tag{6.11}$$

where x is the horizontal distance between the ends of the web. The total vertical force on the web is

$$F_y = \int_0^v q\, dy = qy \tag{6.12}$$

where y is the vertical distance between the ends of the web. The resultant force is qL, where L is the length of the straight line joining the ends of the web, and the resultant force is parallel to this line. Equations 6.11 and 6.12 are independent of the shape of the web, but depend only on the components of the distance between the ends of the web. The moment of the resultant force depends on the shape of the web. The moment of the force $q\,ds$ about any point O, shown in Fig. 6.15(a),

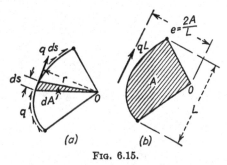

FIG. 6.15.

is the product of the force and the moment arm r. The area dA of the triangle formed by joining point O with the extremities of the element of length ds is $r\,ds/2$. The moment of the shear flows along the entire web can be obtained as the integral of the elementary moments.

$$M = \int qr\, ds = \int 2q\, dA = 2q \int dA$$

The value of this moment will be

$$M = 2Aq \tag{6.13}$$

where A is the area enclosed by the web and by the lines joining the ends of the web with point O, as shown in Fig. 6.15(b). The distance e of the resultant force from point O may be obtained by dividing the moment by the force.

$$e = \frac{2Aq}{qL} = \frac{2A}{L} \tag{6.14}$$

This distance is shown in Fig. 6.15(b). It must be remembered that the shear flow q is assumed constant in the derivation of Eqs. 6.11 to 6.14.

6.5. Shear Center. If a beam cross section is symmetrical about a vertical axis, the vertical loads must be applied in the plane of symmetry in order to produce no torsion on the cross section. If the cross section is not symmetrical about a vertical axis, the loads must be applied at a certain point in the cross section in order to produce no torsion. This point is called the *shear center* and may be obtained by finding the position of the resultant of the shearing stresses on any cross section. The simplest type of beam for which the shear center must be calculated is made of two concentrated flange areas joined by a curved shear web, as shown in Fig. 6.16. The two flanges must lie in the same vertical plane if the beam carries a vertical load. If the web resists no bending, the shear flow in the web will have a constant value q. The resultant of the shear flow will be $qL = V$, and the position of the resultant will be at a distance $e = 2A/L$ to the left of the flanges, as shown in Fig. 6.16(a). All loads must therefore be applied in a vertical plane which is at a distance e from the plane of the flanges.

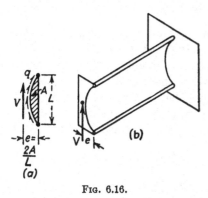

Fig. 6.16.

A beam with only two flanges which are in a vertical plane is not stable for horizontal loads. The vertical location of the shear center would have no significance for this beam. For beams which resist horizontal loads as well as vertical loads, it is necessary to determine the vertical location of the shear center. If the cross section is symmetrical about a horizontal axis, the shear center must lie on the axis of symmetry. If the cross section is not symmetrical about a horizontal axis, the vertical position of the shear center may be calculated by taking moments of the shear forces produced by horizontal loads. The method of calculating the shear center of a beam can best be illustrated by numerical examples.

Example 1. Find the shear flows in the webs of the beam shown in Fig. 6.17(a). Each of the four flange members has an area of 0.5 sq in. The webs are assumed to carry no bending stress. Find the shear center for the area.

Solution. Two cross sections 1 in. apart are shown in Fig. 6.17(b). The increase in bending moment in the 1-in. length is equal to the shear V. The increase of bending stress on the flanges in the 1-in. length is

$$\frac{Vy}{I} = \frac{10,000 \times 5}{50} = 1,000 \text{ psi}$$

The load on each 0.5 sq in. area resulting from this stress is 500 lb and is shown in Fig. 6.17(b). The actual magnitude of the bending stress is not needed in the shear flow analysis, since the shear flow depends only on the change in bending moment, or the shear.   If each web is cut in the spanwise direction, as shown, the shear forces on the cut webs must balance the loads on the flanges.   The force in web *ab* must balance the 500-lb force on flange *a*, and since this spanwise force acts on a 1-in. length the shear flow in the web will be 500 lb/in. in the direction shown.   The shear flow in web *bc* must balance the 500-lb force on flange *b*, and also the 500-lb spanwise force in web *ab*, and consequently has a value of 1,000 lb/in.   The shear flow in web *cd* must balance the 1,000-lb spanwise force in web *bc* as well as the 500-lb force on flange *c*, which is in the opposite direction. The shear flow in web *cd* is therefore 500 lb/in., and is checked by the equilibrium of flange *d*.

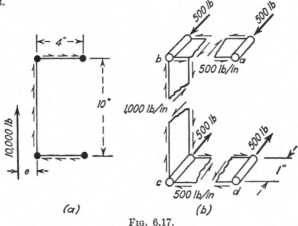

Fig. 6.17.

The directions of the shear flows on the vertical beam cross sections are obtained from the directions of the spanwise forces.   Since each web has a constant thickness the shear flows, like shear stresses, must be equal on perpendicular planes.   The shear flows on a rectangular element must form two equal and opposite couples.   The directions of all shear flows are shown on the perspective sketch of Fig. 6.17(*b*) and are shown for the front cross section in Fig. 6.17(*a*). The shear flow in the vertical web is in the same direction as the external shearing force, since the external force is the resultant of the shear flows.   In some books, the shear flows are shown for the back cross section, and are thus opposite to the external shearing force.   Where there is any possibility of confusion, it is desirable to make a perspective sketch rather than a sketch similar to Fig. 6.17(*a*).

The shear center is found by taking moments about point *c*.   The shear flow in webs *bc* and *cd* will produce no moment, and the moment of the force on web *ab* must equal the moment of the external shearing force.

$$10,000e = 500 \times 4 \times 10$$

or

$$e = 2 \text{ in.}$$

The shear center will be on the horizontal axis of symmetry, since a horizontal force along this axis will produce no twisting of the beam.

Alternate Solution. Equation 6.7 may be used in the solution of this problem. The shear flow is $f_s b$, and is found from Eq. 6.7.

$$q = f_s b = \frac{V}{I} \int y \, dA$$

For web *ab*, the integral represents the moment of the area of flange *a*.

$$q_{ab} = \frac{10,000}{50} (5 \times 0.5) = 500 \text{ lb/in.}$$

For web *bc*, the integral represents the moment of areas *a* and *b*.

$$q_{bc} = \frac{10,000}{50} (5 \times 0.5 + 5 \times 0.5) = 1,000 \text{ lb/in.}$$

The moments of areas *a*, *b*, and *c* are substituted for the integral in finding the shear flow in web *cd*.

$$q_{cd} = \frac{10,000}{50} (5 \times 0.5 + 5 \times 0.5 - 5 \times 0.5) = 500 \text{ lb/in.}$$

Example 2. Find the shear flows in the webs of the beam shown in Fig. 6.18(*a*). Each of the four flanges has an area of 1.0 sq in. Find the shear center for the area.

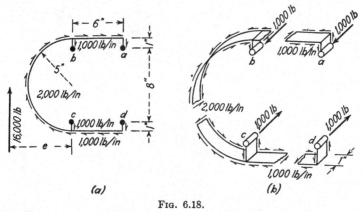

(a)   *(b)*

Fig. 6.18.

Solution. The moment of inertia of the area about the horizontal centroidal axis is

$$I = 4(1 \times 4^2) = 64 \text{ in.}^4$$

The change in axial load in each flange between the two cross sections 1 in. apart is

$$\frac{V}{I} yA = \frac{16,000}{64} \times 4 \times 1 = 1,000 \text{ lb}$$

The axial loads and shear flows are shown in Fig. 6.18(*b*). The shear flows in the webs are obtained by a summation of the spanwise forces on the elements, as in the previous example.

The distance *e* to the shear center is found by taking moments about a point below *c*, on the juncture of the webs. The shear flow in the nose skin produces a moment equal to the product of the shear flow and twice the area enclosed by the semicircle. The shear flow in the upper horizontal web has a resultant force of 6,000 lb and a moment arm of 10 in. The short vertical webs at *a* and *d* each resist forces of 1,000 lb, with a moment arm of 6 in. The resultant forces on the other webs pass through the center of moments. Equating the moment of the 16,000 lb external shearing force to the moment of the shear flows,

$$16,000e = 2 \times 39.27 \times 2,000 + 6,000 \times 10 + 2 \times 1,000 \times 6$$
$$e = 14.32 \text{ in.}$$

Example 3. Find the shear flow distribution and the shear center location for the beam cross section shown in Fig. 6.19(*a*). The rounded corners of the member are approximated by the rectangular areas shown.

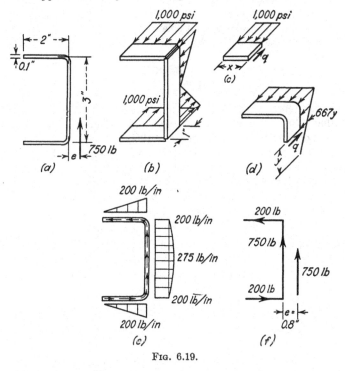

Fig. 6.19.

Solution. In this cross section the webs must resist the bending stresses, as well as the shear stresses. The shear flow *q* will not be constant in any web, but will vary from point to point along the webs. The moment of inertia about a horizontal axis through the centroid is

$$I = 2[2 \times 0.1 \times (1.5)^2] + \frac{0.1 \times (3)^3}{12} = 1.125 \text{ in.}^4$$

The moments of inertia of the horizontal webs about their own centroids are negligible, and the bending stress is assumed constant over the horizontal webs,

and equal to the stress at their centroids. The difference in the bending moments on two cross sections 1 in. apart is equal to the shear V and the difference
in the bending stresses on the two cross sections is Vy/I. For the horizontal
webs,
$$\frac{Vy}{I} = \frac{750 \times 1.5}{1.125} = 1{,}000 \text{ psi}$$

The stress distribution on the vertical webs varies from zero at the neutral axis
to 1,000 psi at the top and bottom, as shown in Fig. 6.19(b).

The shear flow q on the horizontal web at a distance x from the free edge
must balance the stress of 1,000 psi acting over an area $0.1x$, as shown in
Fig. 6.19(c).
$$q = 1{,}000 \times 0.1x = 100x$$

This equation represents a linear variation from zero at the free edge to 200 lb/in.
at the corner, as plotted in Fig. 6.19(e). The shear flow in the vertical web at a
distance y from the neutral axis must balance the force of 200 lb on the horizontal
web, as well as the force on the vertical web above point y, as shown in
Fig. 6.19(d).
$$q = 200 + 1{,}000 \times 1.5 \times \frac{0.1}{2} - 667y \times 0.1\frac{y}{2}$$
or
$$q = 275 - 33.3y^2$$

This equation represents a parabolic distribution of the shear flow with a maximum value of 275 lb/in. at the neutral axis and a value of 200 lb/in. at the top
and bottom of the web, as shown in Fig. 6.19(e).

The resultant force on each horizontal web is obtained as the product of the
average shear flow of 100 lb/in. and the length of 2 in., or as the area of the triangle shown in Fig. 6.19(e). This force of 200 lb is shown on each horizontal
web in Fig. 6.19(f). The resultant force on the vertical shear web is the sum of
the area of the rectangle and of the parabola shown in Fig. 6.19(e).
$$200 \times 3 + 75 \times 3 \times \tfrac{2}{3} = 750 \text{ lb}$$

This checks with the value of the external shear force. The shear center is now
obtained by equating the moment of the resultant shear to the moment of the
web forces about the lower right-hand corner of the section.
$$750e = 200 \times 3$$
or
$$e = 0.8 \text{ in.}$$

Alternate Solution. The shear flow on the horizontal web may be obtained
from Eq. 6.7,
$$q = f_s b = \frac{V}{I} \int y\, dA = \frac{750}{1.125} \times 1.5 \times 0.1x = 100x$$

where the integral represents the moment about the neutral axis of the area
shown in Fig. 6.19(c). The shear flow on the vertical web may also be obtained
from Eq. 6.7.
$$q = f_s b = \frac{V}{I} \int y\, dA = \frac{750}{1.125}\left[1.5 \times 0.1 \times 2 + \int_y^{1.5} y \times 0.1\, dy \right]$$
$$= 275 - 33.3y^2$$

where the integral represents the moment of the area shown in Fig. 6.19(d).
These equations check those previously obtained.

PROBLEMS

6.6. The area shown is symmetrical about both the horizontal and vertical axes through the centroid. Each of the 10 stringers has an area of 0.4 sq in., and the webs carry no bending stresses. Find the shear flows in all webs by two methods, if the vertical shearing force is 12,000 lb.

6.7. Each of the five upper stringers has an area of 0.4 sq in., and each of the five lower stringers has an area of 0.8 sq in. Find the shear flows in all webs by two methods, if the vertical shearing force is 12,000 lb. Note that the cross section is not symmetrical about a horizontal axis and that it is necessary to find the centroid of the area.

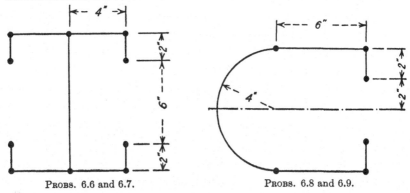

PROBS. 6.6 and 6.7.   PROBS. 6.8 and 6.9.

6.8. Each of the six stringers of the cross section shown has an area of 0.5 sq in. Find the shear flows in all webs and the location of the shear center for a vertical shearing force of 10,000 lb.

6.9. Find the shear flows in all webs for a horizontal shearing force of 3,000 lb. Each stringer has an area of 0.5 sq in.

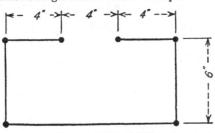
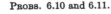
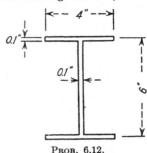

PROBS. 6.10 and 6.11.   PROB. 6.12.

6.10. Find the shear flows in all webs for a vertical shearing force of 12,000 lb. Each stringer has an area of 1.0 sq in.

6.11. Find the shear flows in all webs and the vertical location of the shear center for a horizontal shearing force of 4,000 lb. Each stringer has an area of 1.0 sq in.

6.12. Find the distribution of shear flow in all webs of the area shown if the vertical shearing force is 1,500 lb.

6.6. Torsion of Box Sections. The thin-web beams previously considered are capable of resisting loads which are applied at the shear center, but are not suitable for resisting torsion. In many structures the resultant load is at different positions for different loading conditions and consequently produces torsion. On an airplane wing, for example, the resultant aerodynamic load is farther forward on the wing at high

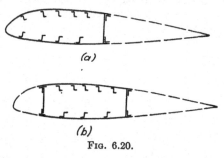

(a)

(b)

FIG. 6.20.

angles of attack than at low angles of attack. The position of this load also changes when the ailerons or wing flaps are deflected. A closed box structure, which is capable of resisting torsion, is consequently used for airplane wings and similar structures. Typical types of wing construction are shown in Fig. 6.20. The wing of Fig. 6.20(a) has only one spar, and the skin forward of this spar forms a closed box which is designed to resist the wing torsion, while the structure aft of the spar is lighter and is not designed to resist bending or torsional loads on the wing. The wing shown in Fig. 6.20(b) has two spars which form a closed box with the top and bottom skin between the spars. The skin forward and aft of the box is not designed to resist the wing bending or torsional loads. In some wings two or more closed boxes may act together in resisting torsion, but such sections are statically indeterminate and will be considered in a later chapter.

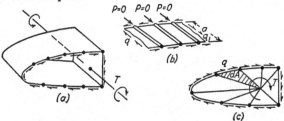

FIG. 6.21.

The box section shown in Fig. 6.21(a) is loaded only by torsional couples T with no shear or bending moment. Since the axial loads in the stringers are produced by wing bending, they will be zero for the condition of pure torsion. If the upper stringers are considered as a free body, as shown in Fig. 6.21(b), the spanwise forces must be in equilibrium. Since there are no loads on the stringers, the force qa on the left side must balance the force q_1a on the right side, or $q_1 = q$. If similar sections containing other stringers are considered, it is obvious that the shear flow at any point must be equal to q. The constant shear flow q around the circumference has no resultant horizontal or vertical force, since in the application of Eqs. 6.11 and 6.12 the horizontal and vertical distances between the end points of the closed web will be zero. The resultant of the shear flow is therefore a couple equal to the applied external couple T, and the moment of the couple will be the same about any axis perpendicular to the cross section. The moment of the shear flow about an axis through point O, shown in Fig. 6.21(c), will be found from Eq. 6.13.

$$T = \int 2q \, dA = 2qA \qquad (6.15)$$

where A is the sum of the triangular areas dA and is equal to the total area enclosed by the box. The area A will be the same regardless of the position of point O, since the moment of a couple is the same about any point. If point O is outside the box, some of the triangular areas, dA, will be negative, corresponding to a counterclockwise moment of the shear flows about point O, but the algebraic sum of all areas, dA, will be equal to the enclosed area A.

6.7. Shear Flow Distribution in Box Beams. A box beam containing only two stringer areas is shown in Fig. 6.22. Since this section is stable

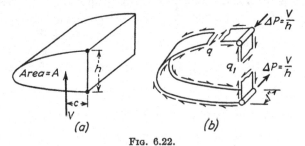

Fig. 6.22.

for torsional moments, the vertical shearing force V may be applied at any point in the cross section. The beam is unstable for a horizontal load, since the two stringers in the same vertical plane cannot resist a bending moment about a vertical axis. If two cross sections 1 in. apart are considered, as shown in Fig. 6.22(b), the difference in axial load on the stringers, ΔP, between the two cross sections is found by dividing the difference in bending moment, V, by the distance between stringers, h. These loads, ΔP, must be balanced by the shear flows shown in Fig. 6.22(b). Considering the equilibrium of the spanwise forces on the upper stringer,

$$q_1 = \frac{V}{h} - q \qquad (6.16)$$

The shear flow q may now be obtained by equating the moment of the shear flow about any point to the moment of the resultant shearing force about that point. Taking moments about the lower stringer,

$$Vc = 2qA$$

or

$$q = \frac{Vc}{2A}$$

where A is the total area enclosed by the box.

Substituting this value in Eq. 6.16,

$$q_1 = \frac{V}{h} - \frac{Vc}{2A} \qquad (6.17)$$

The values of q and q_1 may also be obtained by considering the resultant forces on the webs as obtained by Eqs. 6.11 and 6.12, and as shown in Fig. 6.23. The resultant force on the nose skin is a vertical force qh, and it acts at a distance $2A/h$ from the vertical web. The resultant force on the vertical web is q_1h and acts in the plane of the web. The shearing force V is distributed to the two webs in inverse proportion to its distance from the resultant forces. The force on the nose skin is

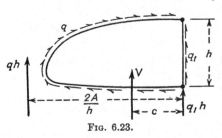

Fig. 6.23.

$$qh = \frac{Vc}{2A/h}$$

or

$$q = \frac{Vc}{2A}$$

The force on the vertical web is

$$q_1h = V - qh$$

or

$$q_1 = \frac{V}{h} - \frac{Vc}{2A}$$

These values of q and q_1 check those previously obtained.

A third method of analyzing this problem is shown in Fig. 6.24. The actual loading, shown in Fig. 6.24(c), may be obtained by a superposition of the load in the plane of the vertical web, shown in Fig. 6.24(a) and a

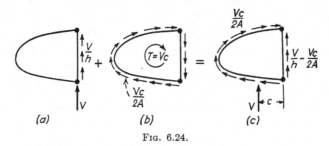

(a)   (b)   (c)

Fig. 6.24.

torsional couple Vc, shown in Fig. 6.24(b). The final shear flow is the result of the superposition of the shear flow V/h in the vertical web and the shear flow $Vc/2A$ around the entire circumference.

The shear flows in box beams with several stringers may be obtained by methods similar to that previously used. From a summation of spanwise loads on various string- ers the shear flows may all be expressed in terms of one un- known shear flow. This shear flow may then be obtained by equating the moments of the shear flows to the external tor- sional moment about a spanwise axis. For the box beam shown in Fig. 6.25, all the shear flows

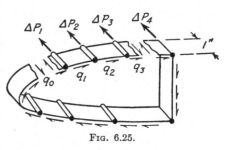

Fig. 6.25.

$q_1, q_2, \ldots, q_n$ may be expressed in terms of the shear flow q_0 by consider- ing the spanwise equilibrium of the stringers between the web o and the web under consideration:

$$q_1 = q_0 + \Delta P_1$$
$$q_2 = q_0 + \Delta P_1 + \Delta P_2$$

or

$$q_n = q_0 + \sum_0^n \Delta P_n \tag{6.18}$$

where $\sum_0^n \Delta P_n$ represents the summation of loads ΔP between web o and any web n. This term is equivalent to the term used in Eq. 6.7

$$\sum_0^n \Delta P_n = -\frac{V}{I} \int_0^n y \, dA$$

where the integral represents the moment about the neutral axis of the areas of all stringers between web o and web n. The negative sign must be introduced if the values of ΔP are positive when they indicate tension increments, since a positive shear produces negative values of ΔP for positive values of y. The flange elements are numbered in a clockwise order around the box, and the shear flows are positive when clockwise around the box. After expressing all shear flows in terms of the unknown q_0, the value of q_0 may be obtained from torsional moments.

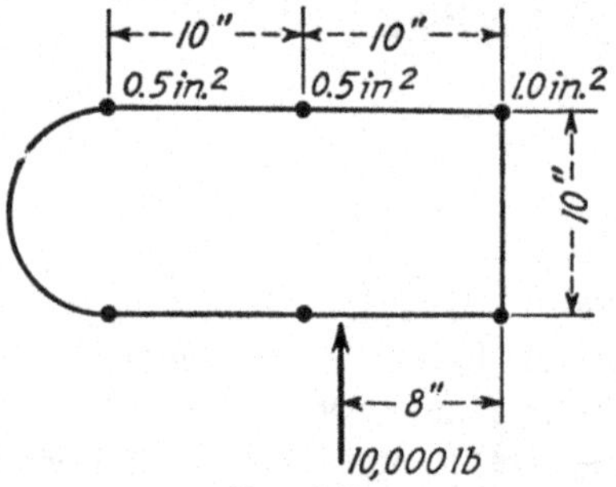

Example 1. Find the shear flows in all webs of the box beam shown in Fig. 6.26. The section is symmetrical about a horizontal axis through the centroid.

Solution. The moment of inertia about the neutral axis is $I = 4 \times 5^2 = 100$ in.[4] The difference in bending stress between the two cross sections 1 in. apart is $Vy/I = 10,000 \times 5/100 = 500$ psi.

This produces loads, ΔP, of -500 lb on the 1-sq in. upper stringer areas and -250 lb on the 0.5-sq in. stringer areas, as shown in Fig. 6.27(a). The shear flow in the leading-edge skin is considered as the unknown q_0, although the shear flow in any other web could be considered as the unknown. The shear flows in all other webs are now obtained in terms of q_0 by considering the spanwise forces on the stringers, and are shown in Fig. 6.27(a).

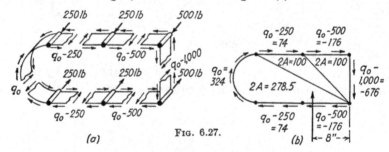

Fig. 6.26.

Fig. 6.27.

The value of q_0 is now obtained by equating the moments of the shear flows to the moment of the external shearing force. If the lower right-hand corner is used as the center of moments, the areas enclosed by lines joining this point to the ends of each web are shown in Fig. 6.27(b). The following moment equation is obtained:

$$10,000 \times 8 = 278.5 q_0 + 100(q_0 - 250) + 100(q_0 - 500)$$
$$478.5 q_0 = 155,000$$

or

$$q_0 = 324 \text{ lb/in.}$$

The shear flows in the other webs are computed in Fig. 6.27(b). Since several shear flows have negative signs they are shown in the correct direction in Fig. 6.28. Extreme care is necessary in using the correct directions of the shear flows. Where only one cross section is shown, both the shear flows and the external shearing forces have been shown for the front face, as is seen by comparing Fig. 6.27(b) and Fig. 6.28. Equal and opposite forces exist on the other

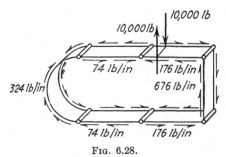

Fɪɢ. 6.28.

cross section, as shown in Fig. 6.28. In referring to other literature on the subject, the student may find that other methods are used, such as showing the shear flows for one cross section to be in equilibrium with the resultant forces on the other cross section. A perspective sketch is required in order to show the direction of the shear flows clearly.

Alternate Solution. The moment equation used above for finding the shear flow q_0 becomes long and cumbersome when it is necessary to find the moments of shear flows for several webs. A tabular solution is much more convenient in such cases. It is seen from Fig. 6.27(a) that the shear flow q_0 appears in each web and that it must be multiplied by the total enclosed area of the box when finding the torsional moment of the shear flows. It is therefore possible to obtain the total shear flows by a superposition of the values shown in Fig. 6.29(a), which were obtained by assuming the nose skin to be cut, and the shear flow q_0

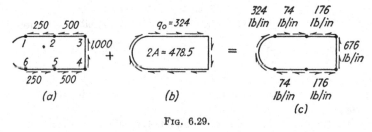

Fɪɢ. 6.29.

around the entire circumference, as shown in Fig. 6.29(b). The value of q_0 is obtained by equating the sum of the moments of the shear flows in Figs. 6.29(a) and (b) to the moment of the external shearing force. These calculations are made in Table 6.1. The values of ΔP are tabulated in column (2), with the negative values indicating an increment of compression load and a positive sign indicating an increment of tension load. The shear flows obtained by con-

TABLE 6.1. SHEAR FLOW ANALYSIS

Flange No. (1)	ΔP (2)	$q' = \Sigma \Delta P$ (3)	$2A$ (4)	$2Aq'$ (5)	$q = q_0 + q'$ $= 324 + q'$ (6)
1	−250				
		−250	100	−25,000	74
2	−250				
		−500	100	−50,000	−176
3	−500				
		−1,000	0	0	−676
4	500				
		−500	0	0	−176
5	500				
		−250	0	0	74
6	500				
		0	278.5	0	324
Total............................			478.5	−75,000	

sidering the beam as an open section with the nose skin cut are shown in Fig. 6.29(a) and are obtained in column (3) by an algebraic summation of the terms in column (2). These shear flows are designated as q' and are assumed positive when they act clockwise around the box. It is also necessary to number the flanges in clockwise sequence from the nose skin. The values of $2A$ tabulated in column (4) represent the double areas used for finding the moments of the shear flows in the webs, as shown in Fig. 6.27(b). The moments of the shear flows are tabulated in column (5) and are obtained by multiplying the values of q' in column (3) by the corresponding terms in column (4). A summation of column (5) gives the total moment of the shear flows q', and the negative sign indicates a counterclockwise moment. The moment of the shear flow q_0 is obtained as the product of q_0 and twice the enclosed area of the box. Equating the moments of the shear flows to the moment of the external shearing force,

$$10,000 \times 8 = -75,000 + 478.5q_0$$

or

$$q_0 = 324 \text{ lb/in.}$$

The positive sign indicates that the shear flow q_0 acts clockwise around the box, as assumed.

The final shear flows are obtained by adding the values of q' in column (3) to the value of q_0, and are tabulated in column (6). The positive signs indicate a clockwise direction around the box. The shear flows are shown in the correct directions in Fig. 6.29(c).

Example 2. Find the shear flow distribution around the circular tube shown in Fig. 6.30. Assume the wall thickness t to be small compared with the radius R of the center of the tube wall.

Solution. An approximate value of the moment of inertia, derived from the assumption that the entire area is concentrated at a distance R from the center of the tube, will be sufficiently accurate in finding the shear flow, although the exact expression should be used when calculating the bending strength of the tube. The cross-sectional area is approximately equal to the product of the wall thickness t and the circumference $2\pi R$. The polar moment of inertia I_p is the product of the area and the square of the distance R.

$$I_p = 2\pi R^3 t$$

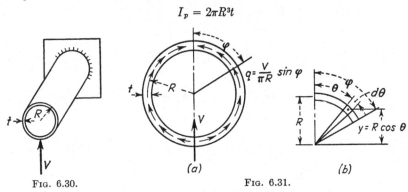

FIG. 6.30.   FIG. 6.31.

Since the circular tube has the same moment of inertia about the horizontal and vertical centroidal axes, and the sum of these moments of inertia is equal to the polar moment of inertia, the following expression applies:

$$I_x = I_y = \frac{I_p}{2} = \pi R^3 t$$

The shear flow will be obtained from Eq. 6.7.

$$q = f_s b = \frac{V}{I} \int y \, dA$$

Since the shear flow will be zero at the top and bottom center line of the tube because of the symmetrical loading, the integral represents the moment of the area between the upper center line and the point considered, as shown in Fig. 6.31(b). The value of dA is $Rt \, d\theta$, and the value of y is $R \cos \theta$. Evaluating the integral for a point defined by the angle ϕ, there is obtained

$$\int y \, dA = \int_0^\phi R^2 t \cos \theta \, d\theta = R^2 t \sin \phi$$

Substituting this value in the previous equation,

$$q = \frac{V}{I} \int y \, dA = \frac{V}{\pi R^3 t} R^2 t \sin \phi = \frac{V}{\pi R} \sin \phi$$

This expression is frequently used for the shear flow in a circular fuselage. While the fuselage stringers are concentrated areas, they have approximately a uniform spacing around the circumference, and the assumption that they represent an area uniformly distributed around the circumference is sufficiently accurate for most purposes.

PROBLEMS

6.13. Find the shear flows in the webs by two methods. Use numerical values from the start of the solution rather than substituting numerical values in the final equations.

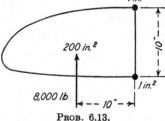

PROB. 6.13.

6.14. Find the shear flow distribution for the section shown by two methods.

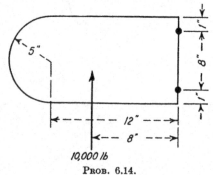

PROB. 6.14.

6.15. Find the shear flows in the webs of the box beam shown by two methods. The area is symmetrical about a horizontal center line.

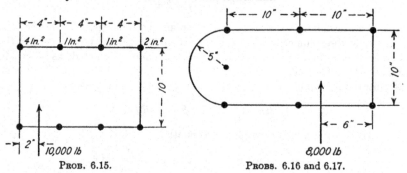

PROB. 6.15.   PROBS. 6.16 and 6.17.

6.16. Find the shear flows in the webs of the beam shown by two methods. All stringers have areas of 1.0 sq in.

6.17. Assume that the two right-hand stringers have areas of 3.0 sq in. and the other stringers have areas of 1.0 sq in. Find the shear flows in the webs by two methods.

6.18. Find the shear flows in all webs by two methods, assuming all stringers to have areas of 1.0 sq in.

6.19. Find the shear flows in all webs by two methods if the two right-hand stringers have areas of 1.5 sq in. and the other stringers have areas of 0.5 sq in.

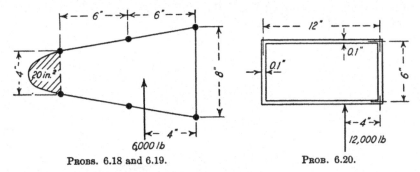

PROBS. 6.18 and 6.19.   PROB. 6.20.

6.20. Find the shear flow distribution in all webs. All parts of the cross section resist bending stresses.

6.8. Tapered Beams. In the preceding analysis of shear stresses in beams it has been assumed that the cross section of the beam remained constant. Since aircraft structures must be as light as possible, the beams are usually tapered so that the bending stresses remain constant. The depth and moment of inertia are greater at points where the bending moment is larger. While this variation in cross section may not cause appreciable errors in the application of the flexure formula for bending stresses, it often causes large errors in the shear stresses determined from Eq. 6.7.

The beam shown in Fig. 6.32(*a*) consists of two concentrated flange areas joined by a vertical web which resists no bending. The flanges are straight and are inclined at angles α_1 and α_2 to the horizontal. The resultant axial loads in the flanges must be in the direction of the flanges and must have horizontal components of $P = M/h$. The vertical components of the loads in the flanges, $P \tan \alpha_1$ and $P \tan \alpha_2$, which are shown in Fig. 6.32(*b*), resist some of the external shear V. Designating this shear resisted by the flanges as V_f and that resisted by the webs as V_w, the following equations apply:

$$V = V_f + V_w \qquad (6.19)$$
$$V_f = P(\tan \alpha_1 + \tan \alpha_2) \qquad (6.20)$$

From the geometry of the beam, as shown in Fig. 6.32, $\tan \alpha_1 = h_1/c$, $\tan \alpha_2 = h_2/c$, and

$$\tan \alpha_1 + \tan \alpha_2 = \frac{h_1 + h_2}{c} = \frac{h}{c}$$

Substituting this value in Eq. 6.20,

$$V_f = P\frac{h}{c} \tag{6.21}$$

Equation 6.21 will apply for a beam with any system of vertical loads. For the loading shown in Fig. 6.32(a), the value of P is Vb/h. Substituting this value for P in Eq. 6.21,

$$V_f = V\frac{b}{c} \tag{6.22}$$

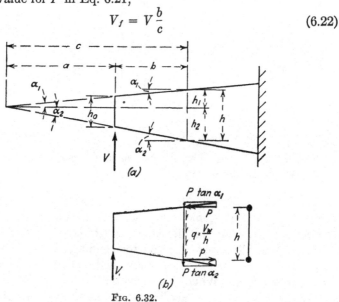

Fig. 6.32.

From Eqs. 6.19 and 6.22, and from the geometry,

$$V_w = V\frac{a}{c} \tag{6.23}$$

Equations 6.22 and 6.23 may be expressed in terms of the depths h_0 and h of the beam, by making use of the proportion $a/c = h_0/h$ obtained from the similar triangles in Fig. 6.32(a).

$$V_w = V\frac{h_0}{h} \tag{6.24}$$

and

$$V_f = V\frac{h - h_0}{h} \tag{6.25}$$

The shear flows in the webs of tapered beams may be obtained by using V_w instead of V in Eq. 6.7. For a beam with two equal flanges of area A and a depth between flanges of h, the shear flow is

$$q = \frac{V_w}{I}\int y\, dA = \frac{V_w}{Ah^2/2}\left(\frac{Ah}{2}\right) = \frac{V_w}{h} \tag{6.26}$$

In case of a beam with several flange areas, such as shown in Fig. 6.33, the shear flows may be obtained by a method similar to that used for the two-flange beam if the areas of the flanges remain constant along the span. If the flange areas vary along the span, and do not all vary in

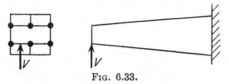

Fig. 6.33.

the same proportion, Eq. 6.7 cannot be applied. When taking moments about a spanwise axis in order to find the shear center of an open beam or the shear flows in a box beam, it is necessary to consider the components of the shear carried by the flanges. The method of obtaining the shear flows in such beams may best be illustrated by numerical examples.

Example 1. Find the shear flows in the web of the beam shown in Fig. 6.34, at 20-in. intervals along the span.

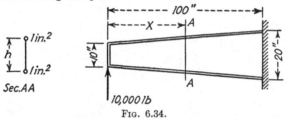

Fig. 6.34.

Solution. The shear flows are obtained by the use of Eqs. 6.24 and 6.26. The solution of these equations is shown in Table 6.2.

TABLE 6.2

x	h	$\dfrac{h_0}{h}$	V_w	$q = \dfrac{V_w}{h}$
0	10	1	10,000	1,000
20	12	0.8333	8,333	694.4
40	14	0.7143	7,143	510.2
60	16	0.6250	6,250	390.6
80	18	0.5555	5,555	308.6
100	20	0.5	5,000	250.0

While slide-rule accuracy is sufficient for shear flow calculations, the values in Table 6.2 are computed to four significant figures for comparison with a method to be developed later.

Example 2. Find the shear flows at section AA of the box beam shown in Fig. 6.35.

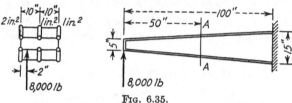

FIG. 6.35.

Solution. The moment of inertia of the cross section at AA about the neutral axis is
$$I = 2(2 + 1 + 1) \times 5^2 = 200 \text{ in.}^4$$
The bending stresses at section AA are
$$f = \frac{My}{I} = \frac{400,000 \times 5}{200} = 10,000 \text{ psi}$$

The horizontal components of the forces acting on the 2-sq in. stringers are 20,000 lb, and the forces on the 1-sq in. stringers are 10,000 lb, as shown in Fig. 6.36. The vertical components are obtained by multiplying the forces by the tangents of the angles between stringers and the horizontal, and are also shown in Fig. 6.36. The sum of the vertical components of forces on all

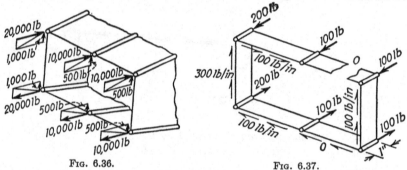

FIG. 6.36.   FIG. 6.37.

stringers V_f is 4,000 lb, and the remaining shear V_w of 4,000 lb is resisted by the shear flows in the webs. If one of the upper webs is cut, as shown in Fig. 6.37, the shear flows in the webs may be obtained from the equation

$$q = \frac{V_w}{I} \int y \, dA \qquad (6.27)$$

where the integral represents the moment of the areas between the cut web and the web under consideration. The change in bending stress on a stringer between two cross sections 1 in. apart is $V_w y/I$ when the effect of taper is considered, and the change in axial load on a stringer of area A_f is

$$\Delta P = \frac{V_w}{I} y A_f \qquad (6.28)$$

These axial loads are shown in Fig. 6.37 in the same way that they were shown previously for beams with no taper. The summation of the terms in Eq. 6.28 gives the values of q in Eq. 6.27. These values are shown in Fig. 6.37.

The shear flow q_0 in the upper web which was assumed to be cut is found by taking moments about a spanwise axis. The sum of the moments of the shear flows and vertical forces shown in Fig. 6.38(a) and the shear flow q_0 around the

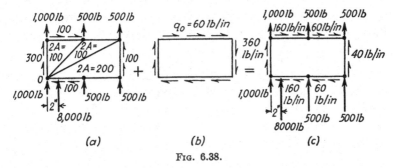

FIG. 6.38.

perimeter must be equal to the moment of the resultant 8,000-lb force. Taking moments about point O in Fig. 6.38(a), the following values are obtained:

$$-8,000 \times 2 = 100 \times 100 - 100 \times 200 - 2 \times 500 \times 10 - 2 \times 500 \times 20 + 400 q_0$$

or

$$q_0 = 60 \text{ lb/in.}$$

The resultant shear flows are obtained by adding algebraically the shear flows shown in Fig. 6.38(a) and the values of q_0. The final values are shown in Fig. 6.38(c).

Alternate Solution. An airplane wing usually has many stringers and is tapered in both depth and width. Each stringer has a different angle with both the horizontal and vertical. A solution for both the vertical and the horizontal components of all the stringer loads would usually require more work than would be justified. An approximate method of obtaining the torsional moment of the horizontal and vertical components of the stringer forces is usually sufficiently accurate. This approximate method consists of taking moments about some point of the cross section about which the stringer forces produce no appreciable torsional moment and of omitting the stringer forces in the moment equation. In this problem the resultant of the vertical forces in the stringers is **7.5** in. from the left web, or through the centroid of the area. In normal wing cross sections, the exact location would be difficult to determine. A torsional axis joining the centroids of the cross sections is usually satisfactory.

The problem will now be solved by the tabular method used in Example 1, Art. 6.7. Point O, shown in Fig. 6.39, will be used as the center of moments. The torsional moment of the external shear about point O is the product of the 8,000-lb force and its moment arm of 5.5 in., or 44,000 in-lb. Since the components of the stringer forces are neglected, only the shear carried by the webs V_w will be assumed to be acting. From Eq. 6.24,

$$V_w = V \frac{h_0}{h} = 8,000 \times \frac{5}{10} = 4,000 \text{ lb}$$

The increments of load in the flanges ΔP in a 1-in. length are found from Eq. 6.28.

$$\Delta P = \frac{V_w}{I} y A_f = \frac{4,000}{200} 5 A_f = 100 A_f$$

where A_f is the area of the flange. These values of ΔP are tabulated in column (2) of Table 6.3 with negative signs indicating compression. The

TABLE 6.3. SHEAR FLOW ANALYSIS

Flange No. (1)	ΔP (2)	$q' = \Sigma \Delta P$ (3)	$2A$ (4)	$2Aq'$ (5)	$q = q_0 + q'$ (6)
1	-100				
		-100	125	$-12,500$	-40
2	$+100$				
		0	50	0	60
3	$+100$				
		$+100$	50	5,000	160
4	$+200$				
		$+300$	75	22,500	360
5	-200				
		$+100$	50	5,000	160
6	-100				
		0	50	0	60
Σ	....	....	400	20,000	

values of q', which are the shear flows shown in Fig. 6.37, are obtained in column (3) by a summation of terms in column (2). Positive values of q' indicate that the shear flows are clockwise around the perimeter of the box. The terms in column (4) represent twice the areas enclosed by the corresponding webs and the lines joining the extremities of the webs and the center of moments, as shown in Fig. 6.39. The moments of the shear flows in the webs are obtained in column (5) as the product of terms in columns (3) and (4). The total moment of the shear flows q' is 20,000 in-lb and is obtained as the sum of terms in column (4). Equating the sum of the moments of the shear flows q' and the shear flows q_0 to the external torsional moment of 44,000 in-lb, the following value of q_0 is obtained:

$$44,000 = 20,000 + 400 q_0$$

or

$$q_0 = 60 \text{ lb/in.}$$

This value of q_0 must be added to the values of q', and the final shear flows are tabulated in column (6). These shear flows are shown in Fig. 6.39 in the correct directions.

It is seen that the shear flows shown in Fig. 6.39 correspond with the values previously obtained in Fig. 6.38(c). Both analyses are exact because Eq. 6.24 gives the correct value of the shear carried in the webs and because the components of the stringer forces have no moment about point O. In the analysis of an actual wing by the tabular solution, it is customary to obtain the center of moments and the value of V_w by approximate methods, and the accuracy of the method would depend on the accuracy of the approximations used.

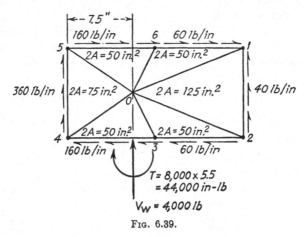

FIG. 6.39.

6.9. Beams with Variable Flange Areas. In the preceding article, beams were considered which varied in depth but which had the same flange areas at all cross sections. In many aircraft beams the cross-sectional area of the flange members varies as well as the depth of the beam. If the areas of all the flange members are increased by a constant ratio, the method of the preceding article may be used, but if the areas at one cross section are not proportional to the areas at another cross section, the method may be considerably in error. The airplane wing shown in Fig. 6.40 represents a structure in which the variation in flange areas must be considered. The flange areas in this wing are designed in such a manner that the bending stresses are constant along the span. In order to resist the larger bending moments near the root of the wing, the bending strength is increased by increasing the depth of the wing and by increasing the area of

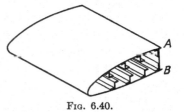

FIG. 6.40.

the spar caps A and B. The stringers, which resist the part of the bending moment not resisted by the spar caps, have the same area for the entire span. Since the stringers have the same area and the same axial stresses at all points, they must also have the same total axial load at all points. The increments of load ΔP will therefore be zero for all the stringers, and will act only on the spar caps A and B. From Eq. 6.18, the shear flow must be constant around the entire leading edge of the wing and will change only at the spar caps. Consequently, the methods of analysis previously used would not be applicable to this problem.

The bending stresses and total stringer loads may be calculated for two cross sections of the beam. The actual dimensions and stringer areas for each cross section are used, so that any changes between the cross sections are taken into consideration. The stringer loads P_a and P_b are shown in Fig. 6.41 for two cross sections a distance a from each

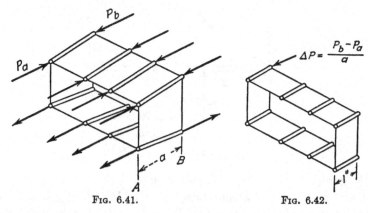

Fig. 6.41. Fig. 6.42.

other. The increase in load in any stringer is assumed to be uniform in the length a. The increase in stringer load per unit length along the span will be

$$\Delta P = \frac{P_b - P_a}{a} \tag{6.29}$$

This force is shown in Fig. 6.42. The shear flows may now be obtained from these values of ΔP as in the previous analyses.

It is seen that the shear force is not used in finding the values of ΔP, and consequently it is not necessary to calculate the vertical components of the stringer loads. The effects of beam taper and changes in flange area are automatically considered when the moments of inertia and bending stresses are calculated. Since it is necessary to calculate the wing bending stresses at frequent stations along the wing in order to design the stringers, the terms P_a and P_b can be obtained without much additional calculation. Consequently, this method of analysis is often simpler and more accurate than the method which considered variations in depth but not variations in flange area.

The distance a between cross sections may be any convenient value. It is common practice to calculate wing bending stresses at intervals of 15 to 30 in. along the span. These intervals are satisfactory for shear flow calculations. If the value of a is too small, the term ΔP in Eq. 6.29

will be small in comparison to P_b and P_a, and small percentage errors in P_b and P_a will result in large percentage errors in ΔP. If the length a is too large, the average shear flow obtained between sections A and B may not be quite the same as the shear flow midway between the sections.

Example 1. Find the shear flows in the beam of Fig. 6.34 by the method of using differences in bending stresses.

Solution. For this two-flange beam the axial load in the flanges has a horizontal component, $P = M/h$. The values of P for various sections are calculated in column (4) of Table 6.4.

TABLE 6.4

x (1)	M (2)	h (3)	$P = \dfrac{M}{h}$ (4)	$P_b - P_a$ (5)	q (6)	Per cent error (7)
10	100,000	11	9,091			
20				13,986	699.3	0.7
30	300,000	13	23,077			
40				10,256	512.8	0.5
50	500,000	15	33,333			
60				7,843	392.1	0.4
70	700,000	17	41,176			
80				6,192	309.6	0.3
90	900,000	19	47,368			

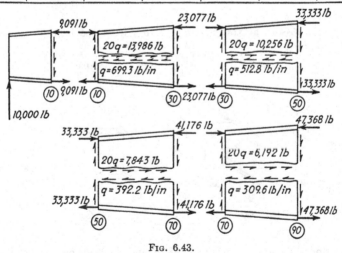

Fig. 6.43.

In computing the shear at any cross section, values of the axial loads at cross sections 10 in. on either side are found. The free-body diagrams are shown in Fig. 6.43. The circled numbers represent stations, or the distance of the cross section from the left end of the beam. The difference in horizontal loads on the upper part of the beam between the cross sections 20 in. apart must be balanced by the resultant of the horizontal shear flow, $20q$. The differences in axial loads are tabulated in column (5), and the shear flows, $q = (P_b - P_a)/20$, are tabulated in column (6). The value of the shear flow at station 20 is thus assumed to be equal to the average horizontal shear between stations 10 and 30. Even though the shear does not vary linearly along the span, the error in this assumption is only 0.7 per cent, as found by comparison with the exact value obtained in Table 6.2. This error is even smaller at the other stations.

Example 2. Find the shear flows at cross section AA of the box beam shown in Fig. 6.35 by considering the difference in bending stresses at cross sections 10 in. on either side of AA.

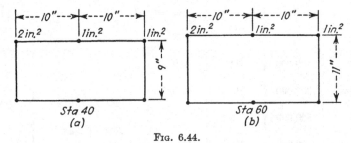

$\xleftarrow{---10''--->}\xleftarrow{---10''--->}$

| $2 in.^2$ | $1 in.^2$ | $1 in.^2$ |

$9''$

Sta 40
(a)

$\xleftarrow{---10''-->}\xleftarrow{---10''-->}$

| $2 in.^2$ | $1 in.^2$ | $1 in.^2$ |

$11''$

Sta 60
(b)

Fig. 6.44.

Solution. The moment of inertia at station 40 (40 in. from the left end) is found from the dimensions shown in Fig. 6.44(a).

$$I = 8(4.5)^2 = 162 \text{ in.}^4$$

The bending stresses at station 40, resulting from the bending moment of 320,000 in-lb are

$$f_b = \frac{My}{I} = \frac{320,000 \times 4.5}{162} = 8,888 \text{ psi}$$

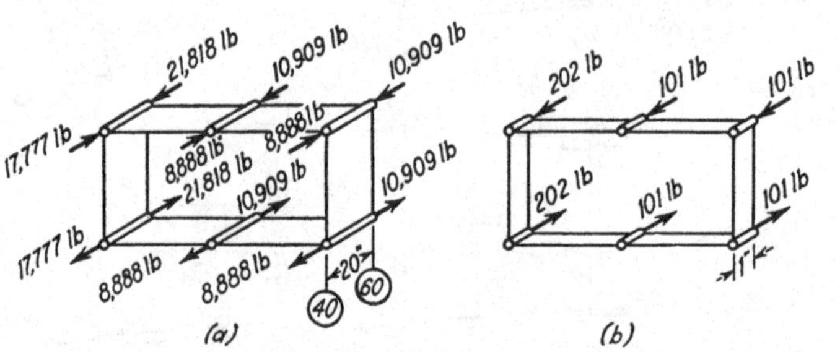

Fig. 6.45.

The loads on the 1-sq in. areas are 8,888 lb, and the loads on the 2 sq-in. areas are 17,777 lb, as shown in Fig. 6.45(a). The moment of inertia at station 60 is found from the dimensions shown in Fig. 6.44(b).

$$I = 8(5.5)^2 = 242 \text{ in.}^4$$

The bending stresses resulting from the bending moment of 480,000 in-lb are

$$f_b = \frac{My}{I} = \frac{480,000 \times 5.5}{242} = 10,909 \text{ psi}$$

The loads on the stringers are 10,909 lb and 21,818 lb, as shown in Fig. 6.45(a).

The increments of flange loads ΔP in a 1-in. length are found from Eq. 6.29. For the area of 2 sq in.,

$$\Delta P = \frac{21,818 - 17,777}{20} = 202 \text{ lb}$$

For the area of 1.0 sq in.,

$$\Delta P = \frac{10,909 - 8,888}{20} = 101 \text{ lb}$$

These values of ΔP are shown in Fig. 6.45(b). The remaining solution is identical with that of Example 2, Art. 6.8, as shown in Table 6.3. The values of ΔP are 1 per cent higher than the exact values shown in Fig. 6.37. The reason for this small discrepancy is that the average shear flow between stations 40 and 60 is 1 per cent higher than the shear flow at station 50. The other assumptions used in the two solutions are identical. The method using differences in bending stresses automatically considers the effects of the shear carried by the stringers, and it is not necessary to calculate the angles of inclination of the stringers. It is, however, necessary to find the torsional moments about the proper axis if the stringer forces are omitted in the moment equation.

PROBLEMS

6.21. Repeat Example 1, Art. 6.8, for a beam in which the depth h varies from 5 in. at the free end to 15 in. at the support.

6.22. Repeat Example 1, Art. 6.9, for a beam in which the depth h varies from 5 in. at the free end to 15 in. at the support.

6.23. Repeat Example 2, Art. 6.8, for a box beam in which the depth varies from 10 in. at the free end to 20 in. at the support. Assume the flange areas, horizontal dimensions, and loads to be as shown in Fig. 6.35.

6.24. Repeat Example 2, Art. 6.9, for a box beam in which the depth varies from 10 in. at the free end to 20 in. at the support.

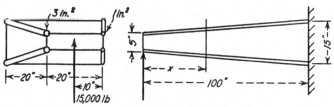

PROBS. 6.25 to 6.28.

6.25. Find the shear flows for the cross section at $x = 50$ in. Consider only this one cross section, but calculate the torsional moments by two methods: (a) by selecting the torsional axis arbitrarily and calculating the in-plane components of the flange loads and (*b*) by taking moments about a torsional axis joining the centroids of the various cross sections.

6.26. Find the increments ΔP of the flange loads by considering cross sections at $x = 40$ in. and $x = 60$ in., as in Example 2, Art. 6.9.   What is the percentage error in these increments?

6.27. Repeat Prob. 6.25 if there is an additional chordwise load of 6,000 lb acting to the left at the center of the tip cross section.

6.28. Repeat Prob. 6.26 if there is an additional chordwise load of 6,000 lb acting to the left at the center of the tip cross section.

CHAPTER 7

BEAMS WITH UNSYMMETRICAL CROSS SECTIONS

7.1. Bending about Both Principal Axes. The simple flexure formula, $f = My/I$, applies only to special cases of beam flexure. The resultant bending moment at the cross section must act about one of the principal axes of the area. In such beams the neutral axis is parallel to the axis of the resultant bending moment. When the resultant bending moment is not about one of the principal axes, the direction of the neutral axis cannot be determined by inspection.

In order to obtain a simple picture of the stress conditions which exist in more general cases of beam flexure, it is convenient to consider first another special case in which the principal axes are axes of sym-

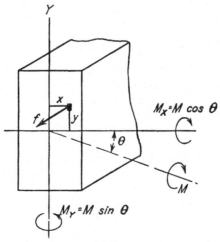

FIG. 7.1.

metry. The beam shown in Fig. 7.1 resists a bending moment M about an axis which is inclined with respect to the principal axes. The bending stresses are readily obtained by superposition of the stresses resulting from the components of the bending moment about the x and y axes. The component bending moments are found from the equations $M_x = M \cos \theta$ and $M_y = M \sin \theta$, where θ is the angle between the x axis and the moment axis. The bending stresses resulting from the couple M_z are shown in Fig. 7.2(a), and are obtained from the equation $f = -M_z y / I_x$. The negative sign indicates compressive stresses for positive values of M_z and y. The positive direction of M_y is that for which compressive stresses are produced at points where x is positive.

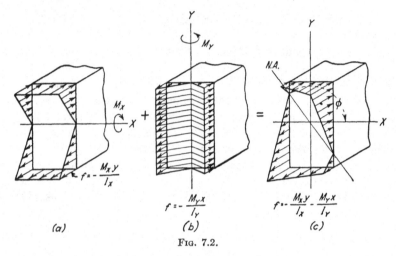

FIG. 7.2.

The bending stresses resulting from the couple M_y are shown in Fig. 7.2(b) and are obtained from the equation $f = -M_y x / I_y$. The resultant bending stresses are obtained by superimposing the stresses for bending about the two principal axes, as shown in Fig. 7.2(c).

$$f = -\frac{M_z y}{I_z} - \frac{M_y x}{I_y} \tag{7.1}$$

The neutral axis is defined as the line on which the bending stresses are zero. The equation of the neutral axis is obtained by substituting $f = 0$ in Eq. 7.1.

$$-\frac{M_z y}{I_z} - \frac{M_y x}{I_y} = 0 \tag{7.2}$$

The neutral axis passes through the centroid of the cross section, since the centroid was used as the origin for the xy coordinates. If the beam resists tension loads in addition to those shown, the neutral axis, or axis of zero stresses, will no longer pass through the centroid, but will be parallel to the line obtained from Eq. 7.2. The angle ϕ designating the slope of the neutral axis is found from Eq. 7.2.

$$\tan \phi = \frac{dy}{dx} = -\frac{M_y/I_y}{M_z/I_z}$$

The axis of the resultant bending moment is designated by the equation, $\tan \theta = M_y/M_z$. The neutral axis therefore coincides with the axis of the resultant bending moment only for cases where M_x or M_y are zero, or where $I_x = I_y$.

Example. Find the bending stresses at section AA of the beam shown in Fig. 7.3.

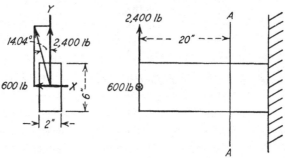

FIG. 7.3.

Solution. The bending moments about horizontal and vertical axes at the cross section are $M_x = 2{,}400 \times 20 = 48{,}000$ in-lb and $M_y = -600 \times 20 = -12{,}000$ in-lb. The corresponding moments of inertia of the cross-sectional area are $I_x = 2 \times 6^3/12 = 36$ in.⁴ and $I_y = 6 \times 2^3/12 = 4$ in.⁴ The bending stresses are obtained from Eq. 7.1.

$$f_b = -\frac{48{,}000}{36} y + \frac{12{,}000}{4} x$$
$$= -1{,}333y + 3{,}000x$$

If values of x and y corresponding to the corners of the area are substituted, the stresses shown in Fig. 7.4(c) are obtained. These values may also be obtained

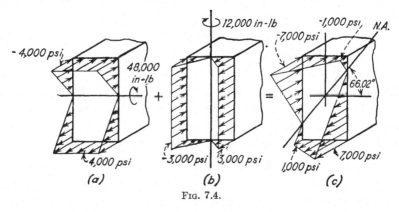

FIG. 7.4.

as the algebraic sum of the stresses resulting from bending about the x axis, shown in Fig. 7.4(a), and the stresses resulting from bending about the y axis, shown in Fig. 7.4(b).

The equation for the neutral axis is obtained by equating the stress to zero.

$$0 = -1,333y + 3,000x$$

or

$$y = 2.25x$$

This is the equation of a straight line through the centroid at an angle of arctan 2.25 or 66.02° with the x axis. The neutral axis is obviously not perpendicular to the resultant load on the beam, as the resultant load is at an angle of 14.04° to the vertical, as shown in Fig. 7.3.

7.2. General Equations for Bending Stress. The general case of bending, in which the bending moments are computed about two arbitrary axes which may not be the principal axes of the cross section, is analyzed by making use of the same assumptions that were used in the case of bending about the principal axes. Any plane cross section of the beam is assumed to remain plane after bending, and stress is assumed to be proportional to strain. As the beam is deformed in bending, a plane cross section rotates about the neutral axis. The deformations of the longitudinal fibers at any point in the cross section are therefore proportional to their distances from the neutral axis. Since the stresses are proportional to the deformations, the stresses must also be proportional to the distance from the neutral axis. The bending stress f_b on an element of area at a distance r from the neutral axis is $f_b = Kr$, where K is a constant to be determined. The force on any element of area dA is $f_b\, dA$. In a case of pure bending, the total axial load on any cross section of the beam is zero.

$$\int f_b\, dA = 0$$

or

$$K \int r\, dA = 0 \tag{7.3}$$

Since the constant K cannot be zero, the integral of Eq. 7.3 must be zero. Since this integral defines a centroidal axis of the area, the neutral axis must pass through the centroid of the cross-sectional area.

If the neutral axis makes an angle ϕ with the x axis, as shown in Fig. 7.5, the distance of any point from the neutral axis is $r = -x \sin \phi + y \cos \phi$. The bending stress may therefore be expressed as follows:

$$f_b = -Kx \sin \phi + Ky \cos \phi$$

or

$$f_b = K_1 x + K_2 y \tag{7.4}$$

where K_1 and K_2 are undetermined constants which are more convenient to use than the related constants K and ϕ.

The external bending moment about the x axis must be equal to the moment of the internal forces $f_b \, dA$ acting on the elements of area of the cross section.

$$M_x = -\int f_b y \, dA \tag{7.5}$$

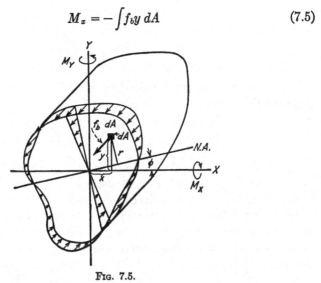

Fɪɢ. 7.5.

The sign convention is the same as that used in Art. 7.1. Substituting values of f_b from Eq. 7.4 into Eq. 7.5,

$$M_x = -K_1 \int xy \, dA - K_2 \int y^2 \, dA \tag{7.6}$$

The integrals in Eq. 7.6 represent the product of inertia of the cross-sectional area I_{xy} and the moment of inertia of the area I_x. Equation 7.6 may then be written

$$M_x = -K_1 I_{xy} - K_2 I_x \tag{7.7}$$

The external bending moment about the y axis must similarly be equal to the sum of the moments of the internal forces $f_b \, dA$ on the elements of area, acting with moment arms x.

$$M_y = -\int f_b x \, dA \tag{7.8}$$

Substituting the value of f_b from Eq. 7.4, and replacing the integrals by more convenient terms I_y and I_{xy},

$$M_y = -K_1 \int x^2 \, dA - K_2 \int xy \, dA$$

or

$$M_y = -K_1 I_y - K_2 I_{xy} \tag{7.9}$$

Equations 7.7 and 7.9 may now be solved simultaneously for K_1 and K_2, with the following results:

$$K_1 = \frac{M_x I_{xy} - M_y I_x}{I_x I_y - I_{xy}^2} \qquad (7.10)$$

$$K_2 = \frac{M_y I_{xy} - M_x I_y}{I_x I_y - I_{xy}^2} \qquad (7.11)$$

The general equation for bending stresses is now obtained by substituting the values from Eqs. 7.10 and 7.11 into Eq. 7.4.

$$f_b = \frac{M_x I_{xy} - M_y I_x}{I_x I_y - I_{xy}^2}\, x + \frac{M_y I_{xy} - M_x I_y}{I_x I_y - I_{xy}^2}\, y \qquad (7.12)$$

In cases where the x and y axes are principal axes of the cross-sectional area, the product of inertia about these axes is zero. The substitution of $I_{xy} = 0$ in Eq. 7.12 yields the expression previously obtained in Eq. 7.1.

In using Eq. 7.12, it is important to observe the sign convention used in the derivation. The bending moments M_x and M_y are assumed to be positive when they produce compressive stresses in the first quadrant of the area, where y and x are positive. In other texts or reference books different sign conventions are used, and consequently some of the signs in Eq. 7.12 will be changed.

In the analysis of aircraft structures, it is often necessary to consider beams with unsymmetrical cross sections. Wing cross sections, for example, are seldom symmetrical about an axis in the plane of the cross section. Even though a symmetrical airfoil might occasionally be used for an airplane wing, it is necessary to provide heavier structural members on the upper surface of the wing than on the lower surface, and the resulting cross section must be considered as unsymmetrical in an analysis for bending stresses. The cross sections of tail surfaces are often symmetrical with respect to the chord line, but must be analyzed as unsymmetrical areas because the thin sheet-metal skin is effective on the tension side of the beam but buckles and becomes ineffective on the compression side of the beam. Most fuselage cross sections are symmetrical about the center line of the airplane and may be analyzed as symmetrical beams.

Bending stresses in beams with unsymmetrical cross sections may be obtained by either of two methods. Equation 7.12 may be used, by substituting bending moments, moments of inertia, product of inertia, and coordinates with respect to arbitrary x and y axes through the centroid of the area. As a second method, the principal axes through the centroid of the area may be obtained by the methods of Chap. 4. The bending moments, moments of inertia, and coordinates of points with respect to these principal axes are then obtained, and the bending stresses are found from Eq. 7.1.

Example 1. Find the bending stresses at points A, B, and C of the beam cross section shown in Fig. 7.6. The properties of this area, obtained in Example 1, Art. 4.6, were as follows:

$I_x = 693.3$ in.^{4}, $I_y = 173.3$ in.^{4}, $I_{xy} = -240$ in.^{4}, $I_p = 787.1$ in.^{4},
$I_q = 79.5$ in.^{4}, $\theta = 21.35°$ (as shown in Fig. 7.7).

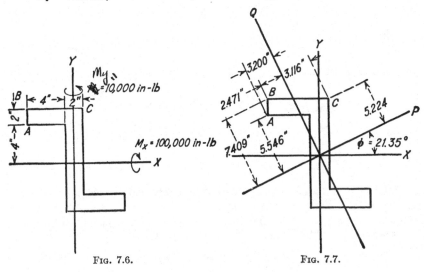

Fig. 7.6.   Fig. 7.7.

Solution by Method 1. The first method of solution will be by the use of Eq. 7.12.

$$f_b = \frac{M_x I_{xy} - M_y I_x}{I_x I_y - I_{xy}^2} x + \frac{M_y I_{xy} - M_x I_y}{I_x I_y - I_{xy}^2} y$$

$$= \frac{(100,000)(-240) - (10,000)(693.3)}{(693.3)(173.3) - (-240)^2} x$$

$$+ \frac{(10,000)(-240) - (100,000)(173.3)}{(693.3)(173.3) - (-240)^2} y$$

$$f_b = -494x - 315y \tag{7.13}$$

The stresses at points A, B, and C are obtained in the following tabular computations by substituting values of x and y into Eq. 7.13. The distribution

TABLE 7.1

Point	x	y	$-494x$	$-315y$	f_b, psi
A	-5	4	2,470	$-1,260$	1,210
B	-5	6	2,470	$-1,890$	580
C	1	6	-490	$-1,890$	$-2,380$

of bending stress over the area of cross section is shown in Fig. 7.8. The equation of the neutral axis may be obtained by substituting $f_b = 0$ in Eq. 7.13. The slope of the neutral axis is obtained from this equation as follows: $\tan \phi = -(^{494}\!/_{315})$, or $\phi = -57.5°$.

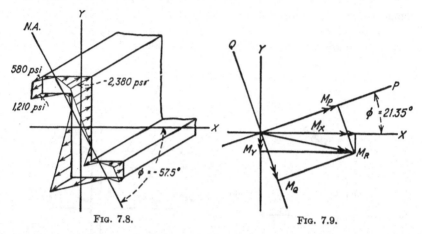

FIG. 7.8.   FIG. 7.9.

Solution by Method 2. In the analysis by the second method, all the section properties are referred to the principal axes, and the bending stresses are obtained from Eq. 7.1. If the coordinates of a point with respect to the principal axes are designated by x_p and y_p, Eq. 7.1 may be written

$$f_b = -\frac{M_p y_p}{I_p} - \frac{M_q x_p}{I_q} \qquad (7.14)$$

where subscripts p and q indicate properties about the principal axes. The bending moments about the P and Q axes are obtained by use of the vector representation of the couples, as shown in Fig. 7.9. Using the left-hand sign convention, the positive value of M_x is shown by a vector to the right along the x axis, and the positive value of M_y is shown by a vector down along the y axis. The vectors M_p and M_q, shown in the positive directions in Fig. 7.9, are obtained by projecting the vectors M_x and M_y along the P and Q axes.

$$\begin{aligned}
M_p &= M_x \cos \theta - M_y \sin \theta \\
&= 100{,}000 \times 0.9313 - 10{,}000 \times 0.3642 = 89{,}490 \text{ in-lb} \\
M_q &= M_y \cos \theta + M_x \sin \theta \\
&= 10{,}000 \times 0.9313 + 100{,}000 \times 0.3642 = 45{,}730 \text{ in-lb}
\end{aligned}$$

Substituting values of M_p, M_q, I_p, and I_q, into Eq. 7.14, there is obtained

$$f_b = -\frac{89{,}490}{787.1} y_p - \frac{45{,}730}{79.5} x_p$$
$$f_b = -113.6 y_p - 575.0 x_p \qquad (7.15)$$

The values of the coordinates of points A, B, and C with respect to the P and Q axes, x_p and y_p, are found from the equations $x_p = x \cos \theta + y \sin \theta$, and

$y_p = y \cos\theta - x \sin\theta$ and are shown in Fig. 7.7. The bending stresses are then obtained by substituting these values into Eq. 7.15, as tabulated below.

TABLE 7.2

Point	x_p	y_p	$-575x_p$	$-113.6y_p$	f_b
A	−3.200	5.546	1,840	−630	1,210
B	−2.471	7.409	1,420	−840	580
C	3.116	5.224	−1,790	−590	−2,380

These bending stresses check those obtained by the first method.

Example 2. Find the bending stress in each stringer of the wing cross section shown in Fig. 7.10. The section properties about the x and z axes through the

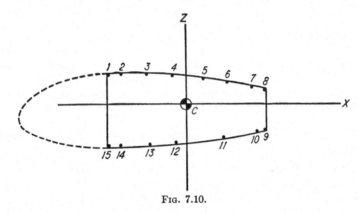

FIG. 7.10.

centroid are $I_x = 71.23$ in.⁴, $I_z = 913.71$ in.⁴, and $I_{xz} = 5.30$ in.⁴ The wing bending moments at the cross section are $M_x = 460,000$ in-lb and $M_z = 42,500$ in-lb. The coordinates of the stringer areas are shown in Table 7.3.

Solution. The bending stresses will be obtained by the use of Eq. 7.12. It is customary to designate the reference axes of wing cross section areas as x and z axes. Equation 7.12 may be written

$$f_b = \frac{M_x I_{zz} - M_z I_x}{I_x I_z - I_{xz}^2} x + \frac{M_z I_{zz} - M_x I_z}{I_x I_z - I_{xz}^2} z$$

Substituting numerical values,

$$f_b = \frac{460,000 \times 5.30 - 42,500 \times 71.23}{71.23 \times 913.71 - (5.30)^2} x + \frac{42,500 \times 5.30 - 460,000 \times 913.71}{71.23 \times 913.71 - (5.30)^2} z$$

or

$$f_b = -9.06x - 6,457z \tag{7.16}$$

The coordinates of each stringer area are substituted in Eq. 7.16 as shown in Table 7.3.

TABLE 7.3

Stringer No.	x	z	$-9.06x$	$-6,457z$	f_b
1	−16.86	4.32	150	−27,890	−27,740
2	−15.76	4.44	140	−28,670	−28,530
3	−9.66	4.19	90	−27,050	−26,960
4	−3.46	3.94	30	−25,440	−25,410
5	2.64	3.64	−20	−23,500	−23,520
6	8.69	3.39	−80	−21,890	−21,970
7	17.19	3.03	−160	−19,560	−19,720
8	18.29	2.74	−170	−17,690	−17,860
9	18.29	−2.25	−170	14,530	14,360
10	16.38	−2.82	−150	18,210	18,060
11	7.54	−3.29	−70	21,240	21,170
12*					
13	−8.31	−4.17	80	26,930	27,010
14	−15.76	−4.43	140	28,600	28,740
15	−16.86	−4.16	150	26,860	27,010

* Stringer 12 ends at this cross section and is not considered in the calculation of bending stresses.

The bending stresses in a wing cross section may also be calculated by referring bending moments and section properties to the principal axes, as was done in the second method of analysis of Example 1. Where it is necessary to find the stresses at several points in the cross section, it becomes rather tedious to calculate the coordinates of these points with respect to the inclined principal axes. This method is occasionally used in the analysis of wing cross sections, however. The coordinates of the stringers would probably be scaled from a drawing of the cross section, rather than computed from the x and z coordinates.

7.3. Unsymmetrical Beams Supported Laterally. The beams considered in the previous article were assumed to be free to deflect in any direction under the imposed loads. In the beam shown in Figs. 7.6 to 7.8, for example, the bending moment produced by vertical loads is ten times as large as the bending moment produced by horizontal loads The horizontal deflection, however, is greater than the vertical deflection since the neutral axis is inclined more than 45° to the horizontal. Any unsymmetrical beam resisting only vertical loads deflects horizontally as well as vertically, since the neutral axis is inclined to the horizontal.

In many structures unsymmetrical beams resist known vertical loads, but are restrained so that they cannot deflect horizontally. As an example, consider the Z-shaped extrusion shown in Fig. 7.11(a), which is riveted to a thin sheet. The sheet is assumed to apply horizontal restraining forces to the extrusion, but to have a negligible effect in resisting the vertical load. Since the beam must deflect vertically, the neutral axis must be horizontal, and the bending stresses must be dis-

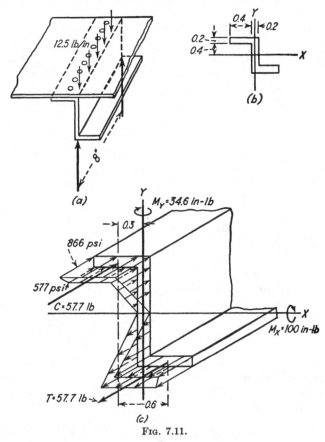

Fig. 7.11.

tributed as shown in Fig. 7.11(c). The bending stresses must resist the external bending moment, and are therefore found from the equation for simple bending

$$f_b = - \frac{M_x y}{I_x} \qquad (7.17)$$

These bending stresses also produce a bending moment about the y axis, since the resultant compressive force on the upper flange of the beam has a moment about the y axis which adds to the moment of the resultant tension on the lower flange of the beam. This bending moment about the y axis is produced by the horizontal restraining forces acting on the beam.　The stresses in the beam may be obtained directly from Eq. 7.17 without obtaining the horizontal forces acting on the beam.

Example.　Find the bending stresses in the beam shown in Fig. 7.11.　The horizontal sheet prevents the beam from deflecting horizontally, but does not resist an appreciable part of the vertical load.　Compare the results with the stresses in a similar beam which is not supported laterally.

Solution.　The dimensions of the cross section shown in Fig. 7.11 are one-tenth the values of the dimensions shown in Fig. 7.6.　The moments and products of inertia are therefore obtained from those used in Example 1, Art. 7.2, by dividing the values previously obtained by 10^4.　Therefore, $I_x = 0.06933$ in.^{4}, $I_y = 0.01733$ in.^{4}, $I_{xy} = -0.0240$ in.^{4}　The bending moment M_x is found as follows:

$$M_x = \frac{WL^2}{8} = \frac{12.5 \times 8^2}{8} = 100 \text{ in-lb}$$

The maximum bending stress is obtained from Eq. 7.17.

$$f_b = -\frac{100 \times 0.6}{0.06933} = -866 \text{ psi}$$

The bending stress at $y = 0.4$ is also obtained from Eq. 7.17.

$$f_b = -\frac{100 \times 0.4}{0.06933} = 577 \text{ psi}$$

The bending moment about the y axis may be obtained by considering the forces on the beam flanges.　The compression force C shown in Fig. 7.11(c) is found by multiplying the average stress by the area of the rectangle.

$$C = \frac{866 + 577}{2} (0.2 \times 0.4) = 57.7 \text{ lb}$$

The moment of this force about the y axis is obtained as the product of the force and the moment arm of 0.3 in.　The bending moment M_y is the sum of the moments of the forces C and T.

$$M_y = -2 \times 0.3 \times 57.7 = -34.6 \text{ in-lb}$$

The horizontal restraining forces on the beam are therefore 0.346 times the vertical loads shown in Fig. 7.11(a).

The value of M_y might also be obtained from Eq. 7.12.　For the neutral axis to be horizontal, it is necessary that the coefficient of x in Eq. 7.12 be zero.

$$\frac{M_x I_{xy} - M_y I_x}{I_x I_y - I_{xy}^2} = 0$$

or

$$M_y = \frac{M_x I_{xy}}{I_x} = \frac{(100)(-0.0240)}{0.06933} = -34.6 \text{ in-lb}$$

The substitution of $M_y = M_x I_{xy}/I_x$ into Eq. 7.12 is found to yield Eq. 7.17.

The stresses in a similar beam which is free to deflect horizontally may be found from Eq. 7.12.   Substituting $M_x = 100$ in-lb, $M_y = 0$, $I_x = 0.06933$ in.⁴, $I_y = 0.01733$ in.⁴, $I_{xy} = -0.0240$ in.⁴, into Eq. 7.12, there is obtained

$$f_b = \frac{M_x I_{xy} - M_y I_x}{I_x I_y - I_{xy}^2} x + \frac{M_y I_{xy} - M_x I_y}{I_x I_y - I_{xy}^2} y$$

$$= \frac{(100)(-0.0240) - 0}{(0.06933)(0.01733) - (-0.0240)^2} x + \frac{0 - (100)(0.01733)}{(0.06933)(0.01733) - (-0.0240)^2} y$$

or

$$f_b = -3,840x - 2,770y \tag{7.18}$$

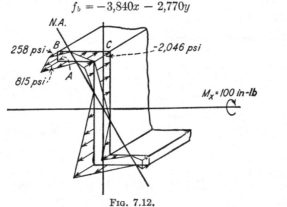

Fig. 7.12.

The stresses at points A, B, and C of Fig. 7.12 are obtained in the following tabular computations by substituting values of x and y into Eq. 7.18.

TABLE 7.4

Point	x	y	$-3,840x$	$-2,770y$	f_b, psi
A	-0.5	0.4	1,920	$-1,108$	812
B	-0.5	0.6	1,920	$-1,662$	258
C	0.1	0.6	-384	$-1,662$	$-2,046$

These stresses are shown in Fig. 7.12, and may be compared with those shown in Fig. 7.11(c).

PROBLEMS

7.1. A cantilever beam 30 in. long carries a vertical load of 1,000 lb at the free end.   The cross section is rectangular and is 6 by 1 in.   Find the maximum bending stress and the location of the neutral axis if (a) the 6-in. side is vertical, (b) the 6-in. side is tilted 5° from the vertical, and (c) the 6-in. side is tilted 10° from the vertical.

7.2. A horizontal beam with a square cross section resists vertical loads. Find the angle of the neutral axis with the horizontal if one side of the beam makes an angle θ with the horizontal.   At what angle should the beam be placed for the bending stress to have a minimum value?

7.3. Find the bending stresses and flange loads for the box beam shown if $M_x = 100{,}000$ in-lb and $M_z = -40{,}000$ in-lb. Assume the areas of the flange members as follows:

    a. $a = b = c = d = 2$ sq in.
    b. $a = b = 3$ sq in. $c = d = 1$ sq in.
    c. $a = d = 3$ sq in. $b = c = 1$ sq in.
    d. $a = c = 3$ sq in. $b = d = 1$ sq in.
    e. $a = c = 1$ sq in. $b = d = 3$ sq in.

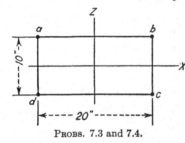

Probs. 7.3 and 7.4.

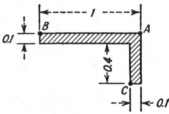

Prob. 7.5.

Check the results by equating the moments of the flange loads to the external moments on the beam.

7.4. Solve parts (*d*) and (*e*) of Prob. 7.3 by a second method.

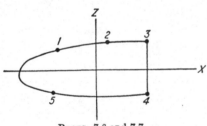

Probs. 7.6 and 7.7.

7.5. A beam with the cross section shown resists a bending moment $M_x = 100$ in-lb. Calculate the bending stresses at points *A*, *B*, and *C* by the use of Eq. 7.12, and check the results by the use of Eq. 7.14.

7.6. The box beam shown resists bending moments of $M_x = 1{,}000{,}000$ in-lb and $M_z = 120{,}000$ in-lb. Find the bending stress in each flange member by the use of Eq. 7.12, if the flanges resist all bending and if the areas and coordinates of the flange members are as follows:

TABLE 7.5

No.	Area, sq in.	x, in.	z, in.
1	1.8	−2.62	8.30
2	0.4	10.81	9.12
3	0.8	24.70	9.75
4	2.3	24.70	−1.30
5	1.0	−2.62	−1.20

7.7. Solve Prob. 7.6 by obtaining the principal axes through the centroid of the area and by the use of Eq. 7.14.

7.4. Beams with Three Concentrated Flange Areas. Aircraft structures are frequently constructed so that all the bending stresses are resisted by three structural members. The area of each of these members may be considered as concentrated at one point. The wing cross section shown in Fig. 7.13, for example, resists bending moments by means of concentrated forces at points A, B, and C. The shear and torsional stresses are resisted by webs AB, BC, and CA, and the shear flows in these webs will be constant. There are six unknown force components at the cross section: the three axial flange loads and the three web shear flows. These six unknowns may be obtained from the six equations of statics, without using the flexure formula.

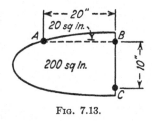

Fig. 7.13.

A box beam with only two concentrated flange areas is unstable for a general condition of loading. If the two flange members are in the same vertical plane, the beam can resist bending moments produced by vertical loads but not those produced by horizontal loads. Three flange members, not located in the same plane and connected by three webs forming a closed box, are necessary and are sufficient for stability under all conditions of loading. Where more than three flange members are used, the structure is statically indeterminate, and the general bending equation which is derived from deflection conditions must be used. In the three-flange beam, the flange loads and web shear flows are independent of the flange areas, since they are obtained from equations of static equilibrium. In the case of beams with more than three flanges, the values of the flange areas must be used in finding the flange loads and the web shear flows.

Three-flange beams are frequently used near splice points, since the three points represent the minimum number of connections necessary for stability. Navy shipboard airplanes with folding wings often use this type of structure near the wing hinge. The hinge line might pass through points A and B of Fig. 7.13, for example, and the latch might be at point C.

In analyzing a wing cross section by the six equations of statics, the outboard section of the wing, shown in Fig. 7.14, is actually used as a free body. However, since aircraft drafting practice makes it customary to show any wing cross section as if the observer is looking inboard, the forces used in further discussion will be those shown on the inboard portion of the wing in Fig. 7.14. These forces are not in equilibrium with the external lift and drag forces on the outboard portion of the wing, but are in the same direction as these forces. The axial loads on the flanges are therefore obtained by equating the internal bending moment to the resultant external bending moment, rather than by equating the sum of the moments to zero.

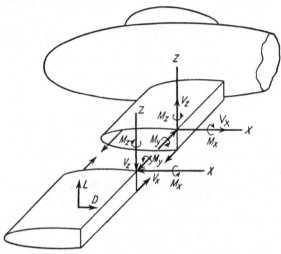

Fig. 7.14.

Example. The beam cross section shown in Fig. 7.13 is subjected to the loads shown in Fig. 7.15. Find the axial load in each flange and the shear flow in each web.

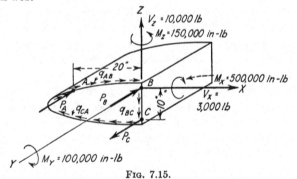

Fig. 7.15.

Solution. The internal forces P_a, P_b, and P_c and those produced by the shear flows q_{ab}, q_{bc}, and q_{ca} have resultants equal to the external couples and forces M_x, M_y, M_z, V_x, and V_z.

Equating the moments of the internal forces to those of the external forces, the following equations are obtained:

$$\Sigma M_z = 150,000 = -20P_a$$

or

$$P_a = -7,500 \text{ lb}$$
$$\Sigma M_x = 500,000 = 10P_c$$

or

$$P_c = 50,000 \text{ lb}$$

The sum of the spanwise forces on the cross section must be zero.

$$\Sigma F_y = P_b - 50,000 - 7,500 = 0$$

or

$$P_b = 57,500 \text{ lb}$$

The shear flows q_{ab}, q_{bc}, and q_{ca} may be obtained from the other three equations of statics. Equations 6.11 to 6.13 are used in finding the resultant horizontal or vertical force, or the resultant moment for the total shear flow on any web. The area A used in Eq. 6.13 is enclosed by the web and the lines joining the center of moments with the ends of the web. If point B, Fig. 7.13, is the center of moments, the enclosed areas are 20 sq in. for web AB and 200 sq in. for web AC. The shear flows in web BC will produce no moment about point B.

$$\Sigma M_y = 2 \times 20q_{ab} + 2 \times 200q_{ca} = 100,000$$
$$\Sigma F_x = 20q_{ab} - 20q_{ca} = 3,000$$
$$\Sigma F_z = -10q_{bc} + 10q_{ca} = 10,000$$

The shear flows are obtained by solving these equations simultaneously.

$$q_{ab} = 363.5 \text{ lb/in.} \qquad q_{bc} = -786.5 \text{ lb/in.} \qquad q_{ca} = 213.5 \text{ lb/in.}$$

The negative signs indicate that the directions of shear flows or forces are opposite to the directions assumed in Fig. 7.15.

7.5. Shear Flows in Unsymmetrical Beams. Shearing stresses in beams with symmetrical cross sections are obtained from Eq. 6.7, $f_s = (V/Ib)\int y\,dA$. This equation was derived from the simple flexure formula, $f = My/I$, and is subject to the same limitations. In the analysis of beams with unsymmetrical cross sections, it is necessary to obtain the bending stress either by means of Eq. 7.12 or else by obtaining section properties about the principal axes and superimposing the stresses from bending about each of the principal axes. Similarly, in analyzing an unsymmetrical beam for shearing stresses, it is necessary either to consider separately the shears along each of the two principal axes or to derive an equation for shearing stress from the general bending equation.

For an unsymmetrical beam with cross sections in the xz plane, Eq. 7.12 may be written as follows:

$$f_b = \frac{M_x I_{zz} - M_z I_x}{I_x I_z - I_{xz}^2} x + \frac{M_z I_{xz} - M_x I_z}{I_x I_z - I_{xz}^2} z \qquad (7.18a)$$

If two cross sections a distance a apart are considered, the change in bending moment about any axis is equal to the product of the shear perpendicular to that axis and the distance a. The change in bending stress Δf_b between the two cross sections is therefore obtained by substituting the terms $M_x = V_z a$ and $M_z = V_x a$ into Eq. 7.18a.

$$\frac{\Delta f_b}{a} = \frac{V_z I_{xz} - V_x I_x}{I_x I_z - I_{xz}^2} x + \frac{V_x I_{xz} - V_z I_z}{I_x I_z - I_{xz}^2} z \qquad (7.19)$$

The change ΔP in the axial load on any flange area between two cross sections a unit distance apart is obtained by multiplying the flange area A_f by the change in stress $\Delta f_b/a$. The shear flows may then be obtained from the axial loads ΔP by the methods previously used for symmetrical beams. The shear flows around the perimeter of a box beam therefore change at each flange area by an amount

$$\Delta q = \Delta P = \left(\frac{V_z I_{xz} - V_x I_x}{I_x I_z - I_{xz}^2} x + \frac{V_x I_{xz} - V_z I_z}{I_x I_z - I_{xz}^2} z \right) A_f \qquad (7.20)$$

where x and z are coordinates of the area A_f.

Example. Find the shear flows in the webs of the beam shown in Fig. 7.16

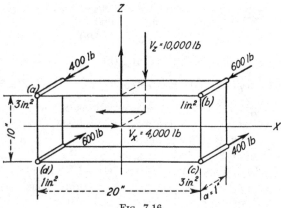

Fig. 7.16.

Solution. The change in bending stress between two cross sections is obtained from Eq. 7.19. The terms to be used in this equation are obtained as follows:

$$I_x = 8 \times 5^2 = 200 \text{ in.}^4$$
$$I_z = 8 \times 10^2 = 800 \text{ in.}^4$$
$$I_{xz} = 2 \times 10 \times 5 - 6 \times 10 \times 5 = -200 \text{ in.}^4$$
$$V_z = 10,000 \text{ lb}$$
$$V_x = 4,000 \text{ lb}$$

The substitution in Eq. 7.19 yields

$$\frac{\Delta f_b}{a} = -23.33x - 73.33z$$

The change in axial load is obtained by multiplying the values given in this equation by the respective areas. The calculations are given in the following table.

TABLE 7.6

Flange	Flange area A_f	Coordinates		$-23.33x$	$-73.33z$	$\dfrac{\Delta f_b}{a}$	ΔP
		x	z				
a	3	-10	5	233.3	-366.7	-133.3	-400
b	1	10	5	-233.3	-366.7	-600	-600
c	3	10	-5	-233.3	366.7	133.3	400
d	1	-10	-5	233.3	366.7	600	600

The terms in the last column represent the change in axial load in the unit length between the two cross sections shown in Fig. 7.16.

The shear flow in each web may now be obtained from the increments of flange load, as was done for symmetrical box beams. The shear flow in the left-hand web is designated as q_0, as shown in Fig. 7.17(a). From the equi-

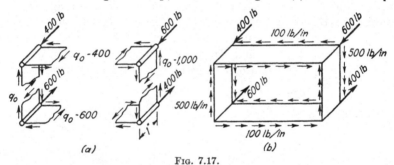

FIG. 7.17.

librium of spanwise forces on the upper left-hand stringer, the shear flow in the upper web is found to be $q_0 - 400$. Similarly, from spanwise forces on the upper right-hand stringer, the shear flow in the right-hand web is $q_0 - 1,000$, as shown in Fig. 7.17(a). The lower web has a shear flow of $q_0 - 600$, as shown. The unknown shear flow q_0 is now obtained from the equilibrium of torsional moments. If the center of moments is taken at the center of the enclosed area, the moments of the shear flows will be zero. The area enclosed by each web and the lines joining its ends to the center of moments is 50 sq in. The equation of torsional moments is written as follows:

$$2 \times 50q_0 + 2 \times 50(q_0 - 400) + 2 \times 50(q_0 - 1,000) + 2 \times 50(q_0 - 600) = 0$$

or

$$400q_0 - 200,000 = 0$$

and

$$q_0 = 500 \text{ lb/in.}$$

The final shear flows are shown in Fig. 7.17(b).

7.6. Beams with Varying Cross Sections. In the derivation of Eq. 7.19, the cross section is assumed to be the same at all points. The equation consequently does not apply to beams tapering in depth or to beams in which the flange areas change from point to point along the span of the beam. Such beams are most conveniently analyzed by calculating the section properties, stresses, and flange loads at two cross sections a short distance apart. The shear flows calculated from the difference in axial flange loads are average values between the two cross sections. Since actual values of dimensions and flange areas at the two cross sections are considered when the section properties are calculated, the assumptions of constant cross section are eliminated.

The method of calculating shear flows for beams with variable cross sections from the differences in flange loads was first published by Shanley and Cozzone.[1] This method seems to be the most convenient to use in the analysis of wings and similar structures, since the bending stresses must be computed at several cross sections for the design of the flange members, and these stresses may then be used in the shear flow analysis of the web members. The method of obtaining the shear flows from the flange loads is the same as the method used for symmetrical beams in Art. 6.9. The bending stresses and flange loads must be obtained by use of Eq. 7.12 in the case of unsymmetrical beams.

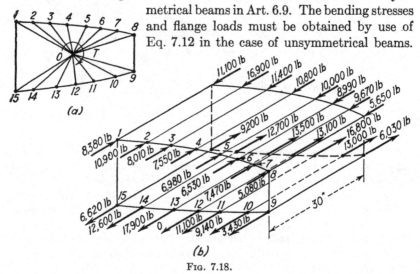

Fig. 7.18.

Example. The wing cross section shown in Fig. 7.10 is analyzed in Example 2, Art. 7.2. The bending stresses thus obtained are multiplied by the stringer areas to find the loads on the flanges. These loads are tabulated in column (3) of Table 7.7 as values of P_0, the flange loads at the outboard section. The flange loads at a cross section 30 in. inboard are tabulated in column (4) of Table 7.7 as values of P_i. Find the shear flows in all the webs, if the torque about point O, Fig. 7.18(a), is 79,000 in-lb and the double areas enclosed by the webs and lines to point O are as tabulated in column (2) of Table 7.7.

TABLE 7.7

Flange No. n (1)	Double enclosed area $2A$ (2)	Axial load outb'd P_0 (3)	Axial load inb'd P_i (4)	Diff. in axial load $P_i - P_0$ (5)	ΔP (6)	q' (7)	Moment $2Aq'$ (8)	Shear flow $q = q' + q_0$ (9)*
				(4) − (3)	(5) ÷ 30	Σ(6)	(2) × (7)	(7) + 496
1		−8,380	−11,100	−2,720	−90.7			
	12.5					−90.7	−1,130	405
2		−10,900	−16,900	−6,000	−200.0			
	38.6					−290.7	−11,230	205
3		−8,010	−11,400	−3,390	−113.0			
	37.3					−403.7	−15,080	92
4		−7,550	−10,800	−3,250	−108.3			
	35.9					−512.0	−18,400	−16
5		−6,980	−10,000	−3,020	−100.7			
	36.6					−612.7	−22,400	−117
6		−6,530	−8,990	−2,460	−82.0			
	71.1					−694.7	−49,400	−199
7		−7,470	−9,670	−2,200	−73.3			
	14.0					−768.0	−10,760	−272
8		−5,080	−5,650	−570	−19.0			
	120.4					−787.0	−94,900	−291
9		3,430	6,030	2,600	86.7			
	21.1					−700.3	−14,770	−204
10		9,140	13,000	3,860	128.6			
	39.1					−571.7	−22,320	−76
11		11,100	16,800	5,700	190.0			
	35.9					−381.7	−13,700	114
12		0	13,100	13,100	436.7			
	28.2					55.0	1,610	551
13		17,900	13,500	−4,400	−146.7			
	30.6					−91.7	−2,800	404
14		12,600	12,700	100	3.3			
	11.4					−88.4	−1,010	408
15		6,620	9,200	2,580	86.0			
	183.6					0(−2.4)	0	496
Total	716	. . . . .	. . . . .	. . . . .	. . . . .	. . . . .	−276,300	

*Calculation of q_0

$$q_0 = \frac{T - \Sigma 2Aq'}{\Sigma 2A} = \frac{79,000 + 276,300}{716} = 496 \text{ lb/in.}$$

Solution. The solution is shown in Table 7.7. This table is similar to Tables 6.1 and 6.3. The values of P_0 and P_i are the axial loads on the stringers at the outboard and inboard cross sections, as shown in Fig. 7.18(b). The difference between these loads, $P_i - P_0$, is obtained in column (5) by subtracting values of P_0 in column (3) from values of P_i in column (4). Since flange 12 ends at the outboard station, it will have no axial load at this point. This stringer is assumed to be fully effective at the inboard station, and consequently the value of ΔP is large, producing large shear flows in the webs adjacent to this stringer. The values of ΔP, the changes in shear flow at each stringer, are obtained in column (6) by dividing the values in column (5) by the distance, $a = 30$ in., between the cross sections. If web 15–1, the front spar web, is first considered as cut, the resulting shear flows are designated as q'. The values of q' are obtained in column (7) by a summation of terms in column (6), which is equivalent to a summation of spanwise forces on a free body consisting of the stringers or flanges between web 15–1 and the web under consideration. The moment of these shear flows about the spanwise reference axis through point O is obtained in column (8) as the product of terms in columns (2) and (7). If point O is inside the area enclosed by the box, all terms in column (2) will be positive, and positive (clockwise) shear flows will produce positive (clockwise) moment increments. If point O is outside the box, the terms in column (2) which indicate a negative moment about O of a positive shear flow will be negative. The summation of column (2) represents double the area enclosed by the box, regardless of the position of point O. The summation of column (8) yields the moment of the shear flows q' about point O.

The values of q' do not represent the true shear flows, since the torsional moment of these shear flows is not equal to the external torsional moment. A constant shear flow q_0 must be added to each shear flow q' to obtain the final shear flows q. The internal moments of the shear flows q' and q_0 must be equal to the external torsional moment on the wing.

$$\Sigma 2Aq' + q_0 \Sigma 2A = T$$

or

$$q_0 = \frac{T - \Sigma 2Aq'}{\Sigma 2A} \tag{7.21}$$

The value of q_0 is obtained from Eq. 7.21 in the lower part of Table 7.7. This value of q_0 must now be added to each value of q' in column (7). The final shear flows are tabulated in column (8).

7.7. The Choice of Wing Reference Axes. In a beam which has no taper in width or depth, the stringers are all parallel, and a cross section which is perpendicular to all the stringers may be used in the analysis. The shears and bending moments for the cross section are then referred to three mutually perpendicular axes parallel and perpendicular to the cross section. The stringer loads obtained from the bending stresses are parallel to the stringers and represent the true axial loads in the stringers.

The more common type of aircraft beam tapers in both depth and width. The stringer loads obtained from the bending stresses are components perpendicular to the cross section, and the true stringer loads must be obtained by dividing the known components by the cosines of the angles between the stringers and the normal to the plane of the cross section. It is usually very tedious to calculate the angle of each stringer and the various components of load in the stringers. Where the angle between the stringer and the normal to the cross section is small, it is customary to assume that the stringer load is equal to the component perpendicular to the cross section. For an angle less than 6°, the error of this assumption will be less than 1 per cent. The beam taper may not be neglected in calculating the components of stringer loads in the plane of the cross section, since these components are equal to the product of the axial load and the sine of the angle between the stringer and the normal to the cross section. For an angle of 6°, the in-plane component is over 10 per cent of the axial load and may change the shear flows by 50 per cent or more.

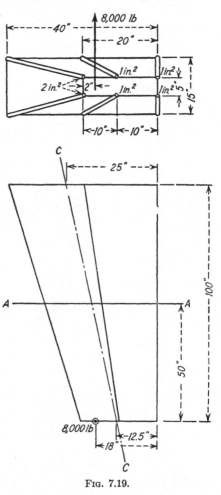

FIG. 7.19.

In Arts. 6.8 and 6.9 it was shown that the in-plane components of the stringer loads might resist a large proportion of the beam shear. In using the method of differences between flange loads on two cross sections, these in-plane components were automatically considered, without the necessity of computing them. It is necessary to select an axis of torsional moments in such a way that the in-plane components of stringer forces have a negligible moment about the axis. In the common type of wing, in which each stringer is at a constant percentage of the wing chord for the entire span, the correct axis for torsional moments is approximately the line joining the centroids of the various wing cross sections. The torsional moment about this axis should be used as the term T in Eq. 7.21.

Example. Find the shear flows at section AA of the beam shown in Fig. 7.19. Solve first by considering the in-plane components and then by selecting reference axes in such a manner that the in-plane components may be neglected.

Solution 1. The shear flows will first be obtained by considering the in-plane components of the stringer loads in computing the torsion. The components of stringer loads perpendicular to the cross section are equal to those computed in Example 2, Art. 6.8, and shown in Fig. 6.36. These loads, and the in-plane components, are also shown in Fig. 7.20. The shear flows, in terms of the unknown shear flow q_0 in the upper right-hand web, are similar to those shown in

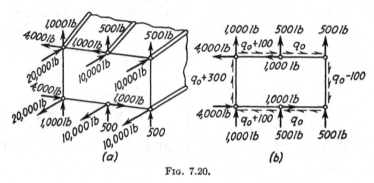

FIG. 7.20.

Fig. 6.37 for the previous example and are shown in Fig. 7.20(*b*). The external torque about a spanwise axis through point *O*, at the lower left-hand corner of section *AA*, is

$$T_0 = -8,000 \times 12 = -96,000 \text{ in-lb}$$

This torque is equal to the internal resisting torque of the in-plane stringer components and the shear flows. The torque resisted by the in-plane components of the stringer forces is as follows:

$$-4,000 \times 10 - 1,000 \times 10 - 2(500 \times 15) - 2(500 \times 30) = -95,000 \text{ in-lb}$$

This obviously represents a large proportion of the torsional resistance. Equating the external torque to the internal torque, the value of q_0 is obtained.

$$-96,000 = -95,000 + 150(q_0 + 100) + 150q_0 + 300(q_0 - 100)$$

or

$$q_0 = 23 \text{ lb/in.}$$

The shear flows in the remaining webs are obtained by substituting the value of q_0 in the terms shown in Fig. 7.20(*b*).

Solution 2. The use of in-plane components of the stringer loads is obviously very tedious in cases where a large number of stringers must be considered. The torsional moment of the in-plane forces about axis *CC*, Fig. 7.21, is negligible. Axis *CC* joins the centroids of the various cross sections. The external torque about this axis is

$$T_c = 8,000 \times 5.5 \cos 7.12° = 43,700 \text{ in-lb}$$

Equating this torque to that produced by the shear flows shown in Fig. 7.22, an equation for q_0 is obtained.

$$43{,}700 = 75(q_0 + 100) + 75q_0 + 187.5(q_0 - 100) + 75q_0$$
$$+ 75(q_0 + 100) + 112.5(q_0 + 300)$$

or

$$q_0 = 23 \text{ lb/in.}$$

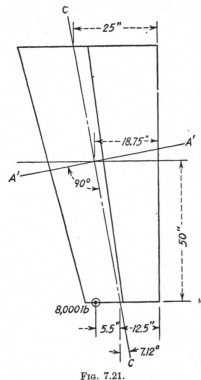

This checks the value previously obtained.

This second method of solution is recommended for most practical problems. The stringer loads may be obtained at various cross sections from the general bending equations. The shear flows may then be obtained from the differences, ΔP, of the stringer loads, as discussed in Art. 7.6. If the torsional axis is then taken as a straight line which approximately joins the centroids of the cross sections, it is not necessary to compute the in-plane components of the stringer loads.

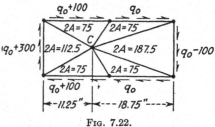

FIG. 7.21.

FIG. 7.22.

7.8. Correction of Wing Bending Moments for Sweepback. In the structural analysis of the wing, it is desirable to obtain the wing bending and torsional moments with respect to a set of mutually perpendicular axes through the cross section, which are oriented in such a manner that the spanwise axis is approximately through the centroids of the cross-sectional areas. Since the centroids are not known at the time preliminary wing bending moments are calculated, it is customary to use first a set of axes in which the spanwise axis is perpendicular to the plane of symmetry of the airplane and to calculate bending and torsional moments with respect to these axes. These moments must then be transferred to the axes through the centroids of the cross section.

The wing shown in Fig. 7.23 is assumed to have a sweepback angle β measured from the y axis, about which the moments are known, to the y_1 axis, about which moments are desired. The wing dihedral is neglected, so that the z and z_1 axes are parallel. The bending moments

M_x, M_y, M_{x1}, and M_{y1} are represented by double arrow couple vectors, according to the left-hand rule of sign conventions. The forces N and C, parallel to the z and x axes, as shown in Fig. 2.23(b), will also affect the

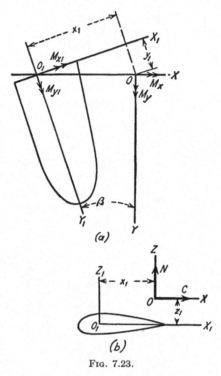

Fig. 7.23.

bending moments. The normal wing bending moment M_{x1} is obtained by taking moments about the x_1 axis shown in Fig. 7.23, as follows:

$$M_{x1} = M_x \cos \beta - M_y \sin \beta + Ny_1 - Cz_1 \sin \beta \qquad (7.22)$$

Similarly, the torsional moment about the y_1 axis is

$$M_{y1} = M_y \cos \beta + M_x \sin \beta - Nx_1 + Cz_1 \cos \beta \qquad (7.23)$$

The chordwise bending moment M_{z1} is not considered here, because for our assumed conditions of no dihedral and no force in the y direction, the moments M_{z1} and M_z will be equal.

In some cases it will be necessary to correct bending moments for the wing dihedral angle. Methods similar to those used in setting up Eqs. 7.22 and 7.23 can be used.

Example. For the wing shown in Fig. 7.19, the x and y reference axes are taken through section AA and the right-hand side of the wing. The moments referred to these axes are $M_x = 400,000$ in-lb and $M_y = 144,000$ in-lb. Transfer these moments to axes $A'A'$ and CC shown in Fig. 7.21.

Solution. The coordinates x_1 and y_1, shown in Fig. 7.22(a), are $x_1 = 18.75 \cos 7.12° = 18.6$ in. and $y_1 = 18.75 \sin 7.12° = 2.32$ in. From Eq. 7.22, $M_{z1} = 400,000 \cos 7.12° - 144,000 \sin 7.12° + 8,000 \times 2.32 = 398,000$ in-lb. From Eq. 7.23 the value of M_{y1} is obtained.

$$M_{y1} = 144,000 \cos 7.12° + 400,000 \sin 7.12° - 8,000 \times 18.6$$
$$= 43,700 \text{ in-lb}$$

This checks the value obtained in the second solution of the example of Art. 7.7.

PROBLEMS

7.8. Find the shear flows in the webs of the cross section shown. Assume no taper. Note that the areas of the flanges do not affect the shear flows.

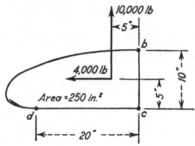

PROBS. 7.8, 7.9, and 7.10.

7.9. Find the shear flows at the cross section for $x = 50$ in. The cross section is the same as that used in the previous problem. Consider only the one cross section, and calculate the in-plane components of the flange loads.

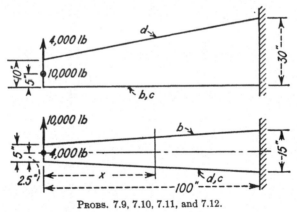

PROBS. 7.9, 7.10, 7.11, and 7.12.

7.10. Repeat Prob. 7.9, using the differences in flange loads at the cross sections for $x = 40$ in. and $x = 60$ in.

7.11. Calculate the shear flows in the webs of the cross section shown at $x =$ 50 in.   Assume the flange areas as follows:

a. $a = b = 3$ sq in. $c = d = 1$ sq in.
b. $a = c = 1$ sq in. $b = d = 3$ sq in.
c. $a = c = 3$ sq in. $b = d = 1$ sq in.

Consider only the one cross section, and calculate the in-plane components of the flange loads.

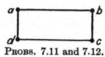

PROBS. 7.11 and 7.12.

7.12. Repeat Prob. 7.11, using the differences in flange loads at the cross sections for $x = 40$ in. and $x = 60$ in.   Use a torsional axis joining the centroids of the cross sections, and do not calculate the in-plane components.

REFERENCE FOR CHAPTER 7

1. SHANLEY, F. R., and F. P. COZZONE: Unit Method of Beam Analysis, *J. Aeronaut. Sci.*, Vol. 8, No. 6, p. 246, April, 1941.

CHAPTER 8

ANALYSIS OF TYPICAL MEMBERS OF
SEMIMONOCOQUE STRUCTURES

8.1. Distribution of Concentrated Loads to Thin Webs. Modern aircraft structures are constructed primarily from sheet metal. The metal is necessary for a covering, and is therefore utilized for structure as well. The thin sheets or webs are very efficient in resisting shear or tension loads in the planes of the webs, but must usually be stiffened by members more capable of resisting compression loads and loads normal to the web. If no stiffening members are used, and the skin or shell is designed to resist all loads, the construction is called *monocoque* or *full monocoque*, from the French word meaning "shell only." It is usually not feasible to have the skin thick enough to resist compression loads, and stiffeners are provided to form *semimonocoque* structures. In such structures the thin webs resist tension and shearing forces in the planes of the webs. The stiffeners resist compression forces in the plane of the web, or small distributed loads normal to the plane of the web.

When semimonocoque structures must resist large concentrated loads, it is necessary to transmit the loads to the planes of the webs. Since the concentrated loads may have components along three mutually perpendicular axes, it is necessary to provide webs in different planes, so that the loads may be applied at the intersection of two planes. A fuselage structure, for example, has closely spaced rings or bulkheads which resist loads in transverse planes, while the fuselage shell resists loads in the fore-and-aft direction. Concentrated loads must be applied at the intersection of the plane of the bulkhead and the shell, or else additional structural members must be provided to span between bulkheads and transfer the loads to two such intersecting planes.

When a concentrated load is applied in the plane of a web, a stiffening member is required to distribute this load to the web, as shown in Fig. 8.1(a). This member should be in the direction of the load, or the load should be applied at the intersection of two stiffeners, so that each stiffener resists the load component in its direction. The load P shown in Fig. 8.1 is distributed to the web by the stiffener AB. The shear flows q_1 and q_2 in the adjacent webs are approximately constant for the length of the stiffener. The axial load in the stiffener therefore varies linearly from P at point B to zero at point A, as shown in Fig. 8.1(c). From the equilibrium of the forces shown in Fig. 8.1(b), $P = (q_1+q_2)d$. The required length d of the stiffener therefore depends on the ability

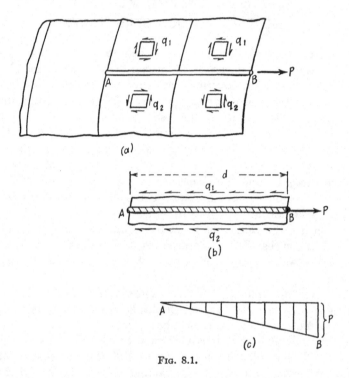

Fig. 8.1.

of the webs to resist shear, since a longer stiffener reduces the shear flows q_1 and q_2. The end of the stiffener, point A, should always be at a transverse stiffener. If a stiffener ends in the center of a web, it produces abrupt changes in the shear flows at the end of the stiffener and undesirable stress concentration conditions.

In this chapter, thin webs will be assumed to resist pure shear along their boundaries. In actual structures, the thin webs may wrinkle in shear, thus introducing tension field stresses in addition to those calculated. The effects of tension field stresses will be calculated in later chapters. It will be found at that time that the tension field stresses can be readily superimposed on those calculated by the methods used here, and that the methods used in this chapter remain valid for obtaining the shear distribution in tension field webs. In some cases the tension field stresses, produced by wrinkling of the webs, will induce additional axial compression loads in stiffeners. These loads would be computed separately and added algebraically to the loads obtained in this chapter.

A study of simple numerical examples will demonstrate the method by which loads are distributed to shear webs. The beam shown in Fig. 8.2(*a*) is similar to a wing rib which is supported by spars at the

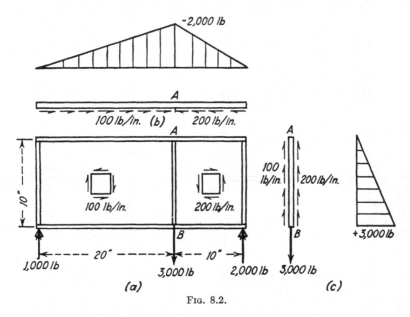

Fig. 8.2.

ends and which resists the load of 3,000 lb as shown. The stiffener *AB* transmits this load to the two webs in inverse proportion to the horizontal lengths of the webs, since the vertical shear at any cross section of the beam must be in equilibrium with the external reaction on the beam. The axial load in *AB* is shown in Fig. 8.2(*c*); it varies from 3,000 lb at *B* to zero at *A*. The axial load in the upper flange of the beam may be obtained from the bending-moment diagram of the beam, or it may be obtained by a summation of the shear flows, as shown in Fig. 8.2(*b*). The compression at point *A* of 2,000 lb may be obtained from the shear flow of 100 lb/in. for 20 in. or from the shear flow of 200 lb/in. for 10 in.

The cantilever beam shown in Fig. 8.3 resists a load *R* which has a horizontal component of 1,500 lb and a vertical component of 3,000 lb. The horizontal stiffener *AB* must be provided to resist the horizontal component of the load, and the vertical stiffener *CBD* must resist the vertical component. The intersection of these stiffeners, point B, should be on the line of action of R. The shear flows q_1 and q_2 may be obtained from the equilibrium of these stiffeners. For stiffener AB to be in equilibrium under the forces shown in Fig. 8.3(c),

$$10q_1 - 10q_2 = 1{,}500 \qquad (8.1)$$

Similarly, for member CBD to be in equilibrium under the forces shown in Fig. 8.3(b),

$$5q_1 + 10q_2 = 3{,}000 \qquad (8.2)$$

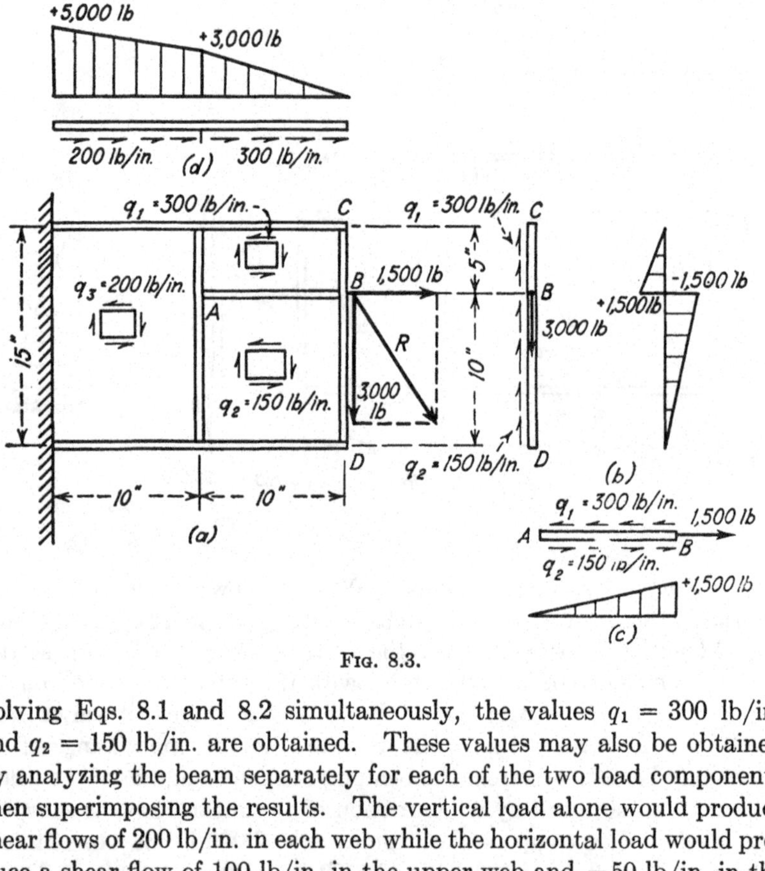

FIG. 8.3.

Solving Eqs. 8.1 and 8.2 simultaneously, the values $q_1 = 300$ lb/in. and $q_2 = 150$ lb/in. are obtained. These values may also be obtained by analyzing the beam separately for each of the two load components then superimposing the results. The vertical load alone would produce shear flows of 200 lb/in. in each web while the horizontal load would produce a shear flow of 100 lb/in. in the upper web and -50 lb/in. in the lower web. The axial loads in the members are shown in Figs. 8.3(b) to (d). The axial loads in the upper flange member, shown in Fig. 8.3(d), could not be readily obtained from a bending-moment diagram of the member.

The loads considered above were assumed to act in the plane of the web. When loads have components along all three reference axes, the structure should preferably be arranged so that the loads act at the intersection of two webs, as shown in Fig. 8.4(a). Here, each of the three components of the force R is distributed to the webs by a stiffener in the direction of the force component. In some cases this is not practical, and a load normal to a web, as shown in Fig. 8.4(b), cannot be

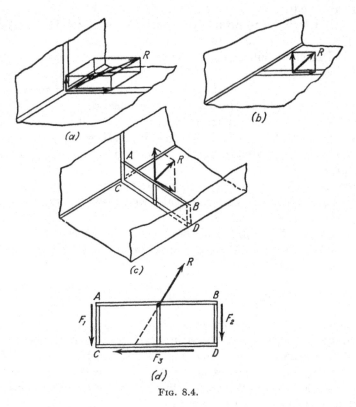

Fig. 8.4.

avoided. If the load is small, the stiffener may be designed to have enough bending strength to resist the load. In many cases the loads are such that it is necessary to provide an additional member, such as web $ABCD$ of Fig. 8.4(c), to resist the load. This member spans between ribs or bulkheads and can resist any load in its plane by means of the three reactions F_1, F_2, and F_3 shown in Fig. 8.4(d). Even small loads, such as those from brackets supporting control pulleys, should not be applied as normal loads to an unsupported web. Such brackets may be attached to stiffeners, or may be located at the intersections of webs.

8.2. Loads on Fuselage Bulkheads. The structural unit which transfers concentrated loads to the shell of an airplane fuselage or wing is commonly called a bulkhead. Bulkheads are attached to the wing or fuselage skin continuously around their perimeters. They may be solid webs with stiffeners or beads, webs with access holes, or truss structures. Fuselage bulkheads are usually open rings or frames, in order that the fuselage interior shall not be obstructed. The chordwise bulkheads in wings are normally called ribs, while fuselage bulkheads are called rings or frames. In addition to transferring loads to the skin, wing and fuselage bulkheads also supply column support to stringers and redistribute shear flows in the skin. The first step in the design of a bulkhead is to obtain the loads which act on the bulkhead and thus hold it in static equilibrium. In the case of fuselage rings, this first step is simpler than the next problem of obtaining the unit stresses from the loads. The analysis for unit stresses in fuselage rings and similar structures will be treated in a later chapter on statically indeterminate structures.

Fuselage shells are normally symmetrical about a vertical center line, and are often loaded symmetrically with respect to the center line. The fuselage bending stresses may then be obtained by the simple flexure formula, $f = My/I$, and the fuselage shear flows obtained from the related expression, which was derived in Chap. 6.

$$q = \frac{V_w}{I} \int y \, dA \qquad (6.27)$$

In applying Eq. 6.27 to a symmetrical box structure, it is usually convenient to consider only half of the structure, since the shear flow must be zero at the top and bottom center lines. Thus, each term of Eq. 6.27 may apply to only half of the fuselage shell. If stringers or longerons are located on the top or bottom center lines, half of their area is considered to act with each side of the structure.

The fuselage ring shown in Fig. 8.5 is loaded by a vertical load P on the center line of the airplane. This vertical load P must be in equilibrium with the running loads q which are applied to the perimeter of the ring, as shown in Fig. 8.5(c). The present problem is to obtain the distribution of the forces q. The fuselage cross section just forward of the ring has an external shear V_a, and the cross section aft of the ring has a shear V_b, as shown in Figs. 8.5(a) and (b). The load P on the ring must be equal to the difference of these shears, as expressed in the following equation:

$$V_a - V_b = P \qquad (8.3)$$

If, for the moment, the shear resisted by the in-plane components of the stringer loads is neglected, the shear flows on the two cross sections adjacent to the ring are obtained as follows:

$$q_a = \frac{V_a}{I} \int y \, dA \tag{8.4}$$

and

$$q_b = \frac{V_b}{I} \int y \, dA \tag{8.5}$$

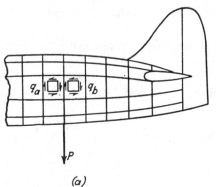

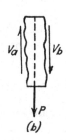

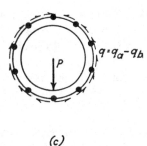

(a)

(b)

(c)

Fig. 8.5.

The load q transmitted to the perimeter of the ring must be equal to the difference between q_a and q_b, or

$$q = q_a - q_b \tag{8.6}$$

From Eqs. 8.3 to 8.6,

$$q = \frac{V_a - V_b}{I} \int y \, dA$$

or

$$q = \frac{P}{I} \int y \, dA \tag{8.7}$$

When the areas resisting bending of the shell are concentrated as flange areas A_f, the integral is replaced by a summation, as in the following equation:

$$q = \frac{P}{I} \Sigma y A_f \qquad (8.8)$$

Equations 8.7 and 8.8 are correct even when the relieving effect of the in-plane components of the stringer forces are considered, since this shear resisted by the stringers must be the same on both fuselage cross sections adjacent to the ring, if the stringers have no abrupt change in direction at the ring. The difference in total shear forces, $V_a - V_b$, must therefore be equal to the difference in the shears resisted by the webs.

In many cases a fuselage structure may be symmetrical but the loads may not be symmetrical. Any unsymmetrical vertical load may be resolved into a vertical load at the center line and a couple. The couple applied to the ring will be resisted by a constant shear flow

$$q_T = \frac{T}{2A} \qquad (8.9)$$

where T is the magnitude of the couple and A is the area enclosed by the fuselage skin in the plane of the bulkhead.

A fuselage ring may also resist loads which have horizontal components. In this case it is not possible to find a web with zero shear flow by inspection, as in the case for symmetrical vertical loads. It is necessary first to obtain all the shear flows in terms of one unknown and then to find this unknown from the equilibrium of moments, as was previously done in the analysis of box beams. The method will be obvious from a study of Example 2.

Example 1. The fuselage bulkhead shown in Fig. 8.6 resists a symmetrical load, as shown. Each stringer has an area of 0.1 sq in., and the coordinates y' of the centroids of the stringers are as shown in Table 8.1. Find the loads which the skin applies to the bulkhead.

Solution. This problem may be solved by the use of Eq. 8.8. Only one-half of the shell is considered, as shown in Fig. 8.6(b). The value of P, resisted on this half of the structure is 500 lb, and the moment of inertia is found for only one-half of the structure. The value of P/I in Eq. 8.8 will of course be the same if both values are obtained for the entire structure, since both will be doubled.

The solution is performed in Table 8.1. The areas A_f listed in column (2) are the total areas of stringers 2, 3, and 4 but only half areas for stringers 1 and 5, since the structure shown in Fig. 8.6(b) is being considered. The centroid $\bar{y}$ is determined from the summations of columns (4) and (2), as follows:

$$\bar{y} = \frac{\Sigma A_f y'}{\Sigma A_f} = \frac{6.2}{0.4} = 15.5 \text{ in.}$$

TABLE 8.1

Str. No. (1)	A_f (2)	y' (3)	$A_f y'$ (4)	y (5)	yA_f (6)	$y^2 A_f$ (7)	$\Sigma y A_f$ (8)	q, lb/in (9)
1	0.05	34.0	1.7	18.5	0.925	17.12		
							0.925	10.20
2	0.10	24.0	2.4	8.5	0.85	7.23		
							1.775	19.55
3	0.10	15.0	1.5	−0.5	−0.05	0.02		
							1.725	19.00
4	0.10	6.0	0.6	−9.5	−0.95	9.02		
							0.775	8.54
5	0.05	0.0	0.0	−15.5	−0.775	12.01		
Σ	0.4		6.2			45.40		

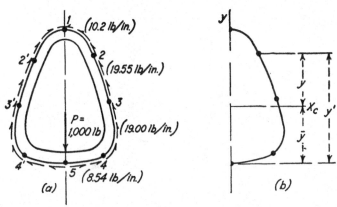

FIG. 8.6.

It is now necessary to obtain coordinates of the stringers with respect to the centroidal axis. These values, $y = y' - \bar{y}$, are obtained by subtracting 15.5 from terms in column (3), and are shown in Column (5). The terms yA_f and y^2A_f are calculated in columns (6) and (7). The summation of column (7) yields the moment of inertia I. Equation 8.8 then becomes

$$q = \frac{P}{I} \Sigma y A_f = \frac{500}{45.4} \Sigma y A_f$$

The values of q are calculated in column (9) and are shown in Fig. 8.6(a).

Example 2. The fuselage bulkhead shown in Fig. 8.7 resists a horizontal load as shown. The stringer coordinates are given in column (3) of Table 8.2. The areas enclosed by the skin segments and the lines to reference point O are indicated by the double areas listed in column (8) of Table 8.2. Find the reactions of the skin on the bulkhead.

TABLE 8.2

Str. No. (1)	A_f (2)	x (3)	xA_f (4)	x^2A_f (5)	ΣxA_f (6)	q' (7)	$2A$ (8)	$2Aq'$ (9)	q (10)
1	0.05	0	0	0					
					0	0	140	0	+16.8
2	0.10	8	0.8	6.4					
					0.8	−15.1	100	−1,510	+1.7
3	0.10	10	1.0	10.0					
					1.8	−34.1	100	−3,410	−17.3
4	0.10	10	1.0	10.0					
					2.8	−53.0	160	−8,480	−36.2
5	0.05	0	0	0					
Σ	0.4			26.4			500	−13,400	

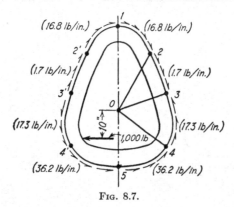

FIG. 8.7.

Solution. The shear flows will first be obtained with the assumption that web 1–2 resists a zero shear flow. The resulting shear flows q' are as follows:

$$q' = \frac{P}{I_y} \Sigma xA_f \tag{8.10}$$

The calculations are performed in Table 8.2. The value of I_y for half the structure is obtained in column (5) as 26.4 in.[4] If the shear flows are positive clockwise around the ring, the force P will be considered as negative in Eq. 8.10, or

$$q' = \frac{-1,000}{2 \times 26.4} \Sigma xA_f$$

The values of q' are calculated in column (7). These values are obviously the same for the left half of the structure because of symmetry. The shear flows q' produce moments of $2Aq'$ about point O. These moments are calculated in column (9). It is now necessary to superimpose a constant shear flow q_0 around the ring such that external moments on the ring are in equilibrium. Taking moments about O and considering clockwise moments as positive,

$$q_0 \Sigma 2A + \Sigma 2Aq' + 1,000 \times 10 = 0$$

or

$$1,000 q_0 - 2 \times 13,400 + 1,000 \times 10 = 0$$

or

$$q_0 = 16.8 \text{ lb/in.}$$

The resulting shear flows q are found by adding 16.8 to each value in column (7). These values are shown in column (10) and in Fig. 8.7.

8.3. Analysis of Wing Ribs. In the simplest type of wing structure, that in which the bending stresses are resisted by only three concentrated flange members, the skin reactions on the ribs may be obtained from the equations of statics. There will be only three unknown shear flows, and these may be readily obtained from the equations for the equilibrium of forces in the vertical and drag directions and for the equilibrium of moments about a spanwise axis.

The internal stresses in the rib are then obtained from the shears and bending moments at the various cross sections. There will normally be axial loads in the rib in addition to the shears and bending moments; therefore it will be necessary to calculate bending moments about a point with a vertical position corresponding to the neutral axis of the rib. For the rib analyzed in the following numerical example, it is assumed that all bending moments are resisted by the rib flange members and that all shears are resisted by the webs. With these assumptions it is more convenient to calculate bending moments about the top or bottom flange of the rib. If the entire depth of the rib resists bending, it is more convenient to calculate bending moments about the neutral axis.

In the more general case of a wing in which the bending moments are resisted by more than three flange members, it is necessary to determine the section properties of the wing cross section before the shear reactions on the ribs may be obtained. The problem is similar to that of calculating reactions on fuselage bulkheads, but differs in the condition that the fuselage cross section is usually symmetrical, whereas the wing cross section is seldom symmetrical. The skin shear flows, and consequently the skin reactions on the ribs, must therefore be obtained by the more general methods which involve the product of inertia of the cross section.

Example 1. Find the shear flows acting on the rib of Fig. 8.8. The wing bending moments are resisted by the three flange areas shown at Figs. 8.8(a) to (c).   Calculate the loads in the rib flanges and the shear flows in the rib webs at a vertical cross section through flange a and at vertical cross sections a short distance to either side of the applied loads.

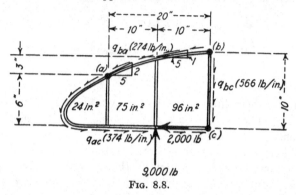

Fig. 8.8.

Solution.   The reactions of the wing skin on the rib must be in equilibrium with the applied loads of 9,000 lb and 2,000 lb.   From a summation of moments about point c and a summation of forces in the vertical and drag directions, the following equations are obtained:

$$\Sigma M_c = 9{,}000 \times 10 - 168 q_{ac} - 222 q_{ba} = 0$$
$$\Sigma F_x = 20 q_{ac} - 20 q_{ba} - 2{,}000 = 0$$
$$\Sigma F_z = 9{,}000 - 10 q_{bc} - 4 q_{ba} - 6 q_{ac} = 0$$

The solution of these equations yields $q_{ba} = 274$ lb/in., $q_{ac} = 374$ lb/in., and $q_{bc} = 566$ lb/in.

At a vertical cross section through flange a, the stresses are obtained by considering the free body shown in Fig. 8.9(a).   The total shear at the cross section is $V = 374 \times 6 = 2{,}244$ lb, and the bending moment is $M = 2 \times 24 \times 374 = 17{,}940$ in-lb.   The horizontal components of the axial loads in the flanges are obtained from the bending moment, as $P_1 = P_2 = M/6 = 2{,}990$ lb.   The lower rib flange is horizontal at this point, but the upper member has a slope of 0.4.   The shear carried by the flange is therefore $V_f = 0.4 \times 2{,}990 = 1{,}195$ lb. The remaining shear, which is resisted by the web, is $V_w = V - V_f = 2{,}244 - 1{,}195 = 1{,}049$ lb.   The shear flow at this section is $q = 1{,}049/6 = 178$ lb/in. These values are shown in Fig. 8.9(a).

The stresses at a vertical section to the left of the applied loads are obtained by considering the free body shown in Fig. 8.9(b).   The shear at the cross section is $V = 274 \times 3 + 374 \times 6 = 3{,}066$ lb.   The bending moment about the lower flange is $M = 2 \times 45 \times 274 + 2 \times 54 \times 374 = 65{,}100$ in-lb.   The horizontal component of the upper flange load is $P_3 = 65{,}100/9 = 7{,}240$ lb. The lower flange load is obtained from a summation of horizontal forces as $P_4 = 7{,}240 - 10 \times 274 + 10 \times 374 = 8{,}240$ lb.   It is obvious that the bending moment in the rib depends on the vertical location of the center of moments, since there is a resultant horizontal load at the section.   If the entire depth of a beam resists bending moment, the centroid of the section is used as the center of moments, and the stresses resulting from the axial load are then uniformly distributed over the area of the cross section. The vertical component of the

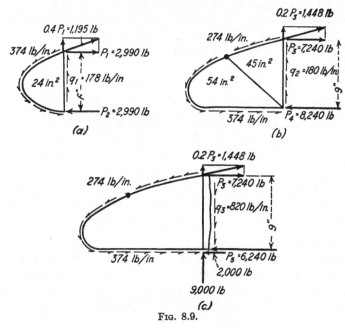

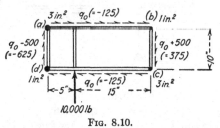

FIG. 8.9.

upper flange load is $0.2 \times 7,240 = 1,448$ lb. The shear flow in the web is therefore $q = (3,066 - 1,448)/9 = 180$ lb/in. These values are shown in Fig. 8.9(b).

The stresses at a vertical cross section just to the right of the applied loads are obtained in a similar manner and are shown in Fig. 8.9(c). Since the bending moment was computed about the intersection of the two applied loads, the value of P_3 is the same as for the previous case. The axial load in the lower flange and the web shear flow differ from those shown in Fig. 8.9(b).

FIG. 8.10.

Example 2. The rib shown in Fig. 8.10 transfers the vertical load of 10,000 lb to the wing spars and to the wing skin. Find the reacting shear flows around the perimeter of the rib, the shear flows in the rib web, and the axial loads in the top and bottom rib flanges.

Solution. The distribution of shear flows depends on the spar-cap areas. The spar areas shown are the same as those used in the example of Art. 7.5 and have the following section properties: $I_x = 200$ in.[4], $I_z = 800$ in.[4], $I_{xz} = -200$ in.[4] The reacting shear flows on the rib are equal to the shear flows in the skin of a box which resists an external shear of 10,000 lb, but have opposite directions. They may be obtained by the method of Art. 7.5. The change in shear flow at each flange area is found from Eq. 7.20.

$$\Delta q = -\left(\frac{V_z I_{xz} - V_x I_x}{I_x I_z - I_{xz}^2}x + \frac{V_x I_{xz} - V_z I_z}{I_x I_z - I_{xz}^2}z\right)A_f$$

assuming q positive clockwise, the following values are obtained:

$$\Delta q = +500 \text{ at flanges } a \text{ and } b$$
$$\Delta q = -500 \text{ at flanges } c \text{ and } d$$

If a shear flow q_0 is assumed in the top skin, the other shear flows are found in terms of q_0, as shown in Fig. 8.10. The final shear flows may now be obtained from the equilibrium of moments about some convenient point, say flange a.

$$200q_0 + 200(q_0 + 500) = 10,000 \times 5$$

or

$$q_0 = -125 \text{ lb/in.}$$

The shear flows in the remaining webs are now obtained from these values of q_0, and are shown in the proper directions in Fig. 8.11. The shear flows in the rib webs are obtained from the equilibrium of forces on the vertical cross sections

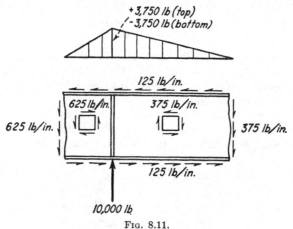

Fig. 8.11.

and are shown in Fig. 8.11 as 625 lb/in. on the left-hand web and 375 lb/in. on the right-hand web. The axial loads in the rib flanges are shown in Fig. 8.11, and are obtained by a summation of the shear flow forces acting on the rib flanges, including shear flows from both the wing skin and the rib web. A comparison of the rib web shears and flange loads shown in Fig. 8.11 with those for a simple beam of the same dimensions shows that the flange loads will be the same in both cases, but the web shears will be different.

PROBLEMS

8.1. Find the shear flow in each web of the beam shown, and plot the distribution of axial load along each stiffening member. Solve for each of the following loading conditions:

$a.$ $P_1 = 3,000$ lb $P_2 = P_3 = 0$
$b.$ $P_2 = 6,000$ lb $P_1 = P_3 = 0$
$c.$ $P_3 = 6,000$ lb $P_1 = P_2 = 0$
$d.$ $P_1 = 3,000$ lb $P_2 = 6,000$ lb $P_3 = 6,000$ lb

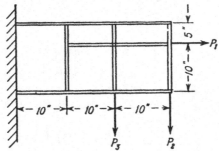

Probs. 8.1 and 8.2.

8.2. Repeat Prob. 8.1 for the following loading condition: $P_1 = 2,400$ lb, $P_2 = 1,200$ lb, and $P_3 = 1,800$ lb.

8.3. The pulley bracket shown is attached to webs along the three sides. Find the reactions R_1, R_2, and R_3 of the webs if $P = 1,000$ lb and $\theta = 45°$.

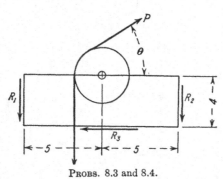

Probs. 8.3 and 8.4.

8.4. Repeat Prob. 8.3 for $P = 2,000$ lb and $\theta = 60°$.

8.5. Find the shear flows applied by the skin to the fuselage ring shown, if $P_1 = 2,000$ lb and $P_2 = M = 0$.

8.6. Find the skin reactions on the fuselage ring shown if $P_2 = 1,000$ lb and $P_1 = M = 0$.

8.7. Find the skin reactions on the fuselage ring if $P_1 = 2,000$ lb, $P_2 = 1,000$ lb, and $M = 10,000$ in-lb.

8.8. Find the skin reactions on the fuselage ring if $P_1 = 1,500$ lb, $P_2 = 500$ lb, and $M = 8,000$ in-lb.

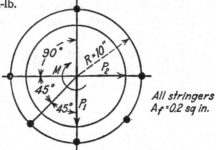

All stringers
$A_f = 0.2$ sq in.

PROBS. 8.5, 8.6, 8.7, and 8.8.

8.9. Find the skin reactions on the rib shown, if the rib is loaded by the distributed load of 20 lb/in. Calculate the shear flows in the rib web and the axial loads in the rib flanges at vertical sections 10 in. and 20 in. forward of the spar.

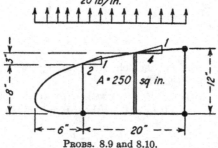

PROBS. 8.9 and 8.10.

8.10. Repeat Prob. 8.9, if the rib is loaded by a concentrated upward force of 600 lb, applied at a point 20 in. forward of the spar, instead of the distributed load.

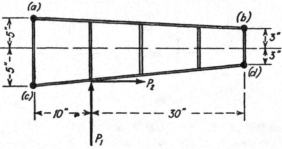

PROBS. 8.11, 8.12, 8.13, 8.14, 8.15, and 8.16.

8.11. Find the skin reactions on the rib shown. Analyze vertical cross sections at 10-in. intervals, obtaining the web shear flows and the axial loads in the rib flanges. Assume the loads $P_1 = 40,000$ lb and $P_2 = 0$. The spar flange areas are $a = b = c = d = 1$ sq in.

8.12. Repeat Prob. 8.11 for $P_1 = 0$ and $P_2 = 8,000$ lb, if the spar flange areas are $a = b = c = d = 1$ sq in.

8.13. Repeat Prob. 8.11 for $P_1 = 20,000$ lb and $P_2 = 8,000$ lb, if the spar flange areas are $a = b = c = d = 1$ sq in.

8.14. Repeat Prob. 8.11 for $P_1 = 40,000$ lb and $P_2 = 0$, if the spar flange areas are $a = 3$ sq in. and $b = c = d = 1$ sq in.

8.15. Repeat Prob. 8.11 for $P_1 = 0$ and $P_2 = 8,000$ lb, if the spar flange areas are $a = 3$ sq in. and $b = c = d = 1$ sq in.

8.16. Repeat Prob. 8.11 for $P_1 = 20,000$ lb, and $P_2 = 8,000$ lb, if the spar flange areas are $a = 3$ sq in. and $b = c = d = 1$ sq in.

8.4. Shear Flow in Tapered Webs. The shear flow in a tapered beam with two concentrated flanges was considered in Art. 6.8, in the discussion of the unit method of shear flow analysis for tapered box beams. The distribution of the shear flow in a tapered web will now be considered in more detail, since a large proportion of the shear webs in an airplane

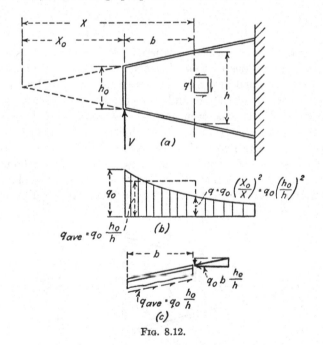

Fig. 8.12.

structure are tapered rather than rectangular. The shear flows in the web of the beam shown in Fig. 8.12(a) were obtained in Art. 6.8. From Eq. 6.24, the shear V_w resisted by the web will be

$$V_w = V \frac{h_0}{h} = V \frac{x_0}{x} \tag{8.11}$$

where the notation corresponds to that shown in Fig. 8.12. The shear flow q may be expressed in terms of the shear flow, $q_0 = V/h_0$, at the free end by the following equations:

$$q = \frac{V_w}{h} = \frac{Vh_0}{h^2} = q_0 \left(\frac{h_0}{h}\right)^2 = q_0 \left(\frac{x_0}{x}\right)^2 \qquad (8.12)$$

The distribution of the shear flow q along the span of the beam is shown in Fig. 8.12(b).

In many problems it is necessary to obtain the average shear flow in a tapered web. The average shear flow between the free end and the point x of the beam shown in Fig. 8.12(a) may be obtained from the spanwise equilibrium of the flange shown in Fig. 8.12(c). The horizontal component of the flange load is found by dividing the bending moment $q_0 h_0 b$ by the beam depth h. The average shear flow in this length, q_{av}, is therefore obtained by dividing this force by the horizontal length b.

$$q_{av} = q_0 \frac{h_0}{h} \qquad \text{or} \qquad q_0 \frac{x_0}{x} \qquad (8.13)$$

If the shear flow on one side of a tapered web is known, the shear flows on the other three sides may be obtained from Eqs. 8.12 and 8.13.

It has been assumed in the derivation of Eqs. 8.12 and 8.13 that the stresses existing on all four boundaries of the tapered plate were pure shearing stresses. It has been shown previously that pure shearing stresses may exist on only two planes, which must be at right angles to each other. Since the corners of the tapered webs do not form right angles, it is necessary for some normal stresses to act at the boundary of the web. In order to estimate the magnitude of these normal stresses, a tapered web in which pure shearing stresses may exist at all the boundaries will be considered.

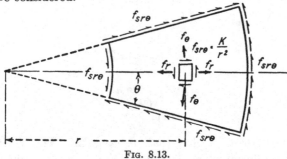

Fɪɢ. 8.13.

It can be shown by the theory of elasticity [1] that a sector such as shown in Fig. 8.13 may have pure shearing stresses on all the boundaries. Under these boundary conditions, any element such as that shown will have no normal stress in the radial direction f_r and no normal stress in the tangential direction, f_θ. The shearing stresses on these radial and tangential faces must satisfy the equation

$$f_{sr\theta} = \frac{K}{r^2} \tag{8.14}$$

where K is an undetermined constant. This equation is similar to Eq. 8.12 if the taper is small.

By comparing the sector of Fig. 8.13 with the tapered web of Fig. 8.12, it is seen that the assumption of pure shear on the top and bottom boundaries of the tapered web was correct. The left and right boundaries must also resist some normal stresses, however. The magnitude

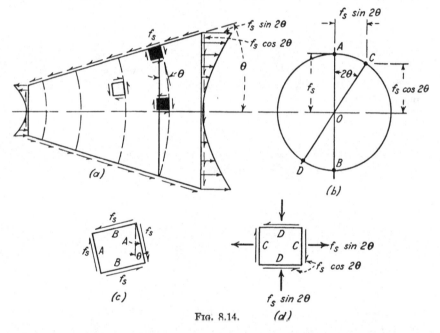

Fig. 8.14.

of these normal stresses may be determined for the Mohr circle of Fig. 8.14(b). The element under pure shearing stresses has faces A and B which are inclined at an angle θ with the vertical and horizontal. The Mohr circle for the pure shear condition will have a center at the origin and a radius f_s. Point A will be at the top of the circle, and point C, representing stresses on the vertical plane, will be clockwise at an angle 2θ from point A. The coordinates of point C represent a tensile stress of $f_s \sin 2\theta$ and a shearing stress of $f_s \cos 2\theta$ on the vertical plane. The normal stresses are obviously negligible for small values of the angle θ.

The equations for shear flow in tapered webs were first derived for the web of a beam with two concentrated flanges and then were shown to be approximately correct for any web which resists no normal loads at its boundaries. It can be shown by examples of other structures containing tapered webs that the shear flows may be applied to the webs by members other than beam flanges. Tapered webs are often used in torque

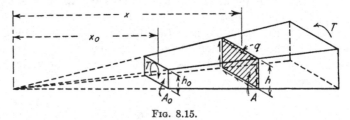

Fig. 8.15.

boxes, such as that shown in Fig. 8.15. For this box, all four sides are tapered in such a way that the corners of the box would intersect if extended. The enclosed area at any cross section varies with x according to the equation

$$\frac{A}{A_0} = \left(\frac{h}{h_0}\right)^2 = \left(\frac{x}{x_0}\right)^2 \tag{8.15}$$

The shear flow at any cross section, for the pure torsion loading condition shown, is obtained from the equation

$$q = \frac{T}{2A} \tag{8.16}$$

From Eqs. 8.15 and 8.16 and from the value of the shear flow at the left end, $q_0 = T/2A_0$, the following expression for q is obtained:

$$q = \frac{T}{2A_0}\left(\frac{x_0}{x}\right)^2 = q_0\left(\frac{x_0}{x}\right)^2 \tag{8.17}$$

This corresponds to the value obtained in Eq. 8.12 for the two-flange beam. The shear flow q given by Eq. 8.17 applies for all four webs of the box. There is consequently no axial load in the flange members at the corners of the box, since the shear flow at the sides of the box will be transmitted directly to the top and bottom webs.

In the structure shown in Fig. 8.15, all four webs are tapered in the same ratio, so that the shear flow obtained from Eq. 8.17 will be the same for all four webs. When the taper ratio for the horizontal webs is not the same as the taper ratio for the vertical webs, the shear flows will not have the same distribution for all webs. If, for example, the top and bottom webs are rectangular and the side webs are tapered, as shown in Fig. 8.16, the shear in the rectangular web must remain constant for the entire length, while the shear in the tapered web must vary according to Eq. 8.12. The shear flow for this structure which has ribs only at the ends cannot be obtained from Eq. 8.16, although Eq. 8.16 is

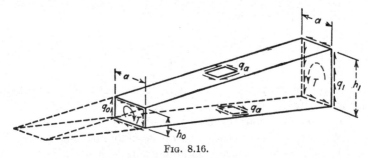

FIG. 8.16.

quite accurate for the common airplane wing structure with closely spaced ribs. The ribs divide the tapered web into several smaller webs and serve to distribute shear flows so that they are approximately equal in the horizontal and vertical webs.

The shear flows in the tapered webs of Fig. 8.16 vary according to Eq. 8.12.

$$q_1 = q_0 \left(\frac{h_0}{h_1}\right)^2 \tag{8.18}$$

Since the flange members at the corners of the box must be in equilibrium for spanwise forces, the shear flows q_a in the top and bottom webs must be equal to the average shear flows for the tapered webs, as obtained from Eq. 8.13.

$$q_a = q_0 \frac{h_0}{h_1} \tag{8.19}$$

and from Eq. 8.18

$$q_a = q_1 \frac{h_1}{h_0} \tag{8.20}$$

The difference in shear flows between two adjacent webs produces axial loads in the flange member between these webs. At any intermediate cross section of the box, the in-plane components of the flange loads must be considered in addition to the web shears in order to check the equilibrium with the external torque on the box. At the end cross sections, the shear flows are in equilibrium with the external torque. For the left end of the box,

$$(q_0 + q_a)ah_0 = T$$

substituting values from Eq. 8.19,

$$q_a = \frac{T}{a(h_1 + h_0)} \tag{8.21}$$

This equation may also be obtained from the equilibrium of forces at the right end of the box.

$$T = ah_1(q_1 + q_a)$$

substituting values from Eq. 8.20,

$$q_a = \frac{T}{a(h_1 + h_0)}$$

which checks the previous value. The denominator of Eq. 8.21 represents the average value of $2A$ for the box, as might be expected from Eq. 8.16. For most conventional wing or fuselage structures, the ribs and bulkheads are closely spaced, and it is seldom necessary to consider the taper of the structure when obtaining torsional shear flows. Equation 8.16 may be used, and the shear flows in all webs will be approximately equal at a cross section. For unconventional structures, however, where the ribs cannot distribute shear flows, it may be necessary to use methods similar to those used for the structure of Fig. 8.16. If the top and bottom webs are also tapered, but have a different taper ratio than the side webs, the shear flows may be obtained by applying Eq. 8.12 to each web, equating the average shear flows for all four webs and then equating the torsional moments of the shear flows at one end to the external torque on the structure.

8.5. Cutouts in Semimonocoque Structures. Typical aircraft structures which consist of closed boxes with longitudinal stiffeners and transverse bulkheads have been analyzed in preceding articles. In actual aircraft structures, however, it is necessary to provide many openings in the ideal continuous structure. Wing structures must usually be interrupted to provide wheel wells for retraction of the main landing gear. Other openings may be necessary for armament installations, fuel tanks, or engine nacelles. Fuselage structures must often be discontinuous for doors, windows, cockpit openings, bomb bays, gun turrets, or landing-gear doors. It is also necessary to provide holes and doors for access during manufacture and for inspection and maintenance in service. These "cutouts" are undesirable from a structural standpoint, but are always necessary. They often occur in regions where high loads must be resisted, and considerable structural weight is often required for reinforcements around the cutouts.

A simplified example of a structure with a large cutout is shown in Fig. 8.17. This corresponds to a wing structure with four flange members in which the lower skin is completely removed. In previous articles it was stated that a closed torque box was necessary in order to provide stability for resisting torsional loads. In order for the structure of Fig. 8.17 to be stable, it is necessary that one end be built in, so that the torsion may be resisted by the two side webs acting independently as cantilever beams, as shown in Fig. 8.17(b). The flange members resist axial loads, which have the values $P = TL/bh$ at the support. The shear flows q in the vertical webs are double the values obtained in a closed torque box with the same dimensions. The horizontal web

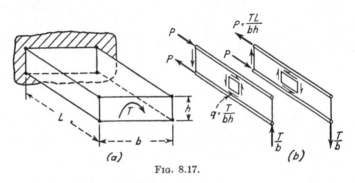

Fig. 8.17.

resists no shear flow in the case of the pure torsion loading, but it is necessary for stability in resisting horizontal loads. The torque box, with webs on all six faces, is capable of resisting torsion with no axial loads in the flange members. It is therefore much more rigid in torsion, since the shear deformations of the web are negligible in comparison to bending deformations of a cantilever beam.

In a full-cantilever airplane wing it is not feasible to have an open structure for the entire span, as the wing tip would twist to an excessive angle of attack under some flight conditions. A closed torque box is necessary for most of the span, but may be omitted for a short length, such as the length of a wheel-well opening. When the lower skin is omitted for such a region, the torsion is resisted by "differential bending" of the spars, as indicated in Fig. 8.17(b). The axial loads in the spar flanges are usually developed at both sides of the opening, since the closed torque boxes inboard and outboard of the opening both resist the warping deformation of the wing cross section. For the torsion loading shown, it would usually be assumed that flange loads were zero at the mid-point of the opening and that loads of $P/2$ were developed at both sides of the opening.

The open box with three webs and four flange areas is stable for any loading, if one or both ends are restrained. The shear flows in the three webs may be obtained from three equations of statics. The method of obtaining the shear flows is obvious from a numerical example such as that indicated by Fig. 8.18.   From the equilibrium of moments about point C of Fig. 8.18(b)

$$10q_1 \times 20 = 10{,}000 \times 10 + 2{,}000 \times 5 + 40{,}000$$

or

$$q_1 = 750 \text{ lb/in.}$$

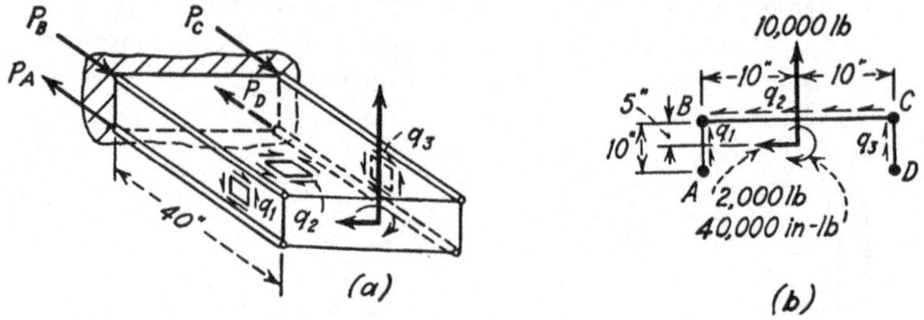

FIG. 8.18.

From the equilibrium of vertical forces,

$$10q_3 + 10 \times 750 = 10{,}000$$

or

$$q_3 = 250 \text{ lb/in.}$$

Similarly, from the equilibrium of horizontal forces,

$$20q_2 = 2{,}000$$

or

$$q_2 = 100 \text{ lb/in.}$$

The axial loads in the flange members may be obtained from a summation of spanwise forces as follows:

$$P_a = 40q_1 = 30{,}000 \text{ lb}$$
$$P_b = 40q_1 + 40q_2 = 34{,}000 \text{ lb}$$
$$P_c = 40q_3 - 40q_2 = 6{,}000 \text{ lb}$$
$$P_d = 40q_3 = 10{,}000 \text{ lb}$$

While these forces satisfy the conditions for static equilibrium, they cannot be obtained from the flexure formula.   For an open beam containing only four flange members, the flange loads are independent of the flange areas.   If the beam has more than four flanges, it is necessary to consider the flange areas when estimating the distribution of axial loads.   This problem is statically indeterminate, and it is usually solved by approximate methods.

A cutout in a short length of the wing structure affects the shear flows in the adjacent sections of the wing which have closed torque boxes.   First considering a case where the wing resists pure torsion, the shear flow in a continuous closed box is

$$q_t = \frac{T}{2A} \tag{8.22}$$

This equation is derived from the assumption that the flange members resist no axial loads. At the edges of the cutout, however, the flange loads resulting from differential bending have their maximum values. These flange loads are distributed to the webs, and at some distance from the cutout the flange loads become zero for the box in pure torsion. The distance along the span required for the distribution of the flange loads depends on the relative rigidities of the members, but it will be approximately equal to the width of the cutout. The shear flows in the torque box are affected considerably by this distribution of load.

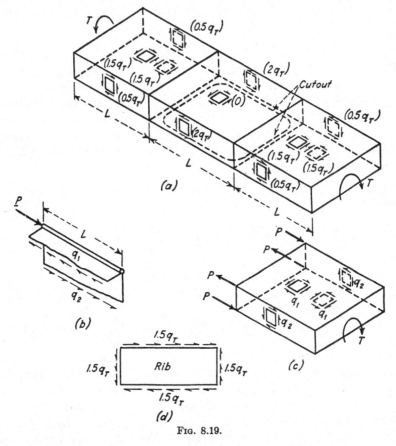

Fig. 8.19.

The rectangular torque box shown in Fig. 8.19(a) resists pure torsion. The lower skin is cut out for the entire width of the box, for a length L. The effect of the cutout is assumed to extend a distance L along the span on either side of the cutout; therefore it is necessary to consider only the length $3L$, which is shown. The shear flows at the section through the

cutout are similar to those obtained in Fig. 8.17(b), or they will be zero
in the upper skin and $2q_t$ in the spar webs, where q_t is the shear flow in a
continuous box, as obtained from Eq. 8.22.   The axial loads P in the
spar flanges are assumed to be equal on the inboard and outboard sides
of the cutout and are therefore half the value shown in Fig. 8.17(b), or

$$P = q_t L \tag{8.23}$$

This axial load must be transferred to the webs adjacent to the flange
in the assumed length L.   From the equilibrium of the flange member
shown in Fig. 8.19(b),

$$q_1 L - q_2 L = P \tag{8.24}$$

or, from Eqs. 8.23 and 8.24,

$$q_1 - q_2 = q_t \tag{8.25}$$

The shear flows q_1 and q_2 must satisfy the conditions of equilibrium of
the structure shown in Fig. 8.19(c).   For the vertical forces to be in
equilibrium at a cross section, the shear flows in the spars must have
equal and opposite values q_2.   For horizontal forces to be in equilibrium,
the shear flows in top and bottom skins must have equal and opposite
values q_1.   For the shear flows in all four webs to react the torque T, the
following condition must be satisfied:

$$q_1 A + q_2 A = T$$

or, from Eq. 8.22,

$$q_1 + q_2 = 2q_t \tag{8.26}$$

Solving Eqs. 8.25 and 8.26,

$$q_1 = 1.5q_t$$

and

$$q_2 = 0.5q_t$$

The values of these shear flows are shown in parenthesis in Fig. 8.19(a).

From these resulting values of shear flows q_1 and q_2 the cutout is seen
to have a serious effect on the shear flows in the closed torque boxes
adjacent to the cutout.   The top and bottom skins have shear flows of
$1\frac{1}{2}$ times the magnitude of those for a continuous box, while the shear
flows in the spars are only one-half as much.   The ribs adjacent to the
cutout also resist high shear flows.   The rib just outboard of the cutout
is shown in Fig. 8.19(d).   The rib receives the shear flow of $1.5q_t$ from
the top and bottom skins of the torque box.   The spars transfer shear
flows of $2q_t$ from the cutout section and $0.5q_t$ from the torque-box section.
Since the shear flows are in opposite directions, the resultant shear flow
applied to the rib is $1.5q_t$.

The problem of a box beam resisting a more general condition of loading is usually analyzed by another method.   The method used for the
structure resisting pure torsion becomes more difficult when the spar

flanges resist axial loads resulting from wing bending in addition to those resulting from the differential bending. The common procedure for the general case is first to analyze the continuous wing structure as if there were no cutout. A system of correcting shear flows must then be obtained and superimposed on the original shear flows obtained for the continuous structure. In finding the correcting shear flows, only a short length on either side of the cutout need be considered, since the loads applied to the wing in obtaining these shear flows are in equilibrium with themselves. One of the established principles of mechanics, formulated by Saint Venant, states that the stresses resulting from such a system of forces will be negligible at a distance from the forces. The distance is approximately equal to the width of the opening.

The method of obtaining correcting shear flows will be illustrated for the wing structure shown in Fig. 8.20. The wing is assumed to have a constant shear of 30,000 lb in the vertical direction and $-9,000$ lb in the chordwise direction for the entire length from station 30 (30 in. from the airplane center line) to station 120. The lower skin is removed for the entire width between the spars from station 60 to station 90. The wing bending moments affect the flange loads but not the shear flows; therefore the bending moments are not considered. The dimensions of the cross section are shown in Fig. 8.20(b). The shear flows in the continuous closed box with no cutout are shown in Fig. 8.20(c). These are computed by the methods discussed in Chap. 7, and the computations will not be discussed here. Since the external shear is constant, the shear flows in all webs between stations 30 and 120 would have the values shown in Fig. 8.20(c) if there were no cutout.

The correcting shear flows are now obtained by applying the loads of 660 lb/in. in the cutout region, as shown in Fig. 8.21(a), and finding the shear flows in the remaining webs. It is obvious that the loads of 660 lb/in. around all four sides of the cutout are in equilibrium with each other. The shear flows at the cross section through the cutout are assumed to be q_1, q_2, and q_3, as shown in Fig. 8.21(b), and must have a resultant equal to the applied load of 660 lb/in. at the lower skin. From a summation of horizontal forces,

$$30q_2 = 30 \times 660$$

or

$$q_2 = 660 \text{ lb/in.}$$

From a summation of vertical forces,

$$10q_1 - 2 \times 660 - 12q_3 = 0$$

From a summation of moments about point O, of Fig. 8.20(b),

$$2 \times 90 \times 660 = 2 \times 75 \times q_1 - 2 \times 200 \times 660 + 2 \times 90 \times q_3$$

Solving these equations simultaneously, $q_1 = 1,340$ lb/in., and $q_3 = 1,010$ lb/in. These correcting shear flows are shown in Fig. 8.21(a) for the structure between stations 60 and 90. The final shear flows in the cutout region are now obtained by superimposing the values shown in Figs. 8.20(c) and 8.21(a). This superposition yields a shear flow of 300 lb/in. in the upper skin, 1,930 lb/in. in the front spar web, and 940 lb/in. in the rear spar web, as shown in parenthesis on Fig. 8.20(a).

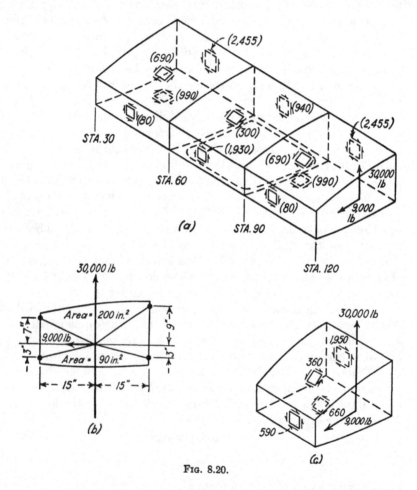

Fɪɢ. 8.20.

The correcting shear flows between stations 90 and 120 are obtained from the equilibrium of the forces on a cross section. The shear flows are shown in Fig. 8.21(c), and the following equations are derived from the equilibrium of the shear flows on the cross section:

$$\Sigma F_x = 30q_5 - 30q_7 = 0$$
$$\Sigma F_z = 10q_4 + 2q_5 - 12q_6 = 0$$
$$\Sigma M_0 = 2 \times 75q_4 + 2 \times 200q_5 + 2 \times 90q_6 + 2 \times 90q_7 = 0$$

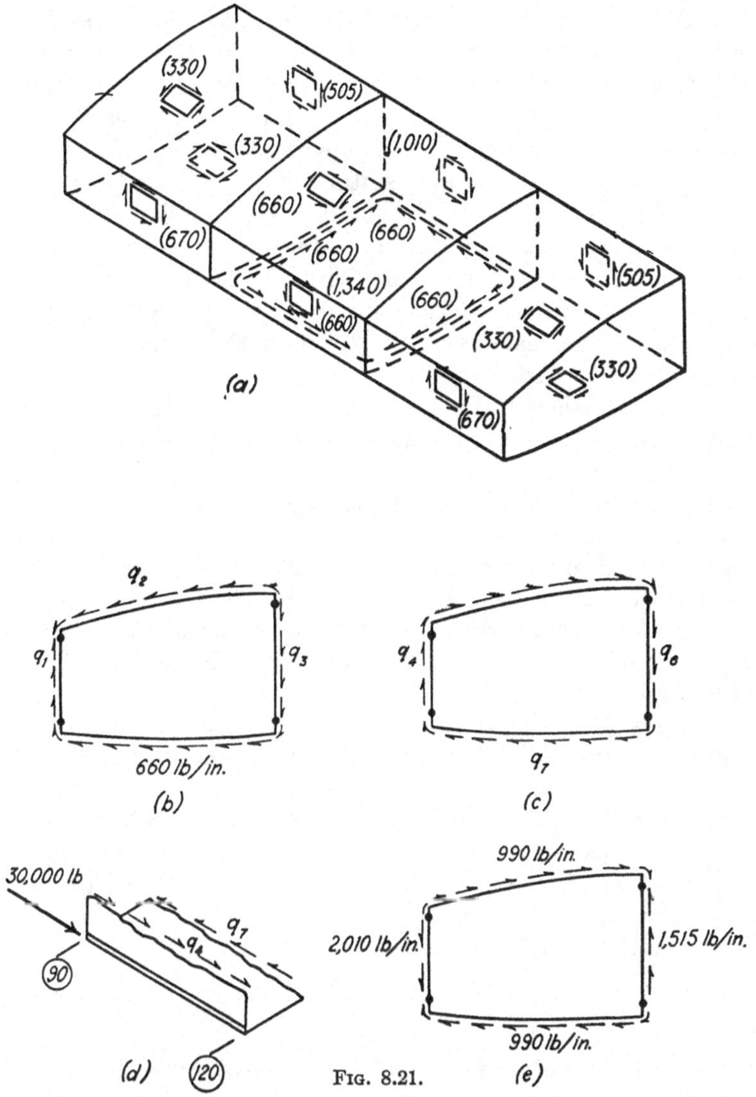

Fig. 8.21.

One additional equation may be derived from the spanwise equilibrium of forces on one of the flange areas. For the flange member shown in Fig. 8.21(d), the axial load at station 90 is obtained by assuming no axial loads at the center of the cutout, station 75. From the shear

flows shown in Fig. 8.21(a), $P = 15(1,340 + 660)$, or 30,000 lb. From Fig. 8.21(d),
$$30q_7 - 30q_4 = 30,000$$

Solving these four equations simultaneously, the values $q_4 = -670$ lb/in., $q_5 = 330$ lb/in., $q_6 = -505$ lb/in., and $q_7 = 330$ lb/in. are obtained. These values of the correcting shears are shown in Fig. 8.21(a). The final shear flows are obtained by superimposing the correcting shear flows and those for the continuous structure, shown in Fig. 8.20(c). The corrected values are shown in parenthesis on Fig. 8.20(a).

The loads acting on the rib at station 90 are obtained from the differences in shear flow on the two sides of the rib and are shown in Fig. 8.2(e). The shear flows transferred to the rib by the wing skin are seen to be greater than the shear flows in the skin, since the skin shears act in the same direction on the rib, and must be added. The rib at station 60 will resist the same loads as the rib at station 90, but the directions of all loads will be reversed.

Cutouts in fuselage structures are treated in essentially the same manner as cutouts in wing structures. Fuselage structures usually have lighter stringers and skin and resist smaller loads, particularly torsional loads. The torsional rigidity of fuselages is not as important as the torsional rigidity of wings, although flutter troubles may develop in high-speed aircraft if the fuselage is too flexible torsionally. Fuselage structures are often open for a large proportion of their length in order to permit long cockpit openings or long bomb bays. These structures are able to resist the torsional loads by differential bending of the sides of the fuselage.

Fuselages of large passenger airplanes often contain rows of windows, as shown in Fig. 8.22. If these windows are equally spaced and have equal sizes, the shear flows in webs adjacent to the windows may be readily obtained in terms of the average shear flow q_0 which would exist in a continuous structure with no windows. If the windows have a spacing w and the webs between them have a width w_1, the shear in these webs, q_1, may be obtained from a summation of forces on a horizontal section through the windows.

$$q_1 = \frac{w}{w_1} q_0 \tag{8.27}$$

Similarly, if the effect of the cutouts is assumed to extend over a vertical distance h, as shown, the shear flows in webs above and below the windows may be obtained by considering a vertical cross section through a window.

$$q_2 = \frac{h}{h_1} q_0 \tag{8.28}$$

The shear flows q_3, shown in Fig. 8.22, may be obtained either by considering a horizontal section through the webs or a vertical cross section through the webs.   The two equations would be

and

$$q_2 w_2 + q_3 w_1 = q_0 w$$
$$q_1 h_2 + q_3 h_1 = q_0 h$$

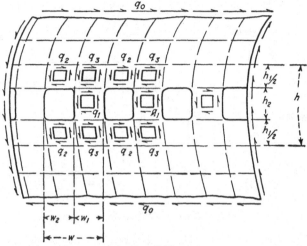

Fig. 8.22.

Either of these two equations, when values from Eqs. 8.27 and 8.28 are substituted, reduces to the following form:

$$q_3 = q_0 \left(1 - \frac{h_2}{h_1}\frac{w_2}{w_1}\right) \qquad (8.29)$$

The notation is shown in Fig. 8.22.

Openings for large fuselage doors may be analyzed in the same manner as wing cutouts.   It is sometimes difficult to provide rigid fuselage bulkheads on either side of an opening, because of interior space limitations. In such cases, a rigid doorframe may be provided so that the doorframe itself resists the shear loads in place of the cutout structure.   If such a structure is provided, it is no longer necessary to have the heavy bulkheads adjacent to the opening.   A fuselage doorframe usually must follow the curvature of the fuselage and hence does not lie in a plane. The structure of the doorframe must therefore be capable of resisting torsion as well as bending, and the frame must be a closed box structure.

PROBLEMS

8.17. The structure of Fig. 8.16 has the dimensions $h_0 = 5$ in., $h_1 = 15$ in., $a = 20$ in., and a length L of 100 in.   For a torque T, of 40,000 in-lb, find q_0,

q_1, and q_a. Find the axial loads in the corner flanges and the shear flows at a cross section 50 in. from one end. Check the values by the equilibrium of torsional moments, including the in-plane components of the flange loads.

8.18. Repeat Prob. 8.17 if $h_1 = 10$ in.

8.19. Find the shear flows and the flange loads for the structure of Fig. 8.18 if only the horizontal load of 2,000 lb is acting.

8.20. Find the shear flows and flange loads for the nacelle structure shown.

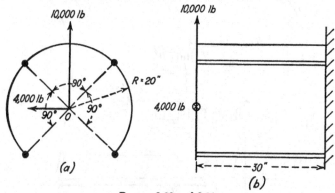

PROBS. 8.20 and 8.21.

8.21. Find the shear flows and flange loads for the nacelle structure, if a clockwise couple load of 200,000 in-lb is acting in addition to the loads shown.

8.22. Find the shear flows in all webs of the structure shown in Fig. 8.20 if a clockwise couple load of 100,000 in-lb is acting in addition to the loads shown.

REFERENCES FOR CHAPTER 8

1. TIMOSHENKO, S.: "Theory of Elasticity," Chap. 3, McGraw-Hill Book Company, Inc., New York, 1934.
2. PEERY, D. J.: Design for Strength at Cut-outs in Aircraft Structures, *Aero Digest*, October, 1947.
3. PEERY, D. J.: Simplified All-metal Wing Structures, *Aero Digest*, January, 1948.

CHAPTER 9

SPANWISE AIR-LOAD DISTRIBUTION

9.1. General Considerations. The entire subject of air-load distribution represents a field which concerns both the aerodynamicist and the stress analyst. The aerodynamicist is usually concerned with properties which affect the performance, stability, or control characteristics of the airplane. He is usually interested in determining the external configuration of the airplane in order to obtain the most desirable flight characteristics. The stress analyst is concerned with the load distributions which will represent the most severe conditions for various parts of the internal structure of the airplane. The aerodynamicist who computes air-load distribution or who obtains experimental aerodynamic data for stress analysis purposes should understand the use which will be made of the data and the accuracy required. If the air-load distribution is calculated by the stress analyst, he should understand the assumptions upon which the calculations are based and the accuracy of the experimental data.

Before obtaining the shear and bending-moment diagrams for use in the design of an airplane wing, it is first necessary to obtain the distribution of loads along the wing span. This spanwise distribution depends on the shape of the wing planform, the airfoil sections used, the distribution of the wing incidence along the span, and the dimensions and position of the wing flaps, or ailerons. It is difficult to measure the pressure distribution along the span of a wind tunnel model, but the spanwise distribution for a wing with any planform or airfoil sections can be accurately computed from airfoil section aerodynamic properties. The aerodynamic characteristics of airfoils are often measured from wind tunnel models which have a uniform airfoil section along the span, and which span the entire width of the wind tunnel test section. These characteristics correspond to those for a wing with an infinite span, since the flow is two-dimensional, with no spanwise velocity component and with the same conditions at all points along the span.

The flow around an airplane wing is not like the flow around a wing with an infinite span. Since the pressure below the wing is above atmospheric and the pressure above the wing is below atmospheric, there is a flow around the wing tips as shown in Fig. 9.1. This produces velocity components in the spanwise direction and introduces an additional downward flow velocity at the wing. The effect of the spanwise velocity components is neglected when calculating air loads. The downward flow velocity, or downwash velocity w combines vectorially with the

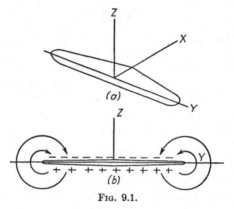

Fig. 9.1.

relative velocity at a distance from the wing V, as shown in Fig. 9.2. The airflow at a great distance from the wing is shown in the horizontal direction as V. The downwash velocity w changes the direction of flow at the airfoil by the angle α_i, which is called the induced angle of attack. Since α_i is a small angle, it may be assumed equal to its tangent or its sine, and the cosine may be assumed to be unity. From Fig. 9.2,

$$\alpha_i = \frac{w}{V} \tag{9.1}$$

A wing of infinite span is assumed to have lift and drag forces of L_0 and D_0, which are perpendicular and parallel to the direction of flow. In

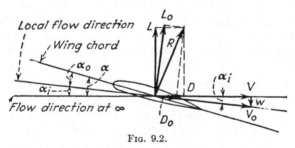

Fig. 9.2.

the wing of finite span, the direction of the local velocity at the wing, V_0, is at an angle α_i with the direction of the flow velocity V at a distance from the wing as shown in Fig. 9.2. The forces L_c and D_0 remain the same as for the wing with an infinite span, provided the local angle of attack α_0 is the same as the angle of attack for the wing with infinite span. The forces L_0 and D_0 for the wing of finite span are perpendicular and parallel to the local flow V_0, but it is desirable to resolve them into forces, L and D, which are perpendicular and parallel to V. Neglecting the vertical component of D_0, and assuming α_i to be a small angle,

$$L = L_0 \tag{9.2}$$

and

$$D = D_0 + \alpha_i L_0 \tag{9.3}$$

The last term of Eq. 9.3 results from the induced angle of attack and is called the induced drag. The spanwise distributions of lift and drag forces are easily obtained from Eqs. 9.1 to 9.3 and from the characteristics of an airfoil with infinite span if the local downwash velocities w are known for various points along the span. Unfortunately, there are no simple equations for obtaining the values of w, and the difficulties of determining the spanwise air-load distribution lie in the calculation of the distribution of the downwash velocities.

9.2. Wing Vortex Systems. The wing downwash velocities must be obtained from a consideration of the circulation theory of lift. A simple type of flow in which circulation exists is a vortex, shown in Fig. 9.3, in which the streamlines are concentric circles and the flow velocity is inversely proportional to the distance r from the center of the vortex. An example of vortex motion is that of the flow around a circular cylinder which is rotating in a fluid for which the velocity at any point has no component parallel to a radial line through the point. The rotation of

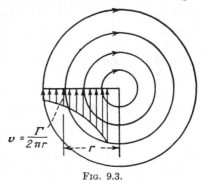

$$v = \frac{\Gamma}{2\pi r}$$

Fig. 9.3.

the cylinder produces the fluid motion, and the fluid velocity at the surface of the cylinder is equal to the peripheral velocity of the cylinder. The tangential velocity at any point in the fluid will be inversely proportional to the distance r of the point from the center of the cylinder, or

$$v = \frac{\Gamma}{2\pi r} \tag{9.4}$$

where Γ is constant for all points and is called the *circulation*. If such a rotating cylinder is placed in a wind tunnel in which the air velocity normal to the axis of the cylinder is V, the lift L per unit length is represented by the equation

$$L = \rho V \Gamma \tag{9.5}$$

Equation 9.5, the Kutta-Joukowski theorem of lift, can be derived theoretically from the equations of flow of an ideal fluid. It might be difficult to check this equation accurately by experiment because of viscosity effects.

In general, the circulation for two-dimensional flow may be defined by the equation

$$\Gamma = \int v \cos \theta \, ds \tag{9.6}$$

in which the integration is performed about any closed path in the plane. The angle θ is measured between the velocity vector v at a point on the path of integration and the tangent to the path which has an increment of length ds. The circulation has dimensions of a velocity times a distance. If Eq. 9.6 is evaluated for any streamline of the vortex shown in Fig. 9.3, the value of $\cos \theta$ is unity, and the circulation Γ is the product of the velocity v and the length of the closed path $2\pi r$. This obviously checks Eq. 9.4 for any streamline. It can be shown that for any type of flow pattern the circulation is zero for any closed path which does not enclose a vortex. For any closed path enclosing a vortex, the circulation is constant regardless of the shape of the path of integration. This constant circulation Γ, corresponding to any vortex regardless of the path of integration, is called the strength of the vortex.

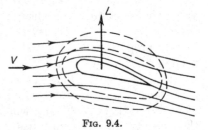

Fig. 9.4.

If Eq. 9.6 is evaluated for the flow pattern around any lifting airfoil, such as that shown in Fig. 9.4, the circulation around any closed path enclosing the airfoil is constant regardless of the path. Thus, the value of Γ for either path shown by the dotted lines in Fig. 9.4 will have the same value. The lift L per unit length of span can also be shown to be equal to that for a rotating cylinder with the same circulation. The Kutta-Joukowski theorem therefore applies to any lifting airfoil, and there is always a circulation corresponding to a lifting surface, with a vortex strength proportional to the lift per unit span.

An airfoil can be represented by a vortex line, which has a strength Γ proportional to the lift per unit span. If the wing has an infinite span and a uniform lift distribution, the circulation Γ is constant. The condition of infinite span is represented by a wind tunnel model which spans the entire width of the tunnel, as shown by the wing in Fig. 9.5. The wing is represented by the vortex line AA' with strength Γ. A vortex line has the property that it can end only at the boundary of the fluid, which is represented by points A and A' of Fig. 9.5. In the airplane wing of finite span, shown in Fig. 9.6, the vortex lines cannot end at the wing tips, but extend downstream as shown for a distance which is considered to be infinite. The strength of the vortex must also be constant at all points along the line. For the "horseshoe" vortex system shown in Fig. 9.6, the downwash velocity at some distance downstream from

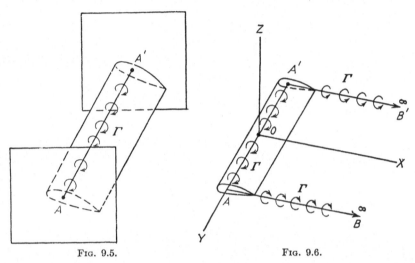

Fig. 9.5.   Fig. 9.6.

the wing is shown in Fig. 9.7. At some distance downstream, the vortex line AB may be considered to have an infinite length in both directions, and the flow velocities correspond to those given in Eq. 9.4 and Fig. 9.3 for a plane vortex. The downwash velocities for the vortex line AB are shown in Fig. 9.7(b), and those for vortex lines AB and $A'B'$ combined are shown in Fig. 9.7(c).

The downwash velocity at a point on the wing span AA' of Fig. 9.6 is only half the value of the downwash velocity shown in Fig. 9.7(c) for a point downstream at a great distance. The downwash at a great distance upstream is zero, and half the final downwash velocity is attained at the wing. In order to prove this relationship and to show the variation of the downwash in the x direction, the effect of a line vortex extending from point A to a point B at infinity is considered. The point P in Fig. 9.8(a) is at a distance r from the vortex line AB. The contribution of the elementary length ds of the vortex line to the downwash velocity w at point P is shown by Glauert[1] to be

$$dw = \frac{\Gamma}{4\pi r} \cos \theta \, d\theta \qquad (9.7)$$

For a vortex line of infinite length, the downwash velocity is

$$w = \frac{\Gamma}{4\pi r} \int_{-\frac{\pi}{2}}^{\frac{\pi}{2}} \cos\theta\, d\theta = \frac{\Gamma}{2\pi r}$$

as given previously. For the semiinfinite vortex line AB, the downwash velocity at point P is

$$w = \frac{\Gamma}{4\pi r} \int_{\theta_P}^{\frac{\pi}{2}} \cos\theta\, d\theta = \frac{\Gamma}{4\pi r}(1 - \sin\theta_P) \tag{9.8}$$

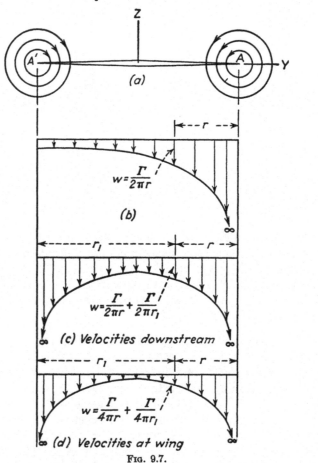

$$w = \frac{\Gamma}{2\pi r}$$

(b)

$$w = \frac{\Gamma}{2\pi r} + \frac{\Gamma}{2\pi r_1}$$

(c) *Velocities downstream*

$$w = \frac{\Gamma}{4\pi r} + \frac{\Gamma}{4\pi r_1}$$

(d) *Velocities at wing*

Fig. 9.7.

If the point P is on the line AA', $\sin\theta_P = 0$, and the downwash velocity for the one vortex line AB is

$$w = \frac{\Gamma}{4\pi r} \tag{9.9}$$

The values of w for other angles θ_P are plotted in Fig. 9.8(b). If the two vortices AB and $A'B'$ are superimposed, the downwash velocities at the wing are distributed as shown in Fig. 9.7(d).

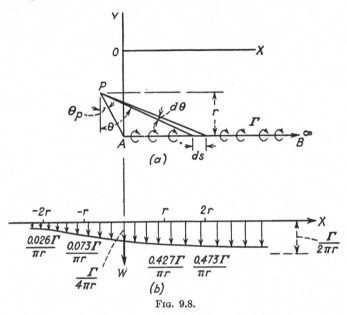

FIG. 9.8.

9.3. Fundamental Equations for Obtaining Downwash Velocities.

The simple horseshoe vortex system shown in Fig. 9.6 cannot represent the actual conditions for an airplane wing. While a rectangular wing might be expected to have a constant spanwise distribution of lift and consequently a constant circulation Γ along the span, this would require infinite downwash velocities at the wing tips and large downwash velocities near the tips, as shown in Fig. 9.7(d). These large downwash velocities produce large induced angles of attack, α_i, near the wing tips. Referring to Fig. 9.2, the effective angles of attack α_0 and consequently the lift forces are decreased near the wing tips. The actual lift distribution along the span of a rectangular wing is consequently represented by the curve shown in Fig. 9.9 rather than by the constant distribution indicated by the vortex system shown in Fig. 9.6.

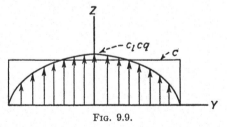

FIG. 9.9.

In order to approximate the lift distribution for a rectangular wing more closely, the three horseshoe vortex systems shown in Fig. 9.10(a) are considered. The strengths of these vortex lines are $\Delta\Gamma_1$, $\Delta\Gamma_2$, and $\Delta\Gamma_3$, as shown. The lift per unit span at any point along the wing is proportional to the total circulation at the point considered. Near the tip of the wing the circulation is $\Delta\Gamma_1$ and the lift per unit length is $\rho V(\Delta\Gamma_1)$, as shown in Fig. 9.10(b). At the center of the span, the total circulation is $\Delta\Gamma_1 + \Delta\Gamma_2 + \Delta\Gamma_3$, and the lift per unit length is

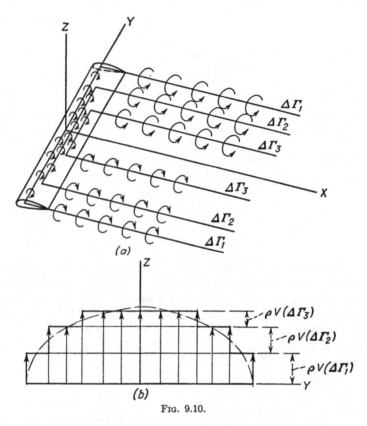

Fɪɢ. 9.10.

$\rho V(\Delta\Gamma_1 + \Delta\Gamma_2 + \Delta\Gamma_3)$ as also shown in Fig. 9.10(b). While this lift distribution composed of three finite steps is a closer approximation to the actual lift condition than that for the single vortex, it still gives conditions of infinite downwash velocities at the wing tips and at the other steps in the lift distribution curve. The true vortex system must therefore consist of an infinite number of simple horseshoe vortices, each having a zero strength. The system of trailing vortices becomes a continuous vortex sheet. The equations for the downwash velocity may be obtained by integrating the effects of these infinitesimal horseshoe systems.

A trailing vortex of infinitesimal strength, $d\Gamma$, is shown at a point y in Fig. 9.11. The downwash velocity at a fixed point y' on the span, resulting from $d\Gamma$, is

$$dw = \frac{d\Gamma}{4\pi r} \tag{9.10}$$

as obtained from Eq. 9.9. The distance r is the distance of the point y' from the vortex, as shown in Fig. 9.11.

$$r = y' - y \tag{9.11}$$

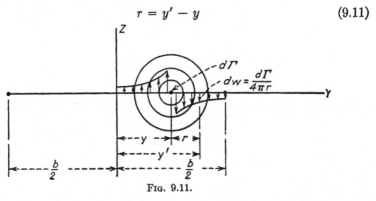

Fɪɢ. 9.11.

The total downwash velocity w at y' is the sum of the downwash velocities resulting from all trailing vortices for the entire span. Substituting the value from Eq. 9.11 into Eq. 9.10 and integrating for the entire span,

$$w = \frac{1}{4\pi} \int_{y=-\frac{b}{2}}^{y=\frac{b}{2}} \frac{d\Gamma}{y' - y} \tag{9.12}$$

The circulation Γ is a function of the distance y.

Another relation between the circulation and the downwash velocity must be satisfied. From Fig. 9.2, it is seen that the effective angle of attack at any point along the span is

$$\alpha_0 = \alpha_a - \frac{w}{V} \tag{9.13}$$

The angle of attack α_a is the absolute angle of attack. It corresponds to the angle α shown in Fig. 9.2 if the wing chord used is the zero-lift chord at the section. The lift coefficient c_l at a point on the span is equal to $m_0\alpha_0$, if m_0 is the slope of the curve of c_l vs. α_0 in radians. Equation 9.13 may therefore be written in terms of c_l as follows:

$$c_l = m_0\left(\alpha_a - \frac{w}{V}\right) \tag{9.14}$$

The coefficient of lift c_l is obtained from a wing of the given airfoil section with an infinite span. The value of c_l varies along the span, and the problem of obtaining the spanwise lift distribution is equivalent to obtaining the distribution of the circulation along the span. Either Γ or c_l may be considered as the unknown term, and the relation between Γ and c_l is obtained by equating the lift on a wing section of unit span and chord c to that given by Eq. 9.5.

$$c_l \frac{\rho V^2}{2} c = \rho V \Gamma$$

or

$$\Gamma = \frac{c_l c V}{2} \tag{9.15}$$

From Eqs. 9.14 and 9.15,

$$\Gamma = \frac{c V m_0}{2} \left(\alpha_a - \frac{w}{V} \right) \tag{9.16}$$

Equations 9.12 and 9.16 represent the conditions which must be satisfied by the spanwise distribution of the circulation Γ. The unknown terms in these equations are the circulation Γ and the downwash velocity w. The downwash velocity may be eliminated from Eqs. 9.12 and 9.16, as follows:

$$\frac{2\Gamma}{m_0 c V} = \alpha_a - \frac{1}{4\pi V} \int_{y=-\frac{b}{2}}^{y=\frac{b}{2}} \frac{d\Gamma}{y' - y} \tag{9.17}$$

Equation 9.17 represents the fundamental equation for the distribution of the circulation Γ. This equation is expressed in terms of the coefficient of lift c_l, if preferred, by substituting values from Eq. 9.15 into Eq. 9.17.

$$\frac{c_l}{m_0} = \alpha_a - \frac{1}{8\pi} \int_{y=-\frac{b}{2}}^{y=\frac{b}{2}} \frac{d(c c_l)}{y' - y} \tag{9.18}$$

The solution of Eq. 9.17 for Γ as a function of y, or the solution of the equivalent equation, Eq. 9.18, for $c c_l$ as a function of y is difficult for most wing planforms. The unknown term Γ (or c_l) occurs on both sides of the equation, and the integral cannot be evaluated in simple terms for most actual wings.

9.4. Wing with Elliptical Planform. The evaluation of Eq. 9.18 is not difficult for the case of a wing which has an elliptical planform, a constant airfoil section, and no aerodynamic twist. For the same airfoil section of all points of the span, the slope of the lift curve m_0 will be constant. If the wing has no aerodynamic twist, the zero-lift chords at all sections lie in a plane; hence the angle of attack α_a is the same at all sections along the span. The requirement of an elliptical planform means that the wing chord lengths c vary as the ordinates of an ellipse, or

$$c = c_s \sqrt{1 - \left(\frac{2y}{b}\right)^2} \tag{9.19}$$

where c_s is the wing chord at the center of symmetry of the wing.

Even for this simple case the value of c_l cannot be determined by a direct solution of Eq. 9.18. It will be assumed that c_l is constant, and this assumption will be verified by substitution. Substituting the value of c from Eq. 9.19 into Eq. 9.18, and assuming c_l to have a constant value of C_L,

$$\frac{C_L}{m_0} = \alpha_a - \frac{C_L c_s}{8\pi} \int_{-\frac{b}{2}}^{\frac{b}{2}} \frac{d[1 - (2y/b)^2]^{1/2}}{y' - y} \tag{9.20}$$

The last term of this equation becomes

$$\frac{C_L c_s}{2b^2 \pi} \int_{-\frac{b}{2}}^{\frac{b}{2}} \frac{y\,dy}{\sqrt{1 - (2y/b)^2}(y - y')} = \frac{C_L c_s}{4b} \tag{9.21}$$

It is more convenient to express the equations for the downwash angle in terms of the wing area S, rather than the root chord c_s. The area of an elliptical wing is found from the equation

$$S = \frac{\pi b c_s}{4} \tag{9.22}$$

From Eqs. 9.20 to 9.22,

$$\frac{C_L}{m_0} = \alpha_a - \frac{C_L S}{\pi b^2}$$

or

$$\frac{C_L}{m_0} = \alpha_a - \frac{C_L}{\pi A} \tag{9.23}$$

where A is the wing aspect ratio b^2/S.

The last term in Eq. 9.23 represents the downwash angle α_i, or w/V, which is seen to be constant along the span of an elliptical wing with no aerodynamic twist. For any other shape of planform, the value of the definite integral obtained as the right side of Eq. 9.21 will be a function of y', and the downwash angle and lift coefficient will vary along the span. The spanwise lift distribution is plotted, as shown in Fig. 9.12, by plotting values of $c c_l$ at all points along the span. If these values are multiplied by the dynamic pressure, $q = \rho V^2/2$, the actual force per unit span is obtained. For the elliptical wing, the spanwise distribution is elliptical for all angles of attack. The circulation Γ is seen from Eq. 9.15 to have an elliptical spanwise distribution also, since Γ is proportional to the lift and to cc_l at any wing section.

The drag for an elliptical wing will also have an elliptical spanwise distribution. From Eq. 9.3, and Fig. 9.2, the drag force at any section along the span is seen to have two components. The drag D_0, which is

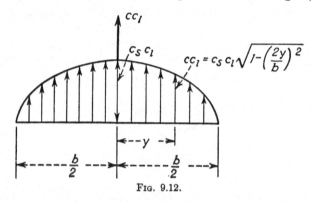

Fig. 9.12.

obtained from a wing with infinite span, depends on the airfoil profile and is called the profile drag. The component $\alpha_i L_0$ depends on the downwash angle, or induced angle α_i, and is called the induced drag. The profile drag for a wing with the same airfoil section along the span is proportional to the chord c and has approximately an elliptical distribution. The induced drag is distributed in the same manner as the lift if the downwash angle is constant, and it also has an elliptical distribution. It can be shown that a wing with an elliptical planform has a smaller induced drag than a wing with any other planform, provided that the lift and span are the same.

9.5. Approximate Lift Distribution for Untwisted Wings. While the exact equations for spanwise airload distribution, Eq. 9.17 or Eq. 9.18, can be solved for any planform and calculation forms are available for solving these equations, the calculations are often quite tedious. A simple approximate solution for the lift distribution has been proposed by Schrenk[2] and has been accepted by the Civil Aeronautics Administration (CAA) as satisfactory for civil airplanes.[3] For wings with no aerodynamic twist and constant airfoil section, the method consists simply of averaging the lift forces obtained from an elliptical lift distribution with those obtained from a planform lift distribution. This approximation is obviously very accurate for wings which approach an elliptical planform. Since the elliptical planform produces a minimum drag, most well-designed wings do not vary greatly from the elliptical planform.

The lift distribution for a wing with a rectangular planform is shown in Fig. 9.13(a). It is convenient to plot the term cc_l, which will yield the lift per unit span when multiplied by q. It is also convenient to plot values corresponding to $C_L = 1.0$ for the entire wing, so that these values may be multiplied by the wing coefficient of lift to obtain the actual values corresponding to any C_L. The values of cc_l corresponding

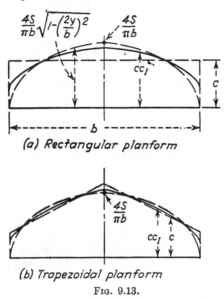

(a) Rectangular planform

(b) Trapezoidal planform

Fig. 9.13.

to a planform distribution are therefore equal to the wing chord c. The values corresponding to the elliptical distribution have a maximum ordinate $4S/\pi b$, as shown. It was shown in Art. 9.3 that a constant spanwise lift distribution could not exist for a wing with a finite span, since an infinite downwash velocity would exist at the wing tip, with a consequent reduction in lift near the tip. The true distribution must be less near the tip, and, for the same total lift must be greater at midspan than the constant lift distribution. It was also shown in Art. 9.4 that an elliptical planform was required for a constant downwash velocity and for an elliptical lift distribution. Where the actual wing chord is greater than the chord of an elliptical wing, the lift is greater than that for an elliptical distribution. Where the chord is less than that of the elliptical wing, the lift must also be less. The assumed average between the elliptical and the planform distributions is therefore seen to satisfy the general requirements for the true lift distribution.

The lift distribution shown in Fig. 9.13(a) is obviously inaccurate at the wing tips, since the lift drops sharply from $c/2$ to zero. Any sharp change in the lift curve corresponds to a trailing vortex of finite strength, with a consequent infinite downwash velocity. At a square wing tip the lift cannot change sharply, but there will be a large change in a comparatively short distance. The downwash velocity is therefore large, and since the induced drag is proportional to the downwash angle, a square wing tip will have a large induced drag. A well-designed wing will therefore have a well-rounded tip, regardless of the planform. For the rounded tip, the lift distribution calculated by the approximate method is much closer to actual conditions than for a wing with a square tip. Even the so-called "exact" method of Eq. 9.18 is inaccurate at the wing tips, and empirical corrections are often applied.

The wing with a trapezoidal planform shown in Fig. 9.13(b) is seen to have a lift distribution which corresponds more closely to the elliptical distribution. If the wing tips are square, however, the same inaccuracy exists at the tips as for the rectangular wing. For a trapezoidal wing with rounded tips, the lift distribution may be obtained very accurately by the approximate method. If the actual wing has square tips or other discontinuities, the lift distribution curve can be faired by comparison with published curves of the actual lift distribution for wings with similar discontinuities.

9.6. Approximate Lift Distribution for Twisted Wings. If the zero-lift chords of all airfoil sections lie in the same plane, the wing has no aerodynamic twist. Most wings, however, have thinner airfoil sections near the tip than near the root, and the zero-lift chords are not quite parallel, even though the reference chords may be parallel. In other cases, wings are designed so that the reference chords are not parallel, in order to obtain better stalling characteristics. Similarly, when wing flaps are deflected, the zero-lift chord for the airfoil section with flaps is at a much larger angle below the wing reference chord than for airfoil sections outboard from the flaps.

The air-load distribution for wings with aerodynamic twist is obtained m two parts. The first part, called the basic lift distribution, is obtained for the angle of attack at which the entire wing has no lift. The second part, called the additional lift distribution, is then obtained by the methods previously used, assuming that the wing has lift but no twist. For the basic condition, some sections of the wing will have a positive lift, and other sections will have a negative lift. The elliptical lift distribution for the entire wing with no lift would correspond to a zero load for the entire span. The average of this zero value, and a value corresponding to the lift at each section for an assumed zero downwash, is just one-half the lift that would be obtained for no downwash. This is equivalent to assuming that the downwash angle of attack is one-half the absolute angle of attack at any section for the condition of zero lift on the wing. The basic lift distribution is therefore obtained from Eq. 9.14, by substituting $\alpha_a/2$ for w/V. The following equation is obtained:

$$cc_{lb} = \tfrac{1}{2}cm_0\alpha_a \tag{9.24}$$

where c_{lb} is the basic lift coefficient at any point on the span and α_a is the angle of attack in radians measured from the zero-lift plane of the entire wing to the zero-lift chord line for the section. There will be large discontinuities in the curve obtained from Eq. 9.24 at the ends of wing flaps. These may be faired to smooth curves by comparison with experimental data for similar wings. The common practice of fairing this curve is shown in Fig. 9.15.

After the basic lift distribution has been obtained, the total lift distribution may be found by adding the values for the basic condition of twist with no lift to the additional condition of lift with no twist. The additional lift coefficient c_{la} is distributed in exactly the same manner as for the untwisted wing discussed in Art. 9.5. In Art. 9.5, a constant slope m_0 of the lift coefficient curve was assumed. A variable slope may easily be considered by use of the following equation:

$$cc_{la1} = \frac{1}{2}\left[\frac{m_0 c}{\overline{m}_0} + \frac{4S}{\pi b}\sqrt{1 - \left(\frac{2y}{b}\right)^2}\right] \tag{9.25}$$

This equation corresponds to the method of Art. 9.5, except that $m_0/\overline{m}_0$ is included. The term c_{la1} is the additional coefficient of lift corresponding to $C_L = 1$ for the entire wing. The total section lift coefficient corresponding to any value of C_L may be found from the equation:

$$c_l = C_L c_{la1} + c_{lb} \tag{9.26}$$

The term $\overline{m}_0$, the average slope of section lift coefficients, may be found from the equation

$$\overline{m}_0 = \frac{\int_0^{\frac{b}{2}} m_0 c\, dy}{S/2} \tag{9.27}$$

It is customary to designate the slope of the section lift coefficient with respect to the angle of attack in radians as m_0 and the slope with respect to the angle of attack in degrees as a_0. Since the symbol α is used in both cases for the angle of attack, the symbol m_0 will be used in all equations. In the numerical examples, the value of a_0 is usually used in place of m_0, and the angle α is in degrees. The wing angle of attack for zero lift is found from the equation

$$\alpha_{w0} = \frac{\int_0^{\frac{b}{2}} m_0 \alpha_{aR} c\, dy}{\int_0^{\frac{b}{2}} m_0 c\, dy} \tag{9.28}$$

where an arbitrary reference plane is assumed and α_{aR} is measured from this plane to the zero-lift chord of each section. α_{w0} is the angle from this reference plane to the plane of zero lift for the wing.

Example 1. Find the spanwise distribution of the section lift coefficient for the wing planform shown in Fig. 9.14 if the flaps are in the neutral position. The section lift coefficient curve has the same slope at all sections, and the wing

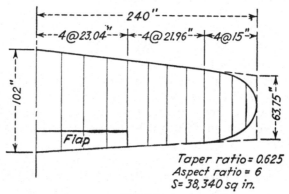

FIG. 9.14.

has no aerodynamic twist. Additional dimensions are given in Table 9.1. This wing is also analyzed by the approximate method in CAMO4,[3] Appendix V, and by an exact method in NACA Technical Report 585.[4]

Solution. Since the slope of the lift curve is the same at all sections, the values of cc_l for a wing C_L of 1.0 will be found as the average of the wing-chord lengths and ordinates of an ellipse. The calculations are shown in Table 9.1. Columns (1) and (3) contain the wing dimensions. Columns (2), (4), and (5) show calculations of the coordinates for an elliptical distribution which will yield $C_L = 1$. The values in column (5) are averaged with the chord lengths of column (3), and the final distribution of load for a unit q and unit C_L is shown in column (6). In column (7) the values of c_l are computed by dividing values in column (6) by the corresponding chord lengths. A tabular solution which is better adapted for calculation purposes is given in CAMO4. The purpose of Table 9.1 is to explain the steps involved.

Example 2. The flap of the wing of Example 1 is deflected 30°. Find the spanwise distribution of c_l for wing lift coefficients of $C_L = 0$, $C_L = 1.00$, and $C_L = 1.72$. The lift curve slope is 0.1 per deg for all sections of the wing. The flap deflection changes the angle of zero lift from $-1.20°$ to $-8.00°$ from the wing reference chord line for the flapped sections.

Solution. In Table 9.1 the elements were chosen in such a way that the wing was divided into three main parts, as shown in Fig. 9.14. The center part, containing the flap, is divided into an even number (in this case four) of divisions, as are the two outer parts. Smaller divisions are used for the tip section, since the planform changes more abruptly. Simpson's rule may be applied in

TABLE 9.1

y (1)	$\dfrac{2y}{b}$ (2)	c (3)	$\sqrt{1-\left(\dfrac{2y}{b}\right)^2}$ (4)	$\dfrac{4S}{\pi b}\sqrt{1-\left(\dfrac{2y}{b}\right)^2}$ (5)	cc_l (6)	c_l (7)
	(1) ÷ 240		$\sqrt{1-(2)^2}$	101.6 × (4)	½[(3) + (5)]	(6) ÷ (3)
0	0	102.0	1.000	101.6	101.8	0.998
23.04	0.0960	98.3	0.995	101.0	99.8	1.016
46.08	0.1920	94.7	0.981	99.6	97.2	1.026
69.12	0.2880	91.0	0.957	97.2	94.1	1.034
92.16	0.3840	87.4	0.924	93.8	90.6	1.038
114.12	0.4755	83.8	0.880	89.3	86.5	1.032
136.08	0.5670	80.3	0.824	83.6	81.9	1.020
158.04	0.6585	76.8	0.752	76.4	76.6	0.998
180.00	0.7500	73.3	0.661	67.1	70.2	0.958
195.00	0.8125	69.2	0.583	59.2	64.2	0.928
210.00	0.8750	62.2	0.484	49.1	55.7	0.896
225.00	0.9375	49.5	0.348	35.4	42.4	0.856
240.00	1.000	0	0	0	0	0

calculating the wing area and other terms when an even number of divisions is used.

The distribution of the basic lift coefficient, the lift coefficient for the wing $C_L = 0$, is obtained in Table 9.2. In order to calculate the wing zero-lift plane, a reference plane is chosen through the zero-lift chords of the outboard sections. The inboard sections with flaps will have an absolute angle of attack of 6.8° from this plane, and the outboard sections will have a zero angle of attack, as shown in column (3). The section for $y = 92.16$, at the end of the flap, will have both angles and is listed twice. Equation 9.28 is used for finding the zero-lift plane, but since m_0 is constant it may be dropped from the equation. The numerator of Eq. 9.28 is evaluated by Simpson's rule, using values in column (4) and increments $\Delta y = 23.04$.

$$\int_0^{\frac{b}{2}} \alpha_{aR}c \, dy = \frac{23.04}{3}\,(694 + 4 \times 669 + 2 \times 644 + 4 \times 619 + 595)$$
$$= 59{,}400$$

The denominator of Eq. 9.28 becomes equal to half the wing area $S/2$ if the slope m_0 is omitted.

$$\alpha_{w0} = \frac{59{,}400}{19{,}170} = 3.10°$$

TABLE 9.2

y (1)	c (2)	α_{aR} from ref. line (3)	$\alpha_{aR}c$ (4)	α_a from wing zero lift (5)	$c_{lb} = \dfrac{m_0}{2}\alpha_a$ (6)	cc_{lb} (7)	cc_{lb} faired (8)	c_{lb} (9)
			$(2)\times(3)$	$(4)-3.1$	$0.05(5)$	$(2)\times(6)$		$(8)\div(2)$
0	102.0	6.80	694	3.70	0.185	18.88	18.87	0.185
23.04	98.3	6.80	669	3.70	0.185	18.18	18.00	0.182
46.08	94.7	6.80	644	3.70	0.185	17.52	14.70	0.155
69.12	91.0	6.80	619	3.70	0.185	16.82	9.70	0.107
92.16	87.4	6.80	595	3.70	0.185	16.16	1.25	0.014
92.16	87.4	0	0	−3.10	−0.155	−13.52	1.25	0.014
114.12	83.8	0	0	−3.10	−0.155	−13.00	−5.15	−0.068
136.08	80.3	0	0	−3.10	−0.155	−12.44	−9.25	−0.115
158.04	76.8	0	0	−3.10	−0.155	−11.90	−11.40	−0.148
180.00	73.3	0	0	−3.10	−0.155	−11.36	−11.36	−0.155
195.00	69.2	0	0	−3.10	−0.155	−10.72	−10.72	−0.155
210.00	62.2	0	0	−3.10	−0.155	−9.65	−9.65	−0.155
225.00	49.5	0	0	−3.10	−0.155	−7.67	−7.67	−0.155
240.00	0	0	0	−3.10	−0.155	0	0	−0.155

The wing zero-lift plane is therefore 3.10° above the zero-lift chords of the outboard sections, and they will have an absolute angle of attack of −3.10° with respect to this plane, as shown in column (5). The sections with flaps have an angle of attack of 6.80 − 3.10, or 3.70° with respect to this plane.

For all sections along the wing, the lift curve has the slope of 0.1 per deg for the airfoil section with infinite span. The angle of attack of 3.70° would therefore correspond to a value of $c_l = 0.370$ and the angle of attack of −3.10° to a value of $c_l = 0.310$, with infinite span. For the zero-lift condition, however, the downwash reduces these values by one-half, and the values of c_l are 0.185 for the inboard section and 0.155 for the outboard section, as shown in column (6). The abrupt change in the curve of cc_{lb} at the outboard end of the flap cannot be accurate, since the circulation must always change gradually. The curve of cc_{lb} is therefore rounded off, or faired, at the end of the flap, as shown in Fig. 9.15. The curve is faired so that the positive area removed is equal to the negative area removed, and the total wing lift remains zero. The calculated curve of cc_{lb} is obtained in column (7), faired in column (8), and then the final values of c_{lb} are calculated in column (9) from points on the faired curve. The total lift coefficients c_l for $C_L = 0$, $C_L = 1.0$, and $C_L = 1.72$, are obtained in Table 9.3 from Eq. 9.26. The values of c_{lb} which were obtained in Table 9.2 must be used in all cases. These

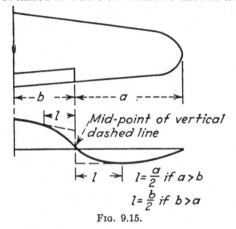

FIG. 9.15.

are repeated in column (2) of Table 9.3 and represent the total c_l for the zero-lift condition $C_L = 0$. The values of c_{la} calculated in Table 9.1 correspond to a wing lift coefficient of unity and are repeated as values of c_{la1} in column (3) of Table 9.3. The total c_l for $C_L = 1.0$ is obtained

TABLE 9.3

(1)	(2)	(3)	(4)	(5)	(6)
	$C_L = 0$		$C_L = 1$	$C_L = 1.72$	
y	c_{lb}	c_{la1}	c_l	$1.72c_{la1}$	c_l
	Table 9.2	Table 9.1	(2) + (3)	1.72(3)	(2) + (5)
0	0.185	0.998	1.183	1.72	1.90
23.04	0.182	1.016	1.198	1.74	1.92
46.08	0.155	1.026	1.181	1.77	1.93
69.12	0.107	1.034	1.141	1.78	1.89
92.16	0.014	1.038	1.052	1.78	1.79
114.12	−0.068	1.032	0.964	1.78	1.71
136.08	−0.115	1.020	0.905	1.76	1.65
158.04	−0.148	0.998	0.850	1.71	1.56
180.00	−0.155	0.958	0.803	1.65	1.50
195.00	−0.155	0.928	0.773	1.60	1.45
210.00	−0.155	0.896	0.741	1.54	1.39
225.00	−0.155	0.856	0.701	1.47	1.32
240.00					

as the sum of values in columns (2) and (3). Similarly, the values of c_l for a wing lift coefficient of 1.72 are obtained in column (6) by multi-

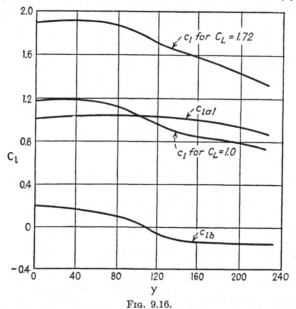

Fig. 9.16.

plying values of c_{la1} from column (3) by 1.72 and adding them to values of c_{lb} from column (2). Values of the distributions of c_l are plotted in Fig. 9.16.

PROBLEMS

9.1. Derive exact equations for the forces L and D of Fig. 9.2 in terms of L_0, D_0, and α_i. Calculate the percentage error of Eqs. 9.2 and 9.3 if $L/D = 10$ and $\alpha_i = 0.05$ rad.

9.2. A long cylinder rotates with its axis perpendicular to an airstream of velocity V. Calculate the lift coefficient c_l, based on a chord equal to the cylinder diameter, if the peripheral velocity of the cylinder is (a) $0.1V$ and (b) $0.2V$.

9.3. Plot the distribution of downwash velocities for the horseshoe vortex system shown in Fig. 9.6. The line AA' has a length b. Compute values of w/Γ at intervals of $0.1b$ along the following lines, all of which are parallel to AA':

$a.$ The line AA'

$b.$ A line $0.1b$ forward of AA'

$c.$ A line $0.2b$ forward of AA'

$d.$ A line $0.4b$ forward of AA'

$e.$ A line $0.8b$ forward of AA'

$f.$ A line $0.1b$ aft of AA'

$g.$ A line $0.2b$ aft of AA'

$h.$ A line $0.4b$ aft of AA'

$i.$ A line $0.8b$ aft of AA'

9.4. Plot all the points calculated in Prob. 9.3 as ordinates from lines parallel to OX.

9.5. Repeat Probs. 9.3 and 9.4 for a wing with an elliptical planform and no aerodynamic twist if Γ is the circulation at the center of the wing span.

9.6. Derive an expression for the slope of the lift curve $m = C_L/\alpha_a$ for an elliptical wing.

9.7. Calculate the spanwise distribution of the section lift coefficient for the wing shown if the flaps are in a neutral position. Assume no aerodynamic twist and a constant value of m_0. Assume $C_L = 1.0$.

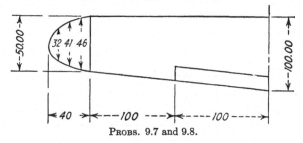

PROBS. 9.7 and 9.8.

9.8. Calculate the spanwise distribution of the section lift coefficient for values of $C_L = 0$, $C_L = 1$, and $C_L = 1.8$, if the flap is deflected so that the zero-lift chord for the flapped portion is 10° from the zero-lift chord for the rest of the wing. Assume that the section lift coefficient has a constant slope of 0.1 per deg. The results of Prob. 9.7 may be used in the solution.

9.7. Spanwise Lift Distribution Obtained by Fourier's Series Method.

An exact solution of the equations for the spanwise lift distribution for a wing with any planform may be obtained by the use of Fourier's series. This method was first proposed by Glauert,[1] and was further developed by Miss Lotz [5] and others. Tabular forms are arranged in ANC–1[6] in such a manner that a computer may obtain and check the spanwise lift distribution with little knowledge of the theory involved.

The circulation at any point in the span of a symmetrically loaded wing could be represented by the series

$$\Gamma = a_1 \cos \frac{\pi y}{b} + a_3 \cos \frac{3\pi y}{b} + a_5 \cos \frac{5\pi y}{b} + \cdots \qquad (9.29)$$

or

$$\Gamma = \sum_{n=1, 3, 5}^{\infty} a_n \cos \frac{n\pi y}{b}$$

where y is the distance of the point from the center of the span, b is the span, and $a_1, a_3, \ldots, a_n$ are constant coefficients as shown in Fig. 9.17. It is possible to represent any symmetrical curve for the circulation by properly choosing the coefficients a_n and by using enough terms. Fortunately, the first few coefficients a_n are usually large in comparison to later terms, and a good approximation to the circulation may be obtained by using only a few terms.

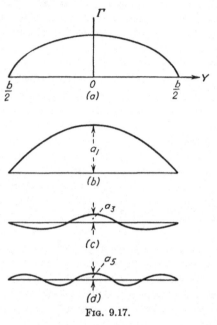

Fɪɢ. 9.17.

A more convenient series for representing the circulation is as follows:

$$\Gamma = \frac{c_s m_s V}{2} \left(A_1 \sin \theta + A_2 \sin 2\theta + A_3 \sin 3\theta + \cdots \right)$$

or

$$\Gamma = \frac{c_s m_s V}{2} \sum_{n=1}^{\infty} A_n \sin n\theta \tag{9.30}$$

where

$$\cos \theta = \frac{2y}{b} \tag{9.31}$$

and c_s and m_s represent constants, equal to the wing chord and the slope of the lift curve for the airfoil at the center of the span. This series can be conveniently used for either symmetrical or unsymmetrical loading. If the loading is symmetrical, the even coefficients, A_2, A_4, A_6, . . ., A_{2n}, will be zero. The first terms in the series of Eq. 9.30 are shown in Fig. 9.18. The angle θ is shown in Fig. 9.18(b), as measured on a circle of radius $b/2$. The first term of Eq. 9.30, $\Gamma_1 = \dfrac{c_s m_s V}{2} A_1 \sin \theta$, is proportional to the ordinates of points on this circle. This first term is seen to represent the ordinates of a semiellipse, since the ordinates of a semiellipse are proportional to the corresponding ordinates of a semi-circle when both are plotted on the same diameter.   Thus, Eq. 9.30 is preferable to Eq. 9.29, because the first term, representing an elliptical lift distribution, approximates the actual wing lift distribution more closely.

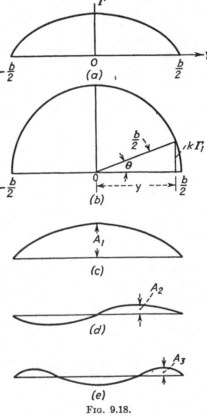

Fig. 9.18.

If the circulation is given by Eq. 9.30, the problem of obtaining the spanwise lift distribution is equivalent to the problem of evaluating the coefficients A_n.   These coefficients may be evaluated by substituting the Fourier series values for Γ into Eq. 9.17, which is repeated below:

$$\frac{2\Gamma}{m_0 cV} = \alpha_a - \frac{1}{4\pi V} \int_{y=-\frac{b}{2}}^{y=\frac{b}{2}} \frac{d\Gamma}{y'-y} \tag{9.17}$$

Differentiating Eq. 9.30, there is obtained

$$d\Gamma = \frac{c_s m_s V}{2} \sum_{n=1}^{\infty} nA_n \cos n\theta \, d\theta \tag{9.32}$$

The last term in Eq. 9.17 is now evaluated, from Eqs. 9.31 and 9.32

$$\frac{c_s m_s}{4\pi b} \int_0^\pi \frac{\sum\limits_{n=1}^\infty n A_n \cos n\theta \, d\theta}{\cos \theta - \cos \theta'} \tag{9.33}$$

where θ' is the particular value of θ corresponding to the coordinate y'. Glauert [1] obtains the value of the integral as follows:

$$\int_0^\pi \frac{\cos n\theta}{\cos \theta - \cos \theta'} \, d\theta = \frac{\pi \sin n\theta'}{\sin \theta'} \tag{9.34}$$

The angle θ' is constant with respect to the integration of Eq. 9.34, but represents any point on the span, so the prime will be dropped in future equations. Substituting the value from Eq. 9.34 into the expression 9.33, the downwash angle becomes

$$\frac{c_s m_s}{4b} \frac{\Sigma n A_n \sin n\theta}{\sin \theta} \tag{9.35}$$

The general equation for the circulation is as follows, when expression 9.35 is substituted for the last term of Eq. 9.17:

$$\frac{m_s}{m_0} \frac{c_s}{c} \Sigma A_n \sin n\theta = \alpha_a - \frac{c_s m_s}{4b} \frac{\Sigma n A_n \sin n\theta}{\sin \theta} \tag{9.36}$$

or

$$\frac{m_s}{m_0} \frac{c_s}{c} \sin \theta \Sigma A_n \sin n\theta = \alpha_a \sin \theta - \frac{c_s m_s}{4b} \Sigma n A_n \sin n\theta \tag{9.37}$$

The terms m_0, the slope of the lift curve, c, the wing chord, and α_a, the angle of attack, vary along the span. In order to represent a wing of any shape, these terms may be expressed by other Fourier series as follows:

$$\frac{m_s}{m_0} \frac{c_s}{c} \sin \theta = \sum_{n=0,1,2,3}^\infty C_{2n} \cos 2n\theta \tag{9.38}$$

and

$$\alpha_a \sin \theta = \sum_{n=1,2,3}^\infty B_n \sin n\theta \tag{9.39}$$

The coefficients C_{2n} and B_n are evaluated from the known wing characteristics, as shown in the following article.

9.8. Evaluation of Fourier Series Coefficients. The terms of the left-hand side of Eq. 9.38 may be computed for any point in the span. In evaluating the coefficients C_{2n}, it is first necessary to decide how many terms of the series are to be used. Most actual calculations are made for 10 terms in the series, and the equations are satisfied for 10 points in the span. Only five points will be considered here, however, since the purpose is to explain the derivation of the equations. The tabular forms for 10-point calculations are readily available in *ANC-1*.

The known terms on the left side of Eq. 9.38 are designated as y_i, or as y_1, y_2, y_3, .. , for various points in the span, as shown in Fig. 9.19. This notation is consistent with that used in *ANC*–1, although it conflicts with the previous use of y to denote coordinates measured along the span. Care must be exercised to avoid confusion with the notation. In this article, points on the span are determined from the value of θ and are not designated by the coordinate y. If five terms of the series are used, Eq. 9.38 may be written as follows:

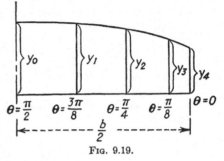

FIG. 9.19.

$$C_0 + C_2 \cos 2\theta + C_4 \cos 4\theta + C_6 \cos 6\theta + C_8 \cos 8\theta = y_i \quad (9.40)$$

Substituting the values of the known coordinates, $\theta = 0$, $y_i = y_4$; $\theta = \pi/8$, $y_i = y_3$; $\theta = \pi/4$, $y_i = y_2$; $\theta = 3\pi/8$, $y = y_1$; and $\theta = \pi/2$, $y = y_0$, the following equations are obtained:

$$
\begin{aligned}
C_0 + C_2 + C_4 + C_6 + C_8 &= y_4 & (a) \\
C_0 + \frac{1}{\sqrt{2}} C_2 + 0 - \frac{1}{\sqrt{2}} C_6 - C_8 &= y_3 & (b) \\
C_0 + 0 - C_4 + 0 + C_8 &= y_2 & (c) \\
C_0 - \frac{1}{\sqrt{2}} C_2 + 0 + \frac{1}{\sqrt{2}} C_6 - C_8 &= y_1 & (d) \\
C_0 - C_2 + C_4 - C_6 + C_8 &= y_0 & (e)
\end{aligned}
\quad (9.41)
$$

These equations may be solved simultaneously for the coefficients C_{2n} in terms of the known ordinates y_i. The solution may be readily obtained by adding and subtracting Eqs. (a) and (e), then Eqs. (b) and (d), and repeating the operations on the resulting equations. The values of C_{2n} are as follows:

$$
\begin{aligned}
C_0 &= \frac{1}{4}\left(\frac{y_4}{2} + y_3 + y_2 + y_1 + \frac{y_0}{2}\right) & (a) \\
C_2 &= \frac{1}{2}\left(\frac{y_4}{2} + \frac{1}{\sqrt{2}} y_3 - \frac{1}{\sqrt{2}} y_1 - \frac{y_0}{2}\right) & (b) \\
C_4 &= \frac{1}{2}\left(\frac{y_4}{2} - y_2 + \frac{y_0}{2}\right) & (c) \\
C_6 &= \frac{1}{2}\left(\frac{y_4}{2} - \frac{1}{\sqrt{2}} y_3 + \frac{1}{\sqrt{2}} y_1 - \frac{y_0}{2}\right) & (d) \\
C_8 &= \frac{1}{4}\left(\frac{y_4}{2} - y_3 + y_2 - y_1 + \frac{y_0}{2}\right) & (e)
\end{aligned}
\quad (9.42)
$$

Numerical calculations for Eqs. 9.42 may be readily made by tabular methods. Table III of ANC-1(1) is a computation form for obtaining the Fourier coefficients C_{2n} when 10 terms of the series and 10 points are considered. The equations are similar to Eqs. 9.42.

The values given in Eqs. 9.42 may also be obtained by the common methods of finding Fourier series coefficients. If Eq. 9.40 is written

$$\frac{a_0}{2} + a_2 \cos 2\theta + a_4 \cos 4\theta + \cdots + a_{2n} \cos 2n\theta + \cdots = y_i \quad (9.43)$$

it is shown in texts on Fourier series that any coefficient may be obtained as follows:

$$a_{2n} = \frac{4}{\pi} \int_0^{\frac{\pi}{2}} y_i \cos 2n\theta \, d\theta \quad (9.44)$$

or, replacing the integral by a summation,

$$a_{2n} = \frac{4}{\pi} \Sigma \cos 2n\theta y_i \, \Delta\theta \quad (9.45)$$

The coefficients a_{2n} correspond to coefficients C_{2n} for all except the first term of the series, as is seen by comparing Eqs. 9.40 and 9.43. To evaluate C_0 by Eq. 9.45, the values $\Delta\theta = \pi/8$, $n = 0$, and $\cos 2n\theta = 1$ are substituted into Eq. 9.45 as follows:

$$C_0 = \frac{a_0}{2} = \frac{2}{\pi} \frac{\pi}{8} \Sigma y_i = \frac{1}{4} \left(\frac{y_0}{2} + y_1 + y_2 + y_3 + \frac{y_4}{2} \right) \quad (9.46)$$

The end ordinates y_0 and y_4 should be multiplied by only one-half the angle $\Delta\theta$ used for other ordinates, and consequently only half the ordinates y_0 and y_4 are used in the summation of Eq. 9.46.

To obtain the coefficient C_6 from Eq. 9.45, the values of $\cos 2n\theta = \cos 6\theta$ are substituted. For increments, $\Delta\theta$, of $\pi/8$, the values of $\cos 6\theta$ will be $1, -1/\sqrt{2}, 0, 1/\sqrt{2}, -1$. The coefficient will have the following value:

$$C_6 = a_6 = \frac{4}{\pi} \frac{\pi}{8} \Sigma \cos 2n\theta y_i$$

or

$$C_6 = \frac{1}{4} \left(\frac{y_4}{2} - \frac{1}{\sqrt{2}} y_3 + \frac{1}{\sqrt{2}} y_1 - \frac{y_0}{2} \right) \quad (9.47)$$

The derivation of such Fourier series coefficients and the tabular methods for calculation have been developed by Runge. An explanation of the procedure is given by Den Hartog,[7] and tabular computation forms are shown in ANC-1.

The Fourier series coefficients B_n, defined in Eq. 9.39, may be obtained in the same manner as the coefficients C_{2n}. For an assumed wing atti tude, the angle of attack α_a is known at all points of the span, and the terms B_n may be calculated to satisfy Eq. 9.39 at a selected number of points.

Example. Calculate the Fourier series terms B_n and C_{2n}, defined by Eqs. 9.28 and 9.39, for the wing of Example 1, Art. 9.6. This wing has no aerodynamic twist, the slope of the lift curve is constant at all points along the span, and the planform is shown in Fig. 9.14.

Solution. Since α_a is a constant, Eq. 9.39 is satisfied by the following values of B_n:

$$B_1 = \alpha_a$$
$$B_n = 0 \qquad n \neq 1$$

In wings with aerodynamic twist, α_a is not constant, and equations similar to Eqs. 9.41 must be solved. A tabular computation form is provided in Table II of ANC–1(1) for use in obtaining values of B_n.

The values for the left-hand side of Eq. 9.38 are calculated in Table 9.4 for five values of θ.

TABLE 9.4

θ	$\dfrac{2y}{b}$	$\dfrac{m_s}{m_0}$	c in.	$\dfrac{c_s}{c}$	$\sin\theta$	$\dfrac{m_s}{m_0}\dfrac{c_s}{c}\sin\theta$	
0	1.00	1			0	0	$= y_4$
$\pi/8$	0.9239	1	53.0	1.922	0.3827	0.735	$= y_3$
$\pi/4$	0.7071	1	75.0	1.360	0.7071	0.961	$= y_2$
$3\pi/8$	0.3827	1	87.4	1.168	0.9239	1.079	$= y_1$
$\pi/2$	0	1	102.0	1.000	1.000	1.0000	$= y_0$

A substitution of these values of y_4, y_3, y_2, y_1, and y_0 into Eqs. 9.42 yields $C_0 = 0.819$, $C_2 = -0.368$, $C_4 = -0.230$, $C_6 = -0.132$, and $C_8 = -0.088$.

9.9. Evaluation of A_n Coefficients. As a further step in solving the general circulation equation, the series of Eqs. 9.38 and 9.39 are substituted into Eq. 9.37. The left-hand side of Eq. 9.37 is then represented by the double series:

$$\Sigma C_{2n} \cos 2n\theta\, \Sigma A_n \sin n\theta \tag{9.48}$$

The following trigonometric relation may be used:

$$2 \sin k\theta \cos l\theta = \sin (k + l)\theta + \sin (k - l)\theta \tag{9.49}$$

A substitution of this relation into expression 9.48 yields

$$\tfrac{1}{2} \sum_{k=1,2,3}^{\infty} \sum_{l=0,2,4}^{\infty} A_k C_l [\sin (k + l)\theta + \sin (k - l)\theta] \tag{9.50}$$

If expression 9.50 is assumed to contain the three terms of the circulation series with coefficients A_1, A_3, and A_5 and six terms of the planform series with coefficients C_0, C_2, C_4, C_6, C_8, and C_{10}, it may be evaluated as follows:

$$\frac{1}{2} \sum_{k=1,3}^{5} \sum_{l=0,2,4}^{10} A_k C_l [\sin (k + l)\theta + \sin (k - l)\theta]$$
$$= \frac{1}{2}(2A_1C_0 - A_1C_2 + A_3C_2 - A_3C_4 + A_5C_4 - A_5C_6) \sin \theta$$
$$+ \frac{1}{2}(A_1C_2 - A_1C_4 + 2A_3C_0 - A_3C_6 + A_5C_2 - A_5C_8) \sin 3\theta$$
$$+ \frac{1}{2}(A_1C_4 - A_1C_6 + A_3C_2 - A_3C_8 + 2A_5C_0 - A_5C_{10}) \sin 5\theta + \cdots$$

$$(9.51)$$

In evaluating the series of Eq. 9.50, the coefficients of 7θ, 9θ, and higher values are neglected, although the two series used in Eq. 9.48 would contain these terms if all values for $n \leqq 5$ are used.

Equation 9.37 may now be written as follows, by substituting from Eqs. 9.38 and 9.39.

$$\Sigma C_{2n} \cos 2n\theta \Sigma A_n \sin n\theta = \Sigma B_n \sin n\theta - \frac{c_s m_s}{4b} \Sigma n A_n \sin n\theta \quad (9.52)$$

Now, by substituting the series of Eq. 9.51 for the left-hand side of Eq. 9.52, equating the coefficients of $\sin \theta$, $\sin 3\theta$, and $\sin 5\theta$, and then rearranging the terms, the following simultaneous equations are obtained:

$$\left.\begin{array}{l} 2P_1A_1 + (C_2 - C_4)A_3 + (C_4 - C_6)A_5 = 2B_1 \\ (C_2 - C_4)A_1 + 2P_3A_3 + (C_2 - C_8)A_5 = 2B_3 \\ (C_4 - C_6)A_1 + (C_2 - C_8)A_3 + 2P_5A_5 = 2B_5 \end{array}\right\} \quad (9.53)$$

where

$$P_1 = C_0 - \frac{1}{2}C_2 + \frac{c_s m_s}{4b}$$

and

$$P_n = C_0 - \frac{1}{2}C_{2n} + n\frac{c_s m_s}{4b} \quad (9.54)$$

The coefficients A_n for the circulation terms may now be obtained by solving Eqs. 9.53 simultaneously.

Example. Calculate the lift distribution for the wing of Example 1, Art. 9.6, for a lift coefficient $C_L = 1.00$. The values of B_n and C_n are obtained in the example problem of Art. 9.8. Assume $C_{10} = 0$. The slope of the section lift curves m_s is 5.73 per rad.

Solution. The following coefficients are obtained in the Example problem of Art. 9.8. First assume an angle of attack α_a of 1 rad.

$$\begin{array}{lll} B_1 = \alpha_a = 1 & C_0 = 0.819 & C_6 = -0.132 \\ B_3 = 0 & C_2 = -0.368 & C_8 = -0.088 \\ B_5 = 0 & C_4 = -0.230 & C_{10} = 0 \end{array}$$

The following terms in Eq. 9.53 are readily obtained:

$$C_2 - C_4 = -0.138$$
$$C_2 - C_6 = -0.236$$
$$C_2 - C_8 = -0.280$$
$$C_4 - C_6 = -0.098$$

From the dimensions shown in Fig. 9.14,

$$\frac{c_s m_s}{4b} = \frac{102 \times 5.73}{4 \times 480} = 0.304$$

From Eq. 9.54,

$$P_1 = 0.819 + 0.5 \times 0.368 + 0.304 = 1.307$$
$$P_3 = 0.819 + 0.5 \times 0.132 + 3 \times 0.304 = 1.797$$
$$P_5 = 0.819 + 0 + 5 \times 0.304 = 2.339$$

Substituting the above values into Eq. 9.53, there is obtained

$$2 \times 1.307 A_1 - 0.138 A_3 - 0.098 A_5 = 2$$
$$-0.138 A_1 + 2 \times 1.797 A_3 - 0.280 A_5 = 0$$
$$-0.098 A_1 - 0.280 A_3 + 2 \times 2.339 A_5 = 0$$

Solving these equations, there is obtained

$$A_1 = 0.767$$
$$A_3 = 0.0308$$
$$A_5 = 0.0179$$

These values of A_1, A_3, and A_5 define the circulation for a hypothetical angle of attack of 1 rad, assuming that the wing does not stall and that the lift curve slope remains constant at this angle of attack. It can be found by integrating the equation for the lift per unit span that the coefficient of lift for the entire wing depends only on A_1 and is defined by the following equation:

$$C_L = \frac{\pi c_s m_s b A_1}{4S} \tag{9.55}$$

Substituting numerical values,

$$C_L = \frac{\pi \times 102 \times 5.73 \times 480 \times 0.767}{4 \times 38,340} = 4.41$$

The Fourier series coefficients A_1, A_3, and A_5 for a lift coefficient of unity may therefore be obtained by proportion from those for a lift coefficient of 4.41, as follows:

$$A_1 = \frac{0.767}{4.41} = 0.1740$$

$$A_3 = \frac{0.0308}{4.41} = 0.0070$$

$$A_5 = \frac{0.0179}{4.41} = 0.0041$$

The lift per unit span cc_l may now be obtained from Eqs. 9.15 and 9.30.

$$cc_l = c_s m_s \Sigma A_n \sin n\theta$$
$$= 102 \times 5.73(0.1740 \sin \theta + 0.0070 \sin 3\theta + 0.0041 \sin 5\theta)$$

Numerical values are computed in Table 9.5 for values of θ at 18° intervals.

TABLE 9.5

1	θ	90°	72°	54°	36°	18°
2	$y = \dfrac{b}{2} \cos \theta$	0	74.2	141.1	194.0	228.2
3	$\sin \theta$	1.000	0.951	0.809	0.588	0.309
4	$\sin 3\theta$	−1.000	−0.588	0.309	0.951	0.809
5	$\sin 5\theta$	1.000	0	−1.000	0	1.000
6	$0.1740 \sin \theta$	0.1740	0.1655	0.1408	0.1022	0.0538
7	$0.0070 \sin 3\theta$	−0.0070	−0.0041	0.0022	0.0066	0.0057
8	$0.0041 \sin 5\theta$	0.0041	0	−0.0041	0	0.0041
9	$\Sigma A_n \sin n\theta$	0.1711	0.1614	0.1389	0.1088	0.0636
10	cc_l	100.0	94.4	81.1	63.6	37.2

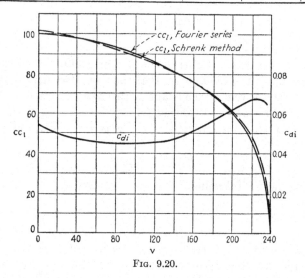

FIG. 9.20.

The values obtained in line 10 of Table 9.5 are plotted in Fig. 9.20, as are the corresponding values obtained in column (6) of Table 9.1. These values correspond closely for most of the span.

9.10. Spanwise Distribution of Induced Drag. The wing of finite span has been shown to have a greater drag per unit span than the corresponding wing of infinite span. This increase in drag because of the span effect has been termed the induced drag, and results from the fact that the forces on the wing are inclined at an additional angle α_i, as shown in Fig. 9.2. The magnitude of the induced drag is obtained as the product of the lift and the induced angle of attack.

$$D_i = L\alpha_i \tag{9.56}$$

where D_i is the induced drag. The spanwise distribution of induced drag therefore depends on both the distribution of lift and induced angle of attack. In the case of an elliptical wing planform, the value of α_i is constant along the span, and the induced drag will be distributed in the same manner as the lift. In most cases, the lift and induced angle of attack will both vary along the span.

It is often convenient to consider the induced drag coefficient c_{di} rather than the total drag. If both sides of Eq. 9.56 are divided by qc, the product of the dynamic pressure and the chord, the following equation is obtained:

$$c_{di} = c_l\alpha_i \tag{9.57}$$

The distribution of c_l was discussed in the previous articles, but the methods of obtaining α_i were not discussed. If the values of c_l are known at various points along the span, the values of α_i are obtained from the following equation, which is obvious from Fig. 9.2.

$$\alpha_i = \alpha_a - \alpha_0$$

or

$$\alpha_i = \alpha_a - \frac{c_l}{m_0} \tag{9.58}$$

where α_a is the angle of attack measured to the zero-lift chord at the section and m_0 is the slope of the section lift coefficient.

The spanwise distribution of the coefficient of induced drag may readily be obtained from Eqs. 9.57 and 9.58, after the distribution for lift has been obtained. The percentage error in the use of the approximate method may be somewhat greater for the drag distribution than for the lift distribution, since Eq. 9.58 requires the subtraction of terms of similar magnitude. This inaccuracy will not seriously affect the structural design of the wing.

It is necessary to know the angle of attack α_a at various sections in order to apply Eq. 9.58. When the spanwise lift distribution was obtained by the Fourier series method, the values of α_a were used in the solution. In the approximate method, however, the distribution of c_l was obtained but the angle of attack corresponding to any wing C_L was not known. It is therefore necessary to obtain a relationship between the wing C_L and the angle of attack. For the elliptical wing, Eq. 9.23 gives this relation.

$$\frac{C_L}{m_0} = \alpha_a - \frac{C_L}{\pi A} \tag{9.23}$$

The lift curve slope, $m = C_L/\alpha_a$, is obtained from Eq. 9.23

$$m = \frac{C_L}{\alpha_a} = \frac{m_0}{1 + \dfrac{m_0}{\pi A}} \tag{9.59}$$

It is customary to use an expression similar to Eq. 9.59 for wings of any planform, as follows:

$$m = \frac{m_0}{1 + (m_0/\pi A)(1 + \tau)} \tag{9.60}$$

where τ is a correction factor which accounts for the deviation of the planform from an ellipse. The values of τ may be computed from the Fourier series method and are given for straight tapered wings in Fig. 9.21.[6] The terms appearing in Fig. 9.21 are shown in Fig. 9.23.

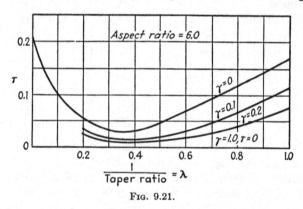

FIG. 9.21.

The coefficient of induced drag for an elliptical wing is readily obtained as the coefficient of lift times the induced angle of attack.

$$C_{Di} = C_L \alpha_i$$

or

$$C_{Di} = \frac{C_L^2}{\pi A} \tag{9.61}$$

A similar equation which applies for any planform is as follows:

$$C_{Di} = \frac{C_L^2}{\pi A}(1 + \sigma) \tag{9.62}$$

The term σ is also a correction factor which accounts for the deviation of the planform from an ellipse. This term is readily computed from the Fourier series terms as follows:

$$1 + \sigma = \frac{\Sigma n A_n^2}{A_1^2} \tag{9.63}$$

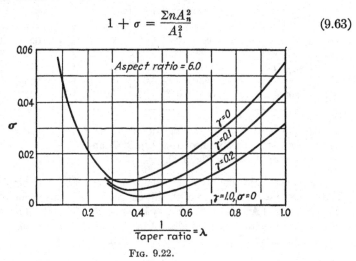

Fig. 9.22.

The values of σ for straight tapered wings may be obtained from Fig. 9.22.[6]

Example. Calculate the spanwise distribution of c_{di} for the wing of the example of Art. 9.9.

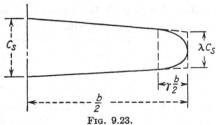

Fig. 9.23.

Solution. The calculations are shown in Table 9.6. The points selected are not the same as those used in Table 9.5; therefore the values of cc_l shown in line 2 are read from the curve of Fig. 9.20. The chord lengths c, corresponding to the points used, are given in line 3, and the values of c_l are computed in line 4 by dividing values in line 2 by values in line 3. The value of α_a is the same at all stations and is obtained by dividing the angle of attack of 1 rad, used in Art. 9.9, by the corresponding lift coefficient $C_L = 4.41$, which was computed by Eq. 9.55. The effective angle of attack at each station, c_l/m_0, is computed in line 6 by dividing the values of c_l in line 4 by the section lift curve slope, $m_0 = 5.73$, which was used in Art. 9.9. The induced angle α_i is computed from

TABLE 9.6

1	y	0	45.1	92.2	136.1	180	195	210	225
2	cc_l	100	97.5	91.8	82.5	70.0	63	54.4	41
3	c	102	94.7	87.4	80.3	73.3	69.2	62.2	49.5
4	c_l	0.98	1.03	1.05	1.03	0.95	0.91	0.87	0.83
5	$\alpha_a = 1/4.41$	0.226	0.226	0.226	0.226	0.226	0.226	0.226	0.226
6	$c_l/m_0 = c_l/5.73$	0.171	0.180	0.183	0.180	0.166	0.159	0.152	0.145
7	$\alpha_i = \alpha_a - c_l/m_0$	0.055	0.046	0.043	0.046	0.060	0.067	0.074	0.081
8	$c_{di} = c_l\alpha_i$	0.054	0.047	0.045	0.047	0.057	0.061	0.064	0.067

Eq. 9.58, in line 7 by subtracting terms in line 6 from terms in line 5. The values of c_{di} are now obtained from Eq. 9.57, and are shown in line 8. These values are also plotted in Fig. 9.20.

The same solution could be obtained from the results of Example 1, Art. 9.6. The values of α_a, or the slope of the lift curve $m = 4.41$, were not available from the equations of Art. 9.6. These may be computed from Eq. 9.60 and Fig. 9.21, as follows: For the wing of Fig. 9.14, $\lambda = 0.625$, $\gamma = 0.25$, $A = 6.0$, and $m_0 = 5.73$. From Fig. 9.21, $\tau = 0.022$. From Eq. 9.60,

$$m = \frac{5.73}{1 + (5.73/6\pi)(1.022)} = 4.38$$

This is slightly less than the value of 4.41 previously obtained. The discrepancy, which is less than 1 per cent, probably results from taking only a few of the series terms in Art. 9.9.

9.11. Other Factors Affecting Distribution of Air Loads. In the preceding discussion only the airplane wing has been considered. The airflow about the complete airplane may be considerably different from the airflow about an isolated wing. The shape of the fuselage and nacelles, and the nature of the wing-fuselage junction, may make a large

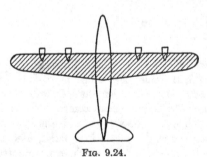

FIG. 9.24.

difference in the air-load distribution. It is customary, however, to assume that the wing area extends straight through the fuselage and nacelles, as shown in Fig. 9.24, and to assume that the shaded wing area shown acts as an isolated wing. For a well-designed wing-fuselage junction, the lift forces acting on the fuselage will approximate those acting on the section of the wing which is inside the fuselage. A mid-wing arrangement is preferable aerodynamically, but other considerations often make it desirable to have a low-wing or a high-wing arrangement. In such cases it may be necessary to provide large fillets at the wing-fuselage junction in order to obtain good lift and low drag characteristics.

When wing flaps end at the side of the fuselage, it is desirable to have the inboard ends of the flaps sealed in such a way that air cannot flow between the flaps and the fuselage when the flaps are deflected. With such an arrangement, it is possible for the high lift of the flapped sections to extend across the fuselage sections. It is usually difficult to prevent leakage between the flaps and the fuselage, however, and for most installations it is necessary to assume that the fuselage section acts as if there were no flaps across the fuselage.

The assumption that the wing acts as a lifting vortex line introduces errors in calculating the lift and drag distribution near the wing tip. A study of Fig. 9.8 indicates that there is a large change in downwash velocity between the leading edge and the trailing edge of the wing in the case of airfoil sections near the tip where the distance r is small. The assumed downwash velocity at the lifting line vortex is therefore in error, and measured air-load distributions near wing tips show that the calculated distributions are in error near the wing tips.

The effects of fuselage interference and wing tip errors cannot be determined theoretically. Wind tunnel tests on models, or flight tests on similar airplanes, are necessary in order to obtain data with which to verify the assumptions of the analysis. Since the assumptions may be in considerable error, the designer is not justified in calculating air loads to a high degree of precision. In most cases the approximate method of obtaining air-load distribution is as accurate as the assumptions on which the calculations are based. The added precision of calculations by the Fourier series method is probably justified only in cases where unusual planforms or unusual flight conditions are investigated.

9.12. Limitations of Lifting Line Theory. The methods of calculating the spanwise air-load distribution have been based on the assumption that the wing could be replaced by a lifting vortex line along the span of the wing. This concept, commonly known as the "lifting line" theory, was originally proposed by Prandtl, and it has been used extensively for conventional unswept wings with aspect ratios greater than 5 or 6. With aircraft designed for transonic or supersonic speeds, however, it is frequently desirable to use wing planforms of radical shapes, and the lifting line theory is no longer applicable. When wings have small aspect ratios, or have pronounced sweep or taper, it is necessary to use a more accurate vortex system for the wing.

The more accurate methods which have been developed for wings with unconventional planforms may also be used in the analysis of conventional wings, in order to correct some of the inherent errors in the assumption of a lifting vortex line. While the Fourier series method provides an accurate solution of Eq. 9.17, several assumptions are made in the derivation of this equation.

In the lifting line theory the wing was replaced by a large number of horseshoe vortex systems similar to that shown in Fig. 9.6. The lifting vortex line AA' was assumed to be at the quarter-chord point of the wing, and the effective angle of attack of the wing was calculated from the downwash velocity at the quarter-chord point. It can be shown that the true effective angle of attack of an airfoil depends on the local angle of attack at the three-quarter-chord point. From Fig. 9.8(*b*), it is seen that the downwash velocity is greater at the three-quarter-chord point than at the one-quarter-chord point and that the effective angle of attack, α_0 of Fig. 9.2, should be smaller than that obtained from the lifting line theory. This effect is large near the wing tip and smaller near the wing root and is therefore more noticeable in wings with small aspect ratios.

The values of the lift curve slopes $m = dC_L/d\alpha$, calculated from the lifting line theory, are too high, but this error does not greatly affect the structural loading conditions. For a given wing angle of attack, the true lift forces are less than those calculated from the lifting line theory, particularly in the outer portion of the span. For a given total lift on the wing, the true spanwise lift distribution is therefore less in the outer portion of the span and greater in the center portion of the span than the calculated values. The calculated distributions therefore yield conservative values for the wing bending moments.

Weissinger[3] has developed a modified lifting line theory in which the effects of aspect ratio and sweepback are considered. He assumes a bound vortex at the one-quarter-chord line and calculates the downwash velocities and local angles of attack at the three-quarter-chord line. The elementary horseshoe vortex of strength $d\Gamma$ is therefore as shown in Fig. 9.25. The effect of sweepback is to increase the lift in the outer portion of the span and to decrease the lift in the center portion. Similarly, if a wing is swept forward the lift is decreased near the tips and increased in the center of the span. These effects are quite pronounced, and the simple lifting line theory is not adequate for a wing with a large angle of sweep.

Certain approximations are made by Weissinger in assuming the lifting line vortex at the quarter-chord points and control points at the three-quarter-chord points. The ideal solution would assume a lifting surface composed of a bound vortex sheet of the true wing planform shape and a trailing vortex sheet in the wing chord plane. Such a solution involves too many mathematical difficulties. Falkner[9] has replaced this lifting surface by a pattern of isolated loads, so that both chordwise and spanwise load distribution may be considered. The number and location of

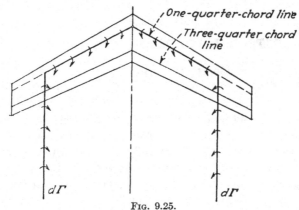

FIG. 9.25.

the isolated loads is carefully chosen in order to obtain a good approximation with a reasonable amount of numerical calculations. For wings of radical shape, it may be necessary to use a lifting surface method.

PROBLEM

9.9. Solve Prob. 9.7 by the Fourier series method.

REFERENCES FOR CHAPTER 9

1. GLAUERT, H.: "The Elements of Aerofoil and Airscrew Theory," Cambridge University Press, London, 1926.
2. SCHRENK, O.: A Simple Approximation Method for Obtaining the Spanwise Lift Distribution, *NACA TM* 948, 1940.
3. "Civil Aeronautics Manual," Part 04, Appendix V, U.S. Department of Commerce, Civil Aeronautics Administration.
4. PEARSON, H. A.: Span Load Distribution for Tapered Wings with Partial Span Flaps, *NACA TR* 585, 1937.
5. LOTZ, IRMGARD: Berechnung der Auftriebsverteilung beliebig geformter Flugel, *Z. Flugtech. u. Motorluftschiffahrt*, Vol. 22, No. 7, Apr. 14, 1931.
6. ANC–1(1), "Spanwise Air-load Distribution," Army-Navy-Commerce Committee on Aircraft Requirements, 1938.
7. DEN HARTOG, J. P.: "Mechanical Vibrations," 3d ed., p. 25, McGraw-Hill Book Company, Inc., New York, 1947.
8. WEISSINGER, J.: The Lift Distribution of Swept-back Wings, *NACA TM* 1120, 1947.
9. FALKNER, V. M.: The Calculation of Aerodynamic Loading on Surfaces of Any Shape, *Reports and Memoranda 1910*, Aeronautical Research Committee (Great Britain), 1943.

CHAPTER 10

EXTERNAL LOADS ON THE AIRPLANE

10.1. General Considerations. Airplanes may be subjected to a wide variety of loading conditions in flight. The pilot may perform a large number of intentional maneuvers, and each maneuver may be performed at various speeds and with various rates of displacement of the controls. In addition to the intentional maneuvers imposed on the airplane, it may also be subjected to high loads when flying in stormy weather.

It is obviously impossible to investigate every loading condition which an airplane might encounter. It is necessary to select a few conditions such that one of these conditions will be critical for every structural member of the airplane. These conditions have been determined from past investigation and experience and are definitely specified by the licensing or procuring agencies. Private or commercial airplanes must be licensed by the Civil Aeronautics Administration (CAA) of the U.S. Department of Commerce. These airplanes are consequently designed in accordance with current CAA specifications. Military airplanes are procured by the Bureau of Aeronautics of the U.S. Department of the Navy or the Air Materiel Command of the U.S. Department of War. These agencies issue detailed specifications, and current specifications are usually classified as confidential for purposes of military secrecy. Many specifications which apply to both military and civil aircraft are standardized by the three agencies and are published as *ANC* (Army-Navy-Civil) publications, which may be obtained from the U.S. Government Printing Office.

Different types of airplanes are designed for different loading conditions. Commercial transport airplanes, for example, are never subjected to violent intentional maneuvers. Air-line pilots have reliable weather information at all times and can plan their flights so as to avoid severe storm centers. These airplanes are consequently designed for comparatively small load factors. A placard is placed in the cockpit to inform the pilot of the maximum permissible values of the gross weight, diving speed, and load factors for the airplane. It would be physically possible for a pilot to cause structural failure of the airplane by maneuvering it violently and exceeding the limitations specified on the placard. The pilot must therefore use care in observing the structural limitations of the aircraft as well as care in avoiding other flight hazards. Military airplanes used in fighter or dive-bomber operations are designed to resist violent maneuvers. The design conditions are usually determined from the maximum acceleration which the human body can withstand, and the pilot will black out and lose consciousness before reaching the load factor which would cause structural failure of the airplane. The maximum load factor which a pilot can withstand is about 8.0, and such airplanes are designed so that they will not be damaged at this load factor and will not completely fail at a load factor of 12.0, since a safety factor of 1.5 is normally used. Even fighter airplanes, however, carry placards showing the maximum permissible diving speed, gross weight, and load factor. Since detailed design specifications for any airplane will be available elsewhere, they are not included here. Only basic flight conditions which are applicable to any conventional type of airplane are considered. These will apply to military or civil land airplanes or seaplanes. Special conditions may apply to unconventional types such as flying wing, tail first, or rotary-wing aircraft.

10.2. Basic Flight Loading Conditions. One of four basic conditions will probably produce the highest stress in any part of the airplane for any flight condition. These conditions are usually called positive high angle of attack, positive low angle of attack, negative high angle of attack, and negative low angle of attack. All of these conditions represent symmetrical flight maneuvers, *i.e.*, there is no motion normal to the plane of symmetry of the airplane.

The positive high angle of attack condition is obtained in a pull-out at the highest possible angle of attack of the wing. The lift and drag forces are perpendicular and parallel to the relative wind, which is shown as horizontal in Fig. 10.1(a). The resultant R of these forces always has an aft component with respect to the relative wind, but will usually have a forward component C with respect to the wing chord line, because of the high angle of attack α. The maximum forward component C will be obtained when α has a maximum value. In order to account for uncertainties in obtaining the stalling angle of attack under unsteady flow conditions, most specifications arbitrarily require that a value of α be used which is higher than the wing stalling angle under steady flow conditions. An angle of attack corresponding to a coefficient of lift of 1.25 times the maximum coefficient of lift for steady flow conditions is often used, and aerodynamic data is extrapolated from data measured for steady flow conditions. Experiments show that these high angles of attack and high lift coefficients may be obtained momentarily in a sudden pull-up before the airflow reaches a steady condition, but it is difficult to obtain accurate lift measurements during the unsteady conditions.

In the positive high angle of attack condition, the bending moments from the normal forces N, shown in Fig. 10.1(a), produce compressive stresses on the upper side of the wing and the moments from the chordwise forces C produce compressive stresses on the leading edge portion of the wing. These compressive stresses will be additive in the upper flange of the front spar and the stringers adjacent to it. The positive

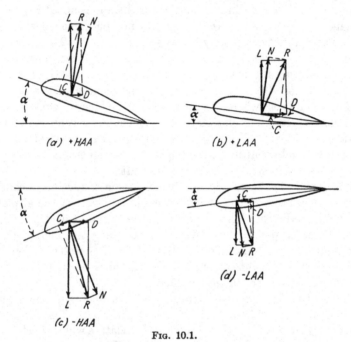

(a) +HAA   (b) +LAA

(c) -HAA   (d) -LAA

Fig. 10.1.

high angle of attack condition will therefore be critical for compressive stresses in the upper forward region of the wing cross section and for tensile stresses in the lower aft region of the wing cross section. For normal wings, in which the aerodynamic pitching moment coefficient is negative, the line of action of the resultant force R is farther forward on the wing in the positive high angle of attack condition than in any other possible flight attitude producing an upload on the wing. The upload on the horizontal tail in this condition will usually be larger than for any other positive flight attitude, since pitching accelerations are normally neglected, and the load on the horizontal tail must balance the moments of other aerodynamic forces about the center of gravity of the airplane.

In the positive low angle of attack condition, the wing has the smallest possible angle of attack at which the lift corresponding to the limit load factor may be developed. For a given lift on the wing, the angle of attack decreases as the indicated airspeed increases, and consequently the positive low angle of attack condition corresponds to the maximum indicated airspeed at which the airplane will be dived. This limit permissible diving speed depends on the type of airplane, but usually is specified as 1.2 to 1.5 times the maximum indicated speed in level flight, depending on the function of the airplane. Some specifications require that the terminal velocity of the airplane, the velocity obtained in a vertical dive sustained until the drag equals the airplane weight, be calculated and the limit diving speed determined as a function of the terminal velocity. Even fighter airplanes are seldom designed for a limit diving speed equal to the terminal velocity, since the terminal velocity of such airplanes is so great that difficult aerodynamic and structural problems are encountered. Airplanes are placarded so that the pilot will not exceed the limit diving speed.

In the positive low angle of attack condition, shown in Fig. 10.1(*b*), the chordwise force C is the largest force acting aft on the wing for any positive flight attitude. The wing bending moments in this condition produce the maximum compressive stresses on the upper rear spar flange and adjacent stringers and maximum tensile stresses on the lower front spar flange and adjacent stringers. In this condition the line of action of the resultant wing force R is farther aft than for any other positive flight condition. The moment of this force about the center of gravity of the airplane has the maximum negative (pitching) value, and consequently the download on the horizontal tail required to balance the moments of other aerodynamic forces will be larger than for any other positive flight condition.

The negative high angle of attack condition, shown in Fig. 10.1(*c*), occurs in intentional flight maneuvers in which the air loads on the wing are down, or when the airplane strikes sudden downdrafts when in level flight. The load factors for intentional negative flight attitudes are considerably smaller than for positive flight attitudes, because conventional aircraft engines cannot be operated under a negative load factor for a very long period of time and because the pilot is in the uncomfortable position of being suspended from his safety belt or harness. Gust load factors are also smaller for negative flight attitudes, since in level flight the weight of the airplane adds to the inertia forces for positive gusts but subtracts from the inertia forces for negative gusts.

In the negative high angle of attack condition, the wing is usually assumed to be at the negative stalling angle of attack for steady flow conditions. The assumption used in the positive high angle of attack condition, that the maximum lift coefficient momentarily exceeds that for steady flow, is seldom used because it is improbable that negative maneuvers will be entered suddenly. The wing bending moments in the negative high angle of attack condition produce the highest compressive stresses in the lower forward region of the wing cross section and the highest tensile stresses in the upper aft region of the wing cross section. The line of action of the resultant force R is farther aft than for any other negative flight attitude, and it will probably produce the greatest balancing upload on the horizontal tail for any negative flight attitude.

The negative low angle of attack condition, shown in Fig. 10.1(d), occurs at the limit diving speed of the airplane. This condition may occur in an intentional maneuver producing a negative load factor or in a negative gust condition. The aft load C is a maximum for any negative flight attitude, and the compressive bending stresses will have a maximum value in the lower aft region of the wing cross section, and the tensile bending stresses will have a maximum value in the upper forward region of the wing cross section. The resultant force R is farther forward than in any other flight attitude, and the download on the horizontal tail will probably be larger than in any other negative flight attitude.

In summary, one of the four basic symmetrical flight conditions will be critical for the design of almost every part of the airplane structure.

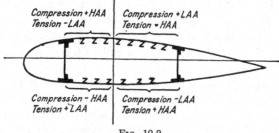

FIG. 10.2.

In the stress analysis of a conventional wing, it will be necessary to investigate each cross section for each of the four conditions. Each stringer or spar flange will then be designed for the maximum tension and the maximum compression obtained in any of the conditions. The probable critical conditions for each region of the cross section are shown in Fig. 10.2.

Some specifications require the investigation of additional conditions of medium high angle of attack and medium low angle of attack, which may be critical for stringers midway between the spars, but usually these conditions are not considered of sufficient importance to justify the additional work required for the analysis. The wing, of course, must be strong enough to resist loads at medium angles of attack, but will normally have adequate strength if it meets the requirements for the four limiting conditions.

For airplanes such as transport or cargo airplanes, in which the load may be placed in various positions in the gross weight condition, it is necessary to determine the balancing tail loads for the most forward and most rearward center of gravity positions at which the airplane may be flown at the gross weight. Each of the four flight conditions must be investigated for each extreme position of the center of gravity. For smaller airplanes in which the useful load cannot be shifted appreciably, there may be only one position of the center of gravity at the gross weight condition. To account for greater balancing tail loads which may occur for another location of the center of gravity, it may be possible to make some conservative assumption and still compute balancing tail loads for only one location.

The gust load factors on an airplane are greater when the airplane is flying at the minimum flying weight than they are at the gross weight condition. While this is seldom critical for the wings, since they have less weight to carry, it is critical for a structure such as the engine mount which carries the same weight at a higher load factor. It is therefore necessary to calculate gust load factors at the minimum weight at which the airplane will be flown.

For airplanes equipped with wing flaps, other high-lift devices, or dive brakes, additional flight loading conditions must be investigated for the flaps extended. These conditions are usually not critical for wing bending stresses, since the specified load factors are not large, but may be critical for wing torsion, shear in the rear spar, or down tail loads, since the negative pitching moments may be quite high. The aft portion of the wing, which forms the flap supporting structure, will be critical for the condition with flaps extended.

Unsymmetrical loading conditions and pitching acceleration conditions for civil airplanes are seldom of sufficient importance to justify extensive analysis. Conservative simplifying assumptions are usually specified by the licensing agency for use in the structural design of members which will be critical for these conditions. The additional structural weight required to meet conservative design assumptions is not sufficient to justify a more accurate analysis. Some military airplanes must perform violent evasive maneuvers such as snap rolls, abrupt rolling pull-outs, and abrupt pitching motions. The purchasing agency for such airplanes would specify the conditions which should be investigated. Such investigations would require the calculation of the mass moment of inertia of the airplane about the pitching, rolling, and yawing axes. The aerodynamic forces on the airplane would be calculated and would be set in equilibrium by inertia forces on the airplane.

10.3. Aerodynamic Data Required for Structural Analysis. Extensive aerodynamic information is required in order to investigate the performance, control, and stability of a proposed airplane. Only the information which is required for the structural analysis will be considered here, although this would normally be obtained as a part of a much more extensive program. The first aerodynamic data required for the structural analysis are the lift, drag, and pitching moment curves for the complete airplane with the horizontal tail removed, through the range of angles of attack from the negative stalling angle to the positive stalling angle. While these data can be calculated accurately for a wing with a conventional airfoil section, similar data for the combination of the wing and fuselage, or the wing, fuselage, and nacelles, are more difficult to calculate accurately from published data, because of the uncertain effects of the aerodynamic interference of various components. It is therefore desirable to obtain wind tunnel data on a model of the complete airplane less horizontal tail. It is often necessary, of course, to calculate these data from published information in order to obtain approximate air loads for preliminary design purposes.

Wind tunnel tests of a model of the complete airplane with the horizontal tail removed will provide values of the lift, drag, and pitching

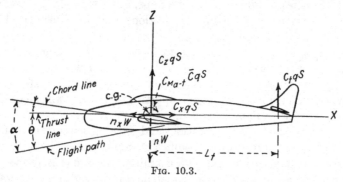

Fig. 10.3.

moment for all angles of attack. Components of the lift and drag forces with respect to airplane reference axes are then obtained. The airplane reference axes are usually chosen parallel and perpendicular to the thrust line, as shown by the x and z axes in Fig. 10.3. The force components are $C_z qS$ and $C_x qS$ along these axes, where $q = \rho V^2/2$ is the dynamic pressure and S is the wing area. The nondimensional force coefficients C_z and C_x are obtained by projecting the lift and drag coefficients for the airplane less horizontal tail along the reference axes by the following equations:

$$C_z = C_L \cos \theta + C_D \sin \theta \tag{10.1}$$
$$C_x = C_D \cos \theta - C_L \sin \theta \tag{10.2}$$

The angle θ is measured from the flight path to the x axis, as shown in Fig. 10.3, and is equal to the difference between the angle of attack α and the angle of wing incidence i.

The pitching moment about the airplane center of gravity is obtained from wind tunnel data and is $C_{m_{a-t}} \bar{c} q S$, where $C_{m_{a-t}}$ is the dimensionless pitching moment coefficient of the airplane less tail and $\bar{c}$ is the mean aerodynamic chord of the wing. The mean aerodynamic chord (MAC) is a wing reference chord which is usually calculated from the wing planform. If every airfoil section along the wing span had the same pitching moment coefficient c_m, the MAC is determined so that the total wing pitching moment is $c_m \bar{c} q S$. For a rectangular wing planform, the value of $\bar{c}$ (the MAC) is equal to the wing chord, and for a trapezoidal planform of the semiwing the value of $\bar{c}$ is equal to the chord at the centroid of the trapezoid. The MAC is actually an arbitrary length, and any reference length would be satisfactory if used consistently in all wind tunnel tests and calculations. For irregular shapes of planforms, the specifications of some procuring or licensing agencies require that the mean chord (wing area divided by wing span) be used as the reference chord.

10.4. Balancing Tail Loads. The balancing air load on the horizontal tail, $C_t q S$, is obtained from the assumption that there is no angular acceleration of the airplane. The moments of the forces shown in Fig. 10.3 about the center of gravity are therefore in equilibrium.

$$C_t q S L_t = C_{m_{a-t}} \bar{c} q S$$

or

$$C_t = \frac{\bar{c}}{L_t} C_{m_{a-t}} \tag{10.3}$$

where C_t is a dimensionless tail force coefficient expressed in terms of the wing area, and L_t is the distance from the airplane center of gravity to the resultant air load on the horizontal tail, as shown in Fig. 10.3. Since the pressure distribution on the horizontal tail is different for various attitudes of the airplane, the value of L_t theoretically varies for different loading conditions. This variation is not great, and it is customary to assume L_t constant, using a conservative forward position of the center of pressure on the horizontal tail. The total aerodynamic force on the airplane in the z direction, $C_{z_a} q S$, is equal to the sum of the force $C_z q S$ on the airplane less tail and the balancing tail load $C_t q S$.

$$C_{z_a} q S = C_z q S + C_t q S$$

or

$$C_{z_a} = C_z + C_t \tag{10.4}$$

For power-on flight conditions the moment of the propeller or jet thrust about the center of gravity of the airplane should also be considered. This will add another term to Eq. 10.3.

The aerodynamic coefficients may now be plotted against the angle of attack α, as shown in Fig. 10.4. If more than one position of the center of gravity is considered in the analysis, it is necessary to calculate the curves for $C_{m_{a-t}}$, C_t, and C_{z_a} for each center of gravity position. The right-hand portions of the solid curves shown in Fig. 10.4 represent

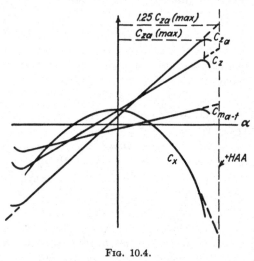

FIG. 10.4.

the aerodynamic characteristics after stalling of the wing. Since stalling reduces the air loads on the wing, these portions of the curves are not used. Instead, the curves are extrapolated as shown by the dotted lines in order to approximate the conditions of a sudden pull-up, in which high lift coefficients may exist for a short time. For the positive high angle of attack (+HAA) condition, the angle of attack corresponding to a force coefficient of 1.25 times the maximum value of C_{z_a} is used, and the curves are extrapolated to this value, as shown in Fig. 10.4.

10.5. Velocity—Load-factor Diagram. The various loading conditions for an airplane are usually represented on a graph of the limit load factor n plotted against the indicated airspeed V. This diagram is often called a V-n diagram, or a V-g diagram, since the load factor n is related to the acceleration of gravity g. In all such diagrams the indicated airspeed is used, since all air loads are proportional to q or $\rho V^2/2$. The value of q is the same for the air density ρ and the actual airspeed at altitude, as it is for the standard sea level density ρ_0 and the indicated airspeed, since the indicated airspeed is defined by this relationship.

The V-n diagram is therefore the same for all altitudes if indicated airspeeds are used. Where compressibility effects are considered, they depend on actual airspeed rather than indicated airspeed, and consequently are more pronounced at altitude. Compressibility effects will not be considered in this chapter.

The aerodynamic forces on an airplane are in equilibrium with the forces of gravity and inertia. If the airplane has no angular acceleration, both the inertia and gravity forces will be distributed in the same manner as the weight of various items of the airplane and will have resultants acting through the center of gravity of the airplane. It is convenient to combine the inertia and gravity forces as the product of a load factor n and the weight W as described in Chap. 3. The z component of the resultant gravity and inertia force is the force nW acting at the center of gravity of the airplane, as shown in Fig. 10.3. The load factor n is obtained from a summation of forces along the z axis.

$$C_{z_a}qS = nW$$

or

$$n = \frac{C_{z_a}\rho S V^2}{2W} \tag{10.5}$$

The maximum value of the normal force coefficient C_{z_a} may be obtained at various airplane speeds. For level flight at a unit load factor, the value of V corresponding to $C_{z_{a\max}}$ would be the stalling speed of the airplane. In accelerated flight, the maximum coefficient might be obtained at higher speeds. For $C_{z_{a\max}}$ to be obtained at twice the stalling speed, a load factor $n = 4$ would be developed, as shown by Eq. 10.5. For a force coefficient of 1.25 $C_{z_{a\max}}$, representing the highest angle of attack for which the wing is analyzed, the value of the load factor n, obtained from Eq. 10.5, may be plotted against the airplane velocity V, as shown by line OA in Fig. 10.5. This line OA represents a limiting condition, since it is possible to maneuver the airplane at speeds and load factors corresponding to points below or to the right of line OA, but it is impossible to maneuver at speeds and load factors corresponding to points above or to the left of line OA because this would represent angles of attack much higher than the stalling angle.

The line AC in Fig. 10.5 represents the limit maximum maneuvering load factor for which the airplane is designed. This load factor is determined from the specifications for which the airplane is designed, and the pilot must restrict maneuvers so that he does not exceed this load factor. At speeds below that corresponding to point A it is impossible for the pilot to exceed the limit load factor in any symmetrical maneuver, because the wing will stall at a lower load factor. For airspeeds between those corresponding to points A and C, it is not practical to design the airplane structure so that it could not be overstressed by violent maneuvers. Some types of airplanes may be designed so that the pilot would have to exert large forces on the controls in order to exceed the limit load factor.

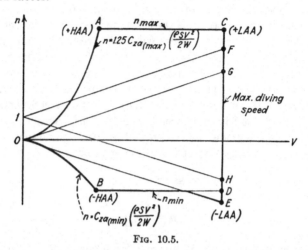

FIG. 10.5.

The line CD in Fig. 10.5 represents the limit permissible diving speed for the airplane. This value is usually specified as 1.2 to 1.5 times the maximum indicated airspeed in level flight. Line OB corresponds to line OA, except that the wing is at the negative stalling angle of attack, and the air load is down on the wing. The equation for line OB is obtained by substituting the maximum negative value of C_{z_a} into Eq. 10.5. Similarly, the line BD corresponds to the line AC, except that the limit load factor specified for negative maneuvers is considerably less than for positive maneuvers.

The airplane may therefore be maneuvered in such a manner that velocities and load factors corresponding to the coordinates of points within the area $OACDB$ may be obtained. The most severe structural loading conditions will be represented by the corners of the diagram, points A, B, C, and D. Points A and B represent the positive high angle of attack ($+$HAA) and negative high angle of attack ($-$HAA) conditions. Point C represents the positive low angle of attack ($+$LAA) condition in most cases, although the positive gust load condition, represented by point F, may occasionally be more severe. The negative low angle of attack ($-$LAA) condition is represented by point D or by the negative gust condition, point E, depending on which condition produces the greatest negative load factor. The method of obtaining the gust load factors, represented by points E and F, is explained in the following article.

10.6. Gust Load Factors. When an airplane is in level flight in calm air, the angle of attack α is measured from the wing chord line to the horizontal. If the airplane suddenly strikes an ascending air current

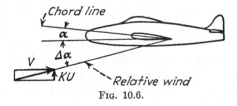

Fig. 10.6.

which has a vertical velocity KU, the angle of attack is increased by the angle $\Delta\alpha$, as shown in Fig. 10.6. The angle $\Delta\alpha$ is small, and the angle in radians may be considered as equal to its tangent.

$$\Delta\alpha = \frac{KU}{V} \tag{10.6}$$

The change in the airplane normal force coefficient C_{z_a}, resulting from a change in angle of attack $\Delta\alpha$, may be obtained from the curve of C_{z_a} vs. α of Fig. 10.4. This curve is approximately a straight line, and it has a slope m which may be considered constant.

$$m = \frac{\Delta C_{z_a}}{\Delta\alpha} \tag{10.7}$$

After striking the gust, the airplane normal force coefficient increases by an amount determined from Eqs. 10.6 and 10.7.

$$\Delta C_{z_a} = \frac{mKU}{V} \tag{10.8}$$

The increase in the airplane load factor Δn may be obtained by substituting the value of ΔC_{z_a} from Eq. 10.8 into Eq. 10.5.

$$\Delta n = \frac{\Delta C_{z_a}\rho S V^2}{2W}$$

or

$$\Delta n = \frac{\rho S m K U V}{2W} \tag{10.9}$$

where ρ is the standard sea level air density (0.002378 slug/ft³), S is the wing area in square feet, m is the slope of the curve of C_{z_a} vs. α in radians, KU is the effective gust velocity in feet per second, V is the indicated airspeed in feet per second, and W is the gross weight of the airplane in pounds.

For purposes of calculation it is more convenient to determine the slope m per degree and the airspeed V in miles per hour. Introducing the necessary constants in Eq. 10.9, there is obtained

$$\Delta n = 0.1 \frac{mKUV}{W/S} \qquad (10.10)$$

where m is the slope of C_{z_a} vs. α per degree, V is the indicated airspeed in miles per hour, and other terms correspond to those in Eq. 10.9.

When the airplane is in level flight, the load factor is unity before striking the gust. The change in load factor Δn, from Eq. 10.10, must be combined with the unit load factor in order to obtain the total gust load factor.

$$n = 1 \pm 0.1 \frac{mKUV}{W/S} \qquad (10.11)$$

Equation 10.11 may be plotted on the V-n diagram, as shown by the inclined straight lines through points F and H of Fig. 10.5. These lines represent load factors obtained when the airplane is in a horizontal attitude and strikes positive or negative gusts. Equation 10.10 is similarly plotted, as shown by the inclined lines through points G and E of Fig. 10.5. These lines represent load factors obtained when the airplane is in a vertical attitude and strikes positive or negative gusts in directions normal to the thrust line.

The gust load factor represented by point F of Fig. 10.5 may be more severe than the maneuvering load factor represented by point C. In the case shown, however, the maneuvering load factor is obviously greater and will represent the positive low angle of attack design condition. The negative gust load factor represented by point E is greater than the negative maneuvering load factor represented by point D and will determine the negative low angle of attack condition. It might seem that the gust load factors should be added to the maneuvering load factors, in order to provide for the possibility of the airplane striking a severe gust during a violent maneuver. While this condition is possible, it is improbable because the maneuvering load factors are under the pilot's control, and the pilot will restrict maneuvers in gusty weather. Both the maneuvering and gust load factors correspond to the most severe conditions expected during the life of the airplane, and there is little probability of a combined gust and maneuver producing a condition which would exceed the limit load factor for the design condition.

The "effective sharp-edged gust" velocity KU is the velocity of a theoretical gust which, if encountered instantaneously, would give the same load factor as the actual gust. Actually, it is impossible for the upward air velocity to change suddenly from zero to its maximum value. There is always a finite distance in which the air velocity changes gradually from zero to the maximum gust velocity, and a short interval of time is required for the airplane to move through this transition region. Most specifications require that the airplane be designed for a gust velocity U of 30 ft/sec with the gust effectiveness factor K of 0.8 to 1.2, depending on the wing loading W/S. Airplanes with higher wing loadings are usually faster and pass through the transition region from calm air to air with the maximum gust velocity in a shorter interval of time, and hence must be designed for larger values of K. The design values of KU are obtained from accelerometer readings for airplanes flying in turbulent air, and they represent the maximum effective gust velocities which will ever be encountered during the service life of the airplane. Some specifications require gust velocities of 50 ft/sec, with corresponding gust reduction factors K of about 0.6. Since the values of KU in this case are also about 30 ft/sec, the net effect is equivalent to a gust velocity U of 30 ft/sec with a K of 1.0. The actual maximum vertical air velocities probably exceed 50 ft/sec, but the transition is gradual, corresponding to the values of $K = 0.6$. High gust load factors exist for only a fraction of a second, and the airplane cannot move far in this interval of time.

In order to understand the effect of gusts, it is necessary to study the motion of the airplane after encountering a gust. If the gust is encountered instantaneously, the factor K is 1.0, and the effective gust velocity is U. The airplane is accelerated upward with an initial acceleration a_0 and attains a variable vertical velocity v. The gust angle of attack ($\Delta\alpha$ of Fig. 10.6) has a maximum value of U/V at the time the gust is encountered, $t = 0$, but this angle of attack is decreased to $(U - v)/V$ after the airplane attains an upward velocity. When the upward velocity v is equal to U, the relative wind is again horizontal, and the airplane is no longer accelerated. The variable vertical acceleration a is therefore

$$a = \frac{dv}{dt} = a_0 \frac{U - v}{U} \tag{10.12}$$

Separating the variables and integrating,

$$\int_0^v \frac{dv}{U - v} = \frac{a_0}{U} \int_0^t dt$$

or

$$\log \frac{U - v}{U} = -\frac{a_0 t}{U}$$

using the exponential form, and substituting the value of a from Eq. 10.12, the following expression for the acceleration a at time t is obtained:

$$\frac{\dot{a}}{a_0} = e^{-a_0 t / U} \tag{10.13}$$

As a numerical example, consider a gust velocity U of 30 ft/sec and an initial acceleration a_0 of $5g$, corresponding to a gust load factor of 6.0. Substituting these values into Eq. 10.13 and plotting a vs. t, the curve of Fig. 10.7 is obtained. The gust acceleration is seen to approach zero asymptotically in an infinite time, but it decreases greatly in the first tenth of a second. Thus, if the airplane has a forward speed of 500 ft/sec

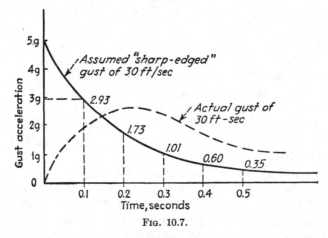

FIG. 10.7.

(340 mph) it would move forward only 50 ft in 1/10 sec. It seems logical to expect that atmospheric conditions are such that it is more than 50 ft from any region of calm air to a region in which the gust velocity is 30 ft/sec. The actual gust acceleration is probably represented more accurately by the dotted line of Fig. 10.7, which would indicate an effectiveness factor K of about 0.6. However, since airplane accelerometer readings have shown effective gust velocities KU of 30 ft/sec, the true conditions are probably represented by gust velocities U of more than 50 ft/sec with effectiveness factors K less than 0.6.

10.7. Numerical Example of Air-load Calculations. The procedures for obtaining air loads, which have been discussed in this chapter, will be illustrated by numerical calculations for a typical airplane. The airplane shown in Fig. 10.8 will be analyzed. The wing of this airplane is shown in Fig. 9.14, and the aerodynamic properties of the wing have been calculated in the various numerical examples in Chap. 9. The following conditions will be specified:

W = airplane gross weight = 8,000 lb

S = airplane wing area = 266 ft² W/S = 30 lb/ft²

KU = effective gust velocity = 34 ft/sec

V_d = design diving speed = 400 mph

n = limit or applied maneuver load factor = +6.00 and −3.00

The aerodynamic characteristics of the airplane with the horizontal tail removed have been obtained from corrected wind tunnel data and are given in Table 10.1. The moment coefficient C_M is about the center of gravity of the airplane and is expressed in terms of the wing area and

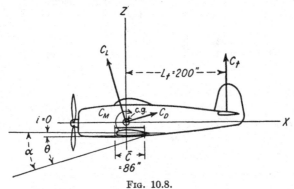

Fig. 10.8.

the mean aerodynamic chord for the wing, $\bar{c} = 86$ in. The stalling angle of the wing is 20°, corresponding to a maximum lift coefficient of 1.67. The aerodynamic data are extrapolated to the angle of attack of 26°. The negative stalling angle is −17°.

TABLE 10.1

$\alpha = \theta$, deg	C_L	C_D	C_M
26	2.132	0.324	0.0400
20	1.670	0.207	0.0350
15	1.285	0.131	0.0280
10	0.900	0.076	0.0185
5	0.515	0.040	0.0070
0	0.130	0.023	−0.0105
−5	−0.255	0.026	−0.0316
−10	−0.640	0.049	−0.0525
−15	−1.025	0.092	−0.0770
−17	−1.180	0.115	−0.0860

The force coefficients acting normal to the thrust line are calculated in Table 10.2. The components of C_L and C_D are calculated in col-

TABLE 10.2

θ, deg (1)	$C_D \sin \theta$ (2)	$C_L \cos \theta$ (3)	C_t (4)	C_{za} (5)
26	0.143	1.918	0.017	2.078
20	0.071	1.570	0.015	1.656
15	0.034	1.240	0.012	1.286
10	0.013	0.887	0.008	0.908
5	0.004	0.512	0.003	0.519
0	0	0.130	−0.004	0.126
−5	−0.002	−0.254	−0.013	−0.269
−10	−0.008	−0.630	−0.022	−0.660
−15	−0.024	−0.990	−0.032	−1.046
−17	−0.034	−1.130	−0.036	−1.200

umns (2) and (3). The tail-load coefficient C_t is calculated in column (4) by means of Eq. 10.3, or the values of C_M in Table 10.1 are multiplied by the ratio $\bar{c}/L_t = 86/200$. The final values of C_{za}, the normal force coefficient for the entire airplane, are obtained in column (5) as the sum of values from columns (2), (3), and (4).

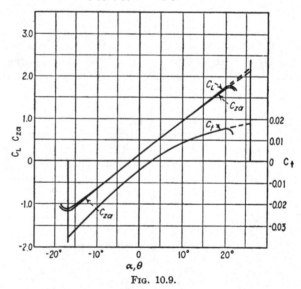

Fig. 10.9.

The curves for C_{za}, C_L, and C_t are plotted as functions of α or θ in Fig. 10.9. These curves are almost identical. The V-n diagram is constructed from the curve for C_{za}. For the portion OA of the curve of Fig. 10.5, the value of C_{za} is assumed to be 1 25 times the value at the stalling angle for the wing, or $C_{za} = 1.25 \times 1.656 = 2.070$. This corresponds with the angle of attack of 26°, within the accuracy of the data, and this angle will be assumed. The equation for the curve OA of Fig. 10.5 is found as follows:

$$n = 2.078 \frac{\rho S V^2}{2W} = 2.078 \times 0.00256 \frac{266}{8,000} V^2$$

$$n = 0.0001772 V^2$$

For point A, $n = 6$, and $V = 184$ mph. The equation for the curve OB of Fig. 10.5 is found as follows:

$$n = -1.200 \frac{\rho S V^2}{2W} = -0.0001024 V^2$$

For point B, $n = -3$, and $V = 172$ mph. The points C and D are plotted with the coordinates 400, 6, and 400, −3. The diagram is shown in Fig. 10.10.

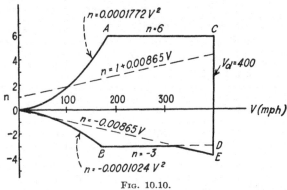

Fig. 10.10.

The gust load factors are now obtained from Eqs. 10.10 and 10.11. The slope m of the curve for C_{za} is obtained from the extreme coordinates, assuming a straightline variation.

$$m = \frac{2.078 + 1.200}{26 + 17} = 0.0763 \text{ per deg}$$

From Eq. 10.10,

$$\Delta n = 0.1 \frac{mKUV}{W/S} = \frac{0.1 \times 0.0763 \times 34}{30} V$$

$$\Delta n = 0.00865 V$$

For $V = 400$ mph, $\Delta n = 3.46$. The points F and E represent gust load factors of 4.46 and −3.46, respectively.

The wing bending moment will now be calculated for the positive high angle of attack condition, which is represented by point A on the V-n diagram. The wing has an angle of attack of 26° at an indicated airspeed of 184 mph. The spanwise distribution of the lift and drag coefficients for this wing were calculated in Chap. 9 for a unit value of the lift coefficient C_L. For the high angle of attack condition, the lift coefficient is 2.132, and the values of c_{la1} shown in Fig. 9.16 must be multiplied by 2.132. The values shown for the induced drag coefficient c_{di} in Fig. 9.20 must be multiplied by C_L^2 and must be added to the profile drag coefficient c_{d0} for the wing sections. The profile drag coefficient at any section will be assumed to have the following value:

$$c_{d0} = 0.010 + 0.12c_{di}$$

The wing drag coefficient at any section is therefore obtained from the following equation:

$$c_d = 0.010 + 1.12c_{di1}(C_L)^2 \tag{10.14}$$

where c_{di1} represents the values of c_{di} shown in Fig. 9.20 for a unit wing lift coefficient. For $C_L = 2.132$, the value of c_d is $0.010 + 5.09\,c_{di1}$, and it is calculated at various stations in columns (2), (3), and (4) of Table 10.3. The wing bending moment is usually calculated about the wing chord line, which in this case is parallel to the thrust line. The components of the drag coefficients in a direction perpendicular to the wing chord are obtained in column (5) by multiplying values in column (4) by sin 26°, or 0.439.

The lift coefficient at any cross section of the wing is obtained by multiplying the values of c_{la1} by 2.132. The components of the lift

TABLE 10.3

Sta. (1)	c_{di1} (2)	$5.09c_{di1}$ (3)	c_d (4)	$c_d \sin \alpha$ (5)	c_{la1} (6)	$c_l \cos \alpha$ (7)	c_n (8)
	Fig. 9.20		0.01+(3)	0.439(4)	Fig. 9.16	1.915(6)	(5) + (7)
240							
220	0.066	0.336	0.346	0.192	0.870	1.667	1.859
200	0.062	0.316	0.326	0.143	0.918	1.760	1.903
180	0.057	0.290	0.300	0.132	0.960	1.840	1.972
160	0.052	0.264	0.274	0.120	1.000	1.915	2.035
140	0.048	0.244	0.254	0.111	1.016	1.942	2.053
120	0.046	0.234	0.244	0.107	1.030	1.970	2.077
100	0.045	0.230	0.240	0.105	1.035	1.980	2.085
80	0.045	0.230	0.240	0.105	1.040	1.990	2.095
60	0.046	0.234	0.244	0.107	1.035	1.980	2.087
40	0.047	0.240	0.250	0.110	1.020	1.950	2.060
20	0.051	0.260	0.270	0.119	1.010	1.930	2.049
0	0.054	0.264	0.274	0.120	0.99	1.900	2.020

coefficients in the direction perpendicular to the wing chord are obtained by multiplying the coefficients by cos 26°. The values in column (7) of Table 10.3 are therefore obtained by multiplying values in column (6) by 2.132 cos 26°, or 1.915. The total force coefficients normal to the wing chord are obtained in column (8) as the sum of values in columns (5) and (7).

The final wing bending moments are calculated in Table 10.4. The load per inch along the span is calculated in column (4) by multiplying the normal force coefficients c_n by the wing chord and by the dynamic pressure $q/144$ in pounds per square inch of wing area. The dynamic pressure is calculated for the indicated airspeed of 184 mph. The final shear forces in pounds are given in column (6) and the final bending moments in 1,000 in-lb are given in column (8). The shears and bending moments are calculated by the method of Art. 5.3.

TABLE 10.4

Sta. (1)	c (2)	cc_n (3)	$\dfrac{cc_nq}{144}$ (4)	ΔV (5)	V (6)	$\dfrac{\Delta M}{1,000}$ (7)	$\dfrac{M}{1,000}$ (8)
240					0		0
				560		5	
220	50	93	56		560		5
				1,320		24	
200	66	126	76		1,880		29
				1,630		54	
180	73.2	144	87		3,510		83
				1,810		88	
160	76.4	156	94		5,310		171
				1,920		125	
140	79.6	163	98		7,230		296
				2,020		165	
120	82.8	172	104		9,250		461
				2,120		206	
100	86.0	179	108		11,370		667
				2,210		250	
80	89.2	187	113		13,580		917
				2,290		295	
60	92.4	193	116		15,870		1,212
				2,350		341	
40	95.6	197	119		18,220		1,553
				2,410		389	
20	98.8	202	122		20,630		1,942
				2,460		437	
0	102.0	206	124		23,090		2,379

If the wing reactions are at the side of the fuselage, station 20, the wing shear is zero inboard of station 20, and the wing bending moment is constant at 1,942,000 in-lb across the fuselage. It is preferable to calculate the shear at station 0 as shown in Table 10.4, in order to check the air loads. The total air load on the airplane normal to the thrust line is nW, or $6 \times 8,000 = 48,000$ lb. The shear at station 0 would therefore be 24,000 lb, if there were no air loads on the tail and fuselage. If one-half of the tail load and the vertical component of the fuselage drag are subtracted from 24,000 lb, the resulting value should check the shear force of 23,090 lb. In this case the wing shear is about 1.5 per cent too small as a result of errors in reading the curves for spanwise distribution, but this error is negligible in comparison with the assumptions involved.

PROBLEMS

10.1. Assume the center of gravity of the airplane analyzed in Art. 10.7 to be moved forward 8 in. without changing the external aerodynamic configuration. The distance L_t will now be 208 in., and the values of the aerodynamic pitching moments about the center of gravity will be $C_M - 8C_z/86$, where values of C_M are given in Table 10.1.

a. Calculate curves for C_t and C_{za}.

b. Construct a V-n diagram, using the conditions specified in Art. 10.7.

c. Calculate the wing bending-moment diagram for air loads normal to the wing chord for the positive high angle of attack condition.

d. Calculate the wing bending-moment diagram for chordwise air loads.

e. Calculate the air-load torsional moments about the wing leading edge, if the leading edge is straight and perpendicular to the plane of symmetry of the airplane. Assume the airfoil at any section to have an aerodynamic center at the quarter chord point and to have a negligible pitching moment about this point.

10.2. Calculate the wing normal and chordwise bending-moment diagrams for the positive low angle of attack condition for the airplane analyzed in Art. 10.7.

10.3. For a power-off condition, calculate the thrust load factors (parallel to the x axis) for the four main loading conditions of the airplane analyzed in Art. 10.7.

10.4. If the airplane wing of Art. 10.7 has a weight of 4.0 lb/ft², which is assumed distributed uniformly over the area, calculate the wing bending moments resulting from gravity and inertia forces normal to the wing chord for the four primary loading conditions.

REFERENCES FOR CHAPTER 10

1. Civil Air Regulations, Part 03, "Airplane Airworthiness—Normal, Utility, Acrobatic, and Restricted Purpose Categories," Civil Aeronautics Board, 1946.
2. "Civil Aeronautics Manual," Part 04, U.S. Department of Commerce, Civil Aeronautics Administration.

CHAPTER 11

MECHANICAL PROPERTIES OF AIRCRAFT MATERIALS

11.1. Stress-Strain Diagrams. The materials used in the airplane power plant and those used in the remaining structure, or airframe, are selected and proportioned by different criteria. Materials used in the airplane power plant are often subjected to high temperatures and must not "flow" or "creep" when stressed at the operating temperatures. The structural members of the power plant may also be subjected to repeated loadings of several millions of cycles during their service life. Such members must not be stressed beyond their "fatigue limit," or the minimum stress at which fatigue failure may occur for an infinitely large number of repetitions of loading. Other obvious considerations which are important in the selection of materials for engine parts are the friction and wearing properties of parts which are in contact and which have relative motion.

Airframe materials are subjected to much different loading conditions. The temperature variations are never sufficient to affect the properties of the materials. The maximum loads occur only a few times during the service life of the airplane, and fatigue failures of the type which are considered in the engine design need not be considered in the airframe design. The materials used for the airframe members are therefore different from those used for the power plant members. Since the design of the power plant is beyond the scope of this book, the future discussion will be confined to airframe materials, and references to aircraft materials or aircraft structures are understood to apply only to the airframe. Almost all the important structural properties of an airframe material are given by the stress-strain diagram for a simple tension or compression specimen of the material.

A tensile stress-strain diagram for a material is obtained by loading a specimen, such as that shown in Fig. 11.1(a), and measuring the total elongation δ in a gage length L for various values of the tension force P. For small loads, the elongation is uniform over the length L and the unit elongation or *strain e* is expressed in dimensionless form

$$e = \frac{\delta}{L} \tag{11.1}$$

where δ and L are both measured in the same units of length. The *stress* f is uniformly distributed over the cross-sectional area A and is obtained as follows:

$$f = \frac{P}{A} \qquad (11.2)$$

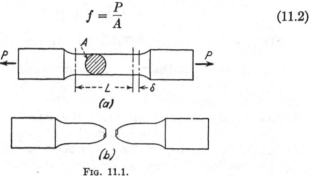

FIG. 11.1.

For common engineering units, the load P is in pounds, the area A is in square inches, and the stress f in pounds per square inch (psi). The stress-strain diagram for a material is obtained by plotting values of the stress f against corresponding values of the strain e, as shown in Fig. 11.2. For small values of the stress, the stress-strain curve is a straight line, as shown by the line OA of Fig. 11.2. The constant ratio of stress to strain for this portion of the curve is called the *modulus of elasticity E*, as defined in the following equation:

$$E = \frac{f}{e} \qquad (11.3)$$

where E has units of pounds per square inch (psi).

The straight-line portion OA of the stress-strain curve is independent of the size and dimensions of the test specimen. As the test specimen is loaded to higher stresses, it tends to "neck down" for a small portion of its length, and failure finally occurs at the reduced area of the necked-down portion, as shown in Fig. 11.1(*b*). After the specimen starts to neck down, the stress and strain are no longer uniformly distributed over the length L. The right-hand portion of the stress-strain curve therefore depends on the size and shape of the test specimen, and test data for this part of the curve do not apply accurately to the actual aircraft structural member. In order to obtain uniform comparative data on stress-strain curves, the American Society for Testing Materials (ASTM) has specified standard sizes of test specimens. Round tensile test specimens are approximately 0.5 in. in diameter and have a gage length of 2 in. Specimens of flat sheet usually have a width of 0.5 in. and a gage length of 2 or 4 in.

A material such as plain low-carbon steel, which is commonly used for bridge and building structures, has a stress-strain diagram such as that shown in Fig. 11.2(a). At point B, the elongation increases with no increase in load. This stress at this point is called the *yield point*, or *yield stress*, F_{ty} and is very easy to detect when testing such materials.

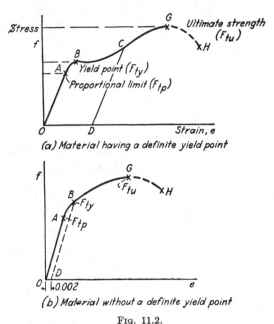

(a) Material having a definite yield point

(b) Material without a definite yield point

FIG. 11.2.

The stress at point A, at which the stress-strain curve first deviates from a straight line, is called the *proportional limit* F_{tp} and is much more difficult to measure when conducting a test. Specifications for structural steel are usually based on the yield stress rather than the proportional limit, because of the ease in obtaining this value.

Structural or plain low-carbon steel is seldom used for aircraft structures because of its low strength in proportion to its weight. Aircraft structures are usually made of aluminum alloys, high-carbon steels, alloy steels, or cold-worked steels. These materials do not have a definite yield point, but have a stress-strain diagram similar to that shown in Fig. 11.2(b). It is convenient to specify arbitrarily the yield stress for such materials as the stress at which a permanent strain of 0.002 is obtained. The point B of Fig. 11.2(b) represents this yield stress and is obtained by drawing the line BD parallel to OA through a point D representing zero stress and 0.002 strain, as shown. When the load is removed from a test specimen which has passed the proportional limit, the specimen does not return to its original length, but retains a permanent strain. For the material represented by Fig. 11.2(a), the load might be gradually removed at point C. The stress-strain curve would then follow line CD, parallel to OA, until at point D a permanent strain equal to OD is obtained for no stress. Upon a subsequent application of load, the stress-strain curve would follow lines DC and CG. Similarly, if the specimen represented by Fig. 11.2(b) were unloaded at point B, the stress-strain curve would follow line BD until a permanent strain of 0.002 is obtained for no stress.

It is customary to use the initial area A of the tension test specimen rather than the actual area of the necked-down specimen when computing the unit stress f. While the true stress, calculated from the reduced area, continues to increase until failure occurs, the apparent stress, calculated from the initial area, decreases as shown by the dotted lines GH of Fig. 11.2. The actual failure occurs at point H, but the maximum apparent stress, represented by point G, is the more important stress to use in design calculations. This value is defined as the *ultimate strength* F_{tu}. In the design of tension members for aircraft structures, it is accurate to use the initial area of the member and the apparent ultimate tensile strength F_{tu}. In using the stress-strain curve to calculate the ultimate bending strength of beams, as shown in a later chapter, the results will be slightly conservative because the beams do not neck down in the same manner as tension members.

Compressive stress-strain curves are more difficult to obtain than tensile stress-strain curves. Compressive specimens must be carefully supported laterally so that they cannot buckle as columns. The supports must be carefully adjusted and lubricated so that they do not resist any of the compression load applied to the specimen. Because of the importance of compressive stress-strain data in the design of columns, methods have been devised for obtaining these data, but the test procedure is not standardized as in the case of tensile stress-strain data.

11.2. Designations for Metal Materials. The properties of metals depend on their chemical composition and also on the methods used in heat-treating, aging, and mechanical working of the metal during manufacture. The material is usually designated by a number which indicates the chemical composition and by a further designation of the heat-treatment. For several years, all alloy steels were classified according to a system developed by the Society of Automotive Engineers (SAE). Under the SAE system of classification, the first two figures indicate the principal alloying element, and the approximate percentage of this element, or, in some cases, the principal alloying elements, as shown in the following table.

TABLE 11.1

Carbon steels	1xxx
Free-cutting carbon steels	11xx
Manganese steels	13xx
Nickel steels	2xxx
Nickel-chromium steels	3xxx
Molybdenum steels	40xx
Chromium-molybdenum steels	41xx
Nickel-chromium-molybdenum steels	43xx
Nickel-molybdenum steels	46xx and 48xx
Chromium steels	5xxx
Chromium-vanadium steels	6xxx
Silicon-manganese steels	9xxx

The last two figures, represented by x's in the table, indicate the approximate carbon content in hundredths of 1 per cent. Thus, 1025 steel represents a plain carbon steel with no alloy and with approximately 0.25 per cent carbon (actually 0.22 to 0.28 per cent). Similarly, 4130 steel represents a chromium-molybdenum steel containing about 0.30 per cent carbon. Almost all aircraft parts were formerly made of X–4130 steel, which was a chrome-molybdenum steel slightly different from 4130 steel.

During the Second World War, the supply of many alloying elements was greatly curtailed in the United States. It was therefore necessary for metallurgists to develop new alloy steels using smaller quantities of these alloying elements, so that the properties would be equivalent to the steels previously used. These steels were designated by a new numbering system, which does not indicate the chemical composition of the alloy. The new steels may permanently replace the X–4130 steel for some applications, and the new numbering system will be retained for these materials.

Aircraft tubing or plate of X–4130 steel is usually received in a "normalized" condition with an ultimate tensile strength of 90,000 to 100,000 psi. The material may then be heat-treated to any desired tensile strength up to 180,000 psi by the aircraft manufacturer. When steel is fabricated by welding, the welding heat reduces the strength near the weld to slightly less than the normalized value. Consequently large welded assemblies such as airplane fuselages, which are too large to heat-treat after welding, are usually fabricated from steel in the normalized condition. Smaller members such as landing gear members are usually heat-treated. The more highly heat-treated steels are less ductile and are harder to machine than steels with lower heat-treatment. Since many other alloy steels have similar properties to X–4130 steel, the structural designer may specify only the ultimate tensile strength of the steel and may use the stress-strain properties of X–4130 steel in his analysis.

Prior to the Second World War, all aluminum alloys manufactured in the United States were made by the Aluminum Company of America (ALCOA). The company designation of these materials was used in specifying aluminum alloys. During the war, other companies produced aluminum alloys, and used different numbers to designate their products. In this book, the ALCOA designations will be used, although other manufacturers produce alloys with similar properties under different designations.

The Aluminum Company of America designates wrought alloys by a number indicating the chemical composition,[1] followed by the letter S, as 2S, 17S, 24S, or 75S. In the case of heat-treatable alloys, this designation is followed by the letters O or T, which designate the condition of heat-treatment. Thus 61S–O is in the soft, or annealed, condition, and 61S–T is heat-treated. Some alloys require artificial aging under special conditions in order to develop their full strength. The type of heat-treating and aging is designated by additional numbers. Thus 61S–O has a strength of 22,000 psi in the annealed condition, 61S–T4 has a strength of 30,000 psi after heat-treatment and aging at room temperature, and 61S–T6 has a strength of 42,000 psi after further artificial aging, although all three have the same chemical composition.

Some wrought alloys are not heat-treatable, but may be hardened by cold working. The full hard temper is designated by H, and other tempers are designated by the fractional symbols $\frac{1}{4}$H, $\frac{1}{2}$H, and $\frac{3}{4}$H, indicating the fraction of the difference between soft and hard tempers. Thus the alloys 52S–$\frac{1}{2}$H and 52S–H have the same chemical composition but different amounts of cold work.

Cast alloys are designated by a number, followed by a designation of the condition of heat-treatment. Thus the alloys 195–T4, 195–T6, and 195–T62 have the same chemical composition but different heat-treatments. A minor variation in the chemical composition of an alloy is indicated by a letter preceding the alloy number; for example, alloy A214 differs from 214 by the addition of zinc to facilitate the pouring of castings in permanent molds. Aircraft rivets are usually made of alloy A17S–T, which is ductile enough to permit the forming of rivetheads, while alloy 17S–T rivets cannot be driven when the material is in the T condition.

In order to protect structural aluminum-alloy sheets from corrosion, they are often manufactured with a thin, integral coating of pure aluminum. This coating is softer and weaker than the alloy, but is lighter and more durable than a similar covering of paint. These sheet materials are designated as Alclad. An Alclad 24S–T sheet, for example, would be similar to a 24S–T sheet, except for the thin coating of pure aluminum. Alclad material is available only in sheets, and not in extrusions, forgings, or castings. The stress-strain curves for Alclad sheet have a slightly smaller slope above a stress of about 10,000 psi, where the aluminum surface material passes its yield stress. The two values of the modulus of elasticity of Alclad materials are designated as primary and secondary moduli. Thus, Alclad 24S-T sheet has a primary modulus of elasticity of 10,500,000 psi and a secondary modulus of elasticity of 9,500,000 psi.

The primary structural material in perhaps 90 per cent of all of the metal airplanes built before 1945 was 24S-T aluminum alloy. The sheet material was either Alclad material or was painted in some manner. Sheets which had only a single curvature, such as wing skins with developable surfaces, were formed from the sheet in the T condition. Members such as wing ribs, which were formed to shapes with a double curvature, were formed from 24S-O material and heat-treated after forming. Stiffeners and stringers were usually made of 24S-T extrusions. The extrusion process consists of forcing the plastic metal through a die of the desired cross section, in the same manner that toothpaste is forced from a tube. About 1945 the newer alloy 75S-T was first used commercially. The 75S-T alloy has an ultimate strength of about 77,000 psi for thin sheets, as compared with about 65,000 psi for 24S-T, and will probably be used extensively in future aircraft construction.

Aluminum-alloy forgings and castings are usually made of materials which are best adapted to the fabrication process. The alloy 14S-T is commonly used for forgings and has an ultimate tensile strength of 65,000 psi. Sand castings are often made from the alloy 220-T4 which has an ultimate strength of 42,000 psi.

11.3. Strength-Weight Comparisons of Materials. In the selection of aircraft structural materials, it is desirable to choose the material which will yield the desired strength for a minimum structural weight. It is, of course, necessary to consider the ease of fabricating the material into the desired structural members. The weights of tension materials required to resist a given load may be compared as the product of the inverse ratio of their ultimate strengths F and the ratio of their weights per cubic inch w.

$$\frac{W_1}{W_2} = \frac{w_1}{w_2} \frac{F_2}{F_1} \tag{11.4}$$

where W_1 and W_2 represent the weights of tension members of different materials resisting the same loads. Members resisting bending or compressive loads cannot be compared in this manner.

As a basis for comparison of materials resisting bending moments, a flat sheet of thickness t will be assumed to resist a bending moment M per unit width, as shown in Fig. 11.3(a). The bending stress will be

$$F = \frac{My}{I} = \frac{6M}{t^2}$$

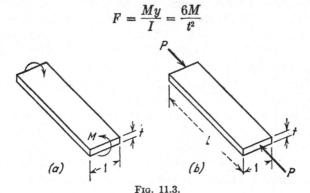

FIG. 11.3.

The required thickness will be

$$t = \sqrt{\frac{6M}{F}}$$

and the weight per square inch will be

$$W = tw = w\sqrt{\frac{6M}{F}} \tag{11.5}$$

The ratio of weights of sheets of two different materials resisting the same bending moment will be

$$\frac{W_1}{W_2} = \frac{w_1}{w_2}\sqrt{\frac{F_2}{F_1}} \tag{11.6}$$

Materials resisting buckling or compression loads will be compared by considering a flat sheet of thickness t which resists a compression load P per unit width, as shown in Fig. 11.3(b). Assuming the strip to buckle as a long column, the buckling load is approximately as follows:

$$P = \frac{\pi^2 EI}{L^2} = \frac{\pi^2 E t^3}{12L^2}$$

The required thickness of sheet is now obtained from this equation.

$$t = \sqrt[3]{\frac{12PL^2}{\pi^2 E}} \tag{11.7}$$

For two materials resisting similar loads, the values of P and L will be the same, and the ratio of weights will be as follows:

$$\frac{W_1}{W_2} = \frac{w_1}{w_2}\sqrt[3]{\frac{E_2}{E_1}} \tag{11.8}$$

Typical aircraft sheet materials are compared in Table 11.2, by means of Eqs. 11.4, 11.6, and 11.8. The weights of the various materials are compared with the aluminum alloy 24S–T. For the materials listed in Table 11.2, the ultimate tensile strengths F and the moduli of elasticity E are almost proportional to the densities w. The weight

TABLE 11.2

Sheet material (1)	F, psi (approx.) (2)	w, lb/in.^{3} (3)	E, 1,000 psi (4)	Ratio of weight to the weight of 24S–T aluminum alloy		
				Tension $\dfrac{w_1}{w_2}\dfrac{F_2}{F_1}$ (5)	Bending $\dfrac{w_1}{w_2}\sqrt{\dfrac{F_2}{F_1}}$ (6)	Compression buckling $\dfrac{w_1}{w_2}\sqrt[3]{\dfrac{E_2}{E_1}}$ (7)
Stainless steel.........	185,000	0.286	26,000	1.23	1.72	2.12
Aluminum alloy 24S–T .	66,000	0.100	10,500	1.00	1.00	1.00
Aluminum alloy 75S–T .	77,000	0.101	10,400	0.87	0.93	1.01
Magnesium alloy.......	40,000	0.065	6,500	1.07	0.83	0.77
Laminated plastic......	30,000	0.050	2,500	1.10	0.74	0.83
Spruce wood..........	9,400	0.0156	1,300	1.09	0.42	0.31

ratios for tension members, shown in column (5), do not vary greatly for the different materials. For members in bending, however, the lower density materials have a distinct advantage, as shown in column (6). Similarly, the lower density materials have an even greater advantage in compression buckling, as shown in column (7). Values of F vary with sheet thickness, and those shown are used only for comparison.

The computations of Table 11.2 indicate that the last three materials, having lower densities, are superior to the aluminum alloys. The aluminum alloys have been used more extensively because of practical fabrication considerations. The magnesium alloys are more subject to corrosion and are more difficult to form than aluminum alloys, although these problems may be solved by further development work. Various wood and plastic materials are available, with a wide variety of strengths and densities. In general, the strength and modulus of elasticity for such materials are approximately proportional to the density and are proportional to those for the typical materials listed in the last two lines of Table 11.2. These materials, which have a lower density than the metals, have much better characteristics for bending and compression members than the metals. Wood and plastic materials usually have a smaller unit elongation at the ultimate tensile strength and are termed *brittle* materials. Brittle materials are undesirable for structures which have numerous bolted connections, or numerous cutouts which produce local high stress concentrations. *Ductile* materials, which have a large unit elongation at the ultimate tensile strength, will yield slightly at points of high local stress and will thus relieve the stress, whereas brittle materials will fail under the same conditions. Wood and plastic materials have been successfully used for aircraft structures, but such structures must be carefully designed to eliminate high stress concentrations.

11.4. Sandwich Materials. The advantages of both high-density and low-density materials may be obtained by using sheets which have a layer of low-density material sandwiched between two layers of high-density material. This type of construction has long been used for plywood, in which the outer veneers are often of a stronger, more durable wood than the inner veneers. Techniques have recently been developed for cementing thin surfaces of metal or plastic materials to low-density core materials of balsa wood, foam rubber, or wood fiber board. While there are many production difficulties in the construction of complete aircraft components of these materials, it is possible that such materials will be used much more extensively in the future.

A cross section of a sheet of sandwich construction is shown in Fig. 11.4. It will be assumed that the low-density filler material, having

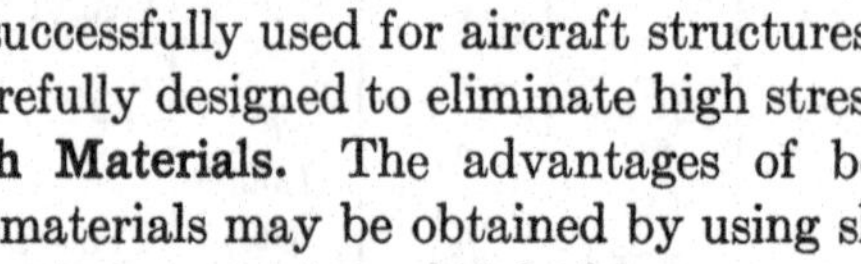

FIG. 11.4.

a weight w_1 lb/in.^{3}, resists no stress. It will also be assumed that the density ratio w_1/w_2 of the two materials is small and that the thickness of the face material kt is small compared with the thickness t. The bending moment resisted by a unit width of the sheet is approximately as follows:

$$M = Fkt^2$$

or

$$t = \sqrt{\frac{M}{kF}} \qquad (11.9)$$

The weight of 1 sq in. of the sheet material is approximately as follows:

$$W = w_1 t + 2w_2 kt \qquad (11.10)$$

Eliminating the term t from Eqs. 11.9 and 11.10, there is obtained

$$W = (w_1 + 2w_2k) \sqrt{\frac{M}{kF}} \qquad (11.11)$$

The value of k for the minimum weight may be obtained by differentiating Eq. 11.11 with respect to k and equating the derivative dW/dk to zero. The following value of k is obtained for a minimum weight:

$$k = \frac{w_1}{2w_2} \qquad (11.12)$$

Equation 11.12 shows that for a sandwich material resisting bending moment, the minimum weight is obtained when the two layers of the face material have approximately the same total weight as the layer of filler material.

For the most efficient sandwich material to resist buckling under compression loads, the face material should not have the same thickness ratio as that given by Eq. 11.12. The buckling load per unit width of sheet is approximately as follows:

$$P = \frac{\pi^2 EI}{L^2} = \frac{\pi^2 Ekt^3}{2L^2}$$

or

$$t = \sqrt[3]{\frac{2PL^2}{\pi^2 Ek}} \qquad (11.13)$$

Substituting the value of t into Eq. 11.10, differentiating the resulting expression, and again equating the derivative dW/dk to zero, the following equation is obtained:

$$k = \frac{w_1}{4w_2} \qquad (11.14)$$

Equation 11.14 shows that the total weight of the two layers of face material should be approximately one-half the weight of the core material, in order to obtain the lightest construction to resist compression buckling loads.

It is now possible to compare the weight of a sandwich material with the weight of a solid sheet of the corresponding face material, assuming that they both resist the same loads. A sandwich material designed to resist bending moments will have a total weight equal to twice the total weight of the face material, if the face and core materials have equal weights.

$$W_s = 4wkt \qquad (11.15)$$

where W_s is the weight of a unit area of the sandwich and w is the unit weight of the face material. From Eqs. 11.9 and 11.15,

$$W_s = 4wk \sqrt{\frac{M}{kF}} \qquad (11.16)$$

The weight W of a solid sheet was obtained in Eq. 11.5, and the ratio of the weight of a sheet of sandwich material to that of a solid sheet of the corresponding face material is obtained from Eqs. 11.5 and 11.16.

$$\frac{W_s}{W} = \frac{4wk\sqrt{M/(kF)}}{w\sqrt{6M/F}}$$

or

$$\frac{W_s}{W} = 1.63\sqrt{k} \tag{11.17}$$

Equation 11.17 applies only to sandwich materials in which the total weight of the face material is equal to the weight of the filler material. In order to compare the weights of sandwich materials with the materials studied in Table 11.2, a sandwich sheet consisting of 24S–T aluminum-alloy face material, and a filler material weighing 0.01 lb/in.³ will be considered. From Eq. 11.12,

$$k = \frac{w_1}{2w_2} = \frac{0.01}{2 \times 0.1} = 0.05$$

From Eq. 11.17,

$$\frac{W_s}{W} = 1.63\sqrt{0.05} = 0.37$$

This sandwich material consequently has only 37 per cent of the weight of a solid sheet resisting the same bending moment. The value of 0.37 is less than any of the other values in column (6) of Table 11.2.

The weight of a sandwich material designed to resist compression buckling loads may be obtained by a method similar to that used in the previous paragraph. The following equation is obtained, by assuming that the thickness of the face material satisfies Eq. 11.14, or that the total weight of the face material is one-half the weight of the filler material:

$$\frac{W_s}{W} = 2.45k^{2/3} \tag{11.18}$$

In the case of 24S–T aluminum-alloy face material and a filler material weighing 0.01 lb/in.³, the thickness ratio is obtained from Eq. 11.14 as follows:

$$k = \frac{w_1}{4w_2} = \frac{0.01}{4 \times 0.01} = 0.025$$

The weight ratio is now obtained from Eq. 11.18 as follows:

$$\frac{W_s}{W} = 2.45 \times (0.025)^{2/3} = 0.21$$

This sandwich material therefore has only 21 per cent of the weight of a corresponding solid sheet of 24S–T which is designed to resist the same buckling load

In the preceding discussion it was assumed that the proportions for sandwich materials were limited only by theoretical considerations. In actual structures, practical considerations will be much more important. The thickness of the face material, for example, would usually be greater than the theoretical value, because it might not be feasible to manufacture and form very thin sheets. The core material was assumed to support the face material sufficiently to develop the same unit stress that was developed in a solid sheet, whereas the actual low-density materials might not provide such support.

In conventional aircraft structures, the skin is supported by closely spaced stringers or stiffeners, and the actual bending or buckling resistance of the sheet itself may not be as important as assumed in the calculations of Table 11.2. In some cases the sheet is stiffened by beads, as shown in Fig. 11.5. The beads provide much greater resistance to bending or buckling than the flat sheet. Another proposed method of stiffening flat sheet is to provide integral ribs, as shown in Fig. 11.5. It is possible to manufacture sheet with such ribs, but at the present time such sheet is not commercially available.

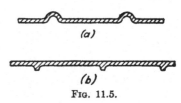

(a)

(b)

FIG. 11.5.

11.5. Typical Design Data for Materials. In the manufacture of materials, it is not possible to obtain exactly the same structural properties for all specimens of a material. In a large number of specimens of the same material, the ultimate strength may vary as much as 10 per cent. In the design of an aircraft structure it is therefore necessary to use stresses which are the minimum values that may be obtained in any specimen of the material. These values are termed the minimum guaranteed values of the manufacturer. The licensing and procuring agencies of the Air Force, Navy, and Civil Aeronautics Administration specify the minimum values which shall be used in the design of aircraft. These values are published by the Air Force-Navy-Civil Committee on Aircraft Design Criteria as *ANC*–5a, "Strength of Aircraft Metal Elements."[2] This publication will be referred to frequently as *ANC*–5. It is revised periodically to include properties of new materials which are developed and to revise properties for old materials for which manufacturing methods are frequently improved. The latest edition of *ANC*–5 should be consulted for all properties of aircraft materials. When tests are made on an actual structural member, it is necessary to obtain the tensile strength of the material in the member and to reduce the test data to correspond to that for material of the minimum guaranteed properties. This reduction of test data is necessary because of the possibility of some production airplanes containing weaker material than that in the test member.

TABLE 11.3.   PROPERTIES OF 24S–T* ALUMINUM-ALLOY SHEET WITH THICKNESS
LESS THAN 0.250 INCH (1,000 psi)

		Tension	
1	F_{tu}	Ultimate stress.............................	L 65, T 64
2	F_{ty}	Yield stress................................	L 48, T 42
3	F_{tp}	Proportional limit..........................	32
4	E	Modulus of elasticity.........	10,500
5		Elongation in 2 in., %	
		Compression	
6	F_{cu}	Ultimate (block) stress	
7	F_{cy}	Yield stress................................	L 40, T 45
8	F_{cp}	Proportional limit	
9	F_{co}	Column yield stress	
10	E_c	Modulus of elasticity.......................	10,700
		Shear	
11	F_{su}	Ultimate stress.............................	40
12	F_{st}	Torsional modulus of rupture	
13	F_{sp}	Proportional limit (torsion)	
14	G	Modulus of rigidity (torsion)..................	4,000
		Bearing	
15	F_{bru}	Ultimate stress† ($e/D = 1.5$).................	98
	F_{bru}	Ultimate stress ($e/D = 2.0$)....................	124
	F_{bry}	Yield stress ($e/D = 1.5$)......................	69
	F_{bry}	Yield stress ($e/D = 2.0$)......................	79

* This table applies only to material supplied by the mill in the "T" temper, and so used without
a reheat-treatment.
† D = hole diameter: e = edge distance measured from the hole centerline in the direction of
stressing. Use value of e/D = 2.0 for all larger values of edge distance.

Typical stress-strain curves for 24S–T aluminum-alloy sheet are shown in Fig. 11.6. The four curves shown represent data from tests of specimens loaded parallel to the direction of rolling of the sheet, and perpendicular to the direction of rolling, often termed longitudinal and transverse properties, and represent tests of tension and compression specimens. The tabular data of minimum guaranteed strength values for this material are shown in Table 11.3, taken from *ANC–5*. For this material, the values of the yield stresses are given in *ANC–5* for all four methods of loading, with L designating longitudinal properties and T

designating transverse properties. It will be noted that the values shown in Table 11.3 apply only to material heat-treated by the manufacturer. The airplane manufacturer, or user, must often form parts of soft material (24S–O) and then heat-treat the part after forming. Since he may not control the heat-treatment as accurately as the material manufacturer, he must use lower allowable stresses for such parts. A different table of allowable stresses for such parts is specified in *ANC–5*.

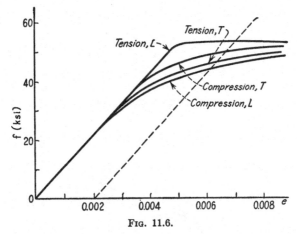

Fig. 11.6.

The values in lines 1 to 10 inclusive of Table 11.3 are obtained from typical stress-strain curves of tension or compression specimens. The remaining values in lines 11 to 15 inclusive will be considered in more detail in a later chapter.

11.6. Equations for Dimensionless Stress-Strain Curves. In the design of aircraft structural members it is necessary to consider the properties of the stress-strain curve at stresses higher than the elastic limit. In other types of structural and machine design, it is customary to consider only stresses below the elastic limit, but weight considerations are so important in aircraft design that it is necessary to calculate the ultimate strength of each member and to provide the same factor of safety against failure for each part of the entire structure. The ultimate bending or compressive strengths of many members are difficult to calculate, and it is necessary to obtain information from destruction tests of complete members. In order to apply the results on tests of members of one material to similar members of another material, it is desirable to obtain an analytical expression for the stress-strain diagrams of various materials.

Ramberg and Osgood[3] have developed a method of expressing any stress-strain curve in terms of the modulus of elasticity E, a stress f_1, which is approximately equal to the yield stress, and a shape factor n. The equation for the stress-strain diagram is

$$\epsilon = \sigma + \tfrac{3}{7}\sigma^n \tag{11.19}$$

where ϵ and σ are dimensionless terms defined as follows:

$$\epsilon = \frac{Ee}{f_1} \tag{11.20}$$

and

$$\sigma = \frac{f}{f_1} \tag{11.21}$$

The curves expressed by Eq. 11.19 are plotted in Fig. 11.7 for various values of n. A material such as mild steel, in which the stress remains almost constant above the yield point, is represented by the curve for $n = \infty$. Other materials, with various types of stress-strain diagrams,

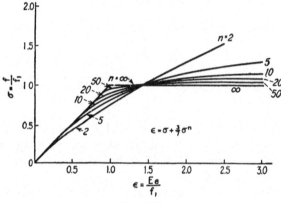

FIG. 11.7.

may be represented by the curves for other values of n. In order to represent the stress-strain diagrams for all materials by the single equation, it is necessary to use the reference stress value of f_1 rather than the yield stress. The value of f_1 is obtained, as shown in Fig. 11.8, by drawing the line $f = 0.7Ee$ from the origin to the stress-strain curve and obtaining the stress coordinate f_1 of this point of intersection. The stress f_1 is approximately equal to the yield stress for typical aircraft materials. The value of n may be determined so that Eq. 11.19 fits the experimental stress-strain curve in the desired region. Ramberg and Osgood show that for most materials the value of n may be accurately determined from the stress f_1 and a similar stress f_2 on the line $f = 0.85Ee$.

11.7. Safety Factors and Margins of Safety. The load factors for which aircraft structures must be designed were discussed in Chaps. 3

and 10. It was shown that the combined inertia and gravity forces acting upon any element of weight in the airplane could be conveniently represented as the product of a load factor and the weight. The load factors were shown to depend on the conditions of acceleration. The *limit* or *applied load factors* represented the worst possible loading conditions the airplane was expected to encounter during its service life.

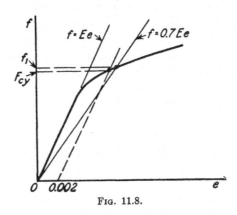

Fig. 11.8.

The *design* or *ultimate load factors* were obtained by multiplying the limit or applied load factors by a *factor of safety* of 1.5. All structural members in the airplane must be designed in such a manner that they do not attain any appreciable permanent deformation at the limit or applied load factors. They must also be designed so that they do not fail at the design or ultimate load factors. It is therefore necessary that all stresses be below the yield point for conditions of applied or limit load and also that all stresses be below the ultimate strength for conditions of design or ultimate load.

For most aircraft materials, the yield point is greater than two-thirds of the ultimate strength, and tension members are designed from the ultimate strength requirement. For some materials, however, as in the case of the 24S-T sheet described in Table 11.3, the yield strength is less than two-thirds the ultimate strength for the transverse loading, and tension members are designed from the limit or applied load conditions. It is obvious, from a simple consideration of the yield and ultimate stresses, which condition will be critical in the design of tension members.

For members stressed beyond the elastic limit in bending or compression, the true stress is no longer proportional to the external loads on the member. It may be necessary to investigate such members more carefully in order to determine whether the yield stresses or the ultimate stresses will be critical in the design of the members. In many types of bending and compression members, it is customary to use fictitious ultimate stresses which are proportional to the external loads. These methods will be treated in later chapters.

The calculated stresses in members are commonly designated by the small letter f with subscripts designating tension, compression, shear, or bending, as f_t, f_c, f_s, or f_b. The allowable stresses are designated by the capital letter F with similar subscripts representing tension, compression, shear, or bending, as F_t, F_c, F_s, or F_b, and sometimes additional subscripts representing ultimate or yield stresses, as F_{tu} and F_{ty}. While an attempt is made to design all members in such a way that the calculated stress f is equal to the allowable stress F, the standard commercial sizes of various materials and parts are usually such that the calculated stresses will be slightly smaller than the allowable stresses. In such cases, calculations are always made for the *margin of safety* (abbreviated MS), which is the ratio of the excess strength to the required strength. The margin of safety is calculated as follows:

$$MS = \frac{F - f}{f}$$

or

$$MS = \frac{F}{f} - 1 \tag{11.22}$$

The margin of safety should never be negative, but should be zero or a small positive value. The margin of safety for each member should be clearly indicated in the stress analysis. If it is later desired to increase the load on any member, or to decrease the size of the member, the MS gives an immediate indication of the permissible load increase.

In the case of members in bending or compression, where the stress is not proportional to the loads, Eq. 11.22 does not apply if true stresses are used. It is customary to work with fictitious stresses which are proportional to the loads, and consequently Eq. 11.22 is usually used. Care must be observed, however, that the stresses used are proportional to the loads.

Additional safety factors must be used for some parts of aircraft structures. Failures are more likely to occur at the end connections of members than in the members themselves because of local stress concentrations at the connections, slight eccentricities of the connections, or more severe vibration conditions. For this reason an additional safety factor is used in the design of all connections or fittings. This factor is called a *fitting factor* and is usually specified as 1.20 for civil airplanes and 1.15 for military airplanes. It is necessary to use the fitting factor for all bolted and welded joints and for the structure immediately adjacent to the joints. It is not necessary to use the fitting factor for continuous lines of rivets at sheet-metal joints.

In bolted joints it is necessary to drill the bolt holes slightly larger than the bolts in order to insert the bolts easily. Where such bolts must resist shock or vibration loading, as in landing-gear members, the bolts tend to hammer back and forth in the holes. This hammering action may enlarge the bolt holes and may eventually cause failure of the members if the bearing stresses are high. Additional safety factors called *bearing factors* are used in calculating the bearing stresses of the bolts against the holes. The bearing factors are 2.0 or more depending on the loading conditions. These bearing factors must be multiplied by the safety factor of 1.5, but need not be multiplied by the fitting factor of 1.15 or 1.2.

The allowable stresses for castings are specified in *ANC-5*, as the minimum stresses obtained from test specimens. The actual cast members, however, occasionally contain air pockets or other hidden imperfections which do not occur in the test specimens. It is therefore necessary to use an additional factor of safety, usually 2.00, for castings. In some cases the factor of safety depends on the method of inspection of the casting. If all castings are x-rayed, the additional safety factor may be smaller than for castings inspected by other methods.

REFERENCES FOR CHAPTER 11

1. "Alcoa Aluminum and Its Alloys," Aluminum Company of America, 1946.
2. *ANC-5a*, "Strength of Metal Aircraft Elements," Subcommittee on Air Force-Navy-Civil Aircraft Design Criteria of the Munitions Board Aircraft Committee, May, 1949.
3. RAMBERG, W., and W. R. OSGOOD: Description of Stress-Strain Curves by Three Parameters, *NACA TN* 902, 1943

CHAPTER 12

JOINTS AND FITTINGS

12.1. Introduction. A complete airplane structure is manufactured from many parts. These parts are made from sheets, extruded sections, forgings, castings, tubes, or machined shapes, which must be joined together to form subassemblies. The subassemblies must then be joined together to form larger assemblies and then finally assembled into a completed airplane. Many parts of the completed airplane must be arranged so that they can be disassembled for shipping, inspection, repair, or replacement, and are usually joined by bolts. In order to facilitate the assembly and disassembly of the airplane, it is desirable for such bolted connections to contain as few bolts as possible. For example, a semi-monocoque metal wing usually resists bending stresses in numerous stringers and sheet elements distributed around the periphery of the wing cross section. The wing cannot be made as one continuous riveted assembly from tip to tip, but must usually be spliced at two or more cross sections. These splices are often designed so that four bolts transfer all the loads across the splice. These bolts connect members called *fittings*, which are designed to resist the high concentrated loads and to transfer them to the spars, from which the loads are distributed to the sheet and stringers. The entire structure for transferring the distributed loads from the sheet and stringers outboard of the splice into a concentrated load at the fitting and then distributing this load to the sheet and stringers inboard of the splice is considerably heavier than the continuous structure which would be required if there were no splice.

Many uncertainties exist concerning the stress distribution in fittings. Manufacturing tolerances are such that bolts never fit the holes perfectly, and small variations in dimensions may affect the stress distribution. An additional margin of safety of 15 per cent for military airplanes and 20 per cent for civil airplanes is used in the design of fittings. A common procedure is to multiply the design loads by a *fitting factor* of 1.15 or 1.20 before calculating the stresses. This fitting factor must be used in designing the entire fitting, including the riveted, bolted, or welded joint attaching the fitting to the structural members. The fitting factor need not be used in designing a continuous riveted joint, although the stress distribution in such a joint is also indeterminate.

The allowable stresses for rivets are rather conservative to account for such uncertainty.

12.2. Bolted or Riveted Joints. Bolted or riveted joints must be investigated for four types of failure: bolt or rivet shear as shown in

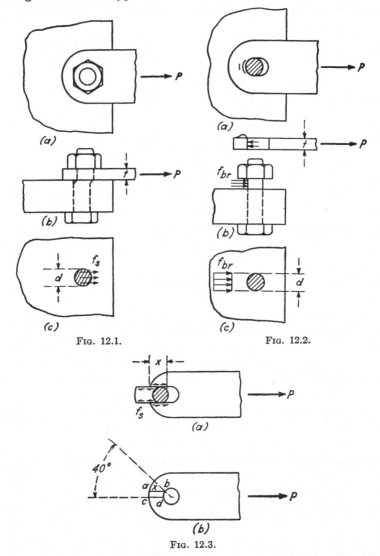

FIG. 12.1.

FIG. 12.2.

FIG. 12.3.

Fig. 12.1, bearing as shown in Fig. 12.2, tear-out as shown in Fig. 12.3, and tension as shown in Fig. 12.4. The true stress distribution is rather complex and will be discussed later. It is customary to assume a simple uniform or average stress distribution in all cases, and the allowable stresses which are used in design are also average stresses which have been obtained from tests of similar joints. It is therefore possible to predict the strength of a joint with an accuracy of a few per cent, although the true maximum stresses may be three or four times as much as the average stresses.

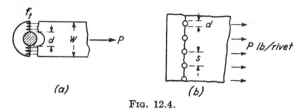

Fig. 12.4.

The average stress for any of the four types of failure is

$$f = \frac{P}{A} \tag{12.1}$$

where f is the average stress, P is the load, and A is the area of the cross section on which failure may occur. The margin of safety (MS) is found from the equation

$$MS = \frac{F}{f} - 1 \tag{12.2}$$

where F is the allowable stress and the stress f is obtained from the load P, which includes the safety factor of 1.5 and usually includes the fitting factor of 1.15 or 1.2. If this fitting factor is included in the stress f, the margin of safety should be zero or a small positive value. Some designers may not include the fitting factor in the stress f, and they must therefore show a minimum margin of safety of 0.15 or 0.20 from Eq. 12.2. In any analysis it should be clearly stated whether the fitting factor is included in the margin of safety. The upper-case letter F always represents an allowable stress, and the lower-case letter f represents a calculated stress. A subscript is used to designate the type of stress, that is, F_s or f_s are shearing stresses, F_{br} or f_{br} are bearing stresses, F_t or f_t are tensile stresses, F_c or f_c are compression stresses, and F_b or f_b are bending stresses.

For investigating the shear strength of a bolt or rivet, the area to be used in Eq. 12.1 is the area of the bolt or rivet cross section, or $A = \pi d^2/4$, where d is the bolt or rivet diameter. The shearing stress is then obtained from Eq. 12.1 as follows:

$$f_s = \frac{4P}{\pi d^2} \tag{12.3}$$

In Figs. 12.1 to 12.4 inclusive, the bolt is shown to be in single shear and one plate is assumed to be rigid in bending, so that the forces on the thin plate are in static equilibrium. The bolt would therefore resist a bending moment $Pt/2$ at the cross section subjected to shear. It will be shown later that this bending moment on the bolt does not exist in most actual single shear connections, and it is customary to disregard this bolt bending moment when the two plates are clamped together by the bolt. In cases where a washer or a filler plate is used between the two stressed plates, the bolt bending must be considered.

The bearing failure of a riveted or bolted joint usually consists of an elongation of the hole in the plate, as shown in Fig. 12.2(a). The allowable bearing stress usually depends on the permissible elongation of the hole. For riveted joints, the allowable bearing stress is determined by arbitrarily specifying a hole elongation equal to a certain percentage of the rivet diameter. The bearing failure is somewhat similar to the tear-out failure shown in Fig. 12.3, and the allowable bearing stress for rivets is reduced when the rivets are too close to the edge of the sheet. The bearing stress is assumed to be uniformly distributed over an area $A = td$, as shown in Fig. 12.2. Substituting this area into Eq. 12.1, the equation for the assumed average bearing stress is obtained.

$$f_{br} = \frac{P}{td} \tag{12.4}$$

Bolt holes must always be slightly larger than the bolt diameter. If the joint is subjected to shock or vibrational loading, as in a landing-gear member, there is a much greater tendency for a bolt hole to elongate than when the joint resists only static loading. Similarly, when relative rotation of the two parts occurs, the bolt hole is more likely to become enlarged. In such cases the bearing stress must be low in order to prevent frequent replacement of the bolt or the hole bushing. The licensing agencies therefore specify that a *bearing factor* of 2.0 or more be used in obtaining the bearing stress when a bolted joint is subject to relative rotation under design loads or to shock or vibration loads. This bearing factor is used in place of the fitting factor and not in addition to the fitting factor.

A tear-out failure of a bolt or rivet hole is shown in Fig. 12.3. The plate material fails in shear on the areas, $A = 2xt$, and the tear-out stress is found from the following equation:

$$f_s = \frac{P}{2xt} \tag{12.5}$$

The distance x is obtained as the length ab in Fig. 12.3, but it is conservative to use the length cd, which is easier to calculate. It is seldom necessary to calculate the tear-out stresses for riveted joints in a sheet of the type shown in Fig. 12.4. From practical considerations, it is desirable to keep the distance from the center of the rivets to the edge of the sheet equal to at least two diameters of the rivet, and there is no danger of tear-out with this edge distance.

A riveted or bolted joint must be investigated for a possible tension failure through the bolt or rivet holes, as shown in Fig. 12.4. The tension stress is assumed to be uniformly distributed over the area $A = (w - d)t$ for the bolted fitting shown in Fig. 12.4(a).

$$f_t = \frac{P}{(w - d)t} \tag{12.6}$$

For the riveted joint shown in Fig. 12.4(b), the tension stress is found from the equation

$$f_t = \frac{P}{(s - d)t} \tag{12.7}$$

where P is the load per rivet, s is the rivet spacing, d the rivet diameter, and t the sheet thickness.

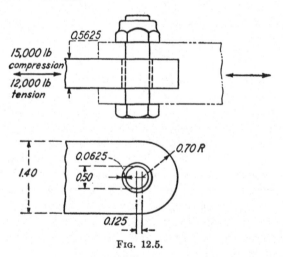

Fig. 12.5.

Example. The fitting shown in Fig. 12.5 is made of a 14S–T forging, for which $F_t = 65,000$ psi, $F_s = 39,000$ psi, and $F_{br} = 98,000$ psi. The bolt and bushing are made of steel for which $F_t = 125,000$ psi, $F_s = 75,000$ psi, and $F_{br} = 175,000$ psi. The fitting resists limit or applied loads of 15,000 lb compression and 12,000 lb tension. A fitting factor of 1.2 and a bearing factor of 2.00 are used. Find the margins of safety for the fitting, for various types of failure.

Solution. The design or ultimate fitting loads are obtained by multiplying the loads given by a safety factor of 1.5 and a fitting factor of 1.2.

Design fitting loads:

$$15,000 \times 1.5 \times 1.2 = 27,000 \text{ lb} \qquad \text{compression}$$
$$12,000 \times 1.5 \times 1.2 = 21,600 \text{ lb} \qquad \text{tension}$$

The bearing of the bolt on the bushing is investigated by using the bearing factor of 2.00 in place of the fitting factor of 1.2.

Design bearing loads:

$$15,000 \times 1.5 \times 2.0 = 45,000 \text{ lb} \qquad \text{compression}$$
$$12,000 \times 1.5 \times 2.0 = 36,000 \text{ lb} \qquad \text{tension}$$

The bolt is in double shear; therefore one-half of the 27,000-lb load must be resisted by each cross section of the bolt in shear. From Eqs. 12.3 and 12.2,

$$f_s = \frac{4 \times 13,500}{\pi (0.5)^2} = 68,600 \text{ psi}$$

and

$$\text{MS} = \frac{75,000}{68,600} - 1 = \underline{0.09} \qquad \text{includes fitting factor}$$

The bearing stress will also be calculated from the larger of the loads for tension and compression. From Eqs. 12.4 and 12.2, the bearing of the bolt on the bushing is investigated.

$$f_{br} = \frac{45,000}{0.5625 \times 0.5} = 160,000 \text{ psi}$$

$$\text{MS} = \frac{175,000}{160,000} - 1 = \underline{0.09} \qquad \text{includes bearing factor}$$

For bearing of the bushing on the forging, it is necessary to use only the fitting factor, because the bushing fits tightly in the hole.

$$f_{br} = \frac{27,000}{0.5625 \times 0.625} = 76,800 \text{ psi}$$

$$\text{MS} = \frac{98,000}{76,800} - 1 = \underline{0.29} \qquad \text{includes fitting factor}$$

The tear-out of the bolt hole is first investigated by assuming that the length x shown in Fig. 12.3 is equal to the length cd rather than the length ab.

$$cd = 0.70 + 0.125 - 0.3125 = 0.5125 \text{ in.}$$

The tension load must be used in calculating the tear-out stress, since the compression load produces no stress on this cross section. From Eqs. 12.5 and 12.2,

$$f_s = \frac{21,600}{2 \times 0.5125 \times 0.5625} = 37,400 \text{ psi}$$

and

$$\text{MS} = \frac{39,000}{37,400} - 1 = \underline{0.04} \qquad \text{includes fitting factor}$$

A more accurate value of the distance x may be calculated from the equation given in Fig. 12.6. The term in brackets may be plotted for various values of r/R in order to reduce the labor of the calculations, where it is necessary to

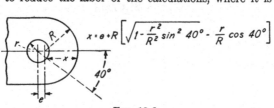

$$x = e + R\left[\sqrt{1 - \frac{r^2}{R^2} \sin^2 40°} - \frac{r}{R} \cos 40°\right]$$

FIG. 12.6.

repeat such calculations frequently. For $R = 0.7$, $r = 0.3125$, and $B = 0.125$, $x = 0.562$.

$$f_s = \frac{21,600}{2 \times 0.562 \times 0.5625} = 34,200 \text{ psi}$$

and

$$MS = \frac{39,000}{34,200} - 1 = \underline{0.14} \qquad \text{includes fitting factor}$$

The tension stress through the bolt hole is obtained from Eq. 12.6.

$$f_s = \frac{21,600}{(1.4 - 0.625)0.5625} = 49,600$$

$$MS = \frac{65,000}{49,600} - 1 = \underline{0.13} \qquad \text{includes fitting factor}$$

12.3. Standard Parts. Since there are many bolts and rivets used in every airplane, it is obviously desirable to standardize on certain sizes, shapes, and materials for these parts. Such standardization not only reduces the original cost of the airplane but also facilitates repair and maintenance. The Army and Navy have adopted standards for hundreds of commonly used aircraft parts such as bolts, rivets, control pulleys, or hydraulic system fillings. The drawings and specifications for these parts are given in an AN Standard Aircraft Parts book. A designer simply lists the AN number on a drawing, rather than drawing and dimensioning a standard part. Standard bolts vary in diameter by increments of $\frac{1}{16}$ in., and standard numbers from AN3 to AN16 indicate bolts whose diameters in sixteenths of an inch are equal to the standard number. These bolts are made from steel which is heat-treated to an ultimate tensile strength of 125,000 psi and which has a shearing strength of 75,000 psi. The shearing strength of standard bolts may be calculated by multiplying the cross-sectional area by the allowable shearing stress. The tensile strength may similarly be calculated by multiplying the bolt area at the root of the thread by the allowable tensile stress. These allowable bolt strengths are tabulated in Table 12.1.

TABLE 12.1. ULTIMATE STRENGTH OF AN BOLTS

AN No.	Nom. diam.	Steel		24S–T Aluminum Alloy	
		Shear $F_s = 75,000$	Tension $F_t = 125,000$	Shear $F_s = 35,000$	Tension $F_t = 62,000$
3	3/16	2,125	2,210	990	1,100
4	1/4	3,680	4,080	1,715	2,030
5	5/16	5,750	6,500	2,685	3,220
6	3/8	8,280	10,100	3,870	5,020
7	7/16	11,250	13,600	5,250	6,750
8	1/2	14,700	18,500	6,850	9,180
9	9/16	18,700	23,600	8,700	11,700
10	5/8	23,000	30,100	10,750	14,900
12	3/4	33,150	44,000	15,500	21,800
14	7/8	45,050	60,000	21,050	29,800
16	1	58,900	80,700	27,500	40,000

The lengths of standard bolts vary by increments of 1/8 in. If the thickness of the parts being joined is slightly greater than the unthreaded length of the bolt, the next length would be used, with washers under the nut, so that the bolt threads will not bear on the structural part. A bolt is specified on a drawing as AN7–5, which indicates a bolt 7/16 in. in diameter and 5/8 in. in length. Standard bolts have holes drilled through the end of the threaded shank, so that a cotter pin may be inserted to safety the nut. Lock washers are not considered satisfactory for safetying aircraft nuts. Self-locking stop nuts of certain types are approved by the licensing or procuring agencies. When a stop nut is used, the bolt is not drilled for a cotter pin, and the standard number is followed by the letter A, as AN7–5A. Aluminum-alloy bolts are seldom used because of the danger of stripping the threads during installation, but 24S–T bolts are accepted as standard parts with the same shapes and sizes of steel bolts. The material is designated by the letters DD in the standard number, as AN7DD5, which represents a 24S–T aluminum-alloy bolt 7/16 in. in diameter and 5/8 in. long.

The most commonly used aircraft rivets are shown in Fig. 12.7. The round-head rivet, AN430, or the flathead rivet, AN442, are used for interior work where the rivetheads are not exposed to the airstream. For external rivets in which the head is exposed to the airstream the countersunk rivets, AN426, are desirable in order to maintain a smooth aerodynamic surface. The countersunk rivets are more expensive to drive, however, and brazier-head rivets, AN456, are sometimes used for external rivets as a compromise between flush rivets and rivets with larger protruding heads.

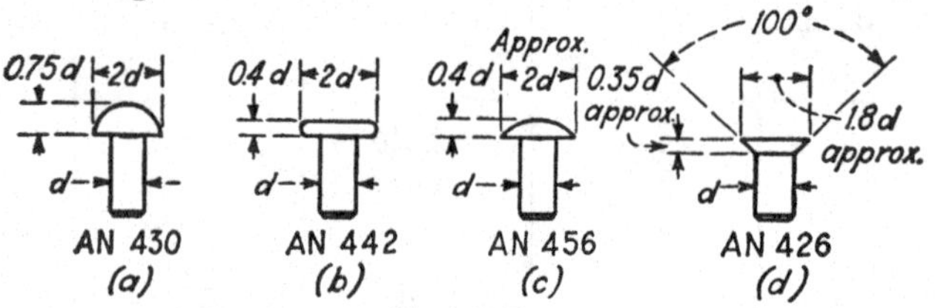

Fig. 12.7.

Aluminum-alloy rivets must be made of a material which has good strength properties but which also is ductile enough to permit driving. The heat-treated alloys do not have sufficient ductility to permit a head to be formed easily without cracking the rivet or damaging the sheet metal by heavy rivet hammer blows. Rivets of the non-heat-treated alloys (SO condition) are not strong enough after driving. A special alloy, A17S-T, is therefore used in almost all of the standard aircraft rivets. This alloy is not quite as strong as 17S-T, but is ductile enough to be easily driven when in the heat-treated and fully aged condition.

Heat-treated alloys do not harden during the heat-treatment, but gradually acquire their full strength and hardness by aging at room temperature for an hour or so after heat-treatment. Rivets of 17S-T or 24S-T could be easily driven within 10 min. after heat-treatment and would then age-harden in place. This procedure is impracticable, but it has been found that the age hardening can be retarded for several days by refrigerating the rivets immediately after heat-treatment. The rivets are therefore stored in refrigerators until immediately before driving. The refrigeration and handling of rivets is costly, and refrigerated rivets are seldom used where it is feasible to use A17S-T rivets, which may be driven as received. For large rivets (over $\frac{3}{16}$ in. diameter) A17S-T rivets require very heavy hammer blows in driving, and the sheet is easily damaged. Refrigerated rivets are more easily driven and must usually be used when these larger rivets are necessary.

The rivet material may be identified by the code shown in Fig. 12.8. Rivets made of the A17S-T alloy have a small dimple in the center of the head and are specified as AD rivets on drawings and in the AN standards book. Rivets made of 17S-T alloy have a small projection on the center of the head and are specified as D rivets. Rivets made of 24S-T material have two raised bars on the head and are specified as DD rivets. Rivet diameters vary by increments of $\frac{1}{32}$ in., and lengths vary by increments of $\frac{1}{16}$ in. A rivet is described on a drawing by the standard number designating the shape, by the letters AD, D, or DD designating the material, and by dash numbers designating the shank diameter and length. Thus, the code number AN426AD4–5

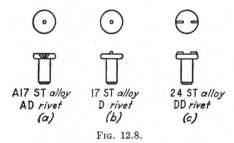

A17 ST *alloy*
AD *rivet*
(a)

17 ST *alloy*
D *rivet*
(b)

24 ST *alloy*
DD *rivet*
(c)

FIG. 12.8.

indicates a 100° countersunk rivet of the shape shown in Fig. 12.7(*d*) and made of A17S–T alloy with a shank diameter of $\frac{4}{32}$ in. and a length of $\frac{5}{16}$ in. Similarly, the code number AN430D10–12 indicates a round-head rivet of refrigerated 17S–T alloy which is $\frac{10}{32}$ in. in diameter and $1\frac{2}{16}$ in. in length. A stress analyst should memorize rivet and bolt codes, as serious mistakes have been made because the wrong type of rivet or bolt was assumed when making the stress analysis.

Rivets may be countersunk to provide flush surfaces by either machine countersinking the sheet, as shown in Fig. 12.9(*a*), or by press counter-sinking or dimpling the sheet, as shown in Fig. 12.9(*b*). The press countersinking or dimpling is preferable structurally, but cannot be

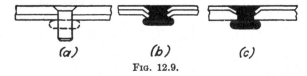

(a) *(b)* *(c)*

FIG. 12.9.

used for thick sheets. When a thin sheet is riveted to the outside of a heavy sheet or extrusion, it is necessary to machine countersink the thick sheet or extrusion and to press countersink the thin sheet, as shown in Fig. 12.9(*c*).

Typical values for the strength of rivets in single shear are shown in Tables 2 and 3 of the Appendix. All types of protruding head rivets, AN430, AN442, or AN456, are analyzed from Table 2 of the Appendix. Table 3 of the Appendix applies to 100° countersunk rivets AN426. The values in Table 2 of the Appendix are calculated by substituting the allowable shear and bearing stresses obtained from *ANC*–5 into Eqs. 12.3 and 12.4, using the nominal hole diameter and correcting the shear strength when the ratio of the hole diameter to sheet thickness d/t exceeds 3.0. Rivet holes are slightly larger than the rivet diameter, and the rivet is upset to fill the hole when it is driven. Most of the values given in Table 2 of the Appendix are determined from the rivet shear strength and are therefore applicable for any sheet material which has a higher allowable bearing stress than that for the Alclad 24S–T alloy. The values shown in Table 2 of the Appendix apply only when the distance from the center of the rivet to the edge of the sheet is two rivet diameters or more, measured in the direction of loading.

The allowable loads for countersunk rivets are determined from tests on the actual riveted joints. The dimpled or press-countersunk rivets are stronger than similar machine-countersunk rivets. For the type of joint shown in Fig. 12.9(c), it may be necessary to conduct special tests, but usually the outside sheet is much thinner than the inside sheet, and the joint may be assumed to have the same strength as the dimpled joint shown in Fig. 12.9(b). Table 3 of the Appendix is taken from *ANC*–5a.

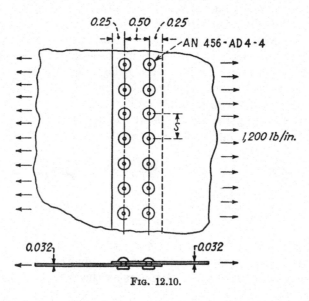

FIG. 12.10.

Example 1. Find the margin of safety for the riveted joint shown in Fig. 12.10 (a) if $s = 0.5$ in. and (b) if $s = 0.375$ in. The sheet is 24S–T Alclad with $F_t = 60,000$ psi.

Solution. a. The allowable rivet loads are obtained from Table 2 of the Appendix for A17S–T rivets as 374 lb. For four rivets per inch,

$$MS = \frac{4 \times 374}{1,200} - 1 = \underline{0.25}$$

The tensile stress in the sheet is obtained from Eq. 12.7, assuming p to be the load on two rivets because of the two rows.

$$f_t = \frac{p}{(s-d)t} = \frac{1,200 \times 0.5}{(0.5 - 0.125)0.032} = 50,000 \text{ psi}$$

$$MS = \frac{60,000}{50,000} - 1 = \underline{0.20}$$

It is not necessary to check the sheet for tear-out of the rivets if the edge distance measured from the center of the rivet is at least two diameters. The margin of safety is therefore 0.20 for the joint.

b. For the spacing of 0.375, the load per rivet is

$$\frac{1,200 \times 0.375}{2} = 225 \text{ lb}$$

For rivet shear,

$$MS = \frac{374}{225} - 1 = \underline{0.66}$$

The tensile stress in the sheet is

$$f_t = \frac{1,200 \times 0.375}{(0.375 - 0.125)0.032} = 56,000 \text{ psi}$$

$$MS = \frac{60,000}{56,000} - 1 = \underline{0.07}$$

The margin of safety for the joint is 0.07. The highest margin of safety for the joint would be obtained with a rivet spacing greater than 0.50.

Example 2. Design a riveted joint to resist the shear load shown in Fig. 12.11, if $q = 900$ lb/in. and the web is 24S–T Alclad with $t = 0.040$ in. Assume that the minimum rivet spacing is $4d$, as shown.

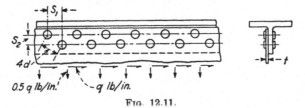

Fig. 12.11.

Solution. As this is an interior joint, either round-head (AN430) or flathead (AN442) rivets will be used, depending upon company practice. From a structural consideration it is desirable for the shear strength and the bearing strength of a rivet to be approximately equal, which results in a rivet diameter of approximately four times the sheet thickness. In the case of sheet gages greater than 0.051 in., the theoretical rivet diameter would be greater than $\frac{3}{16}$ in., for which A17S–T rivets could not be used, and the practical considerations are more important than the theoretical considerations. Consequently, $\frac{3}{16}$-in. A17S–T rivets are usually used in the heavier sheets, although the shear strength is much less than the bearing strength. For the 0.040 gage sheet of this example, AN442AD5 rivets are used, for which the strength is 574 lb. The load per inch is $\sqrt{(900)^2 + (450)^2} = 1,010$ lb/in. The required rivet spacing is as follows:

$$s_1 = \frac{574}{1,010} = 0.568 \text{ in.}$$

The minimum distance between rivets, $4d$, is 0.625 in. The distance between rows, s_2, is now obtained, as indicated in Fig. 12.11.

$$s_2 = \sqrt{(0.625)^2 - (0.568)^2} = 0.27 \text{ in.}$$

The distance s_2 is usually made slightly greater than the calculated value, and the distance s_1 slightly less, so that an even number of spaces will fit between the two fixed end rivets in a line.

12.4. Accuracy of Fitting Analysis. The ultimate strength of a fitting may usually be calculated accurately by the methods previously described. The true stress distribution at stresses below the elastic limit is often much different than the assumed distribution. Before the ultimate strength of the fitting is reached, however, the material yields and the stresses are redistributed so that they usually approach the assumed stress distribution. Because of this plastic yielding of the material and because the allowable shear and bearing stresses are obtained from tests on specimens similar to the actual structure, it is possible to obtain accurate calculated strengths by means of inaccurate assumptions. While the conventional methods are usually satisfactory for making design calculations for fittings, the designer must keep in mind the true stress distribution and must avoid conditions of high local stress wherever possible.

One very common case of stress concentration, shown in Fig. 12.12, is that of a tension plate containing a circular hole. For small loads,

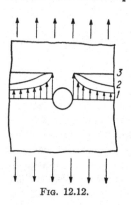

FIG. 12.12.

the tensile stress at the side of the hole is three times the average tensile stress in the plate, as indicated by line 1. As the loads increase, the stress at the side of the hole exceeds the elastic limit, and local plastic yielding of the material occurs near the hole. The stresses near the hole remain almost constant at the yield point while the stresses at a distance from the hole increase with the load, as indicated by line 2. Before failure occurs, yielding has progressed over the entire width of the plate, and the stress is constant over the net section, as shown by curve 3. The customary assumption that failure occurs at a load equal to the product of the ultimate tensile stress and the net area is therefore accurate for ductile materials. Brittle materials,

which fail suddenly with no plastic elongation, should never be used
for aircraft structural members.

Stress concentrations are much more serious in engine parts on which
the loads are repeated millions of times than they are in airframe parts
on which the maximum loads occur only a few times during the life of the
airplane. In airframe design, it is usually safe to consider only average
stresses and to neglect stress concentrations, although certain unfavor-
able conditions, such as radial cracking of sheet around holes when the
holes are press countersunk, may lead to service failures from stress
concentrations.

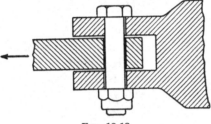

FIG. 12.13.

The double shear connection shown in Fig. 12.13 is assumed to resist
one-half the load by shear on each bolt cross section. Manufacturing
tolerances may permit the hole in the lower lug to be slightly to the left
of the hole in the upper lug, as shown. For small loads, the entire load
is therefore resisted by shear on the upper cross section. As the load
is increased, the parts deflect so that the lower end of the bolt is also
bearing on the lug, but the upper lug continues to resist more than one-
half the load. The fitting factor is intended to account for such eccentric
loading conditions, and in this case the use of a fitting factor of 1.2 is
equivalent to the assumption that one side of the fitting may resist
60 per cent of the total ultimate load.

Most of the bolted and riveted joints in aircraft structures are single
shear joints. For the joints shown in Fig. 12.1 and Fig. 12.2, it was
assumed that one member was rigid, and only the forces acting on the
other member were considered. For this assumed loading the bolt
would resist a bending moment of $Pt/2$, and the heavy member would
resist a larger bending moment. The usual single shear joint has both
members of comparable size. It might first appear that each of the
members shown in Fig. 12.14 could be treated in the same manner as the
upper member of Fig. 12.2(b); in fact many textbooks show the forces
as in Fig. 12.14, and this assumed stress distribution is customary and
satisfactory for design. The forces shown in Fig. 12.14 cannot be in
equilibrium, however, as there is an unbalanced moment Pt on the

plates in Fig. 12.14(a) and a similar unbalanced moment on the pin in Fig. 12.14(b). The correct stress distribution must be as shown in Fig. 12.15. For the forces P to balance, they must act on the same line,

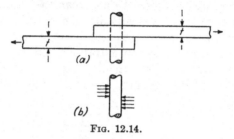

FIG. 12.14.

as shown in Fig. 12.15(a). The stresses in the plate are no longer P/A, but must also include stresses from the bending moment $Pt/2$. If the plate width is b, the plate stress is $P/A \pm My/I$.

$$f = \frac{P}{bt} \pm \frac{Pt}{2}\frac{6}{bt^2} = \frac{P}{bt} \pm \frac{3P}{bt} \tag{12.8}$$

At the inside faces of the plates, the tensile stress from Eq. 12.8 is $4P/A$, and at the outside faces the compressive stress is $2P/A$, as shown.

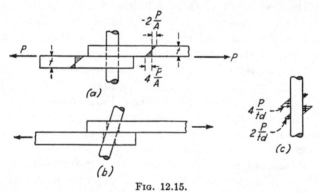

FIG. 12.15.

In order for the pin to be in equilibrium under the bearing stresses, it must bear on opposite corners of the hole, as shown in Fig. 12.15(b). The most optimistic assumption of bearing stresses is the straight-line assumption shown in Fig. 12.15(c), which yields maximum bearing stresses, $4P/tb$ at the inside corner and $2P/tb$ at the outside corners. If the pin is not a tight fit in the hole, the bearing stresses must be higher than those assumed. Thus, for the single-shear pin joint between plates of equal thickness, the maximum plate tension stresses and the bearing stresses are both four times the values assumed in Figs. 12.2 or 12.14.

The bending moment in the pin of Fig. 12.15 is zero at the cross section of maximum shear, and the maximum pin bending moment is $\frac{4}{27}Pt$ at a cross section a distance $t/3$ from the inside of the plates.

The ultimate strength of conventional riveted and bolted joints approaches that assumed in the original simple analysis, because of the clamping action of the rivetheads or boltheads. For a riveted joint between two sheets in tension, the bending and tension stresses in the

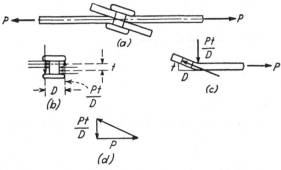

Fig. 12.16.

sheets exceed the elastic limit, and the sheets deform, as shown in Fig. 12.16. The two forces P are almost in the center plane of the sheets, as shown in Fig. 12.16(a), as the ultimate strength is approached. The moment Pt of the bearing forces on the rivet is balanced by the moment of clamping forces under the head of the rivet, as shown in Fig. 12.16(b). These forces on the rivethead have a moment arm D which is slightly less than the diameter of the rivethead. The bending moment in the rivet shank varies from a value of $Pt/2$ at each end of the shank to zero at the plane of rivet shear. After plastic yielding has progressed in the sheet, the bending stresses shown in Fig. 12.15(a) are eliminated, and the sheet is in almost uniform tension at all points. The angular change in the sheet is arctan t/D, as shown in Fig. 12.16(c), and the force exerted by the rivethead on the sheet is just sufficient to keep the resultant tension in the center plane of the sheet. The force triangle at the bend in the sheet is represented by Fig. 12.16(d). The angular change in the sheet is exaggerated in Fig. 12.16. For a rivet-shank diameter of four times the sheet thickness and a rivethead diameter D of twice the shank diameter, the angle is arctan t/D or arctan $\frac{1}{8}$.

Where a tension joint has two lines of rivets, the deformation is shown in Fig. 12.17. If the tension stresses in the sheet were uniform at all points, the sheet would deform as shown in Fig. 12.17(a). Between the rivet lines, however, the sheets have only one-half the average tensile stress that they have at the ends and may therefore resist the bending deformation and assume the deformed shape shown in Fig. 12.17(b).   The forces on the rivets will remain approximately as shown in Fig. 12.16(b), since the clamping forces on the rivethead must balance the moment of the bearing forces.

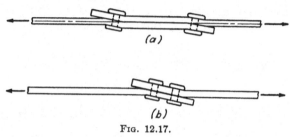

FIG. 12.17.

Any riveted or bolted single-shear joint will have stress conditions which vary between the extreme conditions of Figs. 12.15 and 12.16. At low loads, the sheet must resist bending stresses, as shown in Fig. 12.15, but as local yielding occurs, the stresses are redistributed so that they approach the conditions of Fig. 12.16.   The ultimate strength is predicted accurately from an assumed average tension stress in the sheet and an average bearing stress on the bolt or rivet.   Many types of "blind" rivets or of countersunk rivets do not provide a sufficient amount of clamping action by the rivethead, and strength calculations based on simple stress distributions must be verified by tests.

It is interesting to compare the action of aircraft rivets with the action of hot-driven steel rivets, such as those used in bridges, buildings, boilers, or other steel structures.   The steel rivet is upset when red hot and cools and contracts in place.   The contraction makes the rivet slightly smaller in diameter than the hole, and it also provides a residual tension stress in the rivet approximately equal to the yield stress of the rivet material.   The rivet tension clamps the plates so tightly that small loads are resisted by friction between the plates, and the rivet shank bears on the hole only at higher loads.   This tension does not exist in aircraft rivets, which are driven at room temperatures.

It is common aircraft practice to assume the same allowable bearing stresses for single-shear joints and for double-shear joints of the type shown in Fig. 12.5.   The common practice in bridge or structural steel design is to use higher allowable bearing stresses for joints in double shear.   This practice is logical, since the eccentric distributions as shown in Fig. 12.15 are eliminated in double-shear joints.

When several similar rivets or bolts act together in a joint, it is customary to assume that each rivet or bolt carries a proportionate share of the load. This assumption is very inaccurate when the joint is not highly stressed, but is more accurate as the loads approach the ultimate strength of the joint and local plastic yielding and rivet "slip" have occurred. In Fig. 12.18, the deformations of the various rivets in a double-shear joint are exaggerated in order to show the relative motion between the plates, although the actual plates would be in close contact and the actual deformation would consist of hole elongations as well as rivet shear deformations. It is assumed that the two outside plates

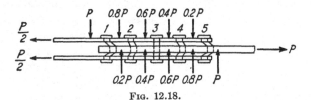

Fig. 12.18.

have the same total area A as the inside plate and that the average stress in all plates is $p = P/A$. If each of the five rivets transfers one-fifth of the total load, as commonly assumed, the stresses in the various plates between rivets will be $0.2p$, $0.4p$, $0.6p$, and $0.8p$, as shown in Fig. 12.18. Between rivets 1 and 2, the outside plates resist tensile stresses of $0.8p$ and the inside plate resists a tensile stress of $0.2p$; therefore the outside plates must elongate four times as much as the inside plate. Thus, rivet 1 must be deformed much more than rivet 2 and must resist a higher shear. Between rivet 2 and rivet 3, the outside plates have 1.5 times the stress and deformation as the inside plate; therefore rivet 2 must resist more load than rivet 3. A study of other deformations shows that the end rivets, 1 and 5, are equally stressed and must resist much higher shears than the other rivets. Rivets 2 and 4 are equally stressed and resist higher shears than rivet 3.

In the case of a longer line of bolts or rivets than that shown in Fig. 12.18, the end bolts or rivets are still more highly stressed in proportion to the bolts and rivets near the center of the line. As the load is gradually applied, the two end rivets must first resist most of the load, until they slip or yield in shear and bearing. The load is then transferred to the next rivets in the line, until they also slip and transfer load to other rivets. The ultimate strength of the joint is accurately predicted as the sum of the strengths of the individual rivets, provided there is enough ductility to permit each rivet to slip considerably and yet still retain its maximum strength after slipping. It is desirable, however, to vary the plate areas in order to obtain approximately constant tension stresses in the plates and thus distribute small loads more equally to all the rivets or bolts. Bolted or riveted joints in brittle materials are undesirable, since the end bolts may fail before they deform enough to redistribute the load and then each bolt in the line will fail in turn. Some spot welds do not have enough ductility to be satisfactory for this type of loading, and occasional rivets are used in most lines of spot welds so that a progressive failure will not extend past the rivet.

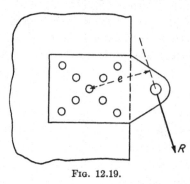

FIG. 12.19.

12.5. Eccentrically Loaded Connections. In many connections the resultant force does not act through the center of the bolt or rivet group. In such cases it is usually convenient to superimpose the effects of an equal parallel force acting at the center of the rivet group and a moment about the center which is equal to the product of the force and its distance from the center. The rivet forces in the typical connection shown in Fig. 12.19 may be obtained by superimposing the forces for the concentric loading of Fig. 12.20(a) and for the moment Re, shown in Fig. 12.20(b).

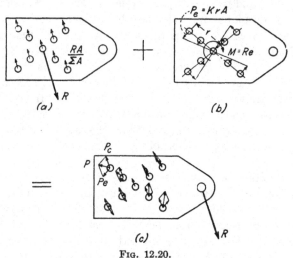

(a)   (b)

(c)

FIG. 12.20.

It will first be assumed that all the rivets are critical in single shear and that all plates are rigid. For the concentric load shown in Fig. 12.20(a), the shearing stresses on all rivets are assumed to be equal. The force P_c on any rivet resulting from this concentric load will be obtained as

$$P_c = \frac{RA}{\Sigma A} \qquad (12.9)$$

where A is the area of the rivet cross section and ΣA is the total cross-sectional area of all the rivets in the group. For a rivet group of n rivets of equal area, Eq. 12.9 reduces to the following form:

$$P_c = \frac{R}{n} \qquad (12.10)$$

The resultant of the forces on the individual rivets passes through the centroid of the areas of the rivet cross sections, and this point must therefore be used as the center of moments for the rivet group.

When the rivets resist a moment, the shearing stresses are assumed to be proportional to the distance r from the centroid of the rivet areas. The force P_e on any rivet of area A resulting from this moment is obtained as follows:

$$P_e = KrA \qquad (12.11)$$

The constant K is obtained by equating the sum of the moments of the individual rivet forces to the external moment.

$$M = \Sigma P_e r = K\Sigma r^2 A \qquad (12.12)$$

The constant K may be eliminated from Eqs. 12.11 and 12.12 and the force P_e obtained.

$$P_e = \frac{MrA}{\Sigma r^2 A} \qquad (12.13)$$

Equation 12.13 is similar in form to the common equations for bending or torsion.

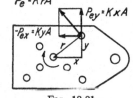

Fig. 12.21.

The resultant force P on any rivet may now be obtained graphically from the component forces P_c and P_e, as shown in Fig. 12.20(c). When an algebraic solution is desired, it is usually more convenient to obtain the horizontal and vertical components of the rivet forces. The distances r do not need to be calculated if the coordinates x and y are used. From Fig. 12.21 and Eq. 12.13, the following equations for the components P_{ex} and P_{ey} are obtained:

$$P_{ex} = \frac{-MyA}{\Sigma x^2 A + \Sigma y^2 A} \qquad P_{ey} = \frac{MxA}{\Sigma x^2 A + \Sigma y^2 A} \qquad (12.14)$$

The method of analysis for an eccentric connection, like methods of analysis for several other types of fittings, must be considered only as a rough approximation. Where bearing stresses are critical in the design of the bolts or rivets, it is customary to substitute P_a, the allowable bearing load for each bolt or rivet, into Eqs. 12.13 and 12.14 in place of the shear area A. In some cases it may be assumed that the loads on all bolts or rivets approach their ultimate strengths, rather than assume that the loads are proportional to r or that bolts or rivets near the center are not highly stressed. Bolts or rivets are frequently attached to members which are more rigid in one direction than another. If the supporting structure is rigid horizontally but flexible vertically, for example, it may be assumed that an applied moment is resisted by horizontal rivet or bolt forces, rather than by forces perpendicular to the radial line. Where standard bolts and rivets are both used in the same connection, it is necessary to design the connection so that either the rivets alone can resist the total load or the bolts alone can resist the total load. Rivets fill the holes completely, but bolts must be slightly smaller than the holes, and consequently the bolts resist no loads until the rivets slip enough to be permanently damaged. Close-tolerance, drive-fit bolts are occasionally used with rivets, and each may be assumed to resist a proportionate share of the load.

Example. Find the resultant force on each rivet of the connection shown in Fig. 12.22. Also find the margin of safety of the most highly stressed rivet, if all rivets are number AN442AD5 in single shear and the sheet is 0.051 gage 24S–T Alclad.

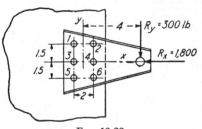

FIG. 12.22.

Solution. The rivet loads are calculated in Table 12.2. The centroid is determined by inspection, and values of x, x^2, y, and y^2 are tabulated in columns (2) to (5). The rivet forces P_{cx} and P_{cy} are obtained by dividing the loads of 1,800 lb and 300 lb by 6, since these loads are resisted equally by each of the six rivets. The values of P_{ex} and P_{ey} are obtained from Eqs. 12.14. Since A is the same for all rivets, the values of A may be omitted from Eqs. 12.14. The moment M is 1,200 in-lb. The values P_x and P_y are each obtained as the sum of terms in the two preceding columns, care being used as to the algebraic signs. The resultant rivet forces P are found as the square root of the sum of the squares of the rectangular components P_x and P_y.

The allowable load for the rivet is obtained from Table 2 of the Appendix as 593 lb. Rivet 6 resists the greatest load, 440 lb. The margin of safety is obtained from these loads.

$$\text{MS} = \frac{593}{440} - 1 = \underline{0.35}$$

It has been assumed that the loads of 1,800 lb and 300 lb were design fitting loads, or that they were obtained by multiplying the applied or limit loads by the safety factor of 1.5 and the fitting factor of 1.2 or 1.15.

TABLE 12.2 ANALYSIS OF RIVETED CONNECTION

Rivet 1	x 2	y 3	x^2 4	y^2 5	P_{ex} 6	P_{ex} 7	P_x 8	P_{cy} 9	P_{ey} 10	P_y 11	P 12
1	-1	1.5	1	2.25	300	-120	180	50	-80	-30	183
2	1	1.5	1	2.25	300	-120	180	50	80	130	221
3	-1	0	1	0	300	0	300	50	-80	-30	302
4	1	0	1	0	300	0	300	50	80	130	327
5	-1	-1.5	1	2.25	300	120	420	50	-80	-30	422
6	1	-1.5	1	2.25	300	120	420	50	80	130	440
Σ			6	9.00							

12.6. Welded Joints. Welding is used extensively for steel-tube truss structures, such as engine mounts and fuselages, and for steel landing gears and fittings. The most common type of welding consists of heating the parts to be joined by means of an oxyacetylene torch and then fusing them together with a suitable welding rod. The grain structure of the material at the weld becomes similar to that of cast metal, and it is more brittle and less able to resist shock and vibration loading than is the original material. Aircraft tube walls are thin and are more difficult to weld than other machine and structural members. All aircraft welding was previously torch welding, but electric arc-welding has now been developed so that it is also satisfactory for the thin aircraft members. In the arc-welding, the welding rod forms an electrode from which current passes in an arc to the parts being joined. The electric arc simultaneously heats the parts and deposits the weld metal from the electrode. The heating is much more localized than in torch welding, and the strength of heat-treated parts is not impaired as much by arc-welding as by torch welding. Design specifications normally require that the same allowable stresses be used for arc-welding and for torch welding.

The strength of welded joints depends greatly on the skill of the welder. The stress conditions are usually uncertain, and it is customary to design welded joints with liberal margins of safety. It is preferable to design joints so that the weld is in shear or compression rather than tension, but it is frequently necessary to have welds in tension. Steel tubes in tension are usually spliced by "fish-mouth" joints, as shown in Fig. 12.23(a), which are designed so that most of the weld is in shear and so that the local heating of the tube at the weld is not confined to one cross section. Where a butt weld must be used, as shown in Fig. 12.23(*b*), the weld is not perpendicular to the center line of the tube.

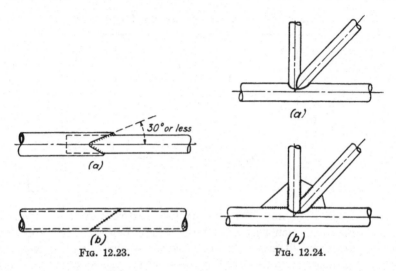

Fig. 12.23.   Fig. 12.24.

Fuselage truss members are often welded as shown in Fig. 12.24(*a*). Only the horizontal member is highly stressed, and the size of the other members is usually determined as a minimum tube size, because they resist small loads.   When these members are highly stressed, it is necessary to insert gusset plates, as shown in Fig. 12.24(*b*).   Steel tubes often have walls as thin as 0.035 in., and the welder must control the temperature to keep from overheating the thin walls and burning holes in them.   It is extremely difficult to weld a thin member to a heavy one, as more heat is required for the heavy member.   The thickness ratio of parts being welded should always be less than 3:1, and preferably less than 2:1.

The structural aluminum alloys cannot be arc-welded or torch-welded without destroying the heat-treatment.   Several of the weaker aluminum alloys are frequently torch-welded for construction of fuel tanks or oil tanks.   Spot welding is frequently used for attaching structural skin to stringers, or for similar secondary structural joints where the loads are not high.   The spot welding is done by clamping the members between two electrodes and passing a high amperage current for a short interval of time.   The electric resistance between the sheets causes heating and fusion of the sheets.   Other types of resistance welding, such as those used in manufacturing steel automobile bodies, can seldom be used in aircraft construction.

The allowable load on the weld metal in welded seams is specified in *ANC*–5 by the following equations:

$$P = 32,000Lt \qquad \text{(low-carbon steel)}$$
$$P = 0.48LtF_{tu} \qquad \text{(chrome-molybdenum steel)} \qquad (12.15)$$

where P = allowable load, lb.
    L = length of welded seam, in.
    t = thickness of thinnest material joined by the weld in the case of lap welds between two steel plates or between plates and tubes, in.
    t = average thickness in inches of the weld metal in the case of tube assemblies. (Cannot be assumed greater than 1.25 times the thickness of the welded stock.)
    F_{tu} = 90,000 psi for material not heat-treated after welding.
    F_{tu} = ultimate tensile stress of material heat-treated after welding, but not to exceed 150,000 psi. Heat-treatable welding rod must be used.

The local heating during welding also reduces the allowable tension or bending stress in the material near the weld. For normalized tubing with no heat-treatment after welding, the allowable tensile stress is 90,000 psi near the weld, for tapered welds making an angle of 30° or less with the axis of the tube, and 80,000 psi for other welds. For tubing which is heat-treated after welding, the allowable tensile stress is F_{tu}.

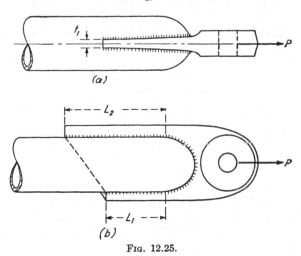

(a)

(b)

Fig. 12.25.

Example. The 1½ by 0.065 chrome-molybdenum steel tube shown in Fig. 12.25 resists a limit or applied tension load of 15,000 lb. Find the margin of safety of the weld and of the tube near the weld if $L_1 = 2.5$ in., $L_2 = 3$ in., $t_1 = 0.20$ in., and the tube area $A = 0.293$ sq in.   Assume (a) that the ultimate tensile stress F_{tu} is 100,000 psi before welding and that there is no subsequent heat-treatment and (b) that the tube assembly is heat-treated to an F_{tu} of 180,000 psi after welding and that the limit load is 22,000 lb.

Solution. a. The weld on the curved end of the tube is neglected, since loads transmitted to this portion of the tube tend only to straighten and flatten the end of the tube and do not increase the tensile strength of the weld appreciably. Because the load P is applied at the center of the tube, one-half of this load is resisted by the weld on each side of the tube.   The two welds of length L_1 must therefore resist the load of $P/2$.   The ultimate or design load is obtained by multiplying the applied load by the safety factor of 1.5.

$$P = 15,000 \times 1.5 = 22,500 \text{ lb}$$

The allowable load P_a will be obtained from Eq. 12.15.   The tube thickness, $t = 0.065$, is critical, since the forging thickness t_1 is more than twice the tube wall thickness.   The length L_1 is welded to the tube on both sides of the forging; therefore a length $L = 5$ in. must resist half of the load.   The allowable load is obtained as follows:

$$\frac{P_a}{2} = 0.48 L t F_{tu} = 0.48 \times 5 \times 0.065 \times 90,000$$

or

$$P_a = 28,000 \text{ lb}$$

The fitting factor of 1.20 must be included in the calculation of the margin of safety.

$$\text{MS} = \frac{P_a}{1.20P} - 1 = \frac{28,000}{1.2 \times 22,500} - 1 = \underline{0.04}$$

This margin appears small for a weld, but was calculated conservatively.   The ultimate tensile stress in the tube near the weld is 90,000 psi since for a slotted tube the weld makes an angle of zero degrees with the tube axis.   The allowable tension in the tube near the weld is

$$P_a = 90,000 \times 0.293 = 26,400 \text{ lb}$$

It is not necessary to use a fitting factor here, since the tube itself, rather than the fitting, is being investigated.

$$\text{MS} = \frac{26,400}{22,500} - 1 = \underline{0.17}$$

b. The allowable load is computed in the same manner as for part (a), but now $F_{tu} = 150,000$ psi.

$$\frac{P_a}{2} = 0.48 \times 5 \times 0.065 \times 150,000$$

or

$$P_a = 46,400 \text{ lb}$$

The design load is

$$P = 22,000 \times 1.5 = 33,000 \text{ lb}$$

The fitting factor is included in calculating the margin of safety

$$MS = \frac{46,400}{33,000 \times 1.2} - 1 = \underline{0.18}$$

The allowable tensile stress in the tube near the weld is F_{tu}.

$$P_a = F_{tu} \times A = 180,000 \times 0.293 = 52,700 \text{ lb}$$
$$MS = \frac{52,700}{33,000} - 1 = \underline{0.60}$$

PROBLEMS

12.1. An end fitting similar to those shown in Figs. 12.5 and 12.6 is made of steel with an ultimate tensile strength F_{tu} of 180,000 psi and has no bushing. It has an AN8 bolt in double shear, a thickness of 0.5 in., and dimensions $R = 0.5$ in. and $e = 0.05$ in., as shown in Fig. 12.6. Find the maximum limit loads in tension and in compression, if the fitting factor is 1.2 and the bearing factor is 1.0. Obtain allowable stresses from *ANC–5*.

12.2. Design a fitting to resist a limit tension load of 15,000 lb and a limit compression load of 20,000 lb. Assume the materials and unit stresses to be the same as those used in the Example problem of Art. 12.2.

12.3. Design an end fitting of steel with an ultimate tensile strength of 125,000 psi. The applied or limit loads are 15,000 lb tension and 20,000 lb compression. Use a fitting factor of 1.2 and a bearing factor of 2.00.

12.4. Find the margin of safety of the riveted joint shown in Fig. 12.10, if the rivets are AN442AD5 and the spacing is $s = 0.625$ in.

12.5. Find the margin of safety of the riveted joint shown in Fig. 12.10 if the rivets are AN426AD4 spaced at $s = 0.55$ in. Assume (a) press-countersunk or dimpled rivets and (b) machine-countersunk rivets.

12.6. Find the margin of safety of the riveted joint shown in Fig. 12.10 if the rivets are AN426AD5 spaced at $s = 0.75$ in. Assume (a) dimpled and (b) machine-countersunk rivets.

12.7. Determine the rivet spacing, s_1 and s_2, for Fig. 12.11 if the web is 24S–T Alclad, t is 0.064 in., and q is 1,500 lb/in.

12.8. Repeat Prob. 12.7 with $t = 0.032$ and $q = 500$ lb/in.

12.9. Find the margin of safety for the joint shown in Fig. 12.22 if $R_x = 3,000$ lb, $R_y = 200$ lb, and the rivets are AN442AD6, in single shear. The plate is 0.072 gage 24S–T Alclad.

12.10. Assume the tube of Fig. 12.25 to be 2×0.083, with $t_1 = 0.2$, $L_1 = 3.0$ in., and $L_2 = 4.0$ in. Find the allowable load, P, if (a) the tube has an allowable stress F_{tu} of 95,000 psi, and (b) the assembly is heat-treated after welding to a tensile strength F_{tu} of 150,000 psi.

REFERENCE FOR CHAPTER 12

ANC–5a, "Strength of Metal Aircraft Elements," Subcommittee on Air Force-Navy-Civil Aircraft Design Criteria of the Munitions Board Aircraft Committee, May, 1949.

CHAPTER 13

DESIGN OF MEMBERS IN TENSION, BENDING, OR TORSION

13.1. Tension Members. Tension members are more readily analyzed and designed than other types of aircraft members. The stress conditions existing in tension members at the ultimate load condition are accurately known and are not subject to the uncertainties which exist in joints, fittings, and other types of structural members. The allowable tensile stress for a structural material is easy to determine, and a single value of the allowable stress will apply to members of any shape. It will be shown later that the allowable stresses for structural members in bending, torsion, or compression depend on the shapes of the members and on other factors which are not considered for tension members.

For a concentric tension load P on a member with a net area A, the tensile stress is found from the equation $f_t = P/A$. The allowable tension stress F_{tu} is the minimum guaranteed value for the material. The margin of safety may be calculated in the usual manner as $F_{tu}/f_t - 1$.

Tension members must frequently resist bending and compression stresses under other loading conditions, and these other conditions often determine the shape of a member even when the tension load is the largest load. Fabric-covered airplanes with structural frameworks may contain tie rods or cables as tension members. Severe vibrations of such members may result in high bending stresses at the ends of the members, and the end fittings must resist these stresses in addition to the direct tensile stresses.

The primary tension structure in a semimonocoque wing consists of the skin, stringers, and spar caps on the under surface. Although the positive bending moments in a wing are about twice as large as the negative bending moments, the compression loads from negative bending determine the design of most of the structure on the under side of the wing. The wing skin resists tension stress but buckles and becomes ineffective for compression stress. The compressive area is therefore less than the tensile area, and the allowable compressive stress is also considerably less than the allowable tensile stress. Even though the compressive loads are smaller, they must always be considered when determining the shape of the stiffening members, and they frequently determine the required areas.

13.2. Plastic Bending. In the previous calculations of bending stresses it has been assumed that the stresses are below the elastic limit. In most types of machine design and structural design, the strength at the yield stress is the important criterion for design and the conventional elastic stress distribution is satisfactory for use in design. In airframe structures, however, the ultimate strength of a member is the criterion used in design. Before failure, the stress exceeds the elastic limit and is said to be in the *plastic* range. The assumptions used in deriving the flexure formula, $f_b = My/I$, no longer apply.

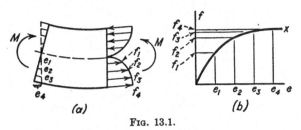

FIG. 13.1.

The initially straight beam shown in Fig. 13.1(a) has bending stresses exceeding the elastic limit. Plane sections remain plane after bending, and the strain distribution is therefore proportional to the distance from the neutral axis, as in elastic beams. If the beam is deflected so that the extreme fiber has an elongation e_4, there will be a stress f_4 at this point, as shown by the stress-strain curve of Fig. 13.1(b). For other strains, e_1, e_2, and e_3, the corresponding stresses f_1, f_2, and f_3 do not vary linearly with the strains above the elastic limit, and the stress-distribution on the beam cross section varies as shown in Fig. 13.1(a).

The ultimate resisting moment of a beam depends both upon the shape of the beam cross section and the shape of the stress-strain curve. Since there is no simple theoretical relationship which applies to a general case, an empirical method is frequently used for determining the ultimate bending strength. A fictitious stress F_b, termed the *bending modulus of rupture*, is defined by the equation $F_b = Mc/I$, where M is the ultimate bending moment as determined from tests of similar beams, I is the moment

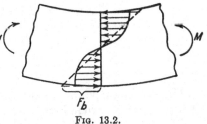

FIG. 13.2.

of inertia of the cross section, and c is the distance from the beam neutral axis to the extreme fiber. The true stress distribution is shown in Fig. 13.2, and the fictitious straight line stress distribution which yields an equal bending moment and has a maximum value of F_b is shown by the dotted line. For geometrically similar sections such as round tubes with the same ratio of outside diameter to wall thickness D/t, the bending modulus of rupture may be obtained for any material by means of tests.

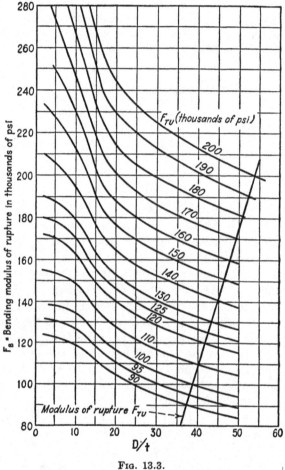

FIG. 13.3.

The bending modulus of rupture for round tubes of chrome-molybdenum steel is shown in Fig. 13.3 for various values of D/t and for various values of F_{tu}, the ultimate tensile stress to which the material is heat-treated. For the larger values of D/t, the tube walls are thin and tend to cripple locally. The local crippling stress of the tube wall, which in itself is difficult to compute theoretically, corresponds to the stress f_4

shown in Fig. 13.1. Thus the tests take into consideration the effects of local crippling, as well as the effects of the shape of cross section and the shape of the stress-strain curve. The bending modulus of rupture is proportional to the bending moment, and the margin of safety may be computed from the usual relation, $F_b/f_b - 1$. The true maximum stress, f_4 of Fig. 13.1, is not proportional to the bending moment and cannot be used in obtaining the margin of safety.

Example. A 1½- by 0.083-in. steel tube resists a bending moment of 25,000 in-lb. What is the margin of safety if the material is heat-treated to an ultimate tensile stress F_{tu} of 180,000 psi?

Solution. From Table 1 of the Appendix, the properties of a 1½- by 0.083-in. tube are obtained as $D/t = 18.08$ and $I/c = 0.1241$ in.³ From Fig. 13.3, $F_b = 220,000$ psi. The fictitious bending stress f_b is obtained from the simple flexure formula, as follows:

$$f_b = \frac{Mc}{I} = \frac{25,000}{0.1241} = 201,000 \text{ psi}$$

The margin of safety is now obtained in the usual manner.

$$MS = \frac{220,000}{201,000} - 1 = \underline{0.09}$$

It will also be necessary to determine a yield margin of safety. For this material the yield stress is 165,000 psi. The applied or limit bending moment is ⅔ × 25,000 = 16,670 in-lb.

$$f_b = \frac{16,670}{0.1241} = 134,000 \text{ psi}$$

or the margin of safety for yielding is

$$MS = \frac{165,000}{134,000} - 1 = \underline{0.23}$$

13.3. Constant Bending Stress. Some materials are such that the stress-strain curve remains almost horizontal after the elongation exceeds

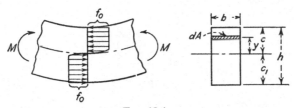

FIG. 13.4.

a value corresponding to the yield point. If a beam of such a material is subjected to bending beyond the yield stress, the bending stresses will approximate those shown in Fig. 13.4. Both the tension and compression stresses may be assumed to have constant values of f_0 over the entire area.   The bending moment is obtained by taking the sum of the moments of infinitesimal forces $f_0 \, dA$ about the neutral axis.

$$M = f_0 \int_{-c_1}^{c} y \, dA \tag{13.1}$$

In the case of a cross-sectional area which is symmetrical with respect to a horizontal axis, the bending moment becomes

$$M = 2Qf_0 \tag{13.2}$$

where Q is defined by the following integral.

$$Q = \int_{0}^{c} y \, dA \tag{13.3}$$

For the symmetrical area, the neutral axis corresponds with the axis of symmetry, as in the case of elastic bending.   For an unsymmetrical area, the neutral axis is not at the centroid, but is located so that the cross-sectional area above the neutral axis is equal to the area below it, since the total tension force must equal the total compression force.

For a rectangular beam of width b and depth h, $Q = bh^2/8$.   Substituting in Eq. 13.2,

$$M = \frac{f_0 b h^2}{4} \tag{13.4}$$

The bending modulus of rupture f_b may be obtained by equating the bending moment of Eq. 13.4 to the expression which defines f_b, or $M = f_b I/c$.   For the rectangular section, $I/c = bh^2/6$, and $f_b = 1.5f_0$.

The parabolic shear stress distribution for a rectangular beam in which the stresses are below the elastic limit was obtained from the bending stress distribution, and does not apply for other distributions of bending stress.   For a rectangular cross section, or for other similar cross sections in which the bending modulus of rupture is considerably larger than the actual stress, the shearing stresses are seldom very high, and may be approximated with sufficient accuracy.

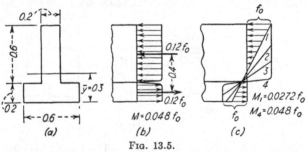

Fig. 13.5.

The plastic bending of a beam in which the cross section is not symmetrical about the neutral axis will be considered by analyzing a numerical example.   The area shown in Fig. 13.5(a) has its centroidal axis at a distance of 0.3 in. above the base. For plastic bending with a constant stress f_0, the neutral axis will be 0.2 in. above the base, in order for the tension area to be equal to the compression area. The tension and compression forces will be equal to $0.12f_0$ and will resist a bending moment of $0.048f_0$, as shown in Fig. 13.5(b). The elastic bending stress for this area is obtained from the equation $M = f_b I/c = 0.0272f_b$. The bending modulus of rupture is therefore equal to $f_b = (0.048/0.0272)f_0 = 1.765f_0$. The stress distribution for various bending moments is shown in Fig. 13.5(c). For bending moments less than $M = 0.0272f_0$, the stresses are below the elastic limit and have a straight-line distribution with the neutral axis at the centroid of the area, as shown by curve 1. For larger values of the bending moment, the stresses will exceed the elastic limit at the upper side of the beam but remain below the elastic limit on the lower side, with the neutral axis shifting downward, as shown by curve 2. For further increases in bending moment, the stresses approach the constant values shown by curve 4, with the neutral axis between the two rectangles.

13.4. Trapezoidal Distribution of Bending Stress. The stress-strain curves for most aircraft materials can be accurately approximated by a trapezoidal curve, as shown in Fig. 13.6(a). Cozzone[1] has proposed this

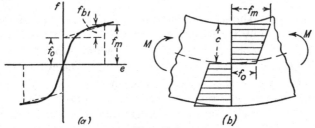

(a) (b)

Fig. 13.6.

approximation for obtaining the bending strength in the plastic range. The bending stresses are distributed as shown in Fig. 13.6(b). The bending moment for the trapezoidal stress distribution is readily obtained as the sum of the bending moment for a constant stress f_0, as given by Eq. 13.2, and the bending moment for a linear stress distribution varying from zero to f_{b1}.

$$M = 2Qf_0 + \frac{f_{b1}I}{c} \tag{13.5}$$

The term f_m may be introduced instead of f_{b1} by substituting $f_{b1} = f_m - f_0$ into Eq. 13.5. Making this substitution and dividing by I/c, the following equation is obtained:

$$F_b = \frac{Mc}{I} = f_m + f_0 \left(\frac{2Q}{I/c} - 1 \right) \tag{13.6}$$

The term in parentheses depends on the shape of the cross section and may vary from zero for concentrated flange areas to 1.0 for a diamond shape. If this term is designated by K, Eq. 13.6 becomes

$$F_b = f_m + K f_0 \qquad (13.7)$$

where

$$K = \frac{2Q}{I/c} - 1 \qquad (13.8)$$

Some values of K for various cross sections are shown in Fig. 13.7.

Section	K	Section	K
	0		0.25 to 0.7
	0.5		1.0
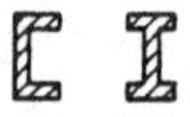	0 to 0.5		⅓

Fig. 13.7.

The value of f_0 should be determined in such a way that the bending moment resisted by the assumed trapezoidal stress distribution is equal to the bending moment resisted by the actual stresses. The correct value of f_0 would therefore depend somewhat on the cross-sectional area. If the value of f_0 were calculated for each area, there would be no advantage in assuming a trapezoidal stress distribution, since it would be necessary to calculate the true resisting moment of the beam in order to calculate f_0. Cozzone has shown that it is sufficiently accurate to calculate f_0 for a rectangular cross section and to use this value for all cross sections.

Example. A beam with the cross section shown in Fig. 13.8 is made of a 14S–T aluminum-alloy forging. The true shape of the forging is shown by the dotted lines, but the trapezoids shown by the solid lines are assumed. Calculate the ultimate bending strength about a horizontal axis, if $f_m = 65{,}000$ psi, $f_0 = 60{,}000$ psi, and the yield stress F_{ty} is 50,000 psi.

Solution. The values of I and Q will be calculated for the assumed area, which is composed of eight of the triangles (1) and four of the rectangles (2).

$$I = \frac{8 \times 0.12 \times 1^3}{12} + \frac{4 \times 0.20 \times 1^3}{3} = 0.347 \text{ in.}^4$$
$$2Q = 8 \times 0.06 \times 0.333 + 4 \times 0.20 \times 0.5 = 0.56 \text{ in.}^3$$

The bending modulus of rupture is now calculated from Eqs. 13.7 and 13.8.

$$K = \frac{2Q}{I/c} - 1 = \frac{0.56}{0.347} - 1 = 0.61$$

$$F_b = f_m + Kf_0 = 65,000 + 0.61 \times 60,000 = 101,600 \text{ psi}$$

The ultimate bending strength is now calculated

$$M = \frac{F_b I}{c} = 101,600 \times \frac{0.347}{1.0} = 35,300 \text{ in-lb}$$

In this case it would not be possible to utilize the full ultimate bending strength because the stress at the applied load condition would exceed the yield stress. The exact amount of permanent set permitted at the applied or limit load is not clearly specified for a member in bending, but depends somewhat on the judgment of the designer. In some cases the bending modulus of rupture is not permitted to exceed the yield stress, or for this problem the value of Mc/I could not exceed 50,000 psi at the applied load, and consequently could not exceed 75,000 psi at the ultimate or design load. Even at the yield stress, however, some plastic

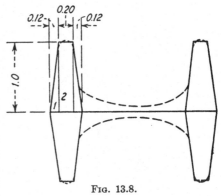

Fig. 13.8.

bending effects may be considered. The stress-strain diagram for a 14S–T forging in which the stress does not exceed the yield stress, shown in Fig. 13.9, may be represented by the trapezoid with $f_0 = 21,200$ psi and $f_m = 50,000$ psi. The

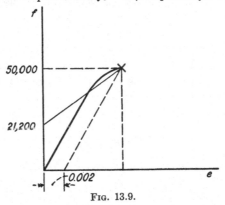

Fig. 13.9.

bending modulus of rupture at the limit load may then be calculated from Eq. 13.7.

$$F_b = f_m + Kf_0 = 50,000 + 0.61 \times 21,200 = 63,000 \text{ psi}$$

The allowable value of Mc/I would then be 63,000 psi for the limit load and $1.5 \times 63,000$, or 94,500 psi, at the ultimate or design load.

13.5. Curved Beams. Most beams in aircraft structures are analyzed by the methods previously considered, in which any initial curvature of the axis of the beam was neglected. However, when the radius of curvature is of the same order of magnitude as the depth of the beam, the stress distribution differs considerably from that for straight beams. The stresses on the concave side of the beam are higher than those for a similar straight beam, and the stresses on the convex side are lower. When the maximum stresses exceed the elastic limit, local yielding occurs

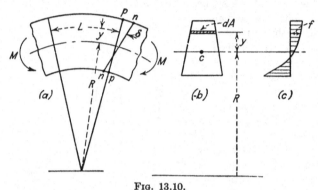

Fig. 13.10.

which permits a redistribution of stress. At the ultimate bending moment, the stresses approach the same distribution as for plastic bending of a straight beam. Thus the beam curvature has the effect of reducing the yield strength but of not appreciably changing the ultimate bending strength.

The beam shown in Fig. 13.10 has an initial radius of curvature R measured to the centroid of the cross section. A plane cross section pp remains plane after bending, and the relative position after bending is shown by nn. A longitudinal fiber of the beam of initial length L is extended a distance δ. Since δ is measured between the straight lines pp and nn, it varies linearly with the distance y from the centroid to the fiber.

$$\delta = k_1 + k_2 y$$

The terms k_1, k_2, and k_3 are undetermined constants. The length of the fiber L is proportional to its distance from the center of curvature.

$$L = k_3(R + y)$$

The unit stress is now obtained as the product of the unit strain δ/L and the elastic modulus E.

$$f = E\frac{\delta}{L} = E\frac{k_1 + k_2 y}{k_3(R + y)}$$

This expression may be simplified by dividing the numerator by the denominator and by grouping the constants into two new undetermined constants a and b.

$$f = a + \frac{b}{R + y} \tag{13.9}$$

This stress distribution is pictured in Fig. 13.10(c).

If a resultant tension force P acts at the centroid of the area, it must equal the sum of the internal forces, that is, $\int f \, dA$. From Eq. 13.9,

$$P = \int f \, dA = aA + b \int \frac{dA}{R + y} \tag{13.10}$$

where A is the total cross-sectional area and a and b are undetermined constants. For the case of pure bending, the force P vanishes.

The external bending moment M about the centroidal axis must be equal to the moment of the internal forces, $\int fy \, dA$.

$$M = \int fy \, dA = a \int y \, dA + b \int \frac{y \, dA}{R + y}$$

The first integral on the right side of the equation is zero because the distance y is measured from the centroidal axis. The second integral may be separated into two terms by division.

$$\int \frac{y \, dA}{R + y} = \int \left(1 - \frac{R}{R + y} \right) dA = A - R \int \frac{dA}{R + y}$$

The bending-moment equation may now be written as follows:

$$M = bA - bR \int \frac{dA}{R + y} \tag{13.11}$$

The unknown constant b may be obtained from Eq. 13.11, since all other terms are known from the geometry and loading of the beam. The other unknown constant a is obtained from Eq. 13.10. The stress distribution is then obtained from Eq. 13.9. The effect of the axial load P is to change the constant a and the stress f by an amount P/A. The same stress distribution can therefore be obtained by superimposing the bending stresses for $P = 0$ and the stresses P/A resulting from the axial load P at the centroid of the area.

The extreme fiber stresses for various curved beams may be determined as ratios of the stresses computed by the flexure formula. These ratios have been computed for various cross sections. The terms k for the equation $f = kMc/I$ are plotted in Fig. 13.11, for a few common cross sections. It is observed that the stresses always become infinite on the concave side when $R/c = 1$, which corresponds to a sharp reentrant angle on the concave surface of the beam. Such reentrant angles should be avoided in any structure or machine part.

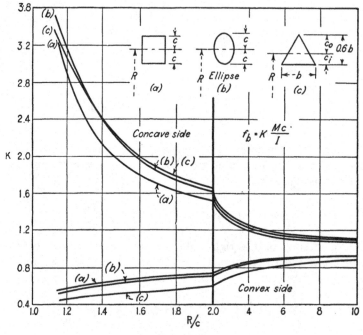

Fig. 13.11.

Another effect of beam curvature which cannot be analyzed by simple theory occurs in beams with thin flanges, as indicated in Fig. 13.12. If the concave side of a beam is in compression, the flanges tend to deflect toward the neutral axis, as shown in Fig. 13.12(*b*). The bending stress is not uniformly distributed along the horizontal beam flange, but is much higher near the web than it is in the outstanding legs, as shown above Fig. 13.12(*b*). When the bending produces compression on the convex side of the beam, the flanges deflect away from the neutral axis, but the stress distribution along the horizontal width of the flanges is essentially the same as shown.

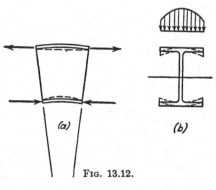

Fig. 13.12.

PROBLEMS

13.1. What is the tensile strength of a 1½- by 0.065-in. steel tube with ends welded so that the welds make angles less than 30° with the axis of the tube? Assume (a) $F_{tu} = 95,000$ psi, (b) $F_{tu} = 150,000$ psi, heat-treated after welding, and (c) $F_{tu} = 150,000$ psi, welded after heat-treatment. Obtain allowable stresses near welds from *ANC*–5.

13.2. What is the ultimate tensile strength of a 24S–T aluminum-alloy tube, 1½ by 0.065 in.? The ends are connected by one line of AN4 bolts. Assume $F_{tu} = 64,000$ psi for the tube.

13.3. Find the ultimate bending moments which may be resisted by steel tubes of dimensions 1¾ by 0.049 in., 1¾ by 0.058 in., and 1¾ by 0.083 in., investigating each tube for heat-treatments of $F_{tu} = 125,000$ psi, $F_{tu} = 150,000$ psi, and $F_{tu} = 180,000$ psi.

13.4. Find the lightest standard steel tube which can resist a bending moment of 10,000 in-lb, if $F_{tu} = 180,000$ psi.

13.5. Design a round steel tube to resist a bending moment of 30,000 in-lb, with $F_{tu} = 125,000$ psi. Calculate the margin of safety for each trial size.

13.6. Find the ultimate resisting moment of a beam with a rectangular cross section, with $b = 0.5$ in. and $h = 2$ in. The material is a 14S–T aluminum-alloy forging. Assume a trapezoidal stress-strain curve with $f_m = 65,000$ psi and $f_0 = 60,000$ psi.

13.7. Repeat the Example problem of Art. 13.4, if the width dimension is changed from 0.2 to 0.4 in. in Fig. 13.8 and the other dimensions remain as shown.

13.8. Calculate the values of K shown in Fig. 13.7, assuming suitable dimensions for the tube, C and I sections.

13.9. Calculate points on curves (a) of Fig. 13.11 for values of R/c of 1.4, 2, and 4.

13.10. Calculate points on curves (c) of Fig. 13.11 for values of R/c of 1.2, 1.6, and 2.0.

13.6. Torsion of Circular Shafts. The stresses resulting from torsional moments acting on elastic cylindrical members of circular cross section may be readily obtained. It has been found experimentally that there is no distortion of any cross section of the shaft either in the direction normal to the plane of the cross section or in the plane of the cross section. Any two cross sections of the shaft, such as those shown in Fig. 13.13(b), have a relative rotation about the axis of the shaft. Since the two cross sections have no relative displacement radially, or along the axis of the shaft, the only stresses are the shearing stresses in the circumferential and axial directions, as shown in Fig. 13.13(b).

The shearing strain γ and consequently the shearing stress f_s must vary linearly in proportion to the distance r from the center of the shaft.

$$f_s = Kr \tag{13.12}$$

The term K is a constant of proportionality which will be determined later. The external torsional moment T must be equal to the sum of the moments of the internal shearing forces on the cross section.

$$T = \int f_s r\, dA = K \int r^2\, dA \qquad (13.13)$$

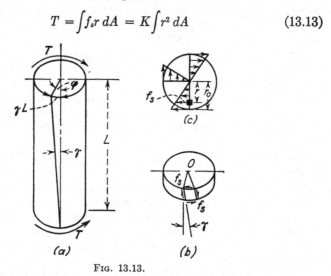

Fig. 13.13.

The integral of Eq. 13.13 represents the polar moment of inertia of the cross-sectional area, and it is usually designated as J or I_p. The value of K may be obtained from Eq. 13.13, as $K = T/J$. Substituting this value into Eq. 13.12, the formula for torsional shear stresses in circular shafts is obtained.

$$f_s = \frac{Tr}{J} \qquad (13.14)$$

It was shown in Eq. 4.14 that $J = I_p = I_x + I_y$. For a circular shaft or tube, $I_x = I_y$ and $r = y$; therefore $J/r = 2I/y$.

The angle of twist of a circular shaft may be obtained from the angle of shearing strain γ. The shearing modulus of elasticity G is defined as the ratio of shear stress to shear strain.

$$G = \frac{f_s}{\gamma} \qquad (13.15)$$

The term G is related to E by the following equation:

$$G = \frac{E}{2(1 + \mu)} \qquad (13.16)$$

For values of Poisson's ratio, μ, of 0.25 to 0.33, G has values of $0.40E$ to $0.375E$.

The shaft of radius r_0 shown in Fig. 13.13 twists through an angle ϕ in a length L. A point on the circumference of the upper cross section moves a distance ϕr_0 during the deformation. This point is also displaced a distance γL, as shown.

$$\phi r_0 = \gamma L$$

The angle of twist ϕ may be expressed in other forms by substituting values from Eqs. 13.15 and 13.14.

$$\phi = \frac{\gamma L}{r_0} = \frac{f_s L}{G r_0} = \frac{TL}{JG} \tag{13.17}$$

These equations apply only to torsion members with solid circular cross sections, or to tubular members with hollow circular cross sections.

13.7. Torsion of a Noncircular Shaft. The stresses in a torsion member of arbitrary cross section cannot be obtained by means of a simple, general equation. A few special cases may be analyzed, however, and some general properties of the shear stress distribution examined. A cylindrical torsion member of arbitrary cross section is shown in Fig. 13.14(a). The small cubical element at the surface of the member

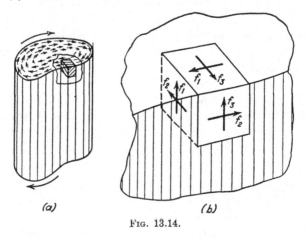

(a)   *(b)*

Fɪɢ. 13.14.

is enlarged in Fig. 13.14(b). In general, three pairs of shearing stresses f_1, f_2, and f_3 exist on any such element, but in the case where one face of this element is a free surface of the member, the shear stresses f_2 and f_3 on the free surface must be zero. The element at the boundary therefore has only the shear stresses f_1, which are parallel to the boundary. The shear stresses on a cross section near any boundary are therefore parallel to the boundary, as shown in Fig. 13.14(a). The resultant of all shearing forces on the cross section must be equal to the external torque.

The exaggerated deformations of a square shaft in torsion are shown in Fig. 13.15(a). An element at a corner of the cross section may be compared to the element shown in Fig. 13.14(b), and since two perpendicular faces of the element can have no shear stresses, all the shear

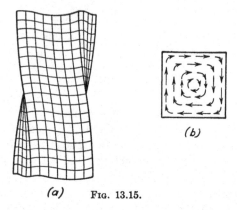

(b)

(a)   Fig. 13.15.

stresses f_1, f_2, and f_3 must be zero at a corner of the member. The shear stresses on the cross section have maximum values at the center of each side and have directions approximately as shown in Fig. 13.15(b). The cubical elements at the corners remain cubical, as indicated in Fig. 13.15(a), but elements near the centers of the sides have rather large shearing deformations. The cross sections therefore do not remain plane but warp as shown.

One type of member which is frequently used in aircraft structures has a narrow rectangular cross section. While such cross sections are inefficient for torsion members, they must frequently resist some torsional stresses. For the cross section shown in Fig. 13.16, of length b

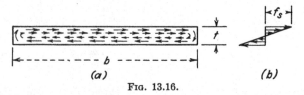

(a)   (b)

Fig. 13.16.

and width t, the shear stresses must be parallel to the boundary. If the length b is large compared to the thickness t, the end effects are small, and the shear stresses may be assumed to be distributed as shown in Fig. 13.16(b) for the entire length b. It can be shown that the shear stress has the following value:

$$f_s = \frac{3T}{bt^2} \tag{13.18}$$

Equation 13.18 is accurate when the width b is large compared to the thickness t. For rectangular cross sections in which the dimensions are of the same order, the maximum stress, which occurs at the middle of the longest side, is found by the following equation:

$$f_s = \frac{T}{\alpha b t^2} \tag{13.19}$$

Values of α are given in Table 13.1. These values have been calculated by theoretical methods,[2] which are not within the scope of the present discussion. For large ratios of b/t, the value of α is 0.333, which corresponds with Eq. 13.18. For smaller values of b/t, the effects of the ends are more noticeable, and the values of α are smaller than 0.333.

TABLE 13.1. CONSTANTS FOR EQUATIONS 13.19 AND 13.20

b/t	1.00	1.50	1.75	2.00	2.50	3.00	4	6	8	10	∞
α	0.208	0.231	0.239	0.246	0.258	0.267	0.282	0.299	0.307	0.313	0.333
β	0.141	0.196	0.214	0.229	0.249	0.263	0.281	0.299	0.307	0.313	0.333

The angle of twist of a rectangular shaft of length L may be obtained from the equation

$$\phi = \frac{TL}{\beta b t^3 G} \tag{13.20}$$

where ϕ is in radians and β is a constant which is given in Table 13.1.

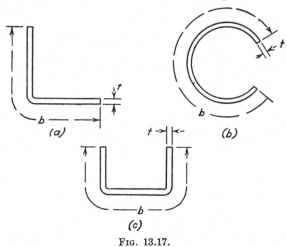

FIG. 13.17.

The torsional properties of a rectangular plate are not appreciably affected if the plate is bent to some cross section, such as those shown in Fig. 13.17, provided that the end cross sections of the member are free to warp. These sections may therefore be analyzed by Eqs. 13.19 and 13.20. The angle of twist for a shaft of any cross section may be expressed in the form

$$\phi = \frac{TL}{KG} \tag{13.21}$$

where K is a constant which depends only on the cross-sectional area. By comparing Eqs. 13.20 and 13.21, the value of K for a rectangular cross section is found to be $\beta b t^3$. For a cross section made up of several rectangular elements, such as that shown in Fig. 13.18, the value of K is given approximately by the following equation:

$$K = \beta_1 b_1 t_1^3 + \beta_2 b_2 t_2^3 + \beta_3 b_3 t_3^3 \tag{13.22}$$

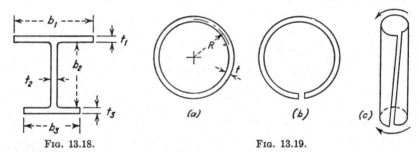

FIG. 13.18.   FIG. 13.19.

Example. The round tube shown in Fig. 13.19(a) has an average radius R and a wall thickness t. Compare the torsional strength and rigidity of this tube with that of a similar tube which is slit for its entire length, as shown in Fig. 13.19(b) and (c). Assume $R/t = 20$.

Solution. Approximate values of the area and moment of inertia will be satisfactory for thin-walled tubes. The area is equal to the product of the circumference $2\pi R$ and the wall thickness t. The polar radius of gyration is approximately equal to R, and the polar moment of inertia is then obtained as follows:

$$J = 2\pi R^3 t \tag{13.23}$$

It will also be sufficiently accurate to use the average radius R in place of the outside radius in computing the maximum shearing stress. The values for the closed tube are obtained from Eqs. 13.14 and 13.17.

$$f_s = \frac{Tr}{J} = \frac{T}{2\pi R^2 t} \tag{13.24}$$

$$\phi = \frac{TL}{JG} = \frac{TL}{2\pi R^3 t G} \tag{13.25}$$

The slit tube will now be analyzed by Eqs. 13.18 and 13.20, assuming $b = 2\pi R$ and $\alpha = \beta = 0.333$.

$$f_s = \frac{3T}{2\pi R t^2} \tag{13.26}$$

$$\phi = \frac{3TL}{2\pi R t^3 G} \tag{13.27}$$

The ratio of shearing stresses for the two members is obtained by dividing Eq. 13.26 by Eq. 13.24. The values of the stress and angle for the closed tube will be designated f_{s0} and ϕ_0.

$$\frac{f_s}{f_{s0}} = \frac{3R}{t} = 60$$

The shearing stresses are therefore sixty times as high in the slit tube as in the closed tube, or the closed tube would be sixty times as strong if the allowable shearing stresses are equal.

The torsional stiffness of the two members may be compared by dividing Eq. 13.27 by Eq. 13.25.

$$\frac{\phi}{\phi_0} = \frac{3R^2}{t^2} = 1,200 \tag{13.28}$$

For a given angle of twist, the closed tube resists 1,200 times the torsion of the open section, or for a given torque, the open section twists through an angle 1,200 times as great as the closed tube. It has been assumed that the open tube was free to distort as shown in Fig. 13.19(c), or that the end cross sections were not restrained against warping.

13.8. End Restraint of Torsion Members. In the previous article it was assumed that the end cross sections of the torsion members were free to warp from their original plane and that there were no stresses normal to the cross sections. It has been pointed out that many aircraft structural members must be constructed with thin webs and that such members are very inefficient in resisting torsional loads unless they form a closed box. In some cases it is necessary to use open sections with thin webs, and in most of these cases the ends should be restrained to provide additional torsional rigidity and strength.

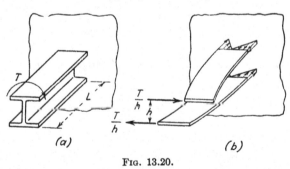

FIG. 13.20.

The I beam shown in Fig. 13.20 resists part of the torsion by means of the shear stresses distributed for the individual rectangles, as shown in Fig. 13.16. The remainder of the torsion is resisted by horizontal bending of the beam flanges, as shown in Fig. 13.20(b). The proportion of the torsion which is resisted by each of the two ways depends on the dimensions of the cross section and the length of the member. This proportion also varies along the member, as more of the torsion is resisted by flange bending near the fixed end than near the free end.

For members in which the webs are thin and the length is not great, all the torsion may be assumed to be resisted by flange bending. This assumption was made in Chap. 8 in the analysis of a wing with the lower skin removed for a wheel well. In the case of long members with thick webs, all the torsion may be assumed to be resisted by the torsional resistance of the rectangular elements. In some cases, however, it may be necessary to calculate the proportion of the torsion resisted by each method.

FIG. 13.21.

A member which is not restrained at the ends twists as shown in Fig. 13.21. The angle of twist will vary uniformly along the length and may be computed from Eq. 13.21. For a member resisting torsion by means of flange bending, as shown in Fig. 13.20(b), the flange bending stresses vary from zero at the free end to maximum values at the fixed end. The angle of twist and the amount of cross section warping vary along the span. For the I-beam cross section, Timoshenko[2] gives the following equations for the maximum flange bending moment, M_{max}, for the maximum torsion resisted by web shears, T'_{max}, and for the angle of twist, θ, in the length L.

$$M_{max} = \frac{T}{h} a \tanh \frac{L}{a} \tag{13.29}$$

$$T'_{max} = T \left(1 - \operatorname{sech} \frac{L}{a}\right) \tag{13.30}$$

$$\theta = \frac{T}{KG}\left(L - a \tanh \frac{L}{a}\right) \tag{13.31}$$

The constant a is a ratio of the relative flange bending rigidity to the torsional rigidity.

$$a = \frac{h}{2}\sqrt{\frac{2I_f E}{KG}} \tag{13.32}$$

The term I_f is the moment of inertia of one flange of the beam about a vertical axis, K is defined by Eq. 13.22, and h is the beam depth between centers of flanges. The analysis for the I beam shown in Fig. 13.20 will also apply for a beam of length $2L$ which has a torque of $2T$ applied at the center and which resists half the torque at each end, if the ends are not restrained against warping. In this case, a cross section at the center of the beam is prevented from warping because of the symmetry of loading.

A channel cross section, such as that shown in Fig. 13.22(a), is frequently used as an aircraft structural member. If a channel member is restrained at the end, it will resist torsion by bending of the flanges in much the same manner as the I-beam flanges shown in Fig. 13.20(b).

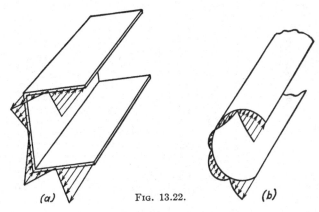

(a)   FIG. 13.22.   (b)

The analysis for flange bending is slightly more complicated than for the I beam, because the vertical web of the channel acts with the flanges in resisting flange bending, as shown in Fig. 13.22(a). The cross section shown in Fig. 13.22(b) acts in the same way as the channel section, but the stress distribution is still more difficult to analyze.

13.9. Torsional Stresses above the Elastic Limit. It has been assumed that the stresses are below the elastic limit in the previous analysis of torsional stresses. In many design applications, the ultimate torsional strength is desired. While there is not much published information concerning stress-strain curves for specimens in pure shear, these curves will have the same general shape as the tension stress-strain curves and will have ordinates approximately 0.6 of those for the tension curves. The torsional stresses in a round bar will therefore be distributed as shown in Fig. 13.23 when the stresses exceed the elastic limit.

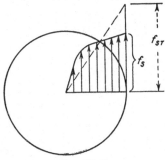

FIG. 13.23.

As in the plastic bending of beams, it is convenient to work with a fictitious stress instead of the exact stress distributions. This stress is designated as the *torsional modulus of rupture* F_{st}, which is defined by the equation

$$F_{st} = \frac{Tr}{J} \tag{13.33}$$

where T is the ultimate torsional strength of the member. For steel tubes, the value of F_{st} depends on the proportions of the cross section. The values of the ratio F_{st}/F_{tu} are shown in Fig. 13.24 for various values of the ratio of outside diameter to wall thickness D/t. These curves are taken from *ANC*–5a.[3]

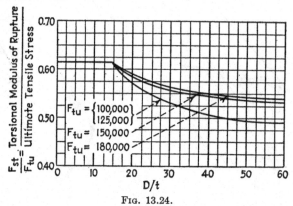

FIG. 13.24.

Equation 13.33 applies only to circular or hollow circular cross sections, and Fig. 13.24 supplies information for allowable stresses on all such cross sections for various aircraft steels. The plastic torsional stress distribution in noncircular sections cannot usually be obtained by any simple analysis. It may often be necessary to make static tests on torsion members of noncircular cross sections in order to determine the allowable torsional moments to use for design.

Example 1. A round steel tube 1 by 0.065 in. resists a design torsional moment of 5,000 in-lb. Find the margin of safety if the ultimate tensile stress F_{tu} is 100,000 psi.

Solution. From Table 1 of the Appendix, the values $D/t = 15.38$ and $I/y = 0.04193$ in. are obtained for this tube. From Fig. 13.24, the ratio F_{st}/F_{tu} = 0.6 is obtained. The margin of safety is therefore calculated as follows:

$$F_{st} = 0.6 \times 100,000 = 60,000 \text{ psi}$$
$$f_{st} = \frac{Tr}{J} = \frac{T}{2I/y} = \frac{5,000}{2 \times 0.04193} = 59,600 \text{ psi}$$
$$MS = \frac{F_{st}}{f_{st}} - 1 = \frac{60,000}{59,600} - 1 = \underline{0.007}$$

Example 2. Design a round tube to resist a torsional moment of 8,000 in-lb. The minimum permissible wall thickness is 0.049 in., and the material has an ultimate tensile stress F_{tu} of 100,000 psi.

Solution. A torsion tube must be designed by a trial-and-error process, because the allowable stress depends on the D/t ratio and cannot be determined exactly until the tube is selected. For tubes of approximately the same weight, the larger I/y values are obtained for larger diameters, but the higher allowable moduli of rupture are obtained for thicker tube walls. Thus several tubes may have about the same weight and strength, although tubes with higher D/t ratios usually have a strength-weight advantage. Assume an average ratio, F_{st}/F_{tu} of 0.6 as a first approximation, or $F_{st} = 60,000$ psi. The required tube properties are now obtained as follows:

$$\text{Required } \frac{J}{r} = \frac{2I}{y} = \frac{T}{F_{st}} = \frac{8,000}{60,000}$$

or

$$\text{required } \frac{I}{y} = 0.0667 \text{ in.}^4$$

From Table 1 of the Appendix, the lightest tube with this value of I/y is a 1½-by 0.049-in. tube, with $D/t = 30.60$, $I/y = 0.07847$, and a weight of 6.32 lb/100 in. From Fig. 13.24, $F_{st}/F_{tu} = 0.53$, or $F_{st} = 53,000$ psi. The margin of safety is calculated below.

$$f_{st} = \frac{T}{2I/y} = \frac{8,000}{2 \times 0.07847} = 51,000 \text{ psi}$$

$$\text{MS} = \frac{F_{st}}{f_{st}} - 1 = \frac{53,000}{51,000} - 1 = \underline{0.04}$$

Other tubes are compared below, in order to select the lightest tube of the required strength. The 1½- by 0.049-in. tube is the only one investigated which has a positive margin of safety. Any other tubes which are lighter than 1½ by 0.049 in. and which have a minimum wall thickness of 0.049 in. are obviously under the required strength.

TABLE 13.2

Tube	$\dfrac{wt}{100 \text{ in.}}$	$\dfrac{D}{t}$	$\dfrac{I}{y}$	F_{st}	f_{st}	MS
1½ × 0.049	6.32	30.60	0.07847	53,000	51,000	0.04
1⅜ × 0.049	5.78	28.05	0.06534	54,000	61,500	−0.12
1¼ × 0.058	6.15	21.55	0.06187	56,000	64,700	−0.12

13.10. Combined Stresses and Stress Ratios. The design of members resisting tension, bending, or torsion has been discussed in the preceding articles. Many aircraft structural members, however, must resist the simultaneous action of two or more of these loading conditions. If stresses are below the elastic limit of the material, the normal and shear stresses at any point may be combined by the methods discussed in Chap. 4. When working stresses based on the elastic limit are used, as in most types of machine design, it is customary to determine the principal stresses and the maximum shearing stresses at a point and to compare these with the allowable working stresses.

When members are designed on the basis of ultimate strength, it is not feasible to calculate the true principal stresses in the case of plastic bending or torsion. Even if the true stresses are known, it is difficult to predict the loads at which failure would occur under combined loading conditions. A tension member fails when the average stress reaches the ultimate tensile strength F_{tu} for the material, but a member resisting combined stresses may fail before the maximum principal stress reaches the value of F_{tu}. Failure in pure shear, for example, occurs when the principal tension and compression stresses are about $0.6F_{tu}$. Various theories of failure for materials under combined loading have been developed, but none of them give a simple method of predicting the failure of all materials.

Shanley[4] has proposed a method which provides a practical means of considering the combination of the fictitious stresses of bending or torsion and of obtaining the allowable ultimate loads for combined loadings. This method consists of using *stress ratios* and has been extensively adopted in the analysis of aircraft structures. The stress ratio method may be applied to almost any combination of two or more types of loading, although in some cases it may be necessary to conduct some tests of specimens in order to apply the method. The method of stress ratios will first be applied to some special cases of loading and will be then stated later in a general form.

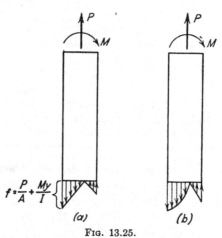

$$f = \frac{P}{A} + \frac{My}{I}$$

(a)   (b)

Fig. 13.25.

One of the simplest types of combined loadings is that of tension and bending, as shown in Fig. 13.25. The stresses may be added algebraically, and for small loads the stress is $P/A + My/I$ at any point in the cross section, as shown in Fig. 13.25(a). When the stresses exceed the elastic limit, however, the distribution becomes similar to that shown in Fig. 13.25(b). The true stress distribution is difficult to calculate, and it is convenient to use the method of stress ratios in predicting the strength. For pure bending with no tension, failure will occur when the bending stress ratio $R_b = f_b/F_b$ approaches unity. Similarly, for tension with no bending, failure will occur when the tension stress ratio $R_t = f_t/F_{tu}$ approaches unity. Since the stresses below the elastic limit add directly, it seems logical to add the stress ratios, and tests substantiate this method. The failure under combined tension and bending therefore occurs under the following condition:

$$R_b + R_t = 1 \tag{13.34}$$

The margin of safety is defined by the following equation, which corresponds with previously used expressions when either of the stress ratios is zero.

$$MS = \frac{1}{R_b + R_t} - 1 \tag{13.35}$$

For round tubes in combined bending and torsion the stresses do not add algebraically. For stresses below the elastic limit, the maximum stress at any point may be readily obtained by methods previously dis-

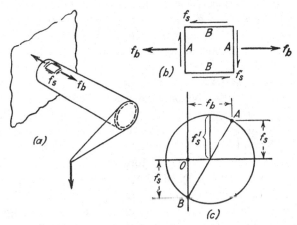

Fig. 13.26.

cussed in Chap. 4. The maximum tension and shearing stresses in the tube shown in Fig. 13.26(a) will occur at the support and on the upper surface as shown. The tension stress in the direction of the axis of the tube f_b is obtained from the bending moment M as follows:

$$f_b = \frac{My}{I} \tag{13.36}$$

The shearing stress f_s on the planes shown is obtained from the torsional moment T.

$$f_s = \frac{Ty}{2I} \tag{13.37}$$

The small element at the top of the tube is enlarged in Fig. 13.26(b). The principal stresses and the maximum shearing stresses at this point on the tube are obtained from the stresses f_b and f_s by means of Mohr's circle construction, as shown in Fig. 13.26(c). The maximum shearing stress $f_{s'}$ is equal to the radius of the circle.

$$f_{s'} = \sqrt{(f_s)^2 + \left(\frac{f_b}{2}\right)^2} \tag{13.38}$$

Substituting values of f_b and f_s from Eqs. 13.36 and 13.37 into Eq. 13.38, there is obtained

$$f_{s'} = \frac{\sqrt{M^2 + T^2}}{2I/y} = \frac{T_e}{2I/y} \tag{13.39}$$

where T_e is an equivalent torque defined by Eq. 13.39. If the torsion is larger than the bending moment, the maximum shearing stress may be used to predict the strength of the tube. If the bending moment is large in comparison to the torsion, the principal stresses are more important than the shearing stresses. In machine design practice, shafts in bending and torsion are designed so as to keep the shearing stress of Eq. 13.39 and the principal stresses smaller than the corresponding allowable working stresses, which are a certain fraction of the yield stress.

When a tube is stressed beyond the elastic limit, Eqs. 13.36 and 13.37 do not yield the true stresses, but yield the fictitious stresses defined as the bending modulus of rupture and the torsional modulus of rupture. Equations 13.38 and 13.39 therefore are not exact when the stresses are beyond the elastic limit. When the torsion is large in comparison to the bending moment, however, Eq. 13.39 may be used to predict the ultimate strength of a tube if the value of F_{st} as obtained from Fig. 13.24 is used for the allowable stress. A more accurate method of designing tubes in torsion or bending is that of stress ratios. The bending and torsional moduli of rupture f_b and f_{st} are calculated from Eqs. 13.36 and 13.37, and the allowable values F_b and F_{st} are found by the same methods as for tubes in bending or torsion only. The stress ratio in bending, $R_b = f_b/F_b$, is combined with the stress ratio in torsion, $R_{st} = f_{st}/F_{st}$, in the same manner that the loads are combined in Eq. 13.39. Failure occurs for the following condition:

$$R_b^2 + R_{st}^2 = 1 \tag{13.40}$$

The margin of safety is defined as follows:

$$MS = \frac{1}{\sqrt{R_b^2 + R_{st}^2}} - 1 \qquad (13.41)$$

Where either the bending moment or the torsional moment is zero, Eqs. 13.40 and 13.41 yield values which were previously obtained for bending or torsion only.

Equations 13.34 and 13.40 represent two ways in which stress ratios may be combined. These equations may be plotted with the two stress ratios as coordinates, as shown in Fig. 13.27. The graph for Eq. 13.34 is shown by the straight line, and that for Eq. 13.40 by the circle. Other combinations of loading conditions may be represented by a more general equation

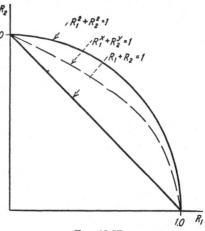

$$R_1^x + R_2^y = 1 \qquad (13.42)$$

FIG. 13.27.

where the exponents x and y must usually be found experimentally by plotting test results, as shown in Fig. 13.27, and by writing an equation for a curve passing through the points.

Example 1. A 2- by 0.095-in. steel tube is heat-treated to an ultimate tensile strength $F_{tu} = 180,000$ psi. (a) Find the margin of safety if the tube resists a design tension load of 50,000 lb and a design bending moment of 30,000 in-lb. (b) Find the margin of safety if the tube resists a bending moment of 30,000 in-lb and a torsional moment of 50,000 in-lb.

Solution. a. From Table 1 of the Appendix, $A = 0.5685$ in.², $D/t = 21.05$, and $I/y = 0.2586$ in.³ From Fig. 13.3, $F_b = 211,000$ psi. The stresses and stress ratios are calculated below.

$$f_b = \frac{My}{I} = \frac{30,000}{0.2586} = 116,000 \text{ psi}$$

$$R_b = \frac{f_b}{F_b} = \frac{116,000}{211,000} = 0.550$$

$$f_t = \frac{P}{A} = \frac{50,000}{0.5685} = 88,000 \text{ psi}$$

$$R_t = \frac{f_t}{F_{tu}} = \frac{88,000}{180,000} = 0.488$$

The margin of safety is obtained from Eq. 13.35.

$$MS = \frac{1}{R_b + R_t} - 1 = \underline{-0.035}$$

The negative margin of safety indicates that the tube is unsatisfactory.

b. The stress ratio for bending is the same as computed in part (a). The torsional modulus of rupture is obtained from Fig. 13.24, for $D/t = 21.05$. $F_{st} = 0.58 \times 180,000 = 104,000$ psi. The stress ratio for torsion is calculated below.

$$f_{st} = \frac{Ty}{2I} = \frac{50,000}{2 \times 0.2586} = 96,500$$

$$R_{st} = \frac{f_{st}}{F_{st}} = \frac{96,500}{104,000} = 0.925$$

The margin of safety is now calculated from Eq. 13.41.

$$MS = \frac{1}{\sqrt{R_b^2 + R_{st}^2}} - 1 = \underline{-0.07}$$

The tube is unsatisfactory.

PROBLEMS

13.11. A 2- by 0.083-in. tube of 24S–T aluminum alloy is 20 in. long and resists a torsional moment T of 8,000 in-lb. Find the maximum shearing stress and the angle of twist.

13.12. An elevator torque tube is made of 24S–T aluminum alloy and has dimensions of 2- by 0.083 in. Find the angle of twist if the shearing stress has a value of 10,000 psi for a length of 80 in.

13.13. A 24S–T extrusion of the shape shown in Fig. 13.18 has dimensions of $b_1 = b_2 = b_3 = 2$ in., and $t_1 = t_2 = t_3 = 0.2$ in. Find the shearing stress and the angle of twist if the section resists a torque of 100 in-lb and has a length of 10 in. Assume (a) both ends free to warp and (b) one end restrained against warping.

13.14. Repeat Prob. 13.13 for a length of 20 in.

In the following problems, all loads are ultimate or design loads.

13.15. Design a round steel tube to resist a torsional moment of 10,000 in-lb. Assume material properties of (a) $F_{tu} = 100,000$ psi, (b) $F_{tu} = 125,000$ psi, and (c) $F_{tu} = 180,000$ psi.

13.16. Design a round steel tube to resist a tension load of 10,000 lb and a bending moment of 5,000 in-lb. Use material properties of (a) $F_{tu} = 100,000$ psi and (b) $F_{tu} = 150,000$ psi.

13.17. Design a round steel tube to resist a bending moment of 6,000 in-lb and a torsional moment of 8,000 in-lb. Use Eq. 13.39 for a preliminary trial, but use the method of stress ratios for the final design. Assume (a) $F_{tu} = 125,000$ psi and (b) $F_{tu} = 180,000$ psi.

13.18. Repeat Prob. 13.17, assuming the cross section to be near a weld, but to be heat-treated after welding.

REFERENCES FOR CHAPTER 13

1. COZZONE, F. P.: Bending Strength in the Plastic Range, *J. Aeronaut. Sci.*, May, 1943, p. 137.
2. TIMOSHENKO, S.: "Strength of Materials," Part I, D. Van Nostrand Company, Inc., New York, 1930.
3. *ANC*–5a, "Strength of Metal Aircraft Elements," Subcommittee on Air Force-Navy-Civil Aircraft Design Criteria of the Munitions Board Aircraft Committee, May, 1949.
4. SHANLEY, F. R., and E. I. RYDER: Stress Ratios, *Aviation Mag.*, June, 1937.

CHAPTER 14

DESIGN OF COMPRESSION MEMBERS

14.1. Beam-deflection Equations. The methods used in the design of compression members, or columns, are based on beam-deflection equations. Columns do not fail as a result of the direct compression stresses only but as a result of the combined compression and bending stresses. Since the magnitudes of the bending stresses depend on the bending deflections, it is necessary to derive the column equations from beam-deflection equations.

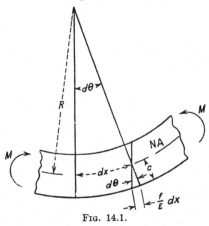

Fig. 14.1.

The equations for beam deflections will be derived from the customary assumptions that stress is proportional to strain and that the deflections are small in comparison to the original dimensions. Only deformations resulting from bending stresses are usually considered, but if shearing deformations are appreciable they may be computed separately and superimposed. An initially straight beam is shown with exaggerated deflections in Fig. 14.1. The two cross sections a distance dx apart are parallel in the unstressed condition, but have a relative angle $d\theta$ in the stressed condition. The angle $d\theta$ may be obtained by considering the small triangle between the neutral axis and a point at a distance c below the neutral axis. The stress f at this point is obtained from the flexure formula:

$$f = \frac{Mc}{I} \tag{14.1}$$

The longitudinal fiber at a distance c from the neutral axis has an elongation $f\,dx/E$ in the length dx. The angle $d\theta$ is obtained by dividing this elongation by the distance c, as shown in Fig. 14.1.

$$d\theta = \frac{f\,dx}{Ec} \tag{14.2}$$

From Eqs. 14.1 and 14.2,

$$d\theta = \frac{M}{EI} dx \qquad (14.3)$$

The deflection curve of the beam will be represented by coordinates x and y, as shown in Fig. 14.2. The beam is assumed to be initially straight and parallel to the x axis. The deflections are small, and the angle θ between a tangent to the deflection curve and the x axis is small enough so that it is sufficiently accurate to assume that the angle in radians, the sine of the angle, and the tangent of the angle are all equal.

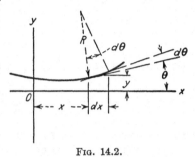

Fig. 14.2.

$$\theta = \sin \theta = \tan \theta \qquad (14.4)$$

$$\theta = \frac{dy}{dx} \qquad (14.5)$$

From Eqs. 14.3 and 14.5,

$$\frac{d^2y}{dx^2} = \frac{M}{EI} \qquad (14.6)$$

All conventional methods of obtaining beam deflections are based on Eq. 14.6. It should be noted that the deflections y are measured positive upward, so that a positive bending moment M produces a positive curvature d^2y/dx^2. Equation 14.6 is often derived from the assumption that y is positive down, and a negative sign is introduced because a positive bending moment would then produce a negative curvature.

14.2. Long Columns. Compression members tend to fail as a result of the lateral bending induced by the compression load, an action which is commonly termed *buckling*. In the case of columns which are long in comparison to their other dimensions, *elastic buckling* occurs, or the columns buckle when the compressive stresses are below the elastic limit. Such columns are termed *long columns*.

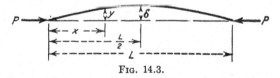

Fig. 14.3.

The initially straight column shown in Fig. 14.3 is assumed to be held in the deflected position by means of the compressive forces P. The bending moment at any cross section is found from the equation

$$M = -Py \qquad (14.7)$$

It is assumed that the material does not exceed the elastic limit at any point, and therefore Eq. 14.6 is applicable. The differential equation of the deflection curve is obtained from Eqs. 14.6 and 14.7.

$$\frac{d^2y}{dx^2} + \frac{P}{EI} y = 0 \tag{14.8}$$

The general solution of Eq. 14.8 is as follows:

$$y = C_1 \sin \sqrt{\frac{P}{EI}} x + C_2 \cos \sqrt{\frac{P}{EI}} x \tag{14.9}$$

This solution may be verified by substitution and must be a general solution of the second-order differential equation because it contains the two arbitrary constants C_1 and C_2. In order to satisfy the end conditions shown in Fig. 14.3, the deflection curve must pass through the points ($x = 0$, $y = 0$) and ($x = L$, $y = 0$). Substituting the first condition into Eq. 14.9, the value $C_2 = 0$ is obtained. The second condition, that $y = 0$ when $x = L$, may be satisfied when $C_1 = 0$, which is a trivial solution, corresponding to the conditions of small loads when the column remains straight. The only solution of interest in column analysis is that for which the column is deflected and C_1 is not zero. This solution, in which the column is buckled, is obtained when the value of P satisfies the condition

$$\sqrt{\frac{P}{EI}} L = \pi, 2\pi, 3\pi, \ldots, \text{ or } n\pi$$

or

$$P = \frac{n^2\pi^2 EI}{L^2} \tag{14.10}$$

The value of P is obviously a minimum when $n = 1$, and higher values of n have no significance in this case, since the column will fail at the smallest value of P which will produce buckling. The *critical*, or buckling, load P_{cr} is therefore defined as follows:

$$P_{cr} = \frac{\pi^2 EI}{L^2} \tag{14.11}$$

It is often more convenient to work with a buckling stress, $F_{cr} = (P_{cr}/A)$. This may be obtained by introducing the radius of gyration of the cross-sectional area, $\rho = \sqrt{I/A}$, into Eq. 14.11.

$$F_{cr} = \frac{\pi^2 E}{\left(\dfrac{L}{\rho}\right)^2} \tag{14.12}$$

The analysis of long columns was first published by the Swiss mathematician Euler. Equation 14.11 (or Eq. 14.12) is commonly called the Euler equation, and the buckling load is often called the Euler load.

The value of C_1 cannot be obtained at the critical load. This value is equal to the maximum deflection δ at the center of the column, which is indeterminate for the assumed conditions. For loads smaller than P_{cr}, the deflection C_1 or δ must be zero, or the column remains straight. At the critical load, any deflection δ, for which the maximum stress is below the elastic limit, will satisfy conditions of equilibrium. This may be shown experimentally by loading a long column in a standard testing machine. As the ends of the column are moved together, the column remains straight until the Euler load is obtained. As the ends continue to move together, the load remains constant at the Euler load, but the lateral deflection δ increases. If the elastic limit is not exceeded, the column will return to its initial shape when the load is removed.

14.3. Eccentrically Loaded Columns. In an actual structure it is not possible for a column to be perfectly straight, or to be loaded exactly at the centroid of the area. The action of a practical long column may be approximated by the member shown in Fig. 14.4, in which the column

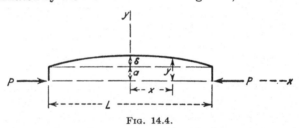

Fig. 14.4.

is initially straight but the loads both have an eccentricity a. The axes of coordinates are taken as shown. The equation of the deflection curve is still represented by Eq. 14.9, since Eqs. 14.7 and 14.8 are applicable, but the constants C_1 and C_2 must be found from the conditions that the deflection curve satisfies the two conditions, $(x = 0, y = \delta + a)$ and $[(x - 0, dy/dx = 0)]$. Substituting these two conditions into Eq. 14.9, there is obtained

$$y = (\delta + a) \cos \sqrt{\frac{P}{EI}}\, x \tag{14.13}$$

The value of δ may now be found from the condition that $y = a$ for $x = L/2$. Substituting these values into Eq. 14.13, the following value of δ is obtained:

$$\delta + a = a\left(\sec \sqrt{\frac{P}{EI}}\, \frac{L}{2}\right) \tag{14.14}$$

The deflection δ of an eccentrically loaded column therefore increases with an increase in the load P. As the value of P reaches the Euler load P_{cr}, as defined by Eq. 14.11, the deflection becomes infinite, since sec $\pi/2 = \infty$. Figure 14.5 shows the relationship between P and δ for various eccentricities a, as determined from Eq. 14.14. All curves are asymptotic to the line $P = P_{cr}$, as this is the theoretical buckling load regardless of the eccentricity of loading. A large deflection may stress the material beyond the elastic limit and cause failure before the Euler load is obtained, since the long-column equations would then no longer apply.

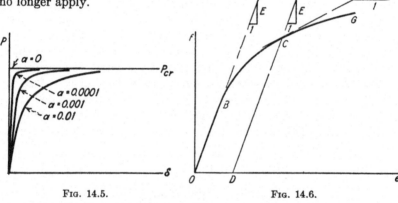

FIG. 14.5.   FIG. 14.6.

14.4. Short Columns. Columns of any specific material are classified according to their slenderness ratio L/ρ. For a slenderness ratio greater than a certain critical value, the column is a long column and is analyzed by Eq. 14.12. Short columns have a slenderness ratio less than this critical value. The critical L/ρ corresponds to the value for which the maximum compressive stress in the column is equal to the stress at which the compressive stress-strain curve deviates from a straight line, as shown by point B in Fig. 14.6. This stress is usually considerably smaller than the yield stress, point C of Fig. 14.6, at which the material has a permanent unit elongation of 0.002.

As stated in Chap. 11, most aircraft materials have stress-strain curves similar to that shown in Fig. 14.6, in which the stress-strain curve has a positive slope at all points. The constant slope of this stress-strain curve below the elastic limit is equal to the modulus of elasticity E, and the variable slope above the elastic limit is termed the tangent modulus of elasticity E_t. Ductile materials, such as mild steel, may have a zero or negative value of E_t near the yield point. If the value of E_t is positive at all points, a short column may remain perfectly straight when loaded to stresses beyond the yield point. If such a column has a slight lateral deflection, the internal resisting moment is found from an equation similar to Eq. 14.6, except that E is replaced by E_t, the tangent modulus for the compressive stress.

$$M = E_t I \frac{d^2y}{dx^2} \tag{14.15}$$

If this internal resisting moment is greater than the bending moment produced by the load P, the column will remain straight when loaded. If the internal resisting moment is not as large as the external bending moment, the deflection will increase, and the column will probably fail. When the bending moment of the load P is equal to the resisting moment defined by Eq. 14.15, the load P may be obtained in the same manner as in the Euler equation, but with E_t substituted for E.

$$P = \frac{\pi^2 E_t I}{L^2} \tag{14.16}$$

This equation is called the tangent modulus equation, or the Engesser equation.

The tangent modulus equation does not quite represent the true conditions for short columns. At point C of the stress-strain diagram of Fig. 14.6, a small increase in the compressive strain produces an increase in compressive stress as determined by the portion CG of the curve, which has a slope E_t. A small decrease in the compressive strain, however, produces a decrease in stress as indicated by line CD, which has a slope E. If a short column deflects laterally in such a way that the compressive strain on the convex side is decreased, the resisting moment will be greater than that given by Eq. 14.15, because the modulus of elasticity for part of the cross section is E rather than E_t. Thus the correct modulus of elasticity should be a value between E and E_t. Values of an effective modulus of elasticity have been derived on the assumption that the column is supported laterally and remains straight until the ultimate load is applied and then buckles with no change in axial load. The column formula obtained by substituting this modulus in the Euler equation is termed the *reduced modulus* equation and is frequently referred to in the literature.

Shanley[1,2] has shown that the correct load resisted by a short column is between the value given by the tangent modulus equation and that given by the reduced modulus equation. The tangent modulus equation yields values which are slightly low, since some strain reversal must take place before the ultimate column load is reached. The reduced modulus equation always yields values which are too high, since the column is not laterally supported when the load is applied. The tangent modulus equation is frequently used, because it corresponds closely to test results and is always conservative.

It is customary to represent column equations by plotting values of the average compressive stress $F_c = P/A$ against values of the slenderness ratio L/ρ. Figure 14.7 shows such curves. The slenderness ratio beyond which this material acts as a long column is about 115. The

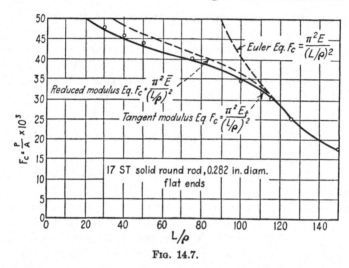

Fig. 14.7.

stress at this point corresponds to the stress at point B of Fig. 14.6. For slenderness ratios less than 115, the compressive stress is higher, and the tangent modulus of elasticity E_t is smaller than E. The points on the column curve are therefore below the Euler curve in the short-column range. The test points are seen to follow the tangent modulus curve very closely. Such test loads will always be slightly lower than theoretical loads because of unavoidable eccentricities of loading. The curves shown in Fig. 14.7 represent values for an actual specimen, whereas similar design curves are based on minimum guaranteed properties of the material, and give somewhat lower stresses.

14.5. Column End-fixity. In the foregoing analysis it has been assumed that the column is hinged at both ends so that it can rotate freely. Such a condition occasionally exists in an aircraft structure when a member is connected by a single bolt at each end. In most cases, however, compression members are connected in such a way that they are restrained against rotation at the ends. If a compression member is rigidly fixed against rotation at both ends, the deflection curve for elastic buckling will have the shape shown in Fig. 14.8(*b*). At the quarter points of the fixed column there will be points of reverse curvature, or points of contraflexure. At points of contraflexure there is no curvature and hence no bending moment. The portion of the column between points of contraflexure may therefore be treated as a pin-ended column.

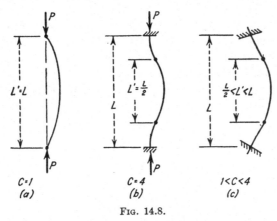

FIG. 14.8.

The length L' between the points of contraflexure is used in place of L in the column equations previously derived, and the slenderness ratio is defined as L'/ρ. An end-fixity term c is often used and is defined in the following equations:

$$F_{cr} = \frac{\pi^2 E}{(L'/\rho)^2} = \frac{c\pi^2 E}{(L/\rho)^2} \qquad \text{for long columns only} \qquad (14.17)$$

or

$$L' = \frac{L}{\sqrt{c}} \qquad \text{for all columns} \qquad (14.18)$$

For the fixed-end condition of Fig. 14.8(*b*), $L' = L/2$ and $c = 4$. The fixed-ended column will therefore resist four times the load of a similar pin-ended column, if both are in the long-column range. This same relation does not hold in the short-column range, because the value of E_t in Eq. 14.16 is smaller for the smaller values of L'. This fact is evident from Fig. 14.7, where it may be seen that a reduction in L'/ρ has a much smaller effect on F_c in the short-column range than it has in the Euler-column range.

In order to obtain complete end-fixity, the compression member must be attached to a structure of infinite rigidity at both ends. This condition is approached less frequently in practice than the condition of hinged ends. Most practical columns have end conditions somewhere between hinged and fixed ends, as shown in Fig. 14.8(*c*). The ends are rigidly attached to a structure which deflects and permits the ends to rotate slightly. The true end-fixity conditions can seldom be determined exactly, and conservative assumptions must be made. Fortunately, short columns are usually used, and the effect of end-fixity on the allowable compressive stress is much smaller than it would be for long columns.

Other common end conditions for columns are shown in Fig. 14.9. For the column fixed at one end and free to both rotate and move laterally at the other end, as shown in Fig. 14.9(a), the length L' is twice the length L, since the column is similar to one-half of the column with two hinged ends. The column with one end fixed and the other free to rotate but not free to move laterally has an effective length $L' \cong 0.7L$, as shown in Fig. 14.9(b).

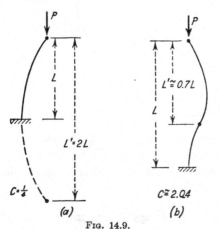

Fig. 14.9.

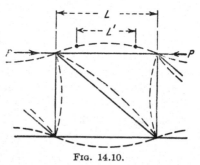

Fig. 14.10.

Welded trusses made up of steel tubes are frequently used in aircraft structures. The ends of a compression member in such a truss cannot rotate without bending all the other members at the end joints. Such a member is shown in Fig. 14.10. The problem of obtaining the true end-fixity of such a compression member is difficult, since the member may buckle either horizontally or vertically and is restrained by the torsional and bending rigidities of many other members. For a steel-tube fuselage truss, it is usually conservative to assume $c = 2.0$ for all members. If a very heavy compression member is restrained by comparatively light members, a smaller end-fixity might be obtained. Similarly, a light compression member restrained by heavy members may approach the fixity condition $c = 4$. If all the members at a joint are compression members, they may all have a tendency to rotate in the same direction, so that none of them helps restrain the others, and all should be designed as pin-ended. Where this rare case exists with members in any plane, the members perpendicular to this plane would probably supply torsional restraint to the joint. Tension members which connect to the ends of compression members supply greater restraint than similar compression members. Steel-tube engine mounts are usually designed with the conservative assumption of pin-ended members with $c = 1.0$.

Stringers which act as compression members in semimonocoque wing or fuselage structures are usually supported by comparatively flexible ribs or bulkheads. Such a stringer is shown in Fig. 14.11. Since the ribs or bulkheads are usually free to twist as shown, their restraining

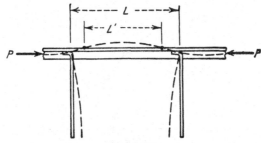

Fig. 14.11.

effect is usually neglected, and the effective column length L' is assumed equal to the length L between bulkheads. Where the bulkheads are rigid enough to provide restraint and clips are provided to attach the stringers to the bulkheads, a value $c = 1.5$, corresponding to an effective length L' of $0.815L$, is sometimes used.

14.6. Other Short-column Formulas. One disadvantage of the tangent modulus formula for short columns is that the relation between the allowable column stress, F_c, and L'/ρ cannot be expressed by a simple equation. It is often more convenient to express this relationship by a simple approximate equation which is reasonably close to the points obtained directly from column tests or from the tangent modulus equation. The short-column curves for many materials approximate the parabola

$$F_c = F_{co} - K \left(\frac{L'}{\rho}\right)^2 \qquad (14.19)$$

The constants F_{co} and K must be chosen so that the parabola fits the test data and is tangent to the Euler curve. Equating the slope of this parabola to that of the Euler curve at the point of tangency and substituting the resulting value of K into Eq. 14.19, the following equation is obtained:

$$F_c = F_{co} \left[1 - \frac{F_{co}(L'/\rho)^2}{4\pi^2 E} \right] \qquad (14.20)$$

The term F_{co} is called the column yield stress. It has little physical significance, since very short columns $(L'/\rho < 12)$ fail by block compression rather than column action, and Eq. 14.20 is not applicable in this range. The value of F_{co} is determined so that Eq. 14.20 will fit short-column test data for values of L'/ρ above the block compression range.

The general second-degree-parabola equation is shown in Fig. 14.12, with the corresponding Euler equation. The value of F_{co} represents the

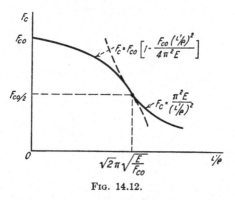

FIG. 14.12.

intercept of this curve at the point $L'/\rho = 0$. The parabola is always tangent to the Euler curve when $F_c = F_{co}/2$, as found by solving Eq. 14.20 simultaneously with the Euler equation. In the same way, the critical slenderness ratio which divides the long-column and short-column ranges is found to be $L'/\rho = \sqrt{2}\pi\sqrt{E/F_{co}}$.

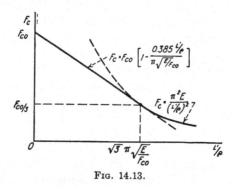

FIG. 14.13.

The short-column curves for most aluminum alloys and for several other materials are represented more accurately by straight lines. A straight line, tangent to the Euler curve, has the equation:

$$F_c = F_{co}\left(1 - \frac{0.385 L'/\rho}{\pi\sqrt{E/F_{co}}}\right) \qquad (14.21)$$

The coordinates of the point of tangency of this curve and the Euler curve are $L'/\rho = \sqrt{3}\pi\sqrt{E/F_{co}}$ and $F_c = F_{co}/3$. The value of L'/ρ at this point is the critical value dividing the short-column and long-column ranges.

Other materials have column curves which may be represented by a semicubic equation. A 1.5-degree equation which is tangent to the Euler curve has the form

$$F_c = F_{co}\left[1 - 0.3027\left(\frac{L'/\rho}{\pi\sqrt{E/F_{co}}}\right)^{1.5}\right] \tag{14.22}$$

The coordinates of the point of tangency are again obtained by solving the short-column and long-column equations simultaneously. The critical slenderness ratio is $L'/\rho = 1.527\pi\sqrt{E/F_{co}}$, corresponding to a stress of $F_c = 0.429F_{co}$.

Equations 14.20 to 14.22 and the Euler equation may be expressed in dimensionless form by using coordinates B and R_a as defined by the following equations:

$$B = \frac{L'/\rho}{\pi\sqrt{E/F_{co}}} \tag{14.23}$$

$$R_a = \frac{F_c}{F_{co}} \tag{14.24}$$

The Euler equation, Eq. 14.12, then becomes

$$R_a = \frac{1}{B^2} \tag{14.25}$$

Equations 14.20 to 14.22 will have the following forms:

$$R_a = 1.0 - 0.25B^2 \tag{14.26}$$
$$R_a = 1.0 - 0.3027B^{1.5} \tag{14.27}$$
$$R_a = 1.0 - 0.385B \tag{14.28}$$

These equations are plotted in Fig. 14.14. The dimensionless form of expressing column curves has the advantage of showing column curves for all materials on a single graph.

14.7. Practical Design Equations. The equations of the preceding article are commonly used for design purposes. The constants are well established for materials in common use, as a result of extensive test information. In the case of new or improved alloys, however, it is necessary to conduct numerous tests in order to establish the constants for use in the equations.

It has been assumed that the column cross section is such that local crippling failures do not occur. Steel tubes, for example, will not fail by local crippling if the ratio of outside diameter to wall thickness D/t is not greater than 50. If this ratio is increased, a point will be reached at which local crippling occurs before primary column failure occurs. The column equations in Table 14.1 will apply to steel tubes with D/t

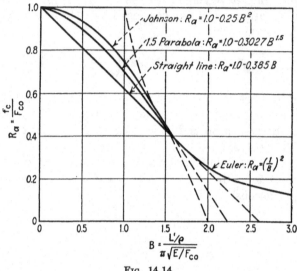

Fig. 14.14.

less than 50, or will apply for any other cross section which does not have thin walls which fail by local crippling. The short-column equations correspond to Eqs. 14.20 and 14.22, with numerical values as specified in *ANC–5*.

The ultimate tensile strengths and tensile yield strengths are listed in Table 14.1 for the various materials. Since tensile stress-strain

TABLE 14.1. COLUMN FORMULAS FOR ROUND STEEL TUBES

Material	F_{tu}, ksi	F_{ty}, ksi	Short columns F_c, psi	Critical L'/ρ	Long columns F_c, psi
1025	55	36	$36,000 - 1.172(L'/\rho)^2$	124	$267 \times 10^6/(L'/\rho)^2$
X–4130	95	75	$79,500 - 51.9(L'/\rho)^{1.5}$	91.5	$286 \times 10^6/(L'/\rho)^2$
X–4130	100	85	$90,100 - 64.4(L'/\rho)^{1.5}$	86.0	$286 \times 10^6/(L'/\rho)^2$
Heat-treated alloy steel	125	100	$100,000 - 8.74(L'/\rho)^2$	75.6	$286 \times 10^6/(L'/\rho)^2$
Heat-treated alloy steel	150	135	$135,000 - 15.92(L'/\rho)^2$	65.0	$286 \times 10^6/(L'/\rho)^2$
Heat-treated alloy steel	180	165	$165,000 - 23.78(L'/\rho)^2$	58.9	$286 \times 10^6/(L'/\rho)^2$

properties are much easier to obtain than compressive stress-strain properties, it is customary to designate a heat-treated alloy steel only by its tensile strength. For example, any heat-treated alloy steel with an ultimate tensile strength of 180,000 psi would have column properties similar to those shown in the last line of Table 14.1. If properties of this particular alloy were questionable the tensile yield stress should be obtained, but compressive stress-strain characteristics would be approximately equivalent to tensile properties.

The critical values for L'/ρ, shown in Table 14.1, represent the values of L'/ρ at the points of tangency of the short-column curve and the Euler curve. At exactly the critical slenderness ratio, either column curve could be used. At larger values of the slenderness ratio the Euler curve must be used, and at smaller values the short-column curve must be used. The long-column equations are seen to be identical for all steels except 1025 steel, which has a slightly lower modulus of elasticity E.

The short-column formulas for round aluminum-alloy tubing, shown in Table 14.2, are straight-line formulas, except in the case of the 75S–T alloy. These may be used for cross sections other than round tubes, provided that the cross sections are stable for local crippling. Even round tubes of aluminum alloy must frequently be investigated for local crippling failure. The effect of local crippling is to flatten the left-hand

TABLE 14.2. COLUMN FORMULAS FOR ROUND ALUMINUM-ALLOY TUBING AND SHAPES

Material	F_{cy}	Short columns	Critical L'/ρ	Long columns
Aluminum alloy (except 75S–T).....		Eq. 14.21, with $F_{co} = F_{cy}\left(1 + \dfrac{F_{cy}}{200,000}\right)$	$1.732\pi\sqrt{E/F_{co}}$	$103.8 \times 10^6/(L'/\rho)^2$
17S–T tubing (re-heat-treated)....	32,000	$37,000 - 269.3L'/\rho$	92.0	$103.8 \times 10^6/(L'/\rho)^2$
17S–T tubing (as received)	36,000	$42,500 - 330.5L'/\rho$	85.7	$103.8 \times 10^6/(L'/\rho)^2$
24S–T tubing..	41,000	$50,000 - 421L'/\rho$	79.2	$103.8 \times 10^6/(L'/\rho)^2$
24SR–T tubing	54,800	$70,000 - 700L'/\rho$	66.7	$103.8 \times 10^6/(L'/\rho)^2$
75S–T........		Eq. 14.20, with $F_{co} = 1.075F_{cy}$	$1.414\pi\sqrt{E/F_{co}}$	
75S–T extrusion....	70,000	$75,300 - 13.70(L'/\rho)^2$	52.5	$103.8 \times 10^6/(L'/\rho)^2$

portion of the column curve at the crippling stress. The column curves for aluminum-alloy tubes given in *ANC*–5 show allowable crippling stresses for various D/t ratios. The calculation of crippling stresses for extrusions and formed-sheet stiffeners will be discussed later.

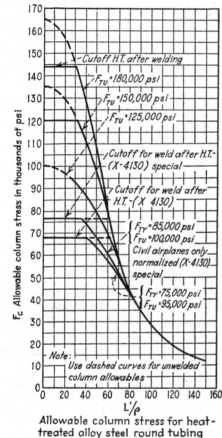

Allowable column stress for heat-treated alloy steel round tubing

Fig. 14.15.

The equations of Tables 14.1 and 14.2 may be conveniently plotted. Since all the steels listed in Table 14.1 except 1025 steel have the same long-column equation, their column curves are plotted on the same graph in Fig. 14.15. Steel-tube columns are normally welded at the ends, and the welding tends to weaken the material. For longer columns, failure occurs at the middle of the member, and the welding of the ends does not affect the strength. For very short columns, however, the allowable column strength at the middle of the member, as determined from the equations of Table 14.1, is greater than the local strength at the weld. The left-hand portions of the column curves of Fig. 14.15 correspond to the strength of the welded end. If the members are connected by some method other than welding, the dotted part of the curves, corresponding to the short-column equations, may be used.

The equations of Table 14.2 are similarly plotted in Fig. 14.16. Aluminum-alloy tubes are not welded at the ends, but have a tendency to fail by local crippling at high values of F_c, or high ratios of diameter to wall thickness D/t. The maximum permissible values of D/t for safety against local crippling failure are shown in Fig. 14.16.

Example 1. Find the column strength of a 1½- by 0.049-in. round alloy steel tube with a length of 30 in. The steel is heat-treated to an ultimate tensile strength of 180,000 psi. Assume the end-fixity coefficient, $c = 2$.

Solution. From Table 1 of the Appendix the properties of this tube are $A = 0.2234$ in.^{2}, $\rho = 0.5132$ in., and $D/t = 30.60$. The effective length is $L' = L/\sqrt{c} = 30/\sqrt{2} = 21.22$ in. The slenderness ratio is $L'/\rho = 21.22/0.5132 = 41.3$. From Table 14.1, the critical slenderness ratio for this material is 58.9, and the short-column equation should therefore be used.

$$F_c = 165{,}000 - 23.78 \left(\frac{L'}{\rho}\right)^2$$

$$= 165{,}000 - 23.78(41.3)^2 = 124{,}400 \text{ psi}$$

The allowable column load is the product of this allowable stress and the cross-sectional area.

$$P = F_c A = 124{,}400 \times 0.2234 = 27{,}800 \text{ lb}$$

Local crippling need not be investigated, since $D/t < 50$.

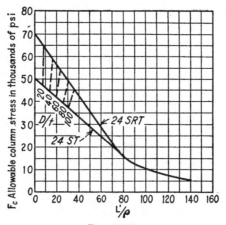

Fig. 14.16.

Alternate Solution. After obtaining the slenderness ratio as above, the value of F_c may be read from Fig. 14.15. For $L'/\rho = 41.3$, read $F_c = 124{,}000$ psi. This checks the value previously obtained with sufficient accuracy.

Example 2. Find the ultimate column strength of a round 24SR–T aluminum-alloy tube, 1×0.049 in., with $L = 20$ in. and $c = 1.5$.

Solution. The properties of the cross section are found from Table 1 of the Appendix. $A = 0.1464$, $\rho = 0.3367$, and $D/t = 20.40$. The effective length is $L' = L/\sqrt{c} = 20/\sqrt{1.5} = 16.34$ in. The slenderness ratio is $L'/\rho = 16.34/0.3369 = 48.5$. From Table 14.2, the critical slenderness ratio is 66.7; therefore the short-column equation must be used.

$$F_c = 70{,}000 - 700 \frac{L'}{\rho}$$

$$= 70{,}000 - 700 \times 48.5 = 36{,}000 \text{ psi}$$

This value of F_c may also be read from Fig. 14.16, by entering the chart with $L'/\rho = 48.5$. The allowable load is the product of this stress and the cross-sectional area.

$$P = F_c A = 36,000 \times 0.1464 = 5,270 \text{ lb}$$

From Fig. 14.16, the permissible value of D/t before local crippling occurs at this stress is over 100.

Example 3. A forging of 14S–T aluminum alloy has a cross section similar to that shown in the figure for Prob. 14.10 if the corner radii and fillets are neglected. The ends of the member are fastened by single-bolt connections, with the bolts parallel to the y axis. The length of the member is 12 in. between bolt centers. Find the allowable column load for the member.

Solution. The area and the moment of inertia about the x axis are computed in the table below.

TABLE 14.3

Element	No.	Area each	Total area	I_x of each element	Total I_x
1	8	0.040	0.320	$\dfrac{0.1(0.8)^3}{12} = 0.00427$	0.0342
2	4	0.096	0.384	$\dfrac{0.12(0.8)^3}{3} = 0.0205$	0.0820
3	1	0.144	0.144	$\dfrac{1.2(0.12)^3}{12} = 0.0002$	0.0002
Total			0.848		0.1164

The radius of gyration about the x axis is

$$\rho_x = \sqrt{\frac{I_x}{A}} = \sqrt{\frac{0.1164}{0.848}} = 0.371 \text{ in.}$$

The radius of gyration about the y axis will probably not be critical and therefore will be estimated conservatively as if the total area were concentrated at the centers of the rectangular elements 2.

$$\rho_y \cong 0.76 \text{ in.}$$

If the end restraints for rotation about both axes were the same, the smaller radius of gyration would always be critical. However, the single-bolt connections act as hinges for end rotation about the y axis, and $c = 1$, whereas they restrain end rotation about the x axis, and it will be assumed that $c = 2$. The slenderness ratios will be

$$\frac{L'}{\rho_x} = \frac{12}{(\sqrt{2} \times 0.371)} = 22.8$$

and

$$\frac{L'}{\rho_y} = \frac{12}{0.76} = 15.8$$

The column will therefore fail by buckling about the x axis, and it is not necessary to calculate ρ_y with any greater accuracy. From ANC–5, the values $F_{cy} = 50,000$ psi and $E = 10,700,000$ psi are obtained for a 14S-T forging. The formula of Table 14.2 yields

$$F_{co} = F_{cy}\left(1 + \frac{F_{cy}}{200,000}\right) = 50,000\left(1 + \frac{50,000}{200,000}\right) = 62,500 \text{ psi}$$

The straight-line formula is specified in ANC–5 for this aluminum alloy. From Eq. 14.21,

$$F_c = F_{co}\left(1 - \frac{0.385L'/\rho}{\pi\sqrt{E/F_{co}}}\right)$$

$$= 62,500\left(1 - \frac{0.385L'/\rho}{\pi\sqrt{\dfrac{10,700,000}{62,500}}}\right)$$

or

$$F_c = 62,500 - 585\frac{L'}{\rho}$$

This is the straight-line formula for a material with $F_{co} = 62,500$ psi and $E = 10,700,000$ psi. The critical slenderness ratio, from Fig. 14.13, is

$$\text{Critical } \frac{L'}{\rho} = \sqrt{3}\,\pi\sqrt{\frac{E}{F_{co}}} = 71.3$$

The value of 22.8 is therefore in the short-column range. The allowable column stress is

$$F_c = 62,500 - 585 \times 22.8 = 49,100 \text{ psi}$$

and the column load is

$$P = F_cA = 49,100 \times 0.848 = 41,700 \text{ lb}$$

14.8. Dimensionless Form of Tangent Modulus Curves. The materials used in aircraft construction are frequently improved. In other types of structural and machine design, weight is relatively unimportant, and materials have become rather standardized. Aircraft designers must adopt new materials and processes which save structural weight, even though the new materials are more expensive than standardized materials. When a new or improved material is introduced, it is difficult to make extensive column tests and crippling tests in order to establish new design allowable stresses. It would be much better to obtain simple compressive stress-strain curves for the new material and to base new column and crippling allowable stresses on these tests than to test numerous built-up column specimens. The Ramberg-Osgood equation for the stress-strain curve, discussed in Art. 11.6, provides the necessary data for comparing similar materials.

The Ramberg-Osgood equation of the stress-strain curve is

$$\epsilon = \sigma + \tfrac{3}{7}\sigma^n \tag{11.19}$$

where σ and ϵ are dimensionless functions of the stress f, the strain e, and the modulus of elasticity E, defined as follows:

$$\epsilon = \frac{Ee}{f_1} \tag{11.20}$$

and

$$\sigma = \frac{f}{f_1} \tag{11.21}$$

The stress f_1 is approximately equal to the yield stress at a permanent strain of 0.002, but must be defined as the stress at a secant modulus of elasticity of $0.7E$, in order that stress-strain curves with equal values of n are geometrically similar.

The tangent modulus of elasticity $E_t = df/de$ is readily obtained from Eqs. 11.19 to 11.21.

$$\frac{E}{E_t} = E\frac{de}{df} = \frac{d\epsilon}{d\sigma} = 1 + \frac{3}{7}n\sigma^{n-1} \tag{14.29}$$

or

$$\frac{E_t}{E} = \frac{1}{1 + \frac{3}{7}n\sigma^{n-1}} \tag{14.30}$$

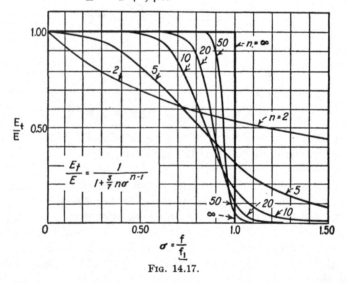

Fig. 14.17.

The tangent modulus equation may now be written as a single expression which includes both the long- and short-column ranges, since the expression 14.30 represents the modulus of elasticity below the elastic limit, as well as above the elastic limit.

$$F_c = \frac{\pi^2 E_t}{(L'/\rho)^2} = \frac{\pi^2 E}{(L'/\rho)^2}\left(\frac{1}{1 + \frac{3}{7}n\sigma^{n-1}}\right) \tag{14.31}$$

For low values of σ, the expression in brackets is approximately unity, and Eq. 14.31 corresponds to the Euler equation. The expression in brackets, corresponding to E_t/E of Eq. 14.30, is plotted in Fig. 14.17 for various values of n.

In order to plot column curves given by Eq. 14.31 in a dimensionless form similar to that shown in Fig. 14.14, similar coordinates will be used, except that the stress f_1 must be used rather than F_{co}. Instead of the coordinates defined by Eqs. 14.23 and 14.24, the following will now be used.

$$B = \frac{L'/\rho}{\pi\sqrt{E/f_1}} \tag{14.32}$$

$$R_a = \frac{F_c}{f_1} \tag{14.33}$$

Substituting these values from Eqs. 14.32 and 14.33 into Eq. 14.31, the following column equation is obtained:

$$R_a = \frac{1}{B^2}\frac{E_t}{E} \tag{14.34}$$

This equation is plotted in Fig. 14.18 for various values of n.

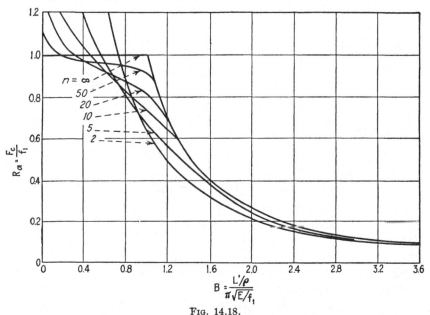

FIG. 14.18.

The form of dimensionless column curve given by Eq. 14.34 and Fig. 14.18 was proposed by Cozzone and Melcon.[3] They also use this same basic diagram for local crippling, initial buckling of sheet in compression and shear, and buckling of sheet between rivets. These further applications will be discussed later. These column curves have a very distinct advantage when analyzing structures of new materials. It is necessary only to obtain the basic compression stress-strain diagram of the material. The shape factor n and the stress f_1, corresponding to the yield stress, supply all the necessary information on the new material. All the information obtained from tests of columns of one material are immediately applicable to a new material.

Some of the test data published by Cozzone and Melcon are given in Table 14.4. The terms n, f_1, and E of Table 14.4 give all the required information for determining the column curve. The additional term F_{cy}, the compression yield stress for a permanent strain of 0.002, is tabulated for reference, since the manufacturer normally guarantees a minimum value of this stress, and values obtained from tests of typical specimens must be reduced to minimum guaranteed stresses.

TABLE 14.4. PROPERTIES OF COMPRESSIVE STRESS-STRAIN CURVES

Material	E, ksi	n	f_1, ksi	f_{cy}, ksi
Aluminum alloys (under 0.250):				
24S–T sheet....................	10,700	10	41	42
24S–T extrusion................	10,700	10	37	38
75S–T extrusion................	10,500	20	71	70
Steel:				
Normalized....................	29,000	20	74	75
$F_{tu} = 100,000$ psi................	29,000	25	94	80
$F_{tu} = 125,000$ psi................	29,000	35	114	100
$F_{tu} = 150,000$ psi................	29,000	40	140	135
$F_{tu} = 130,000$ psi................	29,000	50	165	165

14.9. Effect of Eccentric Loading on Ultimate Strength. In the derivation of the tangent modulus equation for short columns, it was assumed that the column was initially straight and that the load was applied at the centroid of the cross section. As mentioned in Art. 14.3, practical columns are never perfectly straight, and some unavoidable eccentricities always exist. These eccentricities have the effect of reducing the strength of the column, and it would be unconservative to use the tangent modulus equation without a reduction of allowable stresses because of eccentricities. The *ANC*–5 equations for short columns of Tables 14.1 and 14.2 are conservative enough to account for the small misalignments occurring in practical structures.

The approximate effects of column eccentricities may be investigated by considering a material such as that represented by the column curve for $n = \infty$ of Fig. 14.18. Materials such as 1025 steel have this property, since the stress-strain curve has a constant slope E up to the yield point and is almost horizontal above the yield point. A column of such a material is elastic until the stress reaches the yield point, and failure occurs soon after yielding of the material at one point of the cross section.

The deflection of an elastic column resisting a load P with an eccentricity a is given by Eq. 14.14.

$$\delta + a = a\left(\sec \sqrt{\frac{P}{EI}}\,\frac{L'}{2}\right) \tag{14.35}$$

The maximum bending moment at the center of the column is found by multiplying this deflection by the load.

$$M = P(\delta + a) \tag{14.36}$$

The maximum unit stress on the concave side of the column is obtained by combining compression and bending stresses.

$$F_{\max} = \frac{P}{A} + \frac{Mc}{I} \tag{14.37}$$

From Eqs. 14.35 to 14.37,

$$F_{\max} = \frac{P}{A}\left(1 + \frac{acA}{I}\sec\sqrt{\frac{P}{EI}}\,\frac{L'}{2}\right) \tag{14.38}$$

Substituting $K = \dfrac{I}{Ac}$, $F_c = \dfrac{P}{A}$, and $B = \dfrac{L'/\rho}{\pi\sqrt{E/F_{\max}}}$ into Eq. 14.38,

$$F_{\max} = F_c\left(1 + \frac{a}{K}\sec\frac{\pi B}{2}\sqrt{\frac{F_c}{F_{\max}}}\right) \tag{14.39}$$

Values of the ratio $F_c/F_{\max}$ may be substituted into Eq. 14.39 and corresponding values of B obtained for any assumed eccentricity of loading. Curves of these values are shown in Fig. 14.19. These correspond to column curves for ductile materials with eccentric loads. The assumptions made in the derivation of Eq. 14.39, that the material remained elastic until the maximum stress reached a certain constant allowable value $F_{\max}$ and that column failure occurred at this point, apply only to materials such as 1025 steel, with $n = \infty$. Such materials are seldom used in aircraft structures, but the general effects of eccentricities in reducing the allowable column stresses are the same for all materials. Thus all the column curves shown in Fig. 14.18 for ideal straight columns are unconservative for practical columns with unavoidable accidental eccentricities.

The magnitude of the dimension $K = I/Ac$ must be kept in mind.

For a rectangular cross section with width h, $K = h/6$. For a solid circular cross section of diameter D, $K = D/8$. For a thin-wall circular tube, K is approximately $D/4$. Thus, for a column with a square cross section of 0.6 by 0.6 in., the value a/K of 0.1 corresponds to an eccentricity $a = 0.01$ in. Normal manufacturing tolerance would permit a deviation at least this large from an ideal straight column. Similarly, for a 1.25-in. diameter-tube, the value a/K of 0.1 would correspond to

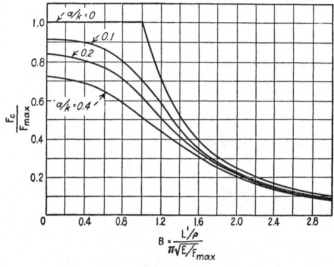

Fig. 14.19.

an eccentricity a of approximately 0.03 in. Such a tube would warp at least this much during welding. The effect of a small initial curvature which produces an eccentricity a at the center of a column is almost equivalent to the effect of the eccentricity a shown in Fig. 14.4.

While the ANC–5 formulas for short columns are conservative enough to account for accidental eccentricities, the long-column formulas are those for ideal columns and are unconservative. From Fig. 14.19 it is seen that the small eccentricity $a/K = 0.1$ may reduce the allowable load as much as 10 per cent in the long-column range. Most aircraft columns are short columns, and most of them have restrained end conditions. For long columns with restrained ends, a conservative assumption for the end-fixity term c will account for accidental eccentricities. For example, the assumption of $c = 2$, where the true value is $c = 2.5$, results in a hidden margin of safety of 25 per cent. For pin-ended long columns, however, the allowable load should be less than the Euler load, because of unavoidable eccentricities.

PROBLEMS

14.1. A long column has an initial curvature defined by the equation, $y_0 = \sigma \sin(\pi x/L)$. Derive an equation for the additional deflection y by integrating Eq. 14.6. Show that the center deflection δ is defined by the equation $\delta + a = \dfrac{a}{1 - P/P_{cr}}$, where P_{cr} is defined by Eq. 14.11. Compare values of $\delta + a$ from this equation with those obtained from Eq. 14.14 for values P/P_{cr} of 0.2, 0.4, 0.6, 0.8, 0.9, 0.95, and 1.00.

14.2. The tangent modulus of elasticity for 24S–T aluminum-alloy extrusions is represented approximately by the following equation:

$$E_t = \frac{10,700,000}{1 + 4.29(f/37,000)^9}$$

From the tangent modulus formula, Eq. 14.16, calculate values of L'/ρ corresponding to values of $f = P/A$ of 40,000, 37,000, 35,000, 30,000, 25,000, and 20,000 psi. Plot the column curve thus obtained, and compare with Fig. 14.16.

14.3. The tangent modulus of elasticity for X–4130 alloy steel heat-treated to an ultimate tensile strength of 100,000 psi is given approximately by the equation below.

$$E_t = \frac{29,000,000}{1 + 10.71(f/94,000)^{24}}$$

Calculate the coordinates of a sufficient number of points, and plot the short-column tangent modulus curve. Compare this curve with that of Fig. 14.15.

14.4. From the expressions $R_a = 1 - KB$ and $R_a = 1/B^2$ derive Eqs. 14.21 and 14.28 by equating the slopes of the two curves at their point of tangency. Find the coordinates of the point of tangency.

14.5. From the expressions $R_a = 1 - KB^2$ and $R_a = 1/B^2$ derive Eqs. 14.20 and 14.26 by equating the slopes of the two curves at their point of tangency. Find the coordinates of the point of tangency.

14.6. Find the column loads which may be resisted by round steel tubes, heat-treated to an ultimate tensile strength of 180,000 psi, with the ends welded before heat-treatment. The dimensions are as follows:

TABLE 14.5

Tube size	L	c
1×0.058	20	2
$1\frac{1}{8} \times 0.049$	20	4
$1\frac{1}{2} \times 0.065$	40	1
$1\frac{1}{4} \times 0.058$	30	2

14.7. Repeat Prob. 14.6 for steel heat-treated to an ultimate tensile strength of 150,000 psi.

14.8. Repeat Prob. 14.6 for steel heat-treated to an ultimate tensile strength of 125,000 psi.

14.9. Repeat Prob. 14.6 for 24S–T round aluminum-alloy tubes. The ends of such tubes are never welded. Assume an end connection which is adequate to develop the strength of the tube.

14.10. A steel forging has dimensions which are approximately as shown. The section properties are calculated in Example 3, Art. 14.7. The ends are

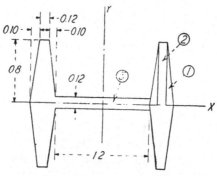

PROB. 14.10.

connected with single bolts, which provide such restraint that $c = 2$ for rotation about the x axis and $c = 1$ for rotation about the y axis. Find the column strength if $L = 15$ in. and $F_{tu} = 180,000$ psi.

14.10. Buckling of Flat Plates in Compression. A flat plate, in which the thickness is small compared to the other dimensions, does not act as a number of parallel narrow beams when resisting bending stresses. The initially flat plate shown in Fig. 14.20(*a*) may be compared to the

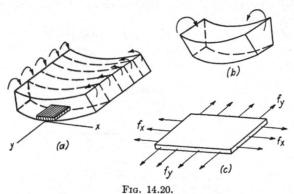

FIG. 14.20.

narrow beam shown in Fig. 14.20(*b*). The initially rectangular cross section of the narrow beam distorts into the trapezoidal cross section, because the compression stresses on the upper face of the beam produce a lateral elongation, while the tensile stresses on the lower face of the beam produce a lateral contraction. The cross sections of the flat plate, however, must remain rectangular.

If the shaded element of Fig. 14.20(a), shown to a larger scale in Fig. 14.20(c), is considered, it will have unit elongations e_x and e_y, as follows:

$$e_x = \frac{f_x}{E} - \mu \frac{f_y}{E} \tag{14.40}$$

and

$$e_y = \frac{f_y}{E} - \mu \frac{f_x}{E} \tag{14.41}$$

where μ is Poisson's ratio. In the plate, the elongation in the y direction must be zero, if the plate is assumed to have no curvature in the y direction. Substituting $e_y = 0$ into Eqs. 14.40 and 14.41, the following relations are obtained:

$$f_y = \mu f_x \tag{14.42}$$

$$e_x = \frac{f_x}{E}(1 - \mu^2) \tag{14.43}$$

Thus, when the flat plate is deflected with single curvature in the x direction, the stresses in the y direction are equal to Poisson's ratio times the stresses in the x direction. Similarly, the unit elongations in the x direction have the ratio of $1 - \mu^2$ to the corresponding elongations in a narrow beam.

Since a flat plate has smaller unit elongations than the corresponding narrow beam, the curvature resulting from an equivalent bending moment will be smaller by the ratio $1 - \mu^2$. Similarly, if the term M/EI of the general beam-deflection relation, Eq. 14.6, is replaced by $M(1 - \mu^2)/EI$, the Euler formula for a flat plate may be obtained as follows:

$$P_{cr} = \frac{\pi^2 EI}{(1 - \mu^2)L^2} \tag{14.44}$$

This equation applies for the condition shown in Fig. 14.21, where the unloaded edges are free and the loaded edges are simply supported, or free to rotate but not free to deflect normal to the plane of the plate. Substituting $I = bt^3/12$, $L = a$, and $P_{cr} = F_{c_{cr}}tb$ into Eq. 14.44, there is obtained

$$F_{c_{cr}} = \frac{\pi^2 E}{12(1 - \mu^2)} \left(\frac{t}{a}\right)^2 \tag{14.45}$$

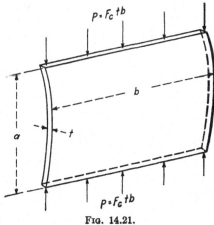

Fig. 14.21.

For a plate simply supported on all four edges, as shown in Fig. 14.22, the buckling compression load is considerably higher. As the plate deflects, both vertical and horizontal strips must bend. The supporting effect of the horizontal strips may be sufficient to cause a vertical strip to deflect into two or more waves, as shown in Fig. 14.22. It can be shown[4] that the buckling load is

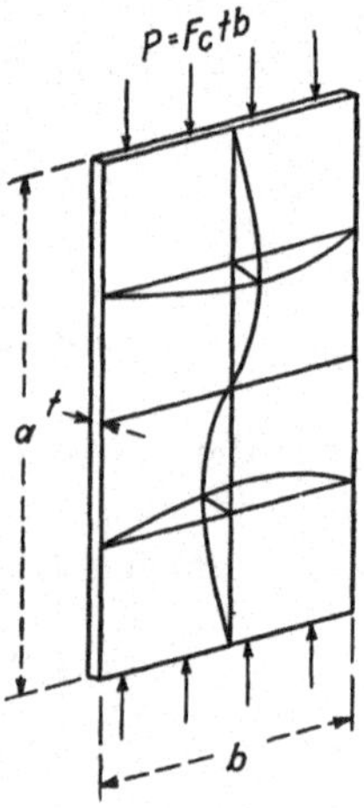

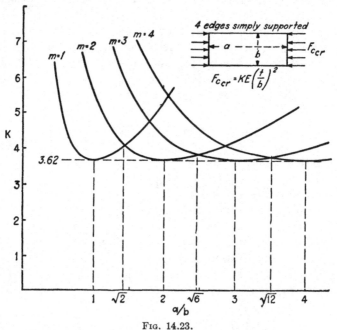

$P = F_c tb$

$$F_{c_{cr}} = \frac{\pi^2 E}{12(1 - \mu^2)} \left(\frac{bm}{a} + \frac{a}{bm}\right)^2 \left(\frac{t}{b}\right)^2 \tag{14.46}$$

where m is the number of waves in the buckled sheet. The value of μ is approximately 0.3 for all metals. Since a large error in μ produces only a small error in $F_{c_{cr}}$, it is seldom necessary to consider the variation of Poisson's ratio. Equation 14.46 may be written as follows:

$$F_{c_{cr}} = KE \left(\frac{t}{b}\right)^2 \tag{14.47}$$

FIG. 14.22.

where K is a function of a/b, and is plotted in Fig. 14.23 for $\mu = 0.3$. Only the curve of Fig. 14.23 which gives the minimum value of K is significant, since the sheet will buckle into the number of waves which

FIG. 14.23.

requires the smallest load. It is seen from Fig. 14.23 that the wave length of the buckles is approximately equal to the width, b, or that $m = 1$ for $a/b = 1$, $m = 2$ for $a/b = 2$, etc. The ratio a/b at which the number of waves changes from m to $m + 1$ is obtained from Eq. 14.46 as $a/b = \sqrt{m(m + 1)}$. The buckling stress obtained from Eq. 14.46 for a square plate with four edges simply supported is four times that obtained from Eq. 14.45 for the plate with sides free and ends simply supported.

The buckling loads for rectangular plates with other edge conditions may also be obtained from Eq. 14.47 by using the correct values of K. Values of K are shown in Fig. 14.25 for various conditions. The loaded edges are termed ends, and the unloaded edges termed sides, as designated on the curves. A free edge may rotate or deflect in a direction normal to the plate. A fixed edge, as shown in Fig. 14.24(a), is prevented from rotating or deflecting. The simply supported edge shown in Fig. 14.24(b) is free to rotate but not to deflect normal to the plane of the plate.

(a)

(b)

Fig. 14.24.

The true edge-fixity conditions for flat plates in an airplane structure cannot be calculated in most cases. It is necessary to estimate the edge-fixity after considering the supporting structure, in a similar manner to the estimates of column end-fixity conditions. The upper skin of an airplane wing, for example, is compressed in a spanwise direction. If the stringers are flexible torsionally, they will rotate as the sheet buckles and will act almost as simple supports for the sheet between the stringers, as shown in Fig. 14.26(a). If the stringers have considerable torsional rigidity, as do the hat sections and the spar flange shown in Fig. 14.26(b), they will rotate only slightly and will provide almost clamped edge conditions. In most structures it is necessary to assume a value for the term K of Eq. 14.47 which will represent a conservative mean between simply supported and clamped-edge conditions.

14.11. Ultimate Compressive Strength of Flat Sheet. The buckling of sheets in compression does not cause the collapse of a semimonocoque structure, because the stiffening members will usually be capable of resisting stresses which are much higher than those at which the initial sheet buckling occurs. It has been previously shown that a long column may resist a compression load when in the buckled condition and that the load is the same for a small lateral deflection as for a large deflection, provided that the stress does not exceed the elastic limit. The compression load resisted by a flat sheet with the sides free also remains constant for any lateral deflection. If the sides are supported, however, the compression load resisted by the sheet will increase as the lateral

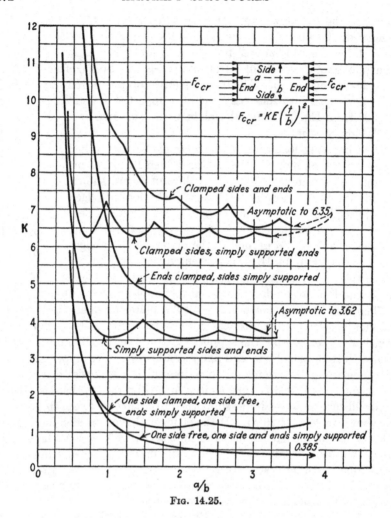

FIG. 14.25.

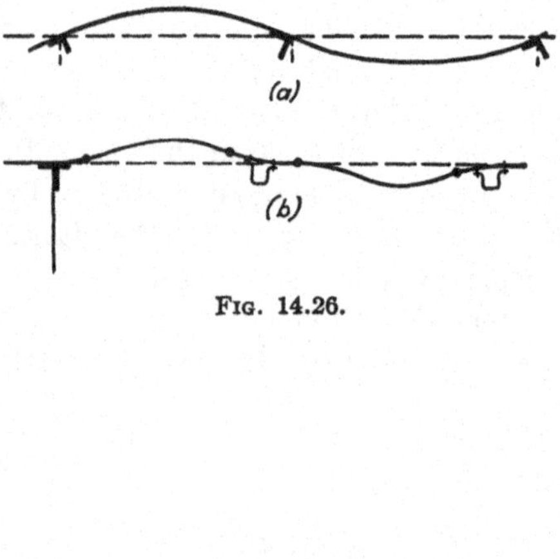

FIG. 14.26.

deflection increases, because the sides of the sheet must remain straight and consequently must be stressed in proportion to the strain in the direction of loading.

The flat plate shown in Fig. 14.27 is simply supported at all four edges and is loaded by a rigid block. The compression stresses are uniformly distributed as shown in Fig. 14.22, if the load is smaller than the buckling load. The stress distribution over the width of the plate is indicated in Fig. 14.28 by lines 1 and 2, with line 2 indicating the stress at initial buckling. As the load is increased beyond the buckling load, the stress distribution is indicated by lines 3, 4, and 5. Near the middle of the cross section the compressive stress remains approximately equal to the buckling stress, or a vertical strip acts in a similar manner to a long column. At the sides of the sheet, buckling is prevented, and the stress increases in proportion to the vertical motion of the loading block. The load may be increased until failure occurs by crushing of the sheet at the sides, although in common aircraft structures the stiffening members supporting the sheet usually fail before the sheet fails.

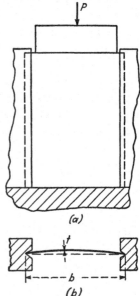

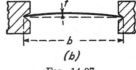

(a)

(b)

Fig. 14.27.

The curve representing the distribution of compressive stress over the width of a sheet is difficult to obtain, and even if it were known, it would

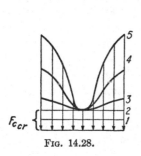

Fig. 14.28.

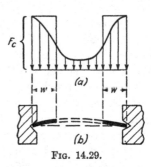

Fig. 14.29.

be tedious to use in analysis. It is more convenient to obtain the total compression load corresponding to a given compression stress at the side of the sheet. It is customary to work with effective widths w, shown in Fig. 14.29, which are defined in such a way that the constant stress F_c acting over the effective widths will yield the total compression load. Thus w is selected so that the area under the two rectangles in Fig. 14.29(a) is equal to the area under the curve of the actual stress distribution. The total compression load P and the edge stress F_c may be readily obtained experimentally, and the widths w may be calculated from the following equation:

$$2wtF_c = P \tag{14.48}$$

An approximate value of w may be obtained by assuming that a long sheet of total width $2w$ will have a buckling stress of F_c. From Eq. 14.47 and Fig. 14.23,

$$F_c = 3.62E \left(\frac{t}{2w}\right)^2$$

or

$$w = 0.95t \sqrt{\frac{E}{F_c}}$$

Test results indicate that this value is too high and that it is more accurate to use the following equation.

$$w = 0.85t \sqrt{\frac{E}{F_c}} \tag{14.49}$$

In obtaining Eq. 14.49, it has been assumed that the sheet is free to rotate at all four edges. In actual structures some degree of restraint always exists and the effective widths may be much greater in many cases. Tests indicate that stringers provide considerable edge-fixity at low stresses, but do not provide much restraint at stresses approaching the ultimate strength of the stringers. Numerous other equations have been used in place of Eq. 14.49, but none of the equations provide accurate correlation with test results under all conditions. Uncertainties regarding the effects of edge restraints in the actual structure, accidental eccentricities in the sheet, and the effects of stresses beyond the elastic limit further complicate the problem. Equation 14.49 yields a smaller effective width than do most other equations and is conservative for use in design. For normal aircraft structure in which the sheet is relatively thin, the weight penalty introduced by using Eq. 14.49 is small, but for high-speed aircraft in which the skin is relatively thick a more accurate analysis may be justified.

The buckling stress for a flat sheet with a large ratio of length to width, with one side simply supported and the other free, may be obtained from Eq. 14.47 and Fig. 14.25. For $K = 0.385$, $F_{c_{cr}} = 0.385E(t/b)^2$. The ultimate load resisted by such a sheet when the supported side is stressed to a value F_c is obtained by considering that an effective width w_1 resists the stress F_c and by obtaining w_1 as b from the above equation.

$$F_c = 0.385E \left(\frac{t}{w_1}\right)^2$$

or

$$w_1 = 0.62t \sqrt{\frac{E}{F_c}}$$

A more conservative value is recommended as follows:

$$w_1 = 0.60t \sqrt{\frac{E}{F_c}} \tag{14.50}$$

The effective sheet widths w and w_1 shown in Fig. 14.30 are obtained from Eqs. 14.49 and 14.50.

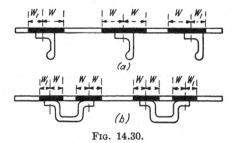

FIG. 14.30.

Example. The sheet-stringer panel shown in **Fig. 14.31** is loaded in compression by means of rigid members. The sheet is assumed to be simply supported at the loaded ends and at the rivet lines and to be free at the sides. Each stringer has an area of 0.1 sq in. Find the total compressive load P for the following conditions:

a. When the sheet first buckles.
b. When the stringer stress F_c is 10,000 psi.
c. When the stringer stress F_c is 30,000 psi.

Assume $E - 10,300,000$ psi for the sheet and stringers.

Solution. *a.* The sheet between the stringers is simply supported on all four edges and has dimensions of $a = 10$ in., $b = 5$ in., and $t = 0.040$ in. From Fig. 14.23, for $a/b = 2.0$, the value $K = 3.62$ is obtained. The buckling stress is $F_c = KE(t/b)^2 = 3.62 \times 10,300,000 \times (0.040/5)^2$, or 2,390 psi. The edge of the sheet has dimensions of $a = 1$ in. and $b = 10$ in. and is simply supported on three edges and free on the fourth edge. From Fig. 14.25, $K = 0.385$. The buckling stress is $F_c = KE(t/b)^2 = 0.385 \times 10,300,000(0.040/1)^2 = 6,200$ psi.

The sheet therefore buckles initially between the stringers. The total area of the sheet is assumed to be effective before buckling occurs. The buckling of a flat sheet in compression is a gradual process, and the load does not drop appreciably when buckling occurs. The load is therefore calculated as follows:

$$A = 3 \times 0.1 + 12 \times 0.040 = 0.78 \text{ in.}^2$$
$$P = F_cA = 2,390 \times 0.78 = 1,865 \text{ lb}$$

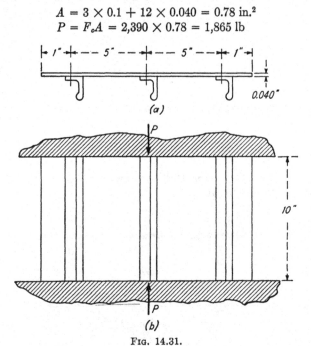

(a)

(b)

Fig. 14.31.

b. The effective sheet widths are obtained from Eqs. 14.49 and 14.50.

$$w = 0.85t \sqrt{\frac{E}{F_c}} = 0.85 \times 0.040 \sqrt{\frac{10,300,000}{10,000}} = 1.09 \text{ in.}$$

$$w_1 = 0.60t \sqrt{\frac{E}{F_c}} = 0.77 \text{ in.}$$

The effective sheet area is

$$A_1 = (4w + 2w_1)t = (4 \times 1.09 + 2 \times 0.77) \times 0.040 = 0.236 \text{ in.}^2$$

The total compressive load is

$$P = F_cA = 10,000(0.3 + 0.236) = 5,360 \text{ lb}$$

c. The solution is similar to that of part (b).

$$w = 0.85t \sqrt{\frac{E}{F_c}} = 0.85 \times 0.040 \sqrt{\frac{10,300,000}{30,000}} = 0.63 \text{ in.}$$

$$w_1 = 0.60t \sqrt{\frac{E}{F_c}} = 0.44 \text{ in.}$$

$$A = 0.3 + (4 \times 0.63 + 2 \times 0.44) \times 0.040 = 0.436 \text{ in.}^2$$

$$F = F_cA = 30,000 \times 0.436 = 13,080 \text{ lb}$$

14.12. Plastic Buckling of Flat Sheet. In the previous discussion of the buckling of sheet elements, it was assumed that the stress did not exceed the proportional elastic limit for the material. This elastic buckling action for flat sheets was similar to the elastic buckling of long columns in that the modulus of elasticity was the only significant material property. Equation 14.47 for sheet buckling is similar to the Euler equation for columns, and in each case the buckling stress is proportional to the modulus of elasticity of the material.

In the case of sheet elements for which the thickness is greater in comparison to the other dimensions, the compressive stresses will exceed the elastic limit before buckling will occur, as is also the case for short columns. Equation 14.47 will be valid in this case if the tangent modulus of elasticity E_t is substituted for the modulus E.

$$F_{c_{cr}} = KE_t \left(\frac{t}{b}\right)^2 \tag{14.51}$$

This equation may be written as follows:

$$F_{c_{cr}} = \frac{KE_t}{(b/t)^2} \tag{14.52}$$

Equation 14.52 is similar to the tangent modulus equation for short columns.

$$F_c = \frac{\pi^2 E_t}{(L'/\rho)^2} \tag{14.53}$$

The tangent modulus curve and other curves for short columns were plotted with values of F_c as ordinates and values of L'/ρ as abscissas. Values of $F_{c_{cr}}$ and b/t could be similarly plotted from Eq. 14.52 for a known value of K. In fact, the column curves can be used for plastic sheet buckling if the values of b/t are multiplied by a constant which is obtained by equating the right side of Eq. 14.52 to the right side of Eq. 14.53, as follows:

$$\text{Equivalent } \frac{L}{\rho} = \frac{\pi}{\sqrt{K}} \frac{b}{t} \tag{14.54}$$

A typical column curve for an aluminum-alloy material is shown in Fig. 14.32. The allowable column stress is obtained from the curve for a known value of L'/ρ. In the short-column range, the theoretical tangent modulus curve may be replaced by the more conservative straight line, in order to account for accidental eccentricities or other unknown conditions. Similarly, the curves of Fig. 14.32 yield the allowable buckling stress for a flat plate in the plastic, or short-column range. The value of $F_{c_{cr}}$ may be obtained for any known value of $(\pi/\sqrt{K})(b/t)$. Either of the short-column curves may be used, depending on the possible initial eccentricities of the sheet element and the degree of conservatism desired.   The value of K is obtained from Fig. 14.23 or Fig. 14.25.

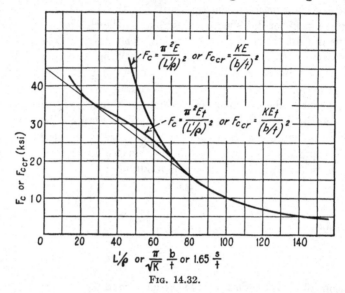

$$F_c = \frac{\pi^2 E}{(L'/\rho)^2} \quad \text{or} \quad F_{c_{cr}} = \frac{KE}{(b/t)^2}$$

$$F_c = \frac{\pi^2 E_t}{(L'/\rho)^2} \quad \text{or} \quad F_{c_{cr}} = \frac{KE_t}{(b/t)^2}$$

$L'/\rho \quad \text{or} \quad \dfrac{\pi}{\sqrt{K}} \dfrac{b}{t} \quad \text{or} \quad 1.65\,\dfrac{s}{t}$

Fig. 14.32.

One common application of plastic buckling is the buckling of compressive skin between rivets attaching the skin to the stringers or spar caps.   A skin element of this type is shown in Fig. 14.33.   The rivets have a uniform spacing s along the stringer, and the restraint is such that the skin element of length s and indefinite width has clamped ends and

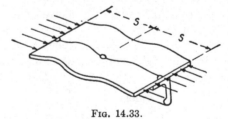

Fig. 14.33.

free sides.   The element therefore resists four times the load of a similar element with hinged ends, which was analyzed by Eq. 14.45.   Substituting $a = s$ and $E = E_t$ into Eq. 14.45, and multiplying the right-hand side by four to account for the end-fixity, there is obtained

$$F_c = \frac{\pi^2 E_t}{3(1 - \mu^2)} \left(\frac{t}{s}\right)^2$$

or, for $\mu = 0.3$

$$F_c = \frac{3.62 E_t}{(s/t)^2} \tag{14.55}$$

The tangent modulus short-column curve may be used in solving Eq. 14.55. An equivalent slenderness ratio may be obtained by equating the right-hand sides of Eqs. 14.53 and 14.55.

$$\text{Equivalent } \frac{L'}{\rho} = \frac{\pi}{\sqrt{3.62}} \frac{s}{t} = 1.65 \frac{s}{t}. \tag{14.56}$$

Example 1. Find the compression buckling stress for a sheet 4 by 4 by 0.125 in., with all four edges simply supported, assuming that the tangent modulus column curve for the material is represented by Fig. 14.32.

Solution. For this sheet, $a = b = 4$ and $t = 0.125$. From Fig. 14.25, for $a/b = 1$ and simply supported edges, $K = 3.62$. From Eq. 14.54, the equivalent L'/ρ is

$$\frac{\pi}{\sqrt{K}} \frac{b}{t} = \frac{\pi}{\sqrt{3.62}} \times \frac{4}{0.125} = 52.8$$

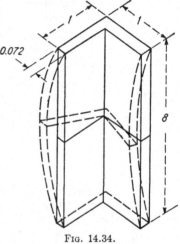

FIG. 14.34.

From Fig. 14.32, $F_{c_{cr}} = 28,000$ psi. If this point had been on the right-hand portion of Fig. 14.32, corresponding to the long-column or elastic range, the buckling stress would correspond to that given by Eq. 14.47.

Example 2. The angle extrusion shown in Fig. 14.34 is loaded in compression. Each leg of the angle buckles as a plate simply supported on the ends and on one side and free on the other side. Find the stress at which this buckling occurs. Assume that Fig. 14.32 represents properties of this material.

Solution. For each leg, $b = 1$, $a = 8$, and $t = 0.072$. For $a/b = 8$, the value of K from Fig. 14.25 is approximately 0.385. The equivalent L'/ρ is

$$\frac{\pi}{\sqrt{K}} \frac{b}{t} = \frac{\pi}{\sqrt{0.385}} \times \frac{1}{0.072} = 70.4$$

From Fig. 14.32, $F_{c_{cr}} = 20,500$ psi.

The type of failure indicated for this section is typical of crippling failures for aluminum-alloy extrusions. The ordinary short-column curves apply only to round tubes, or to stable cross sections which do not cripple locally. Since light extrusions are extensively used as column members in aircraft structures, the subject of crippling failure is very important and will be considered in detail later.

Example 3. An 0.040 sheet is riveted to an extrusion by rivets spaced 1 in. apart. What compressive stress in the extrusion will produce buckling of the sheet between rivets, as shown in Fig. 14.33, if the sheet has column properties as represented by Fig. 14.32?

Solution. From Eq. 14.56, the equivalent L'/ρ is

$$1.65\frac{s}{t} = 1.65 \times \frac{1}{0.040} = 41.2$$

From Fig. 14.32, $F_c = 31,300$ psi.

14.13. Nondimensional Buckling Curves. The plastic buckling stresses in the previous article were obtained from column curves of the type shown in Fig. 14.32. A column curve of this type is applicable to only one material, since the column curve is affected by the shape of stress-strain curve, the modulus of elasticity, and the yield stress of the material. There are numerous advantages to plotting column curves in dimensionless form, as shown in Fig. 14.18, and as discussed in Art. 14.8. When several materials have stress-strain curves of the same general shape, as indicated by the value of n, a single column curve presents the data for all these materials. Test information for any one of the materials is therefore applicable to all of them.

Cozzone and Melcon propose that the nondimensional curves of Fig. 14.18 be used for all problems of plastic sheet buckling, interrivet buckling, and local crippling of compression members. The curves of Fig. 14.18 are represented by the equation

$$\frac{F}{f_1} = \frac{E_t}{E}\frac{1}{B^2} \tag{14.57}$$

where F is the allowable average stress for a column, for sheet buckling, or for crippling, and f_1 is the secant yield stress corresponding to the stress at the intersection between the stress-strain curve and a line through the origin having a slope $0.7E$.

For columns, the term B was defined by Eq. 14.32 as follows:

$$B = \frac{L'/\rho}{\pi\sqrt{E/f_1}} \tag{14.32}$$

For plastic sheet buckling, the value of B is obtained from Eqs. 14.52 and 14.57.

$$B = \frac{b/t}{\sqrt{EK/f_1}} \tag{14.58}$$

For interrivet buckling, the value of B is obtained from Eqs. 14.55 and 14.57.

$$B = \frac{0.525s/t}{\sqrt{E/f_1}} \tag{14.59}$$

The values of B may therefore be calculated from Eqs. 14.32, 14.58, or 14.59, and the value of F_c/f_1 may then be read from the proper curve of Fig. 14.18. Many of the aluminum alloys have stress-strain curves of shapes which may be represented by $n = 10$. The column curve of Fig. 14.18 for $n = 10$ is shown in Fig. 14.35. The column curve in Fig. 14.32 is for $n = 10$, $E = 10,700,000$ psi, and $f_1 = 37,000$ psi, which represents the properties of small 24S–T extrusions with a compressive

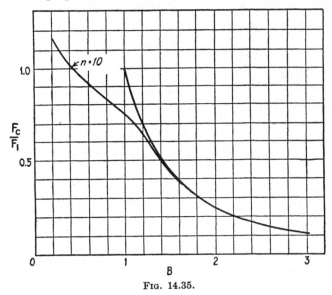

Fig. 14.35.

yield stress of 38,000 psi. The numerical examples of Art. 14.12 will be solved by use of the nondimensional curve of Fig. 14.35, and the results should check those previously obtained.

Example 1. Solve Example 1, Art. 14.12, by use of Fig. 14.35, if $f_1 = 37,000$ psi and $E = 10,700,000$ psi.

Solution. From Eq. 14.58,

$$B = \frac{b/t}{\sqrt{EK/f_1}} = \frac{4/0.125}{\sqrt{10,700,000 \times 3.62/37,000}} = 0.986$$

From Fig. 14.35, $F_c/f_1 = 0.755$.

$$F_c = 0.755 \times 37,000 = 28,000 \text{ psi}$$

Example 2. Solve Example 2, Art. 14.12, by use of Fig. 14.35, if $f_1 = 37,000$ psi and $E = 10,700,000$ psi.

Solution. From Eq. 14.58,

$$B = \frac{b/t}{\sqrt{EK/f_1}} = \frac{1/0.072}{\sqrt{10,700,000 \times 0.385/37,000}} = 1.32$$

From Fig. 14.35, $F_c/f_1 = 0.555$.

$$F_c = 0.555 \times 37,000 = 20,500 \text{ psi}$$

Example 3. Solve Example 3, Art. 14.12, by use of Fig. 14.35, if $f_1 =$ 37,000 psi and $E = 10,700,000$ psi.

Solution. From Eq. 14.59,

$$B = \frac{0.525s/t}{\sqrt{E/f_1}} = \frac{0.525 \times 1/0.040}{\sqrt{10,700,000/37,000}} = 0.77$$

From Fig. 14.35, $F_c/f_1 = 0.845$.

$$F_c = 0.845 \times 37,000 = 31,300 \text{ psi}$$

14.14. Columns Subject to Local Crippling Failure. The column equations previously derived are applicable to closed tubular sections with comparatively thick walls, or to other cross sections which are not subject to local crippling failure.

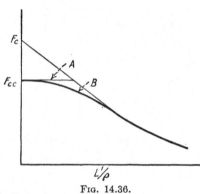

FIG. 14.36.

Many of the columns used in semimonocoque aircraft structures are made of extruded sections or of bent sheet sections and fail by local crippling. In the case of aluminum-alloy tubes, the crippling of thin walls was considered by assuming that for certain D/t ratios the tube walls would not exceed a maximum compressive stress, as given in Fig. 14.16. The assumed column curve is that shown by line A of Fig. 14.36, where the stress F_{cc} is the crippling stress. Tests of columns of extrusions or bent sheet with thin walls subject to local crippling yield values represented by curve B of Fig. 14.36 and indicate that sections subject to crippling failure should be analyzed by different column equations than stable cross sections of the same material.

It is usually desirable to make tests which will cover a range of slenderness ratios, for each thin-walled section which is to be used as a column. This procedure is not always practical for preliminary design, since the designer has a wide choice of cross sections which he may select and since he must be able to make some selection of sections and some prediction of their strength at an early stage of the design. Tests on aluminum-alloy columns subject to crippling failures show that the short-column curve closely approximates a second-degree parabola, as represented by Eq. 14.20 or Eq. 14.26. The crippling stress F_{cc} is substituted for the stress F_{co}, as follows:

$$F_c = F_{cc}\left[1 - \frac{F_{cc}(L'/\rho)^2}{4\pi^2 E}\right] \tag{14.60}$$

As in the case of other short-column curves, Eq. 14.60 does not apply for very short columns ($L'/\rho < 12$) because the end supports increase the crippling stress. The crippling stress F_{cc} may therefore be obtained by testing a column with L'/ρ of about 12. An approximate value of the crippling stress may be obtained as in Example 2, Art. 14.12, by finding the sum of the plastic buckling strengths of the rectangular elements of the cross section.

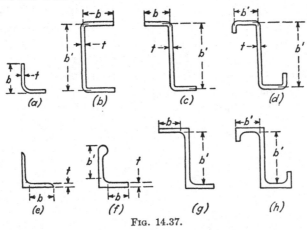

Fig. 14.37.

The column cross sections shown in Fig. 14.37 may be considered as made up of rectangular plates of width b, thickness t, and a length a, which is large in comparison to b. The plates with widths designated as b' are assumed to be simply supported on both sides, and those with widths designated as b are assumed to be free on one side and restrained on the other side. In the case of the angles shown in Fig. 14.37(a) and (e), the plates are assumed to be simply supported on one side, since the two plates buckle at the same stress, and neither plate supplies any edge restraint for the other. In the case of the other cross sections, however, the plates which have one side free have edge conditions between the clamped and simply supported cases for the other side. This difference in edge restraint is seen in comparing the buckled form of the angle shown in Fig. 14.34 with the buckled form of the channel shown in Fig. 14.38. The legs of the angle buckle in one half wave regardless of the length of the column, as is the case for a flat plate with one side free and the other simply supported. The legs of the channel buckle into the same number of half waves as the back of the channel, which buckles in approximately square panels, as shown in Fig. 14.38.

Fig. 14.38.

The initial buckling stress of the plates may be smaller than the stress at which collapse of the member occurs, since the corner resists load after the initial buckling. This effect is considered empirically by assuming the effective width b to be less than the total width, as shown in Fig. 14.37(e). The extrusions resist a greater load at the corners than the bent sheet sections, as indicated by the widths b in Fig. 14.37. The edge conditions for the plates in a bulb-angle extrusion of the type shown in Fig. 14.37(f) depend on the bending stiffness of the bulb, but it is usually assumed that the bulb supports the plate, as indicated.

After obtaining the plastic buckling stress for each element of area by the method of Arts 14.12 and 14.13, the total crippling load on the cross section is found as the sum of the loads on the individual areas. If the areas have dimensions $b_1 t_1$, $b_2 t_2$, and $b_3 t_3$ and buckling stresses F_1, F_2, and F_3, the total crippling stress is obtained as follows:

$$F_{cc} = \frac{F_1 b_1 t_1 + F_2 b_2 t_2 + F_3 b_3 t_3}{b_1 t_1 + b_2 t_2 + b_3 t_3} = \frac{\Sigma Fbt}{\Sigma bt} \qquad (14.61)$$

The denominator of Eq. 14.61 may not be equal to the total area, because the corner areas are not included. The crippling load is obtained by multiplying the stress F_{cc} by the total area, and it may be greater than the numerator of Eq. 14.61 because of the load on the corners.

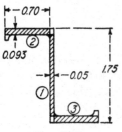

Fig. 14.39.

Example 1. Find the equation of the short-column curve for the extrusion shown in Fig. 14.39, if the material is 24S–T with $E = 10,700,000$, $n = 10$, and $f_1 = 37,000$.

Solution. The column curve for this material is represented by Fig. 14.32. For plastic buckling, the equivalent L'/ρ is obtained from Eq. 14.54, with $K = 3.62$ from Fig. 14.25. For area 1, $b/t = 1.564/0.05 = 31.3$, and for area 2, $b/t = 0.70/0.093 = 7.52$. From Eq. 14.54,

$$\text{Equivalent } \frac{L'}{\rho} = \frac{\pi}{\sqrt{K}} \frac{b}{t} = 1.65 \frac{b}{t}$$

For area 1, $1.65 \times 31.3 = 51.6$.

For area 2, $1.65 \times 7.52 = 12.4$. From Fig. 14.32, $F = 29,000$ for area 1, and $F = 45,000$ for area 2. From Eq. 14.61,

$$F_{cc} = \frac{\Sigma Fbt}{\Sigma bt} = \frac{29,000 \times 1.564 \times 0.05 + 45,000 \times 0.70 \times 0.093 \times 2}{1.564 \times 0.05 + 2 \times 0.70 \times 0.093}$$

$$= 39,000 \text{ psi}$$

The crippling stresses for the individual areas may also be obtained from the nondimensional curve of Fig. 14.35. From Eq. 14.58,

$$B = \frac{b/t}{\sqrt{EK/f_1}} = \frac{b/t}{\sqrt{10,700,000 \times 3.62/37,000}} = 0.0308 \frac{b}{t}$$

For area 1, $B = 0.0308 \times 31.3 = 0.965$. From Fig. 14.35, $F/f_1 = 0.77$, or $F = 28,500$ psi.

For area 2, $B = 0.0308 \times 7.52 = 0.232$. From Fig. 14.35, $F/f_1 = 1.20$, or $F = 45,000$ psi. These check the values obtained from Fig. 14.32.

The short-column curve is now obtained from Eq. 14.60.

$$F_c = F_{cc}\left[1 - \frac{F_{cc}(L'/\rho)^2}{4\pi^2 E}\right]$$
$$= 39,000\left[1 - \frac{39,000(L'/\rho)^2}{4\pi^2 10,700,000}\right]$$
$$= 39,000[1 - 0.0000923(L'/\rho)^2]$$

Example 2. The section shown in Fig. 14.40(a) is made of 24S–RT Alclad sheet with $n = 10$, $E = 9,700,000$, and $f_1 = 46,000$. Find the crippling stress for the cross section.

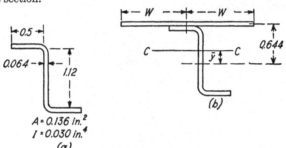

0.5

0.064

1.12

$A = 0.136$ in.^{2}
$I = 0.030$ in.^{4}
(a)

W

W

C

C

$\bar{y}$

0.644

(b)

Fɪɢ. 14.40.

Solution. The web is assumed to be simply supported on both sides, with $K = 3.62$, and buckles into approximately rectangular panels, in a manner similar to the channel section shown in Fig. 14.38. The half waves are approximately 1.12 in. long; therefore the flanges may be considered as simply supported at ends 1.12 in. apart and on one side. From Fig. 14.25 for $a/b = 2.00$, $K = 0.60$. From Eq. 14.58,

$$B = \frac{b/t}{\sqrt{EK/f_1}}$$

For the flanges,
$$B = \frac{0.5/0.064}{\sqrt{9,700,000 \times 0.6/46,000}} = 0.693$$

For the web,
$$B = \frac{1.12/0.064}{\sqrt{9,700,000 \times 3.62/46,000}} = 0.633$$

From Fig. 14.35, $F/f_1 = 0.88$ or $F = 40,500$ psi for the flanges, and $F/f_1 = 0.905$ or $F = 41,600$ psi for the web.

From Eq. 14.61,
$$F_{cc} = \frac{\Sigma Fbt}{\Sigma bt} = \frac{\Sigma Fb}{\Sigma b} = \frac{2 \times 40,500 \times 0.5 + 41,600 \times 1.12}{2 \times 0.5 + 1.12}$$
$$= 41,000 \text{ psi}$$

Example 3. The stringer of Example 2 is used in a wing which has a rib spacing of 18 in. and which is covered with 0.040 in. sheet, with $E = 9,700,000$ psi. Find the ultimate column load which may be resisted by the stringer and the effective skin if the end-fixity coefficient c for the stringer is 1.5.

Solution. The column cross section is composed of the stringer and the effective skin, as shown in Fig. 14.40(b). It is necessary to know the column stress before calculating the effective skin width w from Eq. 14.49, and it is necessary to know the effective skin width in order to calculate ρ for the column equation. A trial solution is therefore necessary. Assuming $F_c = 35,000$ psi, from Eq. 14.49,

$$w = 0.85t\sqrt{\frac{E}{F_c}} = 0.85 \times 0.040\sqrt{\frac{9,700,000}{35,000}} = 0.562 \text{ in.}$$

The effective skin area is $2 \times 0.562 \times 0.040 = 0.045$ in.^{2}

The centroidal distance is

$$\bar{y} = \frac{0.045 \times 0.644}{0.136 + 0.045} = 0.16 \text{ in.}$$

The moment of inertia of the entire area about the centroidal axis is

$$I_c = 0.030 + 0.045(0.644)^2 - 0.181(0.16)^2 = 0.0441$$

The radius of gyration of the area about the centroidal axis is

$$\rho = \sqrt{\frac{I}{A}} = \sqrt{\frac{0.0441}{0.181}} = 0.495 \text{ in.}$$

It should be noted that the column is attached to the sheet so that it is prevented from buckling about any axis but a horizontal axis. If the section were free it would buckle about the axis with the least radius of gyration. The effective column length is

$$L' = L/\sqrt{c} = 18/\sqrt{1.5} = 14.7 \text{ in.}$$

From Eq. 14.60,

$$F_c = F_{cc}\left[1 - \frac{F_{cc}(L'/\rho)^2}{4\pi^2 E}\right]$$

$$= 41,000\left[1 - \frac{41,000(14.7/0.495)^2}{4\pi^2 9,700,000}\right] = 37,200 \text{ psi}$$

The effective width is now recalculated.

$$w = 0.85 \times 0.040\sqrt{\frac{9,700,000}{37,200}} = 0.548$$

The difference in this width and that previously used will not appreciably affect the value of ρ. The total load is

$$P = F_c A = 37,200(0.136 + 2 \times 0.548 \times 0.040) = 6,700 \text{ lb}$$

All calculations of crippling stresses should be substantiated by tests on similar cross sections. The nondimensional curves may be used for obtaining the ratios to be used in correcting crippling test data for use with other materials.

14.15. Curved Sheet in Compression. A thin-walled circular cylinder loaded in compression parallel to its axis may fail by local instability of the thin walls. This type of failure is similar to that which occurs in the compression skin of semimonocoque wing and fuselage structures. The compression buckling of flat sheets was previously considered, but most actual structures are made up of curved sheets, and the curvature has a considerable effect on the buckling and ultimate strengths. A cylinder which is loaded in compression will assume a buckled form similar to that shown in Fig. 14.41. The number of circumferential waves depends on the ratio of R/t, where R is the radius and t is the wall

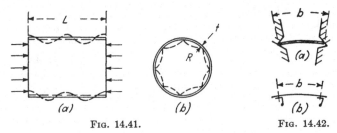

FIG. 14.41.　　　　　　　　FIG. 14.42.

thickness of the cylinder. A large number of waves develops for a large value of R/t. The length of the longitudinal waves is of the same magnitude as the length of the circumferential waves. For the high ratios of R/t which are common in semimonocoque wing and fuselage skins, the wave lengths are so small that a sector of a cylinder, with simply supported edges as shown in Fig. 14.42(a), resists approximately the same buckling stress as the complete cylinder. This sector corresponds to the skin between adjacent stringers, as shown in Fig. 14.42(b). For smaller values of R/t the length of the circumferential waves is greater, and the stringers or edge supports of Fig. 14.42(b) prevent the formation of the waves and thus increase the buckling stress.

　　The compression buckling stress for a thin-walled cylinder may be determined theoretically in a manner similar to that used in obtaining the buckling stresses for flat plates. The classic analysis of cylinders which is based on the assumption of small displacements yields the value

$$F_{c_{cr}} = 0.606E\,\frac{t}{R} \tag{14.62}$$

if Poisson's ratio is 0.3. Test values, however, are much lower than those given by Eq. 14.62, and test results show considerable scatter. This is in contrast to the excellent correlation between theoretical and experimental values for the buckling stresses for flat sheet.

　　Von Kármán, Dunn, and Tsien[5, 6, 7] have shown that the assumptions made in the analysis by the classic theory are in error. In the case of buckling of a flat plate, a longitudinal strip of the plate is supported elastically by lateral strips which exert restraining forces in proportion to their deflection. When the flat plate buckles, it may buckle in either direction, and the load after buckling remains equal to the buckling load, as in a Euler column. In the case of a compressed cylinder, however, the longitudinal strips are supported by circumferential rings which exert restraining forces that are not proportional to their radial deflection. The stiffness of a circular ring increases as it is deflected outward, and it decreases as it is deflected inward. Thus the thin walls of a compressed cylinder buckle inward much more readily than they buckle outward. The buckling is accompanied by a sudden decrease in the load and by a sudden decrease in length if there are no eccentricities of the walls. The buckling load is considerably reduced by small eccentricities of the walls. The buckling stress depends on the rigidity of the testing machine, since any testing machine has some elasticity, and the plates of the machine move together slightly as the resistance of the specimen decreases. The large effects of specimen eccentricity and of testing machine elasticity explain the large scatter of test results.

If the compression load on a cylinder P is plotted against the axial compressive deformation e, curves similar to those shown in Fig. 14.43

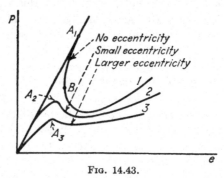

Fig. 14.43.

are obtained. Curve 1 represents a theoretical curve for an ideal cylinder in which the walls are perfectly cylindrical and homogeneous. The point A_1 corresponds to the theoretical buckling stress as obtained from Eq. 14.62, which cannot be obtained experimentally by the most careful testing because the upper branch of the curve is so close to the lower branch. At a deformation e, corresponding to point B_1, the cylinder assumes a buckled form and the load drops. If the deformation has exceeded that corresponding to B_1 before buckling occurs, the cylinder suddenly decreases in length when buckling occurs. Because of the elasticity of the testing machine, the plates of the machine move together, and the cylinder decreases in length when the load drops, even for test specimens with small eccentricities, as represented by curve 2.

Buckling loads obtained experimentally are represented by points A_2 and A_3 of Fig. 14.43. In the case of unstiffened cylinders, these buckling loads represent the ultimate strength of the cylinder in compression. Several empirical equations have been derived from experimental results, and the various equations yield widely divergent values of buckling stress, as might be expected because of the scatter of test values. Kanemitsu and Nojima[3] propose the following equation:

$$\frac{F_{c_{cr}}}{E} = 9\left(\frac{t}{R}\right)^{1.6} + 0.16\left(\frac{t}{L}\right)^{1.3} \tag{14.63}$$

where L is the length of the cylinder. This equation appears to give satisfactory agreement with test values within the ranges of $500 < R/t < 3,000$ and $0.1 < L/R < 2.5$.

Another equation which yields reasonable values of the buckling stress for smaller values of R/t is obtained as approximately one-half of the value of Eq. 14.62.

$$F_{c_{cr}} = 0.3E\frac{t}{R} \tag{14.64}$$

This equation yields results which are much higher than experimental values in cases where R/t is large. Perhaps it is reasonable to use Eq. 14.64 for values of R/t less than 500 and to use Eq. 14.63 for the range in which it applies.

In the case of curved sheet which is stiffened by longitudinal members, as is common in semimonocoque construction, the sheet would resist a buckling stress as given by Eq. 14.47 if there were no curvature and an additional stress as given by Eq. 14.63 because of the curvature. While there is little theoretical justification for adding these buckling stresses, this procedure is substantiated reasonably well by tests.

The compression buckling stress for curved skin on the upper surface of a wing is increased considerably by the negative air pressure on the sheet. Since the curved sheet has a tendency to buckle inward, the aerodynamic forces reduce this tendency. Equation 14.63 will be very conservative in this case. It is very important to prevent the buckling of the wing skin of high-speed aircraft because of the aerodynamic drag of the irregular airfoil section.

The ultimate strength of a stiffened curve sheet panel may be obtained in a similar manner to that used in obtaining the ultimate strength of a flat sheet panel in Art. 14.11. In addition to the compression load resisted by the stringers and by the effective widths of skin acting with the stringers, the sheet between stringers resists load because of its curvature, even though it has buckled. The load resisted by a buckled curved sheet is indicated by the right hand portion of the curves of Fig. 14.43. While this load depends on the elongation e of the stringers, there are many other unknown factors involved. The *ANC*-5 method for calculating this load is to assume that a skin width of $b - 2w$ between stringers resists a stress of $0.25Et/R$, as shown in Fig. 14.44. As an alternate method, this stress might be calculated from Eq. 14.63. Where this buckling stress for the curved sheet exceeds the stringer stress F_c, the entire sheet area is assumed to resist a stress F_c.

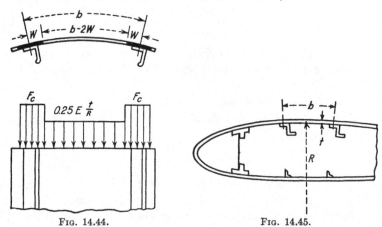

FIG. 14.44.   FIG. 14.45.

Example 1. For the wing shown in Fig. 14.45, $R = 50$ in., $t = 0.064$ in., $b = 6$ in., and the rib spacing $L = 18$ in. Find the compressive stress in the skin at which buckling occurs if $E = 10^7$ psi.

Solution. The buckling stress will be obtained as the sum of the buckling stress for a flat sheet simply supported on four sides, as obtained from Eq. 14.47, and the buckling stress for a cylinder, as obtained from Eq. 14.63. From Eq. 14.47,

$$F'_{c_{cr}} = KE \left(\frac{t}{b}\right)^2 = 3.62 \times 10^7 \times \left(\frac{0.064}{6}\right)^2 = 4,110 \text{ psi}$$

From Eq. 14.63,

$$\frac{F''_{c_{cr}}}{E} = 9 \left(\frac{t}{R}\right)^{1.6} + 0.16 \left(\frac{t}{L}\right)^{1.3}$$

$$= 9 \left(\frac{0.064}{50}\right)^{1.6} + 0.16 \left(\frac{0.064}{18}\right)^{1.3}$$

$$F''_{c_{cr}} = 2,130 + 1,560 = 3,690 \text{ psi}$$

The total buckling stress is the sum of these two values.

$$F_{c_{cr}} = 4,110 + 3,690 = 7,800 \text{ psi}$$

Example 2. Each of the upper stringers shown in Fig. 14.45 has an area of 0.2 sq in. and resists an ultimate stress of 40,000 psi. Find the ultimate compression load resisted by each stringer, by the effective skin acting with each stringer, and by the curved sheet between stringers. Assume $b = 6$ in., $t = 0.064$ in., $R = 50$ in., and $E = 10^7$ psi.

Solution. The effective skin width is obtained from Eq. 14.49.

$$w = 0.85t \sqrt{\frac{E}{F_c}}$$
$$= 0.85 \times 0.064 \sqrt{\frac{10^7}{40,000}} = 0.86 \text{ in.}$$

The assumed stress on the curved sheet between stringers is shown in Fig. 14.44.

$$\frac{0.25 \, Et}{R} = 0.25 \times 10^7 \times \frac{0.064}{50} = 3.200 \text{ psi}$$

The area of curved sheet assumed is

$$(b - 2w)t = (6 - 2 \times 0.86)0.064 = 0.274 \text{ in.}^2$$

The load on this curved sheet is $3,200 \times 0.274 = 880$ lb. The load on the stringer is $40,000 \times 0.2 = 8,000$ lb. The load on the effective skin is $40,000 \times 2 \times 0.86 \times 0.064 = 4,400$ lb. The curved sheet is seen to resist only a small proportion of the ultimate load.

PROBLEMS

14.11. The skin on the upper side of an airplane wing is of 24S–T Alclad material. The stringer spacing is 5 in. and the rib spacing is 20 in. Assuming the edges to be simply supported, find the compression buckling stress for skin gages of (a) 0.020 in., (b) 0.032 in., (c) 0.040 in., and (d) 0.064 in.

14.12. Repeat Prob. 14.11, assuming the values of K to be the average of values for simply supported edges and clamped edges.

14.13. Calculate points on the curve for $m = 1$ of Fig. 14.23, for values of a/b of 0.25, 0.33, 0.5, 1, 2, 3, and 4. Calculate points on the curve for $m = 2$ for values of a/b of 0.50, 0.66, 1, 2, 4, 6, and 8. Note the similarity between the two curves, and devise a system of coordinates which would show all the curves of Fig. 14.23 as a single curve.

14.14. Repeat the Example problem of Art. 14.11 for a skin gage of 0.051 in.

14.15. Repeat the Example problem of Art. 14.11 for a skin gage of 0.064 in.

14.16. Calculate the compression buckling stress for a sheet with $a = 8$ in., $b = 4$ in., and $t = 0.156$ in., if (a) all four edges are simply supported, (b) all four edges are clamped, and (c) if the ends are simply supported and the sides are free. The tangent modulus column curve for the material is shown in Fig. 14.32.

14.17. Solve Prob. 14.16 using a dimensionless buckling curve for $n = 10$. Assume $E = 10,700,000$ psi and $f_1 = 37,000$ psi.

14.18. Repeat Example 1, Art. 14.14, if the depth of the section is changed from 1.75 to 1.50.

14.19. Repeat Examples 2 and 3, Art. 14.14, if the gage of the stringer is 0.051 in. and the area and moment of inertia of the stringer are assumed to be proportional to the gage.

14.20. Repeat Prob. 14.11, including the effect of the skin curvature to a radius R of 50 in.

14.21. Repeat Example 2, Art. 14.15, for $R = 20$ in. and skin gages of (a) 0.072 in., (b) 0.081 in., and (c) 0.091 in.

REFERENCES FOR CHAPTER 14

1. SHANLEY, F. R.: The Column Paradox, *J. Aeronaut. Sci.*, December, 1946, p. 678.
2. SHANLEY, F. R.: Inelastic Column Theory, *J. Aeronaut. Sci.*, May, 1947, p. 261.
3. COZZONE, F. P., and M. A. MELCON: Non-dimensional Buckling Curves— Their Development and Application, *J. Aeronaut. Sci.*, October, 1946, p. 511.
4. TIMOSHENKO, S.: "Theory of Elastic Stability," Engineering Societies Monograph, McGraw-Hill Book Company, Inc., New York, 1936.
5. VON KÁRMÁN, T., L. G. DUNN, and H. S. TSIEN: The Influence of Curvature on the Buckling Characteristics of Structures, *J. Aeronaut. Sci.*, May, 1940, p. 276.
6. VON KÁRMÁN, T., and H. S. TSIEN: The Buckling of Thin Cylindrical Shells Under Axial Compression, *J. Aeronaut. Sci.*, June, 1941, p. 303.
7. TSIEN, H. S.: A Theory for the Buckling of Thin Shells, *J. Aeronaut. Sci.*, August, 1942, p. 373.
8. KANEMITSU, S., and H. NOJIMA: "Axial Compression Tests of Thin Circular Cylinders," Thesis at the California Institute of Technology, 1939.

CHAPTER 15

DESIGN OF WEBS IN SHEAR

15.1. Elastic Buckling of Flat Plates. The buckling of rectangular plates which resist direct compression stresses was discussed in Art. 14.10. Other types of stresses, such as shear stresses and bending stresses, may also produce elastic buckling of thin plates. Only loads in the plane of the plate will be considered here, and components normal to the plane of the plate are assumed to be zero.

The elastic buckling stresses for thin rectangular plates in shear may be calculated theoretically.[1] The analysis is beyond the scope of this book, but the results may be expressed in the same form as Eq. 14.47, if Poisson's ratio, μ, is assumed constant for all materials.

$$F_{s_{cr}} = KE \left(\frac{t}{b}\right)^2 \tag{15.1}$$

The values of K are plotted in Fig. 15.1 for $\mu = 0.3$ and for the two conditions of all four edges clamped and all four edges simply supported. The term t is the plate thickness, and the term E is the elastic modulus of the plate material. For the compressed plate discussed in Art. 14.10, the width b was perpendicular to the direction of loading and the length a parallel to the loads, but since the plate in shear is loaded on all four sides, the dimension b is considered as the smaller of the two plate dimensions. The critical shearing stress $F_{s_{cr}}$ is uniformly distributed along all four sides of the plate.

The rectangular plate which is loaded in pure shear has principal tension and compressive stresses at 45° to the edges. These principal stresses are equal to the shearing stresses. The diagonal compression stresses cause the sheet buckling, and when buckling occurs, the wrinkles form at approximately 45° angles to the edges. The buckling shearing stresses $F_{s_{cr}}$ are considerably higher than the buckling compression stresses $F_{c_{cr}}$ for plates with equal dimensions. This is a result of the restraining effect of the diagonal tension in the plate which is loaded by shearing forces.

The critical buckling stresses in a thin plate loaded in bending as shown in Fig. 15.2 may also be calculated theoretically[1] and expressed in the same form as the equations for compression and shear buckling.

$$F_{b_{cr}} = KE \left(\frac{t}{b}\right)^2 \tag{15.2}$$

where $F_{b_{cr}}$ is the critical maximum bending stress shown in Fig. 15.2 and K is given by the curve of Fig. 15.2, if all four edges are simply supported.

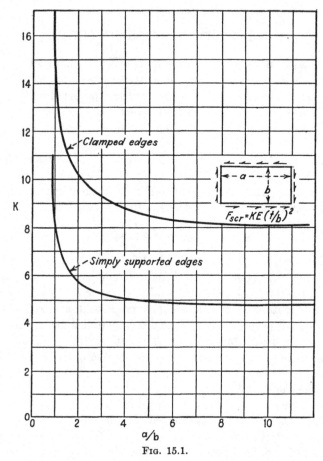

Fig. 15.1.

In the case of buckling of thin plates under the combined action of two of the conditions of compression, shear or bending, the initial buckling stresses have been obtained empirically by the method of stress ratios.[3] The initial buckling occurs when one of the following equations is satisfied:

Compression and bending:

$$R_b^{1.75} + R_c = 1 \tag{15.3}$$

Compression and shear:

$$R_s^{1.5} + R_c = 1 \tag{15.4}$$

Bending and shear:

$$R_b^2 + R_s^2 = 1 \tag{15.5}$$

where the terms, R_b, R_c, and R_s represent the ratios of the stresses in the plate to the critical buckling stresses $f_b/F_{b_{cr}}$, $f_c/F_{c_{cr}}$, and $f_s/F_{s_{cr}}$.

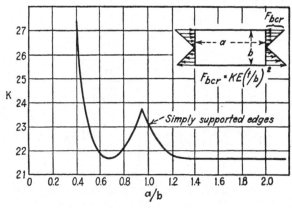

Fig. 15.2.

15.2. Elastic Buckling of Curved Rectangular Plates.

A large part of the structure of a semimonocoque airplane consists of the outer shell, or skin. This skin is usually curved in order to provide the necessary aerodynamic shape, and it must resist tension, compression, shear, and bending stresses. In addition to the conditions of ultimate strength and yield strength, which must be considered in the design of all aircraft structural members, the skin must frequently be designed so that it will not wrinkle under normal flight conditions. Skin wrinkles or other surface irregularities seriously affect the airflow in the case of high-speed aircraft, but may be permissible for aircraft of slower speeds. Unfortunately, both the buckling strength and the ultimate strength of curved plates depend upon many uncertain factors and are difficult to predict accurately. Initial plate eccentricities, air pressure normal to the plate, and conditions of the supports are difficult to evaluate; yet they may have a considerable effect on buckling loads.

The buckling stress for a curved plate in shear, such as shown in Fig. 15.3, is higher than the buckling stress for a flat plate with corresponding dimensions. The buckling stresses obtained experimentally are usually smaller than those calculated theoretically for an ideal plate with small deflections. A condition similar to that described in Art. 14.15 for plates in compression exists for plates in shear; the theoretical buckling stresses for flat plates correspond closely with test results for practical plates, but theoretical buckling stresses for curved plates are usually higher than values obtained experimentally.

The theoretical shear buckling stresses for curved plates have been calculated by Batdorf, Stein, and Schildcrout.[2] For a constant value of Poisson's ratio, $\mu = 0.3$, the shear buckling stress $F_{s_{cr}}$ may be expressed in the form of previous buckling equations.

$$F_{s_{cr}} = K_s E \left(\frac{t}{b}\right)^2 \tag{15.6}$$

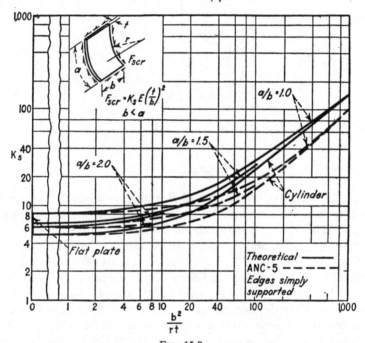

FIG. 15.3.

The term K_s is a function of the ratios a/b and b^2/rt and is plotted in Figs. 15.3 and 15.4. When the circumferential length is greater than the axial length, Fig. 15.3 is applicable, and when the axial length is greater, Fig. 15.4 must be used. The dimension b is smaller than the dimension a in either case. Both figures apply only to plates for which all four edges are simply supported. The points at the left side of the charts, for $b^2/rt = 0$, correspond for the buckling stress coefficients for flat plates, as given in Fig. 15.1.

For design purposes, it is necessary to consider the effects of initial accidental eccentricities, which always cause the buckling stresses to be smaller than the theoretical values. The designer must frequently use his judgment in evaluating these effects for a particular structure. An empirical equation is proposed in *ANC–5* as follows:

$$F_{s_{cr}} = KE\left(\frac{t}{b}\right)^2 + K_1 E\frac{t}{r} \qquad (15.7)$$

where the first term represents the buckling stress for a flat plate, as given by Fig. 15.1, and the last term represents the additional stress

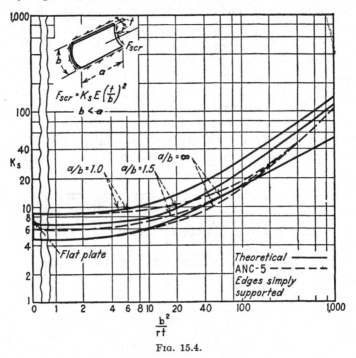

Fig. 15.4.

which can be resisted because of the curvature. The value $K_1 = 0.10$ is recommended. Rewriting Eq. 15.7 and comparing with Eq. 15.6, the following relations are obtained:

$$F_{s_{cr}} = \left(K + K_1\frac{b^2}{rt}\right) E\left(\frac{t}{b}\right)^2$$

$$K_s = K + K_1\frac{b^2}{rt} \qquad (15.8)$$

The values of K_s from Eq. 15.8 are plotted as the dotted lines in Fig. 15.3 and Fig. 15.4, assuming $K_1 = 0.10$. The *ANC–5* values obtained from Eq. 15.7 are seen to represent conservative approximations for all values shown on the chart, except for the case of large values of a/b and b^2/rt shown in Fig. 15.4. Except for this range, the *ANC–5* equation approximates most of the available test information closely and conservatively and may be used in practical design.   While the theoretical curves of Figs. 15.3 and 15.4 apply only to plates with simply supported edges, Eq. 15.7 may be used with Fig. 15.1 for plates with clamped edges, or for other edge conditions by interpolation on Fig. 15.1.

15.3. Pure Tension Field Beams.   The ultimate strength of thin webs in shear is much greater than the initial buckling strength.   In the case of structural members which are not exposed to the airstream, such as wing spars, the shear webs may be permitted to wrinkle at a small fraction of their ultimate loads.   In order to describe the manner in which loads are resisted by shear webs after buckling has occurred, it is convenient to consider a pure tension field beam, in which the web buckles when the shearing forces are initially applied.   Such a web never exists in practice, as even very thin webs have enough buckling resistance to affect the stress distribution appreciably.

The beam shown in Fig. 15.5(a) has concentrated flange areas which are assumed to resist the entire beam bending moment.   The beam web

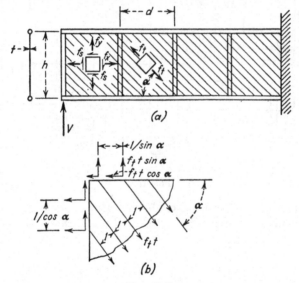

Fig. 15.5.

has a thickness t and a depth h between centroids of the flanges.   The vertical stiffeners are uniformly spaced at a distance d along the span. The shear force V is constant for all cross sections.   The shear flow at all points in the web is therefore equal to V/h, and the shear stress at all points, f_s, is V/th.   If the web is shear resistant, a web element at the neutral axis of the beam will be stressed as shown in Fig. 15.6.   On the vertical and horizontal faces X and Y the element resists only the shearing stresses f_s and no normal stresses. The principal stresses f_t and f_c occur on planes at 45° with the horizontal, as shown in Fig. 15.6(b). The magnitudes of the principal stresses are determined by the Mohr circle construction of Fig. 15.6(c), from which $f_t = f_c = f_s$. The construction of Mohr's circle was described in Art. 4.7.

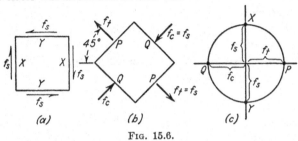

FIG. 15.6.

If the beam web of Fig. 15.5(a) is assumed to be extremely flexible, it will not be capable of resisting the diagonal compressive stress. It will then act as a group of parallel wires, inclined in the direction of the tension diagonal, or at an angle α of approximately 45°, as shown. Such a group of wires cannot resist any of the beam bending moment, and it is customary to assume that every element in a tension field web resists the same stress as an element at the neutral axis. A web element for a pure tension field web is therefore stressed as shown in Fig. 15.7. The

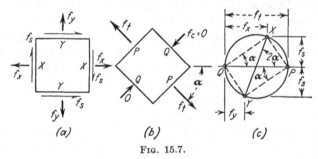

FIG. 15.7.

shearing stresses on the vertical and horizontal faces have the same values, $f_s = V/th$, as for the shear resistant web. These planes, X and Y, also have tensile stresses, f_x and f_y, which are obtained from the Mohr circle construction in Fig. 15.7(c). From the geometry of the circle, the lengths of lines QX and PY are $f_s/\sin \alpha$, and the lengths of lines PX and QY are $f_s/\cos \alpha$. The following stresses are therefore obtained:

$$f_x = f_s \cot \alpha \tag{15.9}$$

$$f_y = f_s \tan \alpha \tag{15.10}$$

$$f_t = \frac{f_s}{\sin \alpha \cos \alpha} = \frac{2f_s}{\sin 2\alpha} \tag{15.11}$$

The relationships expressed by Eqs. 15.9 to 15.11 may be obtained without the use of Mohr's circle construction by referring to Fig. 15.5(b). The web of thickness t, which resists a maximum tensile stress f_t, is assumed to be replaced by wires a unit distance apart, which resist forces of $f_t t$. The vertical component of the wire tension is then $f_t t \sin \alpha$, and the horizontal component is $f_t t \cos \alpha$. Along a horizontal line through the wires, the spacing is $1/\sin \alpha$, corresponding to a horizontal web area of $t/\sin \alpha$. The web tension stress f_y on the horizontal plane is obtained by dividing the vertical component of the wire tension $f_t t \sin \alpha$ by the web area $t/\sin \alpha$.

$$f_y = f_t \sin^2 \alpha \tag{15.12}$$

The shearing stress f_s on a horizontal plane is obtained by dividing the horizontal component of the wire tension $f_t t \cos \alpha$ by the web area $t/\sin \alpha$.

$$f_s = f_t \sin \alpha \cos \alpha \tag{15.13}$$

Equation 15.13 corresponds to Eq. 15.11. Similarly, a vertical line through the wires gives a spacing of $1/\cos \alpha$, corresponding to a web area of $t/\cos \alpha$. The horizontal web stress f_x is obtained by dividing the horizontal component of the wire tension by this area.

$$f_x = f_t \cos^2 \alpha \tag{15.14}$$

Equations 15.9 and 15.10 may be obtained from Eqs. 15.12 to 15.14.

The tension field beam also differs from the shear resistant beam in the manner in which stresses are transferred to the stiffeners, beam flanges, and riveted connections. The vertical stiffeners in a shear resistant beam resist no compression load, but serve only to divide the web into smaller unsupported rectangles and thus to increase the web buckling stress as calculated from Eq. 15.1. In a tension field beam, however, the vertical web tension stresses f_y tend to pull the beam flanges together, and this tendency must be resisted by compression forces in the stiffeners. Each stiffener must resist a compressive force P which is equal to the vertical tension force in the web for a length d equal to the stiffener spacing, as shown in Fig. 15.8(a).

$$P = f_y t d = \frac{Vd}{h} \tan \alpha \tag{15.15}$$

The vertical web tension stresses also tend to bend the beam flanges inward. The flanges act as continuous beams supported by the stiffeners. If the ends of the flanges are assumed to be fixed against rotation, the flange bending-moment diagram will be as shown in Fig. 15.8(b). At the stiffeners the bending moment is

$$M = \frac{f_y t d^2}{12} = \frac{Pd}{12} \qquad (15.16)$$

Midway between the stiffeners, the flange bending moment is

$$M = \frac{f_y t d^2}{24} = \frac{Pd}{24} \qquad (15.17)$$

The direction of the bending moment is such as to produce tension on the outside of the flange at the stiffeners and on the inside of the flange between the stiffeners.

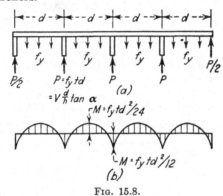

Fig. 15.8.

The horizontal components of the web stresses f_x tend to pull the end stiffeners together with a force $f_x t h = V \cot \alpha$. This force is resisted equally by the two spar flanges, producing compression forces of $V(\cot \alpha)/2$ which must be superimposed on the forces M/h which result from beam bending.

The riveted connections for a shear resistant web must be designed to resist a load $q = f_s t$ per unit length. In a tension field web connection, the horizontal riveted joints must resist shear flows q and also tension forces of $q \tan \alpha$ per unit length in a perpendicular direction. All horizontal riveted joints must therefore resist forces of $q\sqrt{1 + \tan^2 \alpha} = q \sec \alpha$ per unit length. The vertical riveted joints at the ends of the beam or at web splices must resist shear flows of $q = f_s t$ and tensile forces of $f_x t = q \cot \alpha$ per unit length. The joints must therefore be designed for a load of $q\sqrt{1 + \cot^2 \alpha}$ or $q \csc \alpha$ per unit length. This force does not apply for connections between the web and intermediate stiffeners, as no appreciable load is transferred by this connection.

In previous chapters, various shear flow analyses were made in which the webs were assumed to resist pure shear. These analyses remain valid even though the webs are in tension field, since the tension stresses on the X and Y planes may be superimposed on the shearing stresses without affecting the shear flow analysis.

Example. The beam shown in Fig. 15.9 is assumed to have a pure tension field web. Draw free-body diagrams for the stiffeners and flanges, and plot the axial loads in the stiffeners and flanges. Assume $\alpha = 45°$.

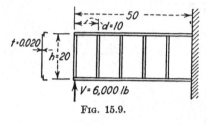

FIG. 15.9.

Solution. The shearing stress on a horizontal or vertical plane of a web element is $f_s = V/th = 6,000/(0.020 \times 20) = 15,000$ psi. The running shear is $q = f_s t = 300$ lb/in. The tension stresses f_x and f_y on these planes, and the tension loads per inch are also equal to f_s and q. The compression load on an intermediate stiffener, $P = Vd/h$, is 3,000 lb. The stiffener at the left end has a compression load of $P/2$ and an additional compression force of 6,000 lb applied at the lower end. Both of the beam flanges have compression loads of $V/2 = 3,000$ lb at the left end. The beam flange loads vary linearly along the span. At the support the flange loads

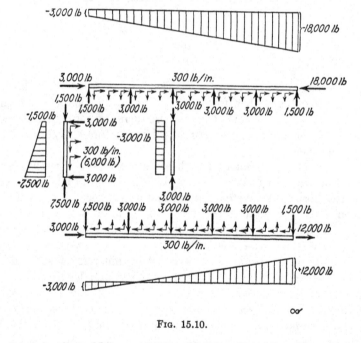

FIG. 15.10.

from beam bending, M/h, are $6,000 \times 50/20 = 15,000$ lb. The compression flange resists a load $- M/h - V/2 = -18,000$ lb, and the tension flange a load of $M/h - V/2 = 12,000$ lb. The free-body diagrams are shown in Fig. 15.10. All intermediate stiffeners resist the same loads.

15.4. Angle of Diagonal Tension in Web. The angle α of the diagonal web tension is usually less than 45°. If the beam framework, consisting of the flanges and stiffeners, has equal stiffness in resisting the horizontal tension f_x and the vertical tension f_y, the two tension stresses will be

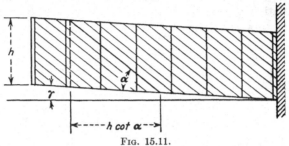

Fig. 15.11.

equal and α will be 45°. In practical beams, the flanges are much more rigid in resisting compression loads than are the stiffeners. The stiffeners deform in compression and permit the flanges to move together, while the stiffeners remain approximately the same distance apart. The horizontal web stress f_x is therefore greater than f_y, and the diagonal tension stress f_t has an angle less than 45°. This angle may be determined from the deflected geometry of the beam.

The beam shown in Fig. 15.11 is initially horizontal, and it has a shearing deformation γ at all cross sections. Bending deflections are not considered here. The deformation γ is caused by axial elongations of the stiffeners and flanges and by the elongation of the web diagonal resulting from the diagonal tension stress. A section of the beam of length $h \cot \alpha$ and depth h will be considered, and the value of α required

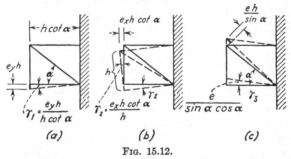

Fig. 15.12.

to produce a minimum deformation γ will be determined. All elongations will be assumed positive as tension in the derivation, although the stiffeners and flanges will always be in compression and have negative elongations. The deformations of the length, $h \cot \alpha$, of the beam in Fig. 15.11 will be the same as those for the truss shown in Fig. 15.12, if the unit elongations in the horizontal, vertical, and diagonal directions are the same for the two structures.

The total elongation of a vertical stiffener is equal to the product of the unit elongation e_y and the length h. This elongation causes a shearing deformation γ_1, as shown in Fig. 15.12(a), which is obtained by dividing the total elongation by the radius $h \cot \alpha$.

$$\gamma_1 = \frac{he_y}{h \cot \alpha} = e_y \tan \alpha \qquad (15.18)$$

The beam flanges have a unit elongation e_x, or a total elongation of $e_x h \cot \alpha$ in the horizontal length considered. The angular deformation γ_2 is obtained by dividing this deformation by the radius h, as shown in Fig. 15.12(b).

$$\gamma_2 = e_x \cot \alpha \qquad (15.19)$$

The diagonal strip of the web has a unit elongation e and a length $h/\sin \alpha$. The angular deformation γ_3 is obtained from the geometry of Fig. 15.12(c).

$$\gamma_3 = \frac{e}{\sin \alpha \cos \alpha} \qquad (15.20)$$

The total shearing deformation for the beam is the algebraic sum of the three components.

$$\gamma = -\gamma_1 - \gamma_2 + \gamma_3$$

Substituting from Eqs. 15.18 to 15.20,

$$\gamma = -e_y \tan \alpha - e_x \cot \alpha + \frac{e}{\sin \alpha \cos \alpha} \qquad (15.21)$$

The angle of the web diagonal tension α will be such that the deformation γ is a minimum. Differentiating Eq. 15.21 and equating $d\gamma/d\alpha$ to zero, the following expression is obtained:

$$\tan^2 \alpha = \frac{e - e_x}{e - e_y} \qquad (15.22)$$

where $e = f_t/E$ is the unit strain along the web diagonal, e_x is the unit strain in the beam flanges resulting from the compression caused by the web tension f_x, and e_y is the unit strain in the vertical stiffeners caused by the compression load P. All strains are positive for tension and negative for compression. The bending of the beam flanges and the slip in the riveted joints at the flanges have the same effect as an elongation e_y and may be included in the analysis.

The unit elongations used in Eq. 15.22 depend on the stresses, which in turn depend on the angle α. It is therefore necessary to solve this equation simultaneously with other equations obtained from the web stress conditions. For normal beam proportions, the flanges do not compress appreciably as a result of the tension field stress, and e_x may be assumed zero. The web diagonal strain e is f_t/E, or $2f_s \csc 2\alpha/E$, from Eq. 15.11. The unit strain e_y is obtained as $-f_s t d \tan \alpha/A_e E$, where A_e is the effective area of a vertical stiffener. Substituting these values into Eq. 15.22, the following expression is obtained:

$$\cot^4 \alpha = \frac{td}{A_e} + 1 \tag{15.23}$$

The effective stiffener area is equal to the true stiffener area A if the stiffener consists of two members symmetrically attached on opposite sides of the web. Where a single stiffener is attached to only one side of the web, it is loaded in bending as well as compression. The compression load P has an eccentricity e measured from the center of the web to the centroid of the stiffener area. The combined bending and compression stress at a distance e from the neutral axis is

$$f_c = \frac{P}{A} + \frac{Me}{I} = \frac{P}{A}\left[1 + \left(\frac{e}{\rho}\right)^2\right] = \frac{P}{A_e}$$

where ρ is the radius of gyration of the stiffener cross section area and A_e is defined as follows:

$$A_e = \frac{A}{1 + \left(\dfrac{e}{\rho}\right)^2} \tag{15.24}$$

The differentiation of Eq. 15.21 may appear questionable, since the elongations e, e_x, and e_y are treated as constant with respect to α. Equation 15.23 may also be obtained by substituting values for the strains as functions of α into Eq. 15.21 before differentiation, which is a more rigorous mathematical procedure but yields the same angle α. Equation 15.22, however, is a general expression for the angle of the principal planes at a point in any structure with two-dimensional stress conditions, when the strains e_x, e_y, and e are known. Wagner[4] first applied this equation to the analysis of tension field webs. Langhaar[6] expressed the strain e in terms of a known distortion γ and equated $de/d\alpha$ to zero in order to find the angle α for the maximum or principal strain. The angle α obtained in this manner is equal to that yielded by Eq. 15.22. Langhaar also expressed the total strain energy as a function of α and equated the derivative of the strain energy to zero in order to find the angle α, which yielded a minimum of the total strain energy. This also gave the same result as that obtained from Eq. 15.22. If flange bending and other deformations are considered, the value of e_y is greater than that used in obtaining Eq. 15.23.

15.5. Semitension Field Beams. In Art. 15.3 it was assumed that the beam web was perfectly flexible and was not capable of resisting any diagonal compressive stress. In practical beams the webs resist some diagonal compressive stress after buckling, and thus act in an intermediate range between shear resistant webs and pure tension field webs. Such beams are termed semitension field beams, partial tension field beams, or incompletely developed diagonal tension field beams. The pure tension field theory is conservative for the design of all parts of a practical beam, but may yield stiffener loads or flange bending moments which are as much as five times the true values. A more accurate theory is therefore necessary for design purposes.

The theory of pure tension field beams was first published by Wagner in 1929. Since then, many investigators have studied the problem of semitension field beams. Lahde and Wagner[5] published empirical data in 1936 which were based on strain measurements of buckled rectangular sheets. These data provided information for the practical design of beams, but the test points had considerable scatter because of the difficulty in making strain measurements of buckled sheet. Many aircraft manufacturers have conducted tests and developed empirical design formulas, but such tests have usually been made on one particular type of beam, and the equations must be used with caution in designing beams of different materials or different proportions than those on which the tests were conducted. The most extensive test program has been conducted by the NACA under the direction of Paul Kuhn. Kuhn and Peterson[7] have measured strains in the vertical stiffeners of a large number of beams and have derived empirical equations from these measurements. The stiffener stresses supply information required for the stiffener design and also for the design of the flanges to resist secondary bending. A theoretical analysis of the stresses in a buckled rectangular sheet has been made by Levy.[8, 9] While some simplifying assumptions are made in the analysis, it provides valuable information regarding the true stress distribution in the web. Levy's analysis shows that the stress conditions vary considerably at different points in the web and that any practical analysis in which the same stress conditions are assumed at all points will have some discrepancies with observed test conditions.

In Kuhn's analysis of semitension field beams, it is assumed that part of the shear load kV is resisted by pure tension field action and that the remaining load $(1 - k)V$ is resisted by the beam acting as a shear-resistant beam. All points of the web are assumed to have the same stress distribution, except for the web adjacent to the vertical stiffeners. This portion of the web is riveted to the stiffeners and does not wrinkle.

The stresses in the web may be obtained by multiplying the values shown for a shear-resistant web in Fig. 15.6 by $1 - k$, and those shown for a pure tension field web in Fig. 15.7 by k and then superimposing them. The values shown in Fig. 15.13(c) represent the total stresses on horizontal and vertical planes and are obtained by superimposing the conditions shown in Figs. 15.13(a) and (b). The angle α is obtained from Eq. 15.22, with sufficient accuracy. There is a conservative error involved in using Eq. 15.22, for partial tension field webs, since this equation yields the angle of the principal stress. The principal tension stress for the element of Fig. 15.13(c) has an angle which is between the 45° angle for the principal stress of the element of Fig. 15.13(a) and the angle α for the element of Fig. 15.13(b).

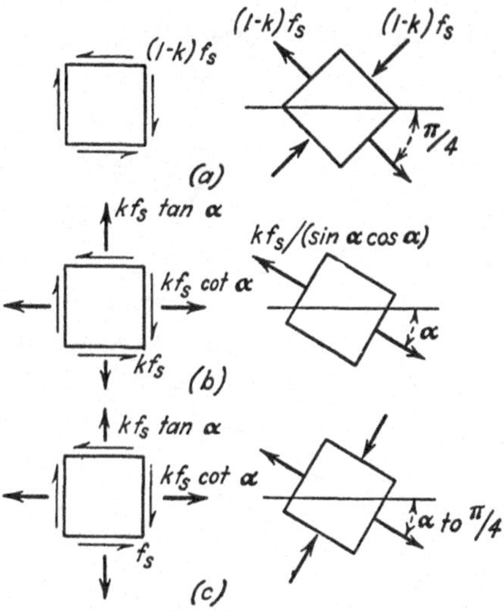

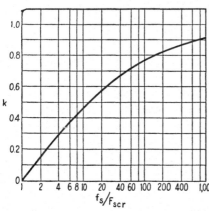

FIG. 15.13.

The diagonal tension factor k is given by Kuhn's empirical equation

$$k = \tanh\left(0.5 \log_{10} \frac{f_s}{F_{s_{cr}}}\right) \tag{15.25}$$

which is plotted in Fig. 15.14. The angle α is obtained from Eq. 15.22, after substituting functions of k, α, and A_e for the strains. Values of $\tan \alpha$ are plotted as functions of k and td/A_e in Fig. 15.15. Since Eq. 15.22 must be solved by trial for $\tan \alpha$, it is much more convenient to use Fig. 15.15 than to solve the equation for each particular case.

The stress conditions for any web element are known after k

FIG. 15.14.

and α are found and are as shown in Fig. 15.13(c). The stiffener compression forces and the flange bending moments are proportional to the vertical component of the web tensile stress f_y, which is shown in Fig. 15.13.

$$f_v = k f_s \tan \alpha \qquad (15.26)$$

The stiffener compression load P is obtained as follows:

$$P = f_v t d = k f_s t d \tan \alpha \qquad (15.27)$$

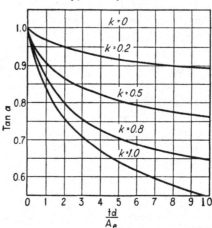

Fig. 15.15.

The flange bending moments are obtained from the stress f_y, as in Eqs. 15.16 and 15.17 for pure tension field webs, but the values of f_y and P are smaller for semitension field webs.

$$M = \frac{f_y t d^2}{12} = \frac{Pd}{12} \qquad \text{at stiffeners} \qquad (15.16)$$

$$M = \frac{f_y t d^2}{24} = \frac{Pd}{24} \qquad \text{between stiffeners} \qquad (15.17)$$

The compression load on a vertical stiffener is resisted by the stiffener and by the effective web which is riveted to the stiffener. If sufficient rivets are provided so that the web wrinkles do not extend through the riveted joint at the stiffener, as in the customary construction, the web must have the same vertical compression strain and approximately the same compression stress as the stiffener at the rivet line. An effective width of web equal to $0.5(1 - k)d$ is assumed by Kuhn to act with the stiffener, and the stiffener compression stress will then have the following value:

$$f_c = \frac{P}{A_e + 0.5(1 - k)td} \qquad (15.28)$$

Since the values of k are obtained empirically from measurements of f_c, an approximate expression for the effective width of web will yield an accurate value of f_c if the same effective width is assumed in calculating k from experimentally determined values of f_c. It seems probable that the true effective widths of the web are less than those assumed and that the empirical values of k thus yield conservative values for f_y and the flange bending moments.

The riveted joints between the webs and the beam flanges must be designed for the resultant of the shearing stress f_s and the tension stress f_y or for a running load of $\sqrt{(f_s)^2 + (f_y)^2}\, t$. Similarly, vertical web splices must be designed for a running load of $\sqrt{(f_s)^2 + (f_x)^2}\, t$. The rivets connecting the vertical stiffeners to the beam flanges should be designed to transfer the load $P_u = A_e f_c$, according to the above theory. In actual beams, it is frequently impractical to provide this strength. The theoretical analysis by Levy, and many tests, indicate that the stiffener load decreases near the end of the stiffener and that as much as half of this load is transferred to the web near the end of the stiffener rather than to the beam flange. In practical beams it is customary to provide a total strength in the rivets connecting the stiffener to the flange and the rivets connecting the end of the stiffener to the web to resist the load P_u. The stiffener-web rivets which are assumed to transfer part of this load P_u are spaced as close to the end of the stiffener as possible.

The allowable strength for beam webs has sometimes been obtained by equating the calculated diagonal tension stresses to the allowable tension stress for the web material, with empirical corrections for rivet holes and various stress concentration factors. More accurate web strength predictions can probably be obtained by equating the total web shearing stress f_s to an allowable stress F_{sw} obtained from tests of beams of common proportions. These allowable stresses are plotted in Fig. 15.16 as functions of $f_s/F_{s_{cr}}$. The curves of Fig. 15.16 were obtained by analyzing data from extensive tests of semitension field beams. The tests were conducted under the supervision of S. A. Gordon of the Glenn L. Martin Company.

It is assumed that the web-flange rivet spacing is in the normal range of three to five times the rivet diameter. A smaller rivet spacing reduces the net area excessively, while a very large spacing may permit the web wrinkles to extend through the rivet lines. The allowable web stresses F_{sw} are often plotted as functions of the sheet thickness, with the heavier sheet gages resisting higher stresses. This practice is permissible because stiffener spacings are usually kept to a maximum of about 8 in., even for very heavy sheets, and the thicker webs therefore have lower ratios of $f_s/F_{s_{cr}}$. For geometrically similar webs, however, the allowable stress would be independent of the web thickness. Where beam flanges and web stiffeners are attached symmetrically to both sides of a web, the edge of the web is better supported and resists higher stresses than those given in Fig. 15.16 for flanges attached to only one side of the web.

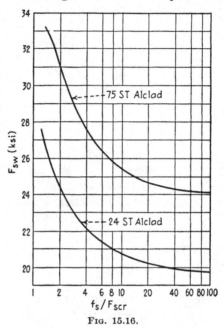

Fig. 15.16.

Example. Determine the margin of safety for the web, intermediate stiffeners, and riveted joints of the beam shown in Fig. 15.17. The web is 24S–T Alclad sheet, and the flanges and stiffeners are 24S–T extrusions.

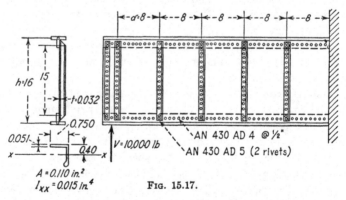

Fig. 15.17.

Solution. The buckling stress for the web in shear is obtained from Eq. 15.1 and Fig. 15.1. The web dimensions for computing buckling stresses are measured between rivet lines, as $a = 15$ in and $b = 8$ in. Entering Fig. 15.1 with

$1/b = 1.875$, the values $K = 10.3$ for clamped edges and $K = 5.9$ for simply supported edges are obtained. An average value, $K = 8.1$, will be used because of the restraining effects of the flanges and stiffeners. From *ANC-5*, $E = 9,700,000$ psi. Substituting in Eq. 15.1,

$$F_{s_{cr}} = KE\left(\frac{t}{b}\right)^2 = 8.1 \times 9,700,000 \times \left(\frac{0.032}{8}\right)^2 = 1,260 \text{ psi}$$

The shear stress is obtained as follows:

$$f_s = \frac{V}{ht} = \frac{10,000}{16 \times 0.032} = 19,500 \text{ psi}$$

The web depth h is always the distance between centroids of the flange areas, rather than the distance between rivet lines, when used in shear stress or shear flow calculations.

The effective stiffener area is computed from Eq. 15.24, since the stiffeners are attached only to one side of the web and are loaded eccentrically. The stiffener radius of gyration is $\rho = \sqrt{I/A} = \sqrt{0.015/0.110} = 0.37$ in. The stiffener eccentricity is $e = 0.40 + t/2 = 0.416$ in. The effective stiffener area is obtained from Eq. 15.24.

$$A_e = \frac{0.11}{1 + (0.416/0.37)^2} = 0.049 \text{ in.}^2$$

The ratio of web area to effective stiffener area is

$$\frac{td}{A_e} = \frac{8 \times 0.032}{0.049} = 5.22$$

This ratio and the ratio of $f_s/F_{s_{cr}}$ determine the stress distribution in the beam.

$$\frac{f_s}{F_{s_{cr}}} = \frac{19,500}{1,260} = 15.5$$

From Fig. 15.14, $k = 0.53$, and from Fig. 15.15, $\tan \alpha = 0.79$. From Fig. 15.16, the allowable web stress is $F_{sw} = 20,400$ psi. The margin of safety for the web in shear is obtained as follows:

$$\text{Web MS} = \frac{F_{sw}}{f_s} - 1 = \frac{20,400}{19,500} - 1 = \underline{0.04}$$

The vertical component of the web tension f_y is obtained as follows:

$$f_y = f_s k \tan \alpha = 19,500 \times 0.53 \times 0.79 = 8,160 \text{ psi}$$

The stiffener compression load is obtained from Eq. 15.25.

$$P = f_y td = 8,160 \times 0.032 \times 8 = 2,090 \text{ lb}$$

The load per inch in the web-flange rivets is

$$q_r = \sqrt{f_s^2 + f_y^2}\, t = \sqrt{(19,500)^2 + (8,160)^2} \times 0.032 = 660 \text{ lb/in.}$$

The allowable load for one $\frac{1}{8}$-in. A17S–T rivet is 375 lb shear and 477 lb bearing on an 0.032 gage sheet, as obtained from *ANC*-5. For the $\frac{1}{2}$-in. spacing, the allowable rivet load is 750 lb/in., as determined from the shear strength.

$$\text{Rivet MS} = \frac{750}{660} - 1 = \underline{0.14}$$

The maximum compression stress in the stiffener, resulting from the eccentric compression load, is obtained from Eq. 15.28.

$$f_c = \frac{2,090}{0.049 + 0.5(1 - 0.53) \times 0.032 \times 8} = 19,200 \text{ psi}$$

This stress exists in the leg attached to the web and decreases to zero or a tension stress in the outstanding leg. The allowable stress is the compression crippling stress for a leg with $b = 0.70$, $t = 0.051$, or $b/t = 13.7$, as computed by the methods of Art. 14.12. The allowable crippling stress is approximately 22,000 psi.

$$\text{MS} = \frac{22,000}{19,200} - 1 = \underline{0.15}$$

The allowable compression stress in the stiffener depends on many factors and is only roughly approximated here. The attached leg of a stiffener is usually made at least one gage thicker than the web in order to prevent a forced crippling from the web wrinkles. Since the outstanding leg is not highly stressed in compression, it supplies torsional rigidity, and the attached leg can probably be assumed to have one side clamped and the other side free in computing the crippling stress.

The two rivets connecting the stiffener to the flange are $\frac{5}{32}$-in. A17S–T rivets with a single shear strength of 596 lb each. They must transfer the compression load in the stiffener.

$$P_u = f_c A_s = 19,200 \times 0.049 = 940 \text{ lb}$$

The remaining part of the force P is resisted by compression in the effective sheet and is not transferred by the rivets. The margin of safety of the two stiffener-flange rivets is as follows:

$$\text{MS} = \frac{2 \times 596}{940} - 1 = \underline{0.27}$$

15.6. Curved Tension Field Webs. The entire external skin of an airplane normally resists shear flows and shearing stresses, and it is curved to provide the desired aerodynamic shape. The skin normally wrinkles in shear before the ultimate strength is reached, in the same manner as flat tension field webs in the interior structure. The buckling stress for curved webs is usually a larger fraction of the ultimate stress than in the case of flat webs. This condition is partly by choice of the designer and is partly an inherent property of curved webs. The designer may keep the buckling stress high for aerodynamic reasons by using rectangular skin panels which are small in comparison to the sheet thickness. A comparison of flat and curved sheet panels with the same dimensions also shows that the curved webs have higher buckling stresses than the flat webs but lower ultimate stresses, because when the curved webs wrinkle in diagonal tension the median plane of the sheets must be stretched severely, whereas the flat webs wrinkle to developable surfaces without stretching the sheets.

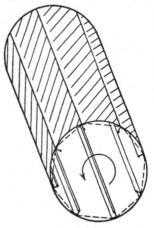

Wagner[11] proposed a tension field theory for curved webs, in which pure tension field was assumed. For a structure such as that shown in Fig. 15.18, the skin could be assumed to be replaced by parallel wires, as shown. The wires would be stretched to a plane containing the adjacent stringers, or the originally curved sheets would deform to flat sheets, and the original circular cross section would become polygonal. The conditions observed in tests of practical curved webs differ from this assumed pure tension field condition con-

Fig. 15.18.

siderably more than practical flat webs differ from pure tension field webs. For a curved web to wrinkle as shown in Fig. 15.18, the sheet must be deformed rather severely, particularly when the sheet panels between stiffeners have a large angle of curvature. For the curved

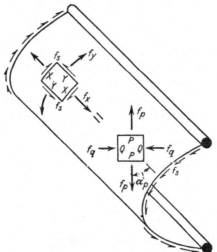

Fig. 15.19.

web shown in Fig. 15.19, the pure tension field theory obviously does not apply, and designers have often erroneously assumed the ultimate strength of such webs to be equal to the buckling strength.

The stress conditions for an element of a curved semitension field web may be represented in the same manner as for a flat semitension field web. For the web elements shown in Fig. 15.19, the stresses on the X and Y planes are represented in the same manner as for the element of a flat web. While some investigators represent the stresses on inclined planes as shown in Fig. 15.13, it seems preferable to consider the principal stresses instead, in order to visualize the true web stress conditions.

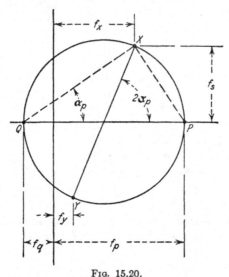

Fig. 15.20.

The inclined web element of Fig. 15.19 shows the principal stresses f_p and f_q on principal planes inclined at an angle α_p. Each of the elements shown in Fig. 15.19 represents the total stress conditions at any point in the web, and the two elements therefore represent equivalent conditions rather than conditions to be superimposed as in Fig. 15.13. The relationship between the stresses is determined by Mohr's circle construction in Fig. 15.20. The following equations are obtained from the geometry of the circle construction:

$$f_x = f_s \cot \alpha_p - f_q \tag{15.29}$$
$$f_y = f_s \tan \alpha_p - f_q \tag{15.30}$$
$$f_s = (f_p + f_q) \sin \alpha_p \cos \alpha_p \tag{15.31}$$

The stress f_q is assumed positive in compression, since it is always a compressive stress in this analysis.

At stresses below the buckling stress, the conditions $f_s = f_p = f_q$, $f_x = f_y = 0$ and $\alpha = 45°$ are applicable, as in flat webs. The tendency for the diagonal compression f_q to increase the web curvature is exactly balanced by the tendency of the diagonal tension f_p to flatten the curvature, and the stress conditions are therefore not affected by the curvature. The tendency of web stresses to change the curvature depends on the circumferential stress f_y, which is zero before buckling and which tends to decrease the curvature if tension and to increase the curvature if compression. At stresses higher than the buckling stress, the action is considerably different from that of flat webs. In flat webs, the stresses f_x and f_y increase in almost the same proportions after buckling occurs, and the angle of the principal stresses, α_p, remains about 45°. Curved webs, however, are capable of resisting stresses f_x in the direction of the axis of the cylinder, but can resist only negligible stresses f_y in the circumferential direction. The angle of the principal stress α_p is therefore much smaller than 45°.

For webs which are almost flat, such as those shown in Fig. 15.18, some circumferential stresses f_y are obtained, and the stress conditions approach those for a plane web. For webs with considerable curvature, such as that shown in Fig. 15.19, the circumferential stress f_y can be assumed zero. The diagonal compressive stress f_q for wrinkled flat webs is considerably higher than the diagonal compression stress $F_{s_{cr}}$ when buckling first occurs, because the increasing diagonal tension stress f_p tends to support the sheet. Perhaps a similar situation occurs in curved webs, but it is conservative to assume that the diagonal compression stress after buckling is equal to that at which buckling first occurs. Substituting $f_y = 0$ and $f_q = F_{s_{cr}}$ into Eqs. 15.29 to 15.31, the following equations are obtained for buckled webs with large curvature:

$$\tan \alpha_p = \frac{F_{s_{cr}}}{f_s} \tag{15.32}$$

$$f_p = \frac{(f_s)^2}{F_{s_{cr}}} \tag{15.33}$$

$$f_x = \frac{(f_s)^2}{F_{s_{cr}}} - F_{s_{cr}} \tag{15.34}$$

The allowable stresses for curved webs of various types have been determined experimentally, but will not be treated here because there is no simple expression which will yield the allowable stress for the general case of a buckled curved web. Kuhn and Griffith[10] present semiempirical methods of analysis and design which will be satisfactory for most design problems.

A tension field web analysis in which the total stresses on horizontal and vertical planes and on principal planes are obtained has several advantages over the method in which the stresses for two assumed conditions are obtained separately and superimposed, as in Fig. 15.13. The principal stresses, and the angle to the principal planes for the stress condition of Fig. 15.13(c), cannot be expressed by any simple equation. In order to represent the stress conditions for curved webs by the method of Fig. 15.13, it is necessary to assume that $\alpha = 0$, when $f_y = 0$. It is not readily apparent from Fig. 15.13 that a horizontal stress f_x permits the web to resist shearing stresses by tension field action, whereas Eqs. 15.32 to 15.34 show this to be the case.

Example. The cylinder shown in Fig. 15.21 is stiffened by rings at 8-in. intervals along its length and by four longerons which are equally spaced around the circumference. Find the compressive forces P in the longerons, the stresses f_p and f_x in the web, and the angle α_p of the principal web stress. Use $E = 10^7$ psi and (a) $f_s = 5{,}000$ psi, (b) $f_s = 10{,}000$ psi, (c) $f_s = 15{,}000$ psi, and (d) $f_s = 20{,}000$ psi.

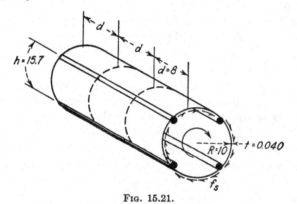

FIG. 15.21.

Solution. The shear buckling stress for the web will be obtained from Eq. 15.7. The dimensions of the sheet for use in the buckling stress equation are $a = 15.7$, $b = 8$, $t = 0.040$, and $r = 10$. From Fig. 15.1, for $a/b = 1.98$, $K = 10.0$ for clamped edges and $K = 5.8$ for simply supported edges. Using an average value $K = 7.9$ and a value $K_1 = 0.1$, the following value is obtained from Eq. 15.7:

$$F_{s_{cr}} = KE \left(\frac{t}{b}\right)^2 + K_1 E \frac{t}{r}$$

$$= 7.9 \times 10^7 \times \left(\frac{0.040}{8}\right)^2 + 0.1 \times 10^7 \times \frac{0.040}{10}$$

$$= 1{,}970 + 4{,}000 = 5{,}970 \text{ psi}$$

a. For the stress of 5,000 psi the web is shear resistant, or $f_p = f_q = f_s = 5{,}000$ psi, $f_x = f_y = P = 0$, and $\alpha = 45°$.

b. For the stress of 10,000 psi, the following values are obtained from Eqs. 15.32 to 15.34:

$$\tan \alpha_p = \frac{F_{s_{cr}}}{f_s} = \frac{5,970}{10,000} = 0.597$$

$$\alpha_p = 30.8°$$

$$f_p = \frac{(f_s)^2}{F_{s_{cr}}} = \frac{(10,000)^2}{5,970} = 16,770 \text{ psi}$$

$$f_x = \frac{(f_s)^2}{F_{s_{cr}}} - F_{s_{cr}} = \frac{(10,000)^2}{5,970} - 5,970 = 10,800 \text{ psi}$$

The longeron load P is equal to $f_x th$.

$$P = 10,800 \times 0.040 \times 15.7 = 6,800 \text{ lb}$$

c. Substituting $f_s = 15,000$ psi into the equations used in part (*b*), the following results are obtained: $\tan \alpha = 0.398$, $\alpha = 21.7°$, $f_p = 37,670$ psi, $f_x = 31,700$ psi, and $P = 20,000$ lb.

d. Substituting $f_s = 20,000$ psi the results are $\tan \alpha = 0.298$, $\alpha = 16.6°$, $f_p = 67,200$ psi, $f_x = 61,230$ psi, and $P = 38,400$ lb.

It is observed that the principal tension stress f_p increases as the square of the shearing stress f_s. The principal tension stress obtained in part (*d*) is obviously greater than the allowable tensile stress in webs of normal materials. The web of part (*c*) may also be overstressed, since the diagonal tension is approximately equal to the allowable diagonal tensile stress for a plane web.

PROBLEMS

15.1. The skin on a fuselage is supported by stringers which are spaced at 5 in. and by rings spaced at 20 in. Find the shear buckling stresses for the flat sheet if (*a*) $t = 0.020$ in., (*b*) $t = 0.032$ in., (*c*) $t = 0.040$ in., and (*d*) $t = 0.064$ in. Assume $E = 10^7$ psi and an average between simply supported and clamped edge conditions.

15.2. Repeat Prob. 15.1 if the skin is curved with a radius of 20 in.

15.3. Plot the axial loads in the flanges and stiffeners of a pure tension field beam similar to that shown in Fig. 15.5, with $h = 10$ in., $d = 10$ in., and $V = 10,000$ lb. Compute the flange bending moments and the load per inch on all rivets. Assume $\alpha = 45°$.

15.4. Repeat Prob. 15.3 for an angle α as determined from Eq. 15.23 with $td/A_e = 5.0$.

15.5. The beam of Fig. 15.9 has a web of 24S–T Alclad material. The stiffeners are attached to one side of the web and have properties $A = 0.08$ sq in., $\rho = 0.3$ in., and $e = 0.35$ in. The webs are riveted to the flanges by AD4 rivets at 0.5-in. spacing. The stiffeners are connected to the flanges by a single AD5 rivet at each end. The allowable maximum compressive stress in the stiffeners is 25,000 psi. Find the margins of safety of the web, stiffeners, and web-flange and stiffener-flange rivets by use of the equations for semitension field beams.

15.6. A beam with the dimensions and rivet spacing shown in Fig. 15.17 has a web of 0.040 gage of 24S–T Alclad material and resists a shear V of 12,000 lb. Find the margins of safety of the web, stiffeners, and riveted joints.

15.7. Design a semitension field beam with $h = 10$ in. and $d = 6$ in., to resist a shear force of $V = 8,000$ lb. Determine the web gage, stiffener cross section, and rivet spacing. As a trial value, a stiffener area $A = 0.5td$ is usually a close approximation.

15.8. Design a semitension field beam with $h = 12$ in., $d = 8$ in., and $V = 12,000$ lb.

15.9. A curved fuselage skin with a radius of 20 in. is supported by stringers with a 5-in. spacing and rings with a 20-in. spacing. The shearing stress is 10,000 psi. Find the principal tension stress in the curved web if (a) $t = 0.020$ in., (b) $t = 0.032$ in., (c) $t = 0.040$ in., and (d) $t = 0.064$ in.

REFERENCES FOR CHAPTER 15

1. TIMOSHENKO, S.: "Theory of Elastic Stability," McGraw-Hill Book Company, Inc., New York, 1936.
2. BATDORF, S. B., M. STEIN, and M. SCHILDCROUT: Critical Shear Stress of Curved Rectangular Panels, *NACA TN* 1348, 1947.
3. *ANC-5*, "Strength of Aircraft Elements," Army-Navy-Civil Committee on Aircraft Design Criteria, Amendment 2, 1946.
4. WAGNER, H.: Flat Sheet Metal Girders with Very Thin Metal Web, Parts I, II, and III, *NACA TM* 604, 605, and 606, 1931.
5. LAHDE, R., and H. WAGNER: Tests for the Determination of the Stress Condition in Tension Fields, *NACA TM* 809, 1936.
6. LANGHAAR, H. L.: Theoretical and Experimental Investigations of Thin-webbed Plate-girder Beams, *Trans. ASME*, Vol. 65, 1943.
7. KUHN, P., and J. P. PETERSON: Strength Analysis of Stiffened Beam Webs, *NACA TN* 1364, 1947.
8. LEVY, S., K. L. FIENUP, and R. M. WOOLLEY: Analysis of Square Shear Web above Buckling Load, *NACA TN* 962, 1945.
9. LEVY, S., R. M. WOOLLEY, and J. N. CORRICK: Analysis of Deep Rectangular Shear Web above Buckling Load, *NACA TN* 1009, 1946.
10. KUHN, P., and G. E. GRIFFITH: Diagonal Tension in Curved Webs, *NACA TN* 1481, 1947.
11. WAGNER, H., and W. BALLERSTEDT: Tension Fields in Originally Curved, Thin Sheets during Shearing Stresses, *NACA TM* 831, 1937.

CHAPTER 16

DEFLECTIONS OF STRUCTURES

16.1. Use and Limitations of Calculated Deflections. The most important applications of the methods for calculating deflections will be in the analysis of statically indeterminate structures. The deflections of most engineering structures are small in comparison to the original dimensions of the structures, and the flexibility of the structures is seldom an important design criterion. However, the relative rigidity of various members of statically indeterminate structures affects the stress distribution in the structures; therefore it is necessary to consider the deflections in the analysis of such structures.

Most of the methods of calculating deflections are applicable to both elastic and inelastic deformations. Elastic deformations are those due to stresses which are below the elastic limit, so that all stresses and deflections are proportional to the applied loads. Inelastic deformations include all deformations which are not proportional to the applied loads, such as deformations at stresses above the elastic limit, slip in bolted or riveted joints, and temperature deformations.

The assumption that deflections are small in proportion to the initial dimensions of the structure is used in all the derivations of this chapter. The moment arms of all forces are therefore assumed to be the same for the deflected structure as they are for the undeflected structure. This assumption is quite accurate for all except a few special types of structures, such as flexible beams in tension or compression, which will be given special consideration.

16.2. Strain Energy in an Axially Loaded Member. The deflections of structures are usually obtained from a consideration of the work done by forces acting on the structures. Elastic deformations, in which the stress and strain are proportional, will first be considered. The external work done by a force acting through a deformation of a structure will be stored in the structural material as potential energy of deformation, or strain energy. This energy is used in restoring the structure to its original position if the load is gradually removed.

An axially loaded elastic member is shown in Fig. 16.1(a). The tension force S is gradually applied, and the elongation of the member Δ will be proportional to the load. The load and the elongation therefore increase simultaneously from zero to their maximum values S_a and Δ_a, as shown in Fig. 16.1(*b*). The total strain energy U stored in the member is therefore equal to the shaded triangular area.

$$U = \frac{S_a \Delta_a}{2}$$ (16.1)

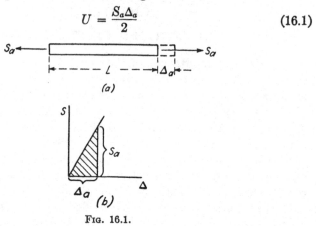

FIG. 16.1.

Equation 16.1 may be expressed in terms of either Δ_a or S_a for a uniform homogeneous bar in which $\Delta_a = S_a L/AE$.

$$U = \frac{S_a^2 L}{2AE} = \frac{\Delta_a^2 A E}{2L}$$ (16.2)

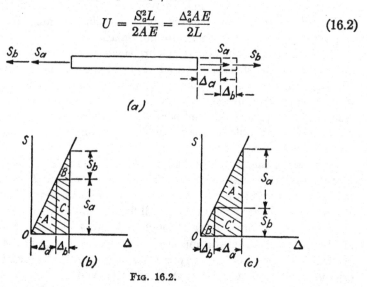

FIG. 16.2.

When two or more axial loads are gradually applied to a member, the total work done on the member will be independent of the order in which the loads are applied, but the total work will be greater than the sum of the work done by the loads acting separately.   If the loads S_a and S_b are applied to the member shown in Fig. 16.2(a), the total strain energy stored in the member is equal to the shaded area shown in Fig. 16.2(b) or Fig. 16.2(c).

$$U = \tfrac{1}{2}(S_a + S_b)(\Delta_a + \Delta_b) \tag{16.3}$$

If the load S_a is first applied, it will do an amount of work indicated by the area of the triangle A of Fig. 16.2(b), or $\tfrac{1}{2}S_a\Delta_a$. When the load S_b is subsequently applied, the increase in strain energy is represented by the sum of the areas B and C, or $\tfrac{1}{2}S_b\Delta_b + S_a\Delta_b$. The work represented by area C results from the total force S_a acting through the total deformation Δ_b. The total strain energy may now be expressed by the following equation.

$$U = \tfrac{1}{2}S_a\Delta_a + \tfrac{1}{2}S_b\Delta_b + S_a\Delta_b \tag{16.4}$$

If the load S_b is first applied, it will do an amount of work represented by area B of Fig. 16.2(c). When the load S_a is subsequently applied, it will do work as represented by areas A and C' of Fig. 16.2(c). The total strain energy may then be expressed as follows:

$$U = \tfrac{1}{2}S_b\Delta_b + \tfrac{1}{2}S_a\Delta_a + S_b\Delta_a \tag{16.5}$$

The values of the strain energy given by Eqs. 16.3 to 16.5 are seen to be equal if the areas C and C' are equal, or if $S_a\Delta_b$ is equal to $S_b\Delta_a$. This is obviously true, since the loads S_a and S_b are proportional to the corresponding deformations Δ_a and Δ_b from the original assumption of an elastic member.

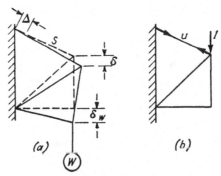

FIG. 16.3.

16.3. Truss Deflections. The truss shown in Fig. 16.3(a) resists an external load W. The upper member is assumed to be elastic and is assumed to have an elongation Δ and a tension force S, which are produced by the application of W. At present, only the one member is assumed to deform, the others remaining inextensible. The external work done by the gradually applied load W acting through the deflection

δ_w must therefore be equal to the strain energy induced in the elastic member by the gradually applied tension force S, acting through the deformation Δ.

$$\frac{W\delta_w}{2} = \frac{S\Delta}{2} \tag{16.6}$$

Equation 16.6 is adequate for finding the deflection δ_w in the direction of the applied load, since the other terms may be obtained from the dimensions of the truss and the known loading. This equation is not adequate when the deflection of some other point is desired, or when several loads act at different points of the truss. A more general method will therefore be developed.

If the deflection δ of the truss shown in Fig. 16.3(a) is desired, a unit load is applied at the point of the desired deflection, and in the direction of the desired deflection, as shown in Fig. 16.3(b). This load is termed a "virtual" load and has no relationship to the actual loads on the structure, but is simply a hypothetical load assumed for the deflection analysis. The unit virtual load produces a tension force u in the elastic member, as shown. If the virtual load is acting at the time W is applied, the additional deflections produced by W will be the same as the real deflections produced by W when the unit load was not acting. The total external work of W and the virtual load will be equal to the internal strain energy in the elastic member. In addition to the terms of Eq. 16.6, the external work will include the product of the unit load and the deflection δ and the strain energy will include the product of the force u and the deformation Δ.

$$\underbrace{\frac{W\delta_w}{2}}_{\substack{\text{(Real} \\ \text{work)}}} + \underbrace{1 \cdot \delta}_{\substack{\text{(Virtual} \\ \text{work)}}} = \underbrace{\frac{S\Delta}{2}}_{\substack{\text{(Real} \\ \text{work)}}} + \underbrace{u \cdot \Delta}_{\substack{\text{(Virtual} \\ \text{work)}}} \tag{16.7}$$

If desired, the work done by the virtual loads before the real loads are applied may be included in Eq. 16.7; but this would add equal terms to each side of the equation, since the external and internal work would be equal during this virtual load application. Subtracting Eq. 16.6 from Eq. 16.7, the following equation is obtained:

$$1 \cdot \delta = u \cdot \Delta \tag{16.8}$$

If all the truss members are elastic, the strain-energy terms and the virtual-work terms must include summations of similar terms for all the truss members. The following expression is obtained for the deflection of a truss:

$$\delta = \Sigma u \cdot \Delta \tag{16.9}$$

The term u represents the force in any member resulting from a unit virtual load acting at the point of the required deflection δ and in the direction of the deflection. As Eq. 16.9 is written, u is dimensionless, since it has units of pounds per pound of virtual load, or tons for a 1-ton virtual load. As Eq. 16.8 is written, both sides of the equations have dimensions of force times distance; thus u may be considered as having dimensions of force before dividing both sides of the equation by the unit force. The summation includes all members of the structure.

Equation 16.9 remains valid for inelastic deformations. From physical considerations of a structure, such as that shown in Fig. 16.3, a small elongation Δ of any member will contribute the same amount to the deflection δ, regardless of whether the member elongates elastically or inelastically. The terms u therefore represent dimensionless ratios which depend only on the geometry of the structure. In investigating the energy relationships for a structure which does not deform elastically, it is seen that Eqs. 16.6 and 16.7 no longer apply. No simple relationship such as Eq. 16.6 exists for the real work of deformation, when rivet slip, plastic deformations, and similar items are included. The virtual work, however, must still satisfy Eqs. 16.8 and 16.9, because the virtual loads are acting at the time the real deformations occur. The unit virtual load therefore acts through the total deformation δ, and the internal virtual forces u act through the total deformations Δ. The real work terms will remain the same as if the virtual loads were not acting, and therefore they will be equal on each side of the equation, as in Eq. 16.7. Thus Eq. 16.9 may be written directly, and it is applicable for small elongations Δ resulting from any cause. When the elongations Δ result from elastic deformations of uniform bars, the following equations apply:

$$\Delta = \frac{SL}{AE} \tag{16.10}$$

$$\delta = \sum \frac{SuL}{AE} \tag{16.11}$$

Example 1. Find the displacement of point C, Fig. 16.4(a), in a direction at 45° with the horizontal, as indicated by the unit load in Fig. 16.4(b). The lengths of all members are 30 in., the areas are 1 sq in., and E is 10,000 ksi. Assume that all displacements result from elastic deformations of the members.

Solution. The forces S in the members, resulting from the applied loads, are shown in Fig. 16.4(a). All values of S are in kips, or kilopounds (1,000 lb). The values of u, the forces in the members resulting from a unit virtual load at point C in the direction of the desired deflection, are shown in Fig. 16.4(b). These have units of pounds per pound or kips per kip and are dimensionless. The deflection

δ is now computed from Eq. 16.11, and the calculations are made in Table 16.1. The elongations Δ and the terms $u \cdot \Delta$ are calculated for all members, and the final deflection is obtained as the sum of the terms $u \cdot \Delta$ for all members. The values of $u \cdot \Delta$ may have negative signs in some problems and must, of course, be added algebraically.

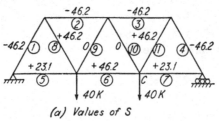

(a) Values of S

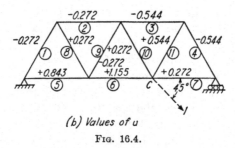

(b) Values of u

Fig. 16.4.

TABLE 16.1

Member	S, kips	$\dfrac{L}{AE}$, in./kip	$\Delta = \dfrac{SL}{AE}$, in.	u, lb/lb	$u \cdot \Delta$, in.
1	-46.2	0.003	-0.1386	-0.272	$+0.0377$
2	-46.2	0.003	-0.1386	-0.272	$+0.0377$
3	-46.2	0.003	-0.1386	-0.544	$+0.0754$
4	-46.2	0.003	-0.1386	-0.544	$+0.0754$
5	$+23.1$	0.003	$+0.0693$	$+0.843$	$+0.0584$
6	$+46.2$	0.003	$+0.1386$	$+1.115$	$+0.1600$
7	$+23.1$	0.003	$+0.0693$	$+0.272$	$+0.0188$
8	$+46.2$	0.003	$+0.1386$	$+0.272$	$+0.0377$
9	0	0.003	0	-0.272	0
10	0	0.003	0	$+0.272$	0
11	$+46.2$	0.003	$+0.1386$	$+0.544$	$+0.0754$

$$\delta = \Sigma u \cdot \Delta = 0.5765$$

Example 2. Members 2 and 3 of the truss of Example 1 are assumed to be fabricated with an initial length of 30.1 in. and to be subjected to a temperature rise of 40°F. Find the displacement of point C resulting from the change in length and from the change in temperature. The other members retain the initial length of 30 in., and none of the members are stressed. Assume a temperature coefficient ϵ of 10^{-5} per degree Fahrenheit.

Solution. The elongations of members 2 and 3 are as follows.

$$\Delta_2 = \Delta_3 = 0.1 + \epsilon Lt = 0.1 + 10^{-5} \times 30 \times 40 = 0.112 \text{ in.}$$

The displacement of point C is found from Eq. 16.9.

$$\delta = u_2 \Delta_2 + u_3 \Delta_3 = -0.272 \times 0.112 - 0.544 \times 0.112 = \underline{-0.0914} \text{ in.}$$

The negative sign indicates that the deflection is opposite in direction to the virtual load. The values of δ obtained in Examples 1 and 2 represent the component of the deflection in the direction of the virtual load and indicate nothing regarding the displacement in the perpendicular direction. If the direction of the total displacement is desired, it is necessary to obtain other components. In such cases, horizontal and vertical components would probably be obtained.

16.4. Strain Energy of Bending. The deflections of beams, rigid frames, and similar structures which resist bending moments are normally obtained by methods of virtual work which are similar to those used in truss analysis. The general methods of virtual work are applicable to both elastic and inelastic deformations. The special applications of energy methods for elastic deformations will first be considered, however.

An element of an initially straight beam is shown in Fig. 16.5. A length dx of the beam deflects through an angle $d\theta$, as shown.

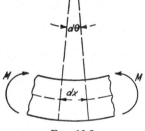

Fig. 16.5.

If the beam is elastic, the angular deflection is proportional to the bending moment, as shown in Art. 14.1.

$$d\theta = \frac{M}{EI} dx \tag{16.12}$$

As the bending moment is gradually applied to the beam, it does work which is equal to the product of the average moment $M/2$ and the final angular deformation $d\theta$. If all elements of the beam are considered, the total work U is obtained by an integration over the length of the beam.

$$U = \int \frac{M}{2} d\theta = \int \frac{M^2}{2EI} dx \tag{16.13}$$

If a beam is deformed by two couples, m and M, as shown in Fig. 16.6, the total work of deformation, or strain energy, is greater when the two couples are acting simultaneously than the sum of the work for the two

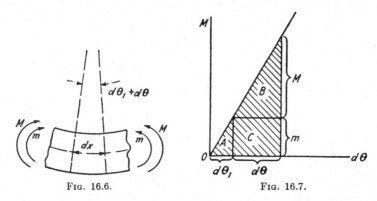

Fig. 16.6.   Fig. 16.7.

couples acting independently. If the couple m deforms the beam through an angle $d\theta_1$ in a length dx, and the couple M deforms the beam element an angle $d\theta$, the total work is represented by the area of Fig. 16.7. The following equation determines the angular distortion produced by the bending moment m:

$$d\theta_1 = \frac{m\,dx}{EI} \tag{16.14}$$

Assuming that the couple m is first gradually applied and that the couple M is then gradually applied, the total strain energy in the length dx is represented by the following equation:

$$dU = \tfrac{1}{2}m\,d\theta_1 + \tfrac{1}{2}M\,d\theta + m\,d\theta \tag{16.15}$$

Substituting values of the angular distortion from Eqs. 16.12 and 16.14 and integrating for the length of the span, the total strain energy is obtained.

$$U = \int \frac{m^2}{2EI}\,dx + \int \frac{M^2}{2EI}\,dx + \int \frac{Mm}{EI}\,dx \tag{16.16}$$

16.5. Beam-deflection Equations. The energy relationships for a deflected beam are similar to those for a deflected truss. It will first be assumed that all deflections are elastic. The beam shown in Fig. 16.8(a) has bending moments M, as shown in Fig. 16.8(b). The external work done by the gradually applied load is $P\delta_p/2$, and it is equal to the internal work defined by Eq. 16.13.

$$\frac{P\delta_p}{2} = \int \frac{M^2}{2EI}\,dx \tag{16.17}$$

This relationship may be used for finding the deflection under the load, but it cannot be used to find the deflection at another point. If the deflection at point A is desired, a unit virtual load may be assumed to act at point A at the time the load P is applied. The unit virtual load

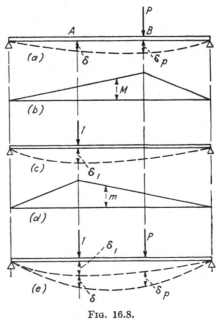

FIG. 16.8.

deflects the beam, as shown in Fig. 16.8(c), and produces bending moments m, as shown in Fig. 16.8(d). The following energy relationship exists for the gradual application of the unit load:

$$\frac{1 \cdot \delta_1}{2} = \int \frac{m^2}{2EI}\, dx \tag{16.18}$$

If the unit load is first applied at point A and the load P is subsequently applied at point B, the additional deflections resulting from P will be equal to the deflections resulting from P acting alone, as shown in Figs. 16.8(a) and (e). The external work done by the two loads will be equal to the strain energy defined by Eq. 16.16.

$$\frac{1 \cdot \delta_1}{2} + \frac{P\delta_p}{2} + 1 \cdot \delta = \int \frac{m^2}{2EI}\, dx + \int \frac{M^2}{2EI}\, dx + \int \frac{Mm}{EI}\, dx \tag{16.19}$$

Equations 16.17 and 16.18 may now be subtracted from Eq. 16.19.

$$\delta = \int \frac{Mm}{EI}\, dx \tag{16.20}$$

The term m represents the beam bending moment for a unit load acting at the point of the deflection δ.

Equation 16.20 represents the virtual work condition for beams which compares with Eq. 16.11 for trusses. Equation 16.19 is similar to Eq. 16.7, but it contains an additional term on each side which represents the work done during the application of the unit load. A similar term could have been included in the energy equation for trusses, but it would not affect the final equation. If the beam distortions are inelastic, the virtual work conditions may be used to find the deflections. A unit virtual load is assumed to be acting on a beam at the time a section of length dx is distorted through an angle $d\theta$. The external virtual work caused by the unit load acting through the real deflection δ will be equal to the internal virtual work of the couple m acting through the angle $d\theta$.

$$1 \cdot \delta = \int m \, d\theta \qquad (16.21)$$

The integration extends over the span of the beam. The equation applies for small angular deformations $d\theta$, which may result from any cause.

Example 1. Find the deflection of the uniform beam shown in Fig. 16.9(*a*), at the point of application of the load. Assume elastic deflections.

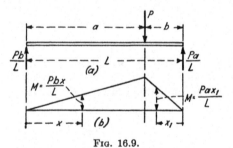

Fig. 16.9.

Solution. The deflection at the load may be obtained from Eq. 16.17 or from Eq. 16.20. Substituting $M = Pm$ into Eq. 16.17, Eq. 16.20 is obtained. The integral of Eq. 16.20 may be most conveniently evaluated by using separate origins of coordinates for the portions of the beam to either side of the load P. For the left portion of the beam, the origin of coordinates is taken as the left support, with x varying from 0 to a. For the right side the origin is taken as the right support with x_1 varying from 0 to b. From values shown in Fig. 16.9(*b*), the following results are obtained:

$$\delta = \int \frac{Mm}{EI} \, dx = \frac{Pb^2}{EIL^2} \int_0^a x^2 \, dx + \frac{Pa^2}{EIL^2} \int_0^b x_1^2 \, dx_1 = \frac{Pa^2b^2}{3EIL}$$

Example 2. Find the deflection of the beam shown in Fig. 16.10 at the point of application of the 1,000-lb load. Assume elastic conditions, with $EI = 10^6$ lb-in.[2]

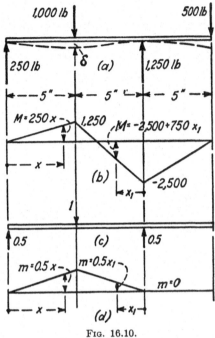

FIG. 16.10.

Solution. The deflection is determined from Eq. 16.20, with values of M and m as shown in Fig. 16.10(b) and (d). The integrals must be evaluated separately for the three parts of the beam.

$$EI\delta = \int Mm \, dx = \int_0^5 0.5 \times 250x^2 \, dx + \int_0^5 (-2{,}500 + 750x_1) \times 0.5x_1 \, dx_1$$
$$= 5{,}208$$
$$\delta = \frac{5{,}208}{10^6} = 0.005208 \text{ in.}$$

Example 3. The semicircular arch shown in Fig. 16.11 is supported in such a manner that the external reactions are vertical. Find the horizontal motion of the right-hand support if the deflections are elastic and EI is constant.

Solution. The derivation of Eq. 16.20 applies to any structure in which other types of deflections are negligible in comparison to bending deflections. Since shearing and axial deformations are negligible in the arch, Eq. 16.20 may be used directly. The length dx of Eq. 16.20 will be replaced by $ds = R \, d\phi$, an increment of length along the arch. The loading is symmetrical with respect to the center line; therefore the integral of Eq. 16.20 will be evaluated for one-half of the structure and then multiplied by 2. Bending moments will be considered positive when they produce compression on the outside of the ring. From Fig. 16.11(a),

$$M = \frac{PR}{2}(1 - \cos \phi)$$

for the ring to the left of the load. Similarly, the values of m are obtained from Fig. 16.11(b), the free-body diagram for the structure loaded with the unit virtual load in the direction of the desired deflection.

$$m = R \sin \phi.$$

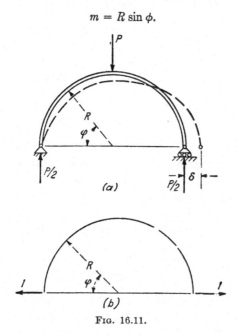

Fig. 16.11.

These values are now substituted into Eq. 16.20.

$$\delta = \int \frac{Mm}{EI} ds = 2 \int_0^{\frac{\pi}{2}} \frac{PR^3}{2EI}(1 - \cos \phi) \sin \phi \, d\phi = \frac{PR^3}{2EI}$$

16.6. Semigraphic Integration. In many practical problems the bending-moment diagrams may be readily plotted, but they cannot be expressed as simple algebraic expressions. The integrals involving such bending moments may often be evaluated more readily by semigraphic methods than by direct integration. The expression $\int M_1 M_2 \, dx$, in which either M_1 or M_2 is a linear function of x, will be evaluated. Assuming M_1 to be linear between two points a and b, it may be expressed as follows:

$$M_1 = M_a + \frac{x}{L}(M_b - M_a) \qquad (16.22)$$

where L is the distance between a and b and M_a and M_b are the values of M_1 at these points, as shown in Fig. 16.12. Substituting the value of M_1 into the integral and expanding, the following values are obtained·

$$\int M_1 M_2 \, dx = M_a \int_a^b M_2 \, dx + \frac{M_b - M_a}{L} \int_a^b M_2 x \, dx \qquad (16.23)$$

The first integral on the right side represents the area of the M_2 diagram and will be designated as A. The second integral represents the moment

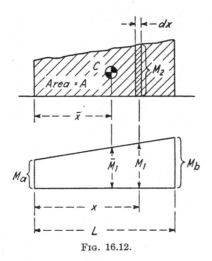

FIG. 16.12.

of the M_2 diagram about point a and will be designated as $A\bar{x}$, where $\bar{x}$ is the distance to the centroid of the M_2-diagram area, as shown in Fig. 16.12. The integral becomes

$$\int_a^b M_1 M_2 \, dx = A \left[M_a + \frac{\bar{x}}{L} (M_b - M_a) \right] \qquad (16.24)$$

where the term in brackets may be compared with Eq. 16.22 and is seen to represent the value of M_1 at a point opposite the centroid of the area of the M_2 diagram. This value is designated as $\overline{M}_1$ in Fig. 16.12. The integral is therefore evaluated by the following equation:

$$\int_a^b M_1 M_2 \, dx = A\overline{M}_1 \qquad (16.25)$$

where the M_1 diagram is a straight line between a and b, A is the area of the M_2 diagram between a and b, and $\overline{M}_1$ is the value of M_1 opposite the centroid of the area of the M_2 diagram.

Example. Find the horizontal displacement δ of the right-hand support of the frame shown in Fig. 16.13(a). The members are elastic and have constant values of $I = 80$ in.^{4} and $E = 10^7$ psi.

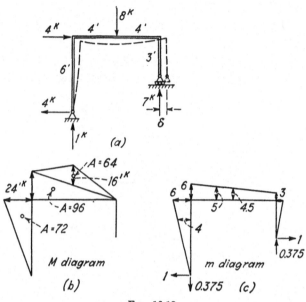

Fig. 16.13.

Solution. The bending-moment diagram is plotted in Fig. 16.13(b) on the compression side of the frame. The areas of the moment diagram are shown in units of ft^2-kips. A unit virtual load is then applied in the direction of the desired deflection. This load produces reactions and bending moments m, as shown in Fig. 16.13(c). The values of m at points opposite the centroids of the M-diagram triangles are indicated. The dimensions of m are feet. The deflection will now be obtained by evaluating the integral of Eq. 16.20 by means of Eq. 16.25.

$$EI\delta = 72 \times 4 + 96 \times 5 + 64 \times 4.5 = 1,056 \text{ ft}^3\text{-kips}$$

The constants 12^3 and $1,000$ are necessary for conversion to inch and pound units.

$$\delta = \frac{1,056}{80 \times 10^7} \times (12)^3 \times 1,000 = 2.28 \text{ in.}$$

16.7. Angular Deflections of Structures. If the rotation at a point of a structure is desired, a virtual loading of a unit couple may be considered as acting at the point during the deformation. For a beam or other structure in which bending deformations predominate, the deflections will be as indicated in Fig. 16.14. The external virtual work is equal to the product of the unit couple and the angle of rotation in radians θ. This is equated to the internal virtual work.

$$1 \cdot \theta = \int \frac{Mm}{EI}\, dx \tag{16.26}$$

This equation as stated is applicable to elastic deformations, but it may be varied to cover inelastic deformations. The values for m are obtained as bending moments resulting from the unit couple, as shown in Fig. 16.14(c). The values of M are the bending moments resulting from the actual loads on the structure.

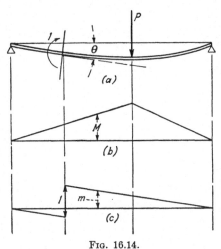

The rotation of a truss member may also be obtained by the

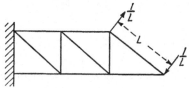

FIG. 16.14.

FIG. 16.15.

method of virtual work. A unit virtual couple is applied as loads of $1/L$ at the ends of the member for which the rotation is desired, as shown in Fig. 16.15. This unit virtual couple causes forces of u in the truss members. The rotation θ is now obtained by equating the external virtual work to the internal virtual work.

$$1 \cdot \theta = \Sigma u \cdot \Delta \tag{16.27}$$

The deformations Δ of the members may be elastic or inelastic, as in previous truss-deflection analysis.

16.8. Deflections Resulting from Torsional Deformations. The deflections of a structure as a result of the torsional deformations of members may be obtained by the method of virtual work in a similar manner to the deflections resulting from axial and bending deformations. The torsion member shown in Fig. 16.16 is subject to loads which deform an element of length dx through an angle $d\phi$, as shown in Fig. 16.16(b). It is desired to obtain the deflection δ of a point A, resulting from this torsional deformation. A unit virtual load is assumed to act at point A in the direction of the desired deflection during the deformation. This virtual load produces a torque, m_t, at any cross section, and this virtual torque acts through the real displacement $d\phi$. Equating the virtual work terms, the following deflection is obtained.

$$1 \cdot \delta = \int m_t \, d\phi \tag{16.28}$$

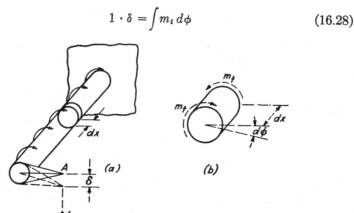

FIG. 16.16.

The integration extends over the length of the member. This equation applies for any elastic or inelastic torsional deformation. For an elastic deformation of a round bar or round tube,

$$d\phi = \frac{T}{JG} \, dx \tag{16.29}$$

where T is the applied torque, G is the shearing modulus of elasticity, and J is the polar moment of inertia of the cross section. Equation 16.28 becomes

$$\delta = \int \frac{Tm_t}{JG} \, dx \tag{16.30}$$

for elastic deformations of round bars or round tubes. For noncircular cross sections, constants depending on the cross section may be substituted for J.

Example 1. The structure shown in Fig. 16.17(a) consists of a tube curved to a 90° arc in a horizontal plane, with one end fixed and the other end resisting a vertical load P. For $P = 1,000$ lb, $I = 1.5$ in.⁴, $E = 10,000,000$ psi, $R = 20$ in., and Poisson's ratio $\mu = 0.25$, find (a) the downward deflection of P and (b) the angle of rotation θ_y.

Solution. a. The bending moment and torque at any cross section are indicated in Fig. 16.17(b). The load P has a moment arm of $R \sin \beta$ in producing bending moment and $R(1 - \cos \beta)$ in producing torsion. The values of m and m_t are for a unit load in the direction of P and are equal to M/P and T/P.

$$M = PR \sin \beta \qquad m = R \sin \beta$$
$$T = PR(1 - \cos \beta) \qquad m_t = R(1 - \cos \beta)$$
$$ds = R d\beta$$

Substituting these values into Eqs. 16.20 and 16.30 and obtaining the total deflection as the sum of the torsional and bending deflections, the following value is obtained:

$$\delta = \int \frac{Mm}{EI} ds + \int \frac{Tm_t}{JG} ds$$

$$\delta = \frac{PR^3}{EI} \int_0^{\frac{\pi}{2}} \sin^2 \beta \, d\beta + \frac{PR^3}{JG} \int_0^{\frac{\pi}{2}} (1 - \cos \beta)^2 \, d\beta$$

$$\delta = \frac{PR^3\pi}{4EI} + \frac{PR^3}{JG} \left(\frac{3\pi}{4} - 2 \right)$$

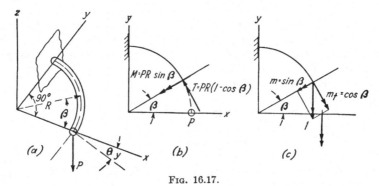

Fig. 16.17.

The numerical value for the shearing modulus is found as follows.

$$G = \frac{E}{2(1 + \mu)} = \frac{10,000,000}{2 \times 1.25} = 4,000,000 \text{ psi}$$

Since $J = 2I$ and $G = 0.4E$, the relation $GJ = 0.8EI$ is approximately correct in all round bars or tubes. A variation in Poisson's ratio causes only a small variation in G, for example, if $\mu = 0.3$, $G = 0.385E$. The numerical value of δ is found by substitution.

$$\delta = \frac{1,000 \times (20)^3 \times \pi}{4 \times 1.5 \times (10)^7} + \frac{1,000 \times (20)^3 \times 0.3565}{3.0 \times 4 \times (10)^6}$$
$$= 0.418 + 0.238 = 0.656 \text{ in.}$$

b. The angle of rotation of the free end about the y axis is found by the following equation:

$$\theta_y = \int \frac{Mm}{EI} ds + \int \frac{Tm_t}{JG} ds$$

The values of M and T are the same as for part (a), since they result from the actual loading. The values of m and m_t result from a unit couple about the *y* axis, applied at the free end.   The couple vector is represented in Fig. 16.17(*c*), and has the following components:

$$m = \sin \beta$$
$$m_t = -\cos \beta$$

The positive directions of couple vectors are arbitrarily assumed as the directions of *M* and *T*, shown in Fig. 16.17.   For this type of structure there is no well-established sign convention such as the sign convention for bending moments in horizontal beams.

$$\theta_y = \frac{PR^2}{EI} \int_0^{\frac{\pi}{2}} \sin^2 \beta \, d\beta - \frac{PR^2}{JG} \int_0^{\frac{\pi}{2}} \cos \beta (1 - \cos \beta) \, d\beta$$
$$= \frac{PR^2 \pi}{4EI} - \frac{PR^2}{JG} \left(1 - \frac{\pi}{4}\right)$$
$$= 0.0209 - 0.0071 = 0.0138 \text{ rad}$$

Example 2.   The structure shown in Fig. 16.18 is made of a round tube with $I = 1.5$ in.^{4}, $E = 10^7$ psi, and $G = 4 \times 10^6$ psi.   Each section is straight and parallel to one of the coordinate axes.   Find the vertical deflection of point *a*.

Solution.   The bending-moment diagram of Fig. 16.19(*a*) shows the bending moments about the coordinate axes.   In evaluating the integral of Eq. 16.20, the bending moment about each axis will be considered separately, since a virtual couple *m* about one axis does no work during an angular deformation about a per-

Fig. 16.18.   Fig. 16.19.

pendicular axis.   The bending moments are plotted on the compression sides of the members.   The torsion diagram is shown in Fig. 16.19(*b*), with the directions indicated by the curved arrows.   The values of *m* and m_t, the bending moments and torques resulting from a unit load, are shown in Fig. 16.19(*c*) and (*d*).

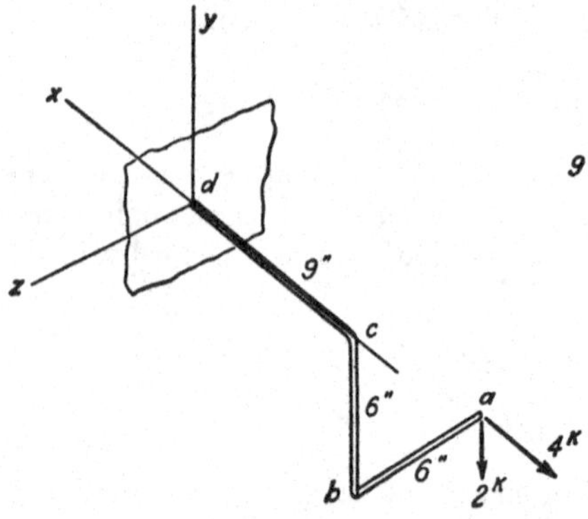

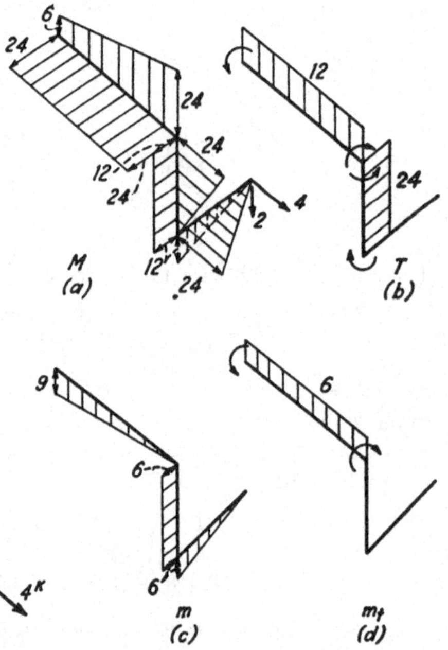

In cases where M and m are opposite in direction, their product is negative. The values of the integrals for Eqs. 16.20 and 16.30 are evaluated by semi-graphic methods.

The torsional distortion of member cd only contributes to the deflection, since $m_t = 0$ for the other members.

$$\int T m_t \, ds = 12 \times 6 \times 9 = 648$$

The bending-moment terms appear for all members. For member ab, only the bending moment in the vertical plane contributes to the vertical deflection.

$$\int M m \, ds = \frac{12 \times 6}{2} \times 4 = 144$$

For member bc, only the bending moment in the yz plane is considered, since m is zero for bending in the xy plane, or bending in the xy plane would not affect the vertical deflection of a.

$$\int M m \, ds = 12 \times 6 \times 6 = 432$$

For member cd, only the bending moment in the vertical plane is considered. The M diagram is assumed to be made up of a rectangular area of $6 \times 9 = 54$ and a triangular area of $18 \times \frac{9}{2} = 81$. The integral will be negative, since M and m have opposite signs.

$$\int M m \, ds = -54 \times 4.5 - 81 \times 3 = -486$$

The total deflection is now found as the sum of the contributions of all the members.

$$\delta = \int \frac{Mm}{EI} \, ds + \int \frac{Tm_t}{JG} \, ds$$

$$\delta = \frac{144 + 432 - 486}{EI} + \frac{648}{JG}$$

$$\delta = \frac{90}{1.5 \times 10^4} + \frac{648}{1.2 \times 10^4} = 0.060 \text{ in.}$$

16.9. Relative Displacements. In the previous analyses, the displacements relative to rigid supports have been obtained. It is frequently necessary to find the displacement of some part of the structure relative to another part, both of which may be displaced relative to the supports. In such cases, the supports for the structure resisting the virtual loads are considered to be different from the actual supports for the structure.

The beam of Fig. 16.20(a) is horizontal in the unstressed condition and deflects as shown. The deflection δ of point a relative to a tangent at point b is desired. The structure resisting the virtual load is therefore assumed to be as shown in Fig. 16.20(b), and the virtual bending-moment diagram is as shown in Fig. 16.20(c). Considering only the distortion between points a and b, and the virtual work of the relative displacements, the virtual-work equation may be written in the same form as previously.

$$\delta = \int m \, d\theta = \int_a^b \frac{Mm}{EI} \, dx \qquad (16.31)$$

The values of δ and m are shown in Fig. 16.20.

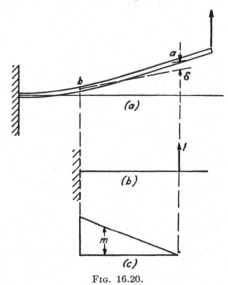

Fig. 16.20.

Angular displacements relative to other parts of the structure may also be obtained by assuming the structure which resists the virtual load to be fixed at the reference point. If the angular displacement of a tangent at a relative to a tangent at b for the beam of Fig. 16.20 is calculated, a unit virtual couple will be applied in place of the unit load of Fig. 16.20(b). The diagram for m will then have a unit value between points a and b and a zero value at other points. The values of the real loads and real distortions $d\theta$ are calculated for the true reaction conditions for the beam.

If the angular displacement of member 1 of the truss of Fig. 16.21 with respect to member 2 is desired, the reactions for the virtual loads are such that member 2 does not rotate. Only the members 1, 2, and 3 will resist the virtual loads, and the values of u for all other members will be zero. The rotation will be

$$\theta = \Sigma u \cdot \Delta \qquad (16.32)$$

where the values of Δ are the true elongations of the members and the values of u are calculated for the structure shown in Fig. 16.21(b). It can be shown that relative deflections for other types of structures may be calculated by adding fictitious supports at the reference member for

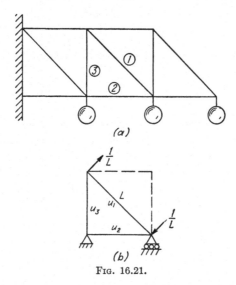

(a)

(b)

Fig. 16.21.

the virtual loading condition. The proof follows directly from the principles of virtual work and will not be discussed because the procedure is usually obvious for most problems which are encountered.

16.10. Shearing Deformations. In the analysis of semimonocoque aircraft structures, the shear stress distribution is of great importance. Much of the classical theory of statically indeterminate structures has been developed for the analysis of heavy structures in which shearing deformations are of minor importance. Consequently much of the published work on structural deflections and indeterminate structures does not treat shearing deformations. The deflections caused by shearing deformations may be determined by the method of virtual work, in the same manner that other types of deflections were analyzed.

The shearing deformation of an elastic rectangular plate with thickness t, width a, and length b is indicated in Fig. 16.22(a). The shear strain γ is obtained from the relation

$$\gamma = \frac{f_s}{G} = \frac{q}{tG} \tag{16.33}$$

where G is the shearing modulus of elasticity, f_s is the shearing stress, and q is the shear flow $f_s t$. The strain energy of shearing deformation may be obtained from the work done by the shearing stresses during the deformation. Only the force on the upper face does work in the condition shown. The force $f_s at$ acts through a displacement γb. The force is gradually applied, and the strain energy is obtained as follows.

$$U = \frac{f_s \gamma abt}{2} \qquad (16.34)$$

The strain energy may be expressed in other forms by substituting values from Eq. 16.33.

$$U = \frac{f_s^2}{2G} abt = \frac{\gamma^2 G}{2} abt = \frac{q^2}{2tG} ab \qquad (16.35)$$

In finding the deflections of a structure resulting from shearing deformations of the webs, it is convenient to use a unit virtual load applied

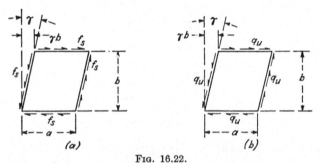

Fig. 16.22.

at the point of the desired deflection δ, as in structures previously analyzed. The virtual load produces a system of shear flows q_u in the webs, which are acting with their full magnitude during the entire deformation. For the web shown in Fig. 16.22(b), the force $q_u a$ acts through the displacement γb. The external virtual work must be equal to the sum of the internal virtual work for all webs.

$$1 \cdot \delta = \Sigma q_u \gamma ab$$

Substituting the value of γ from Eq. 16.33, the following equation is obtained:

$$\delta = \sum \frac{q_u q ab}{tG} \qquad (16.36)$$

The summation includes all webs of the structure which affect the deflection. The terms q_u are shear flows due to the unit virtual load, and the terms q are the real shear flows which produce the deformation. Equation 16.36 applies only to elastic deformations which satisfy Eq. 16.33.

PROBLEMS

16.1. Find the horizontal and vertical components of the deflection of point C of the truss shown in Fig. 16.4(a). For all members, use $L = 30$ in., $A = 1$ sq in., and $E = 10,000$ ksi. Check results by use of the deflection obtained in Example 1, Art. 16.3.

16.2. Find the horizontal and vertical deflections of the joint at the upper left-hand corner of the truss in Fig. 16.4(a). Use the values $L = 30$ in., $A = 1$ sq in., and $E = 10,000$ ksi for all members.

16.3. Repeat Prob. 16.2 if one 40-kip load is removed at point C but the other 40-kip load is acting.

16.4. Find the deflection of the right end of the beam of Fig. 16.10(a). Use $EI = 10^6$ lb-in.^{2}

16.5. Repeat Example 1, Art. 16.5, using semigraphic integration.

16.6. Repeat Example 2, Art. 16.5, using semigraphic integration.

16.7. Find the slope of the tangent to the elastic curve at the load for the beam of Fig. 16.9(a). Find the slopes of the tangents to the elastic curve at the supports. Use the procedure for semigraphic integration.

16.8. Find the slopes of tangents to the elastic curve at both loads and at both supports for the beam of Fig. 16.10(a). Use $EI = 10^6$ lb-in.^{2}

16.9. Find the rotation at the right support of the arch of Fig. 16.11(a).

16.10. For the truss of Fig. 16.4(a), calculate the rotation of member 10 relative to the supports. Also calculate the rotation of member 10 relative to member 6. Discuss the results.

16.11. Find the displacement δ of the frame of Fig. 16.13 if the only load is a uniformly distributed load of 1 kip/ft over the entire length of the horizontal member. Use $EI = 8 \times 10^8$ lb-in.^{2}

16.12. Find the displacement δ of the frame of Fig. 16.13 if the only load is a uniformly distributed horizontal load of 1 kip/ft over the length of the left-hand vertical member.

16.13. For the structure shown in Fig. 16.17(a), find the deflections of the free end parallel to the x and y axes. Find the rotations of the free end with respect to the x and z axes.

16.14. For the structure of Fig. 16.18 and of Example 2, Art. 16.8, find the components of the deflection of point a along x and z axes, and find the rotation at point a about all three axes.

16.15. Assume that the structure of Fig. 16.18 resists an additional load of 3 kips at a, acting toward b. Find the three components of linear displacement and the three components of angular displacement of point a. The structure is made of a round tube with $I = 1.5$ in.^{4}, $E = 10^7$ psi, and $G = 4 \times 10^6$ psi.

16.16. A cantilever wing spar is 10 in. deep between centroids of the flange areas. The bending stresses in the flanges are 30,000 psi, and the shear stresses in the web are 15,000 psi at all points. Find the deflection resulting from shear and bending deformations, and find the percentage of the deflection contributed by the shear at (a) 20 in. from the fixed support, (b) 40 in. from the fixed support, and (c) 100 in. from the fixed support. Use $E = 10^7$ psi and $G = 3,000,000$ psi.

16.17. A landing gear resists a drag force of 4,000 lb at a point 40 in. below the lower skin of a wing. Find the deflection of the point of application of the force in the direction of the force if the wing has dimensions as shown in Fig. 16.25 and the landing gear is 60 in. outboard from the fixed support. Consider only the shear deformation of the four webs. Use $G = 3,000,000$ psi.

16.11. Twist of Box Beam. One of the most common applications of Eq. 16.36 is in finding the angle of twist of a box beam, such as that

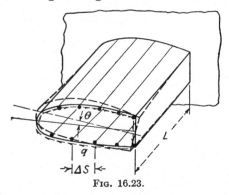

Fig. 16.23.

shown in Fig. 16.23. The shear flows q may result from any condition of loading, and they are obtained by the methods used in Chap. 6. Since an angular deflection is required, a unit virtual couple will be applied as shown in Fig. 16.24, and the resulting virtual shear flows are

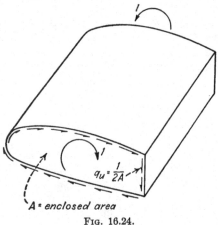

Fig. 16.24.

$q_u = \dfrac{1}{2A}$, where A is the enclosed area of the box. The webs are assumed to have dimensions $a = \Delta s$ and $b = L$. The angle of twist is obtained by substituting these values into Eq. 16.36.

$$\theta = \sum \frac{qab}{2AtG} = \sum \frac{q\,\Delta s L}{2AtG} \qquad (16.37)$$

The summation includes all webs of the structure.

Example. The box beam shown in Fig. 16.25 has front spar-flange areas which are three times the rear spar-flange areas. Find the angle of twist at the free end. Assume $G = 4 \times 10^6$ psi.

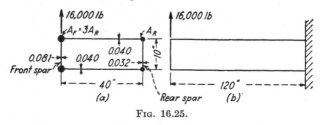

Fig. 16.25.

Solution. The shear flows q are shown in Fig. 16.26. All these shear flows are positive except the shear flow in the right-hand web, which tends to produce a counterclockwise rotation. The values obtained from Figs. 16.25 and 16.26 are substituted into Eq. 16.37.

$$\theta = \sum \frac{q\,\Delta s L}{2AtG} = \frac{L}{2AG}\sum \frac{q\,\Delta s}{t}$$

$$= \frac{120}{2 \times 400 \times 4 \times 10^6}\left(\frac{1{,}400 \times 10}{0.081} + \frac{200 \times 40 \times 2}{0.040} - \frac{200 \times 10}{0.032}\right) = 0.020 \text{ rad}$$

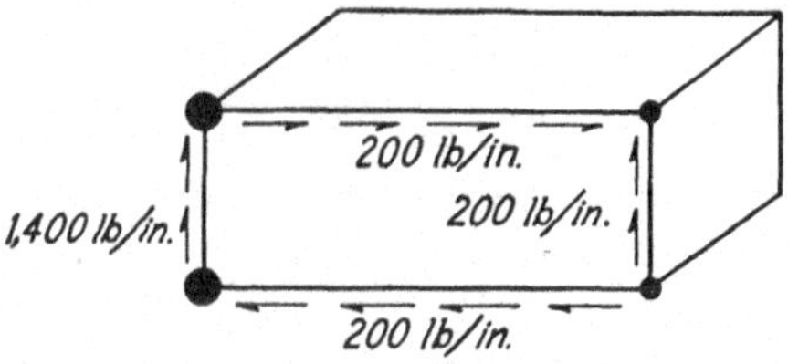

200 lb/in.

200 lb/in.

1,400 lb/in.

200 lb/in.

Fig. 16.26.

16.12. Accuracy of Analysis Methods. Several assumptions were made in Chap. 6 for the analysis of box beams. The approximate errors involved in these assumptions may be determined by calculating deflections of typical structures. In the conventional analysis of the box beam of Fig. 16.25, it is assumed that both of the upper flange areas resist the same bending stresses at all points along the span. This assumption depends on the condition that the box construction permits very little torsional deflection and that the front spar and rear spar therefore have approximately the same bending-deflection curves. The torsional deflection has been calculated for the structure, and it is necessary to compare it with the bending deflection.

It will be assumed that the flange areas of Fig. 16.25 vary along the span in such a manner that the bending stress has a constant value of 40,000 psi in all four flanges and that $E = 10^7$ psi. The bending deflection of the free end is found from Eq. 16.20. Substituting $f = My/I$,

$$\delta = \int \frac{Mm}{EI}\, dx = \int \frac{fm}{Ey}\, dx$$

$$\delta = \frac{40,000 \times 60 \times 120}{10^7 \times 5} = 5.76 \text{ in.}$$

This will be the deflection at the shear center of the area, which is approximately 10 in. from the front spar. The angle of twist is 0.020 rad, as computed in the example of Art. 16.11. This causes an additional

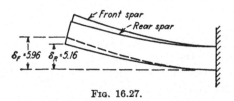

Fig. 16.27.

deflection of the front spar of $10 \times 0.020 = 0.20$ in. The rear spar is 30 in. from the centroid, and it will have a smaller deflection than δ by $30 \times 0.020 = 0.60$ in. The deflections of the spars are shown in Fig. 16.27.

$$\delta_F = 5.76 + 0.20 = 5.96 \text{ in.}$$
$$\delta_R = 5.76 - 0.60 = 5.16 \text{ in.}$$

The deflection of the front spar is therefore about 3 per cent greater and the deflection of the rear spar about 10 per cent less than calculated from bending deflections only. This error is probably typical of most semi-monocoque structures of this type. The bending stresses calculated by the simple flexure formula have approximately the same error as the deflections, but it will be shown later that the bending stresses will be considerably higher in the loaded spar near the supported end but will be approximately equal to those calculated by the flexure formula for most of the span.

16.13. Warping of Beam Cross Sections. When a rectangular box beam is subjected to torsional moments, it deforms as shown in Fig. 16.28. If the cross section is square and the web thickness is the same on all sides, the cross sections will remain plane after the box is twisted. Similarly, if the box beam is subjected to bending with no torsion, the plane cross sections will remain plane after bending. In the usual case, however, the box is rectangular and resists some torsion; therefore the cross sections do not remain plane but "warp." In the analysis of box beams of Chap. 6 it was assumed that torsional moments did not affect the distribution of bending stresses, or that cross sections were not restrained against warping. The shear flows computed from these assumptions are accurate for all cross sections except those which are very close to a fixed cross section.

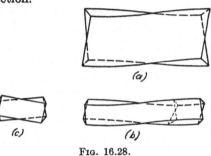

FIG. 16.28.

The amount of warping of a cross section may be measured by the angle θ between spar cross sections, as shown in Fig. 16.29(a). This angle is calculated by applying unit virtual couples, as shown, and calculating the relative rotation from Eq. 16.36. The warping of cross sections of the beam of Fig. 16.25 will be calculated, assuming that the shear flows q are as shown in Fig. 16.26. A unit spanwise length of the

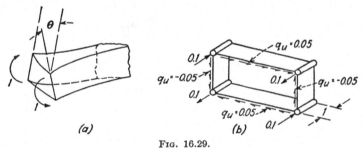

FIG. 16.29.

beam will be considered as shown in Fig. 10.29(b). The unit couples acting on the spars are represented by forces of 0.1 at distances 10 in. apart, and the values of q_u must be 0.05 for all webs, as shown, in order to satisfy all conditions of static equilibrium. Substituting values from Figs. 16.26 and 16.29(b) into Eq. 16.36, the angle θ is obtained.

$$\theta = \sum \frac{q_u q a b}{tG} = \frac{0.05 \times 200 \times 40 \times 1 \times 2}{0.040 \times 4 \times 10^6}$$

$$+ \frac{0.05 \times 200 \times 10 \times 1}{0.032 \times 4 \times 10^6} - \frac{0.05 \times 1400 \times 10 \times 1}{0.081 \times 4 \times 10^6} = \underline{0.00362 \text{ rad}}$$

This warping of the cross section is the same for all cross sections on which the shear flows are as shown in Fig. 16.26. The 1-in. length along the span was selected arbitrarily, but any other length b might be used. The values of q_u shown in Fig. 16.29(b) would be divided by an assumed

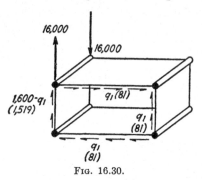

FIG. 16.30.

length b; then the terms in the above summation would be multiplied by b instead of the unit length in order to yield the same final result.

At the fixed support shown in Fig. 16.25(b), the warping of the cross section is obviously prevented; therefore the values of q cannot be as shown in Fig. 16.26. The values of q which are required to prevent warping will now be calculated for this beam. At any cross section the shear flows must be as shown in Fig. 16.30 in order to satisfy conditions of equilibrium. From the equilibrium of moments about a spanwise axis through one corner, and from the equilibrium of horizontal and vertical shearing forces, the shear flows q_1 must be equal and in the directions shown for three webs. For the front spar web the shear flow is $1,600 - q_1$. Substituting the values of q from Fig. 16.30, and the values of q_u from Fig. 16.29(b) into Eq. 16.36, the following equation is obtained:

$$\theta = \sum \frac{q_u qab}{tG} = \frac{0.05q_1 \times 40 \times 1 \times 2}{0.040 \times 4 \times 10^6} + \frac{0.05q_1 \times 10 \times 1}{0.032 \times 4 \times 10^6}$$
$$- \frac{0.05(1,600 - q_1)10}{0.081 \times 4 \times 10^6} = 0.0000304q_1 - 0.00247$$

At the fixed support, $\theta = 0$, or $q_1 = 81$ lb/in. The final shear flows are shown in parentheses in Fig. 16.30. These are seen to be considerably different from those shown in Fig. 16.26.

The shear flows of Fig. 16.26 apply a running load of 400 lb/in. to the rear spar flanges and a running load of 1,200 lb/in. to the front spar flanges. Near the support the shear flows apply a running load of 162 lb/in. to the rear spar flanges and a running load of 1,438 lb/in. to the front spar flanges. Near the support the bending stresses are therefore higher in the front spar than in the rear spar, the axial strains in the front spar flanges are greater, and the cross sections change from plane sections to warped cross sections. The spanwise distance required for the transition depends on the flange areas and web gages. In this problem the shear flows at a section 30 in. from the support have approximately the values shown in Fig. 16.26. The bending stresses outboard of this cross section are approximately equal for the two spars, since all cross sections warp the same amount. The effect of a fixed cross section is to increase the bending stresses and shear flows in the loaded spar for a spanwise distance which is approximately equal to the average of the cross section dimensions.

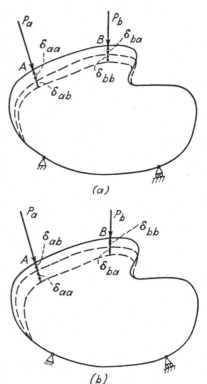

Fig. 16.31.

16.14. Maxwell's Theorem of Reciprocal Displacements. An interesting and useful relationship between deflections resulting from two loads was stated by Maxwell. The relationship will be derived by considering the structure of Fig. 16.31, which represents any elastic body. The loads P_a and P_b are applied at any two points A and B, and the deflections of a point are measured in the direction of the load acting at the point. In Fig. 16.31(a), the load P_a is first applied and causes deflections δ_{aa} and δ_{ba} at points A and B. The load P_b is then applied and causes deflections δ_{ab} and δ_{bb} at points A and B. The first subscript indicates the point at which the deflection is measured, and the second subscript indicates the load producing the deflection. The strain energy is obtained as follows:

$$U = \frac{P_a \delta_{aa}}{2} + \frac{P_b \delta_{bb}}{2} + P_a \delta_{ab} \tag{16.38}$$

If the load P_b is first applied and the load P_a is subsequently applied, the deflections are shown in Fig. 16.31(b). The strain energy may now be expressed as follows:

$$U = \frac{P_b \delta_{bb}}{2} + \frac{P_a \delta_{aa}}{2} + P_b \delta_{ba} \tag{16.39}$$

The total strain energy for the body after the loads are applied must be independent of the order of application. Equating the values from Eqs. 16.38 and 16.39, the following expression is obtained:

$$P_a \delta_{ab} = P_b \delta_{ba} \tag{16.40}$$

This expression is known as Maxwell's theorem of reciprocal displacements. The forces P_a and P_b may have any magnitudes, but it is easier to state the theorem in words if a special case of unit loads is assumed. The theorem may be stated as follows:

The deflection of an elastic structure at point A due to a unit force at point B is equal to the deflection at point B due to a unit force at point A.

The theorem may be extended to include rotations by considering the work done by a couple acting through an angular deflection of an elastic structure. Either a couple and a force or two couples may be considered instead of the forces P_a and P_b. For unit couples or forces, the theorem may be stated as follows:

The angular deflection of an elastic structure at point A due to a unit couple at point B is equal to the angular deflection at point B due to a unit couple at point A.

The deflection of an elastic structure at point A due to a unit couple at point B is equal to the rotation at point B due to a unit force at point A.

In all cases, the component of the deflection at a point in the direction of the load acting at that point is considered.

Maxwell's theorem for any particular type of structure may also be proved by the method of virtual work. In a truss, for example, the deflection at point A due to a unit load at point B is found from Eq. 16.11 by substituting $S = u_b$.

$$\delta = \sum \frac{u_b u_a L}{AE}$$

This is obviously equal to the deflection of point B due to a unit load at point A, since $S = u_a$ for this condition.

16.15. Elastic Axis, or Shear Center. The elastic axis of a wing is defined as the axis about which rotation will occur when the wing is loaded in pure torsion. For the wing shown in Fig. 16.32(a), in which the cross section is uniform along the span, the elastic axis is a straight line. Points on the elastic axis do not deflect in the torsion loading, but points forward of the elastic axis are deflected upward and points to the rear of the elastic axis are deflected downward. It is necessary to calculate the position of the elastic axis in order to make a flutter analysis of the wing.

The shear center of a wing cross section is defined as the point at which the resultant shear load must act to produce a wing deflection with no rotation. The shear force shown in Fig. 16.32(b) deflects the wing in translation, but causes no rotation of the cross section about a spanwise axis. If the wing is an elastic structure, the shear center of a cross section must lie on the elastic axis, since a force at the shear center produces no rotation at the point of application of the couple, and the couple must therefore produce no vertical deflection at the point of application of the force, as a result of the application of Maxwell's theorem. Practical wings deviate slightly from conditions of elasticity because the skin wrinkles and becomes ineffective in resisting compression loads, but for practical purposes the elastic axis may be assumed to coincide with the line joining the shear centers of the various cross sections.

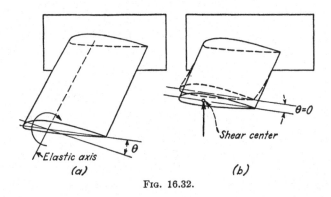

Fig. 16.32.

The shear center of a cross section may be calculated from Eq. 16.37 by finding the position of the resultant shear force which yields a zero angle of twist. The shear-center location depends on the distribution of the flange areas and the thickness of the shear webs. The procedure can best be studied by means of an illustrative example.

Example. Find the shear center for the wing cross section shown in Fig. 16.33. Web 3 has a thickness of 0.064 in. and the other webs have thicknesses of 0.040 in. Assume G constant for all cross sections. The cross section is symmetrical about a horizontal axis.

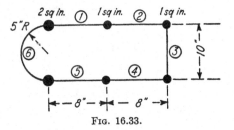

Fig. 16.33.

Solution. The position of the shear center does not depend on the magnitude of the shear force. A shear force $V = 400$ lb is therefore assumed arbitrarily. The shear flow increments are obtained by the methods used in Chap. 6 and are as shown in Fig. 16.34. The shear flow in web 1 is assumed to have a value q_0, and the remaining shear flows are expressed in terms of q_0, as shown. In previous problems, the shear flow q_0 was obtained from the equilibrium of torsional moments, but the external torsional moment is not known in this problem. It will be assumed that the 400-lb shear force acts at a distance $\bar{x}$ from the right side, as shown, and that this point is the shear center. The shear flow q_0 is found from the condition that the angle of twist θ is zero. From Eq. 16.37,

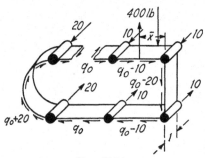

FIG. 16.34.

$$\theta = \sum \frac{q \, \Delta s L}{2 A t G} = 0$$

where $L = 1$ and $2AG$ is constant for all webs and may be taken outside the summation sign and then cancelled.

$$\theta = \sum \frac{q \, \Delta s}{t} = 0 \qquad (16.41)$$

The total shear flow q is the sum of the component q_0 and the component q', where q' is the shear flow if web 1 is cut.

$$q = q_0 + q'$$

Substituting this value into Eq. 16.41 and taking q_0 outside the summation sign because it is constant for all webs, the following equation is obtained:

$$q_0 \sum \frac{\Delta s}{t} + \sum q' \frac{\Delta s}{t} = 0 \qquad (16.42)$$

The numerical solution is made in Table 16.2. Column (1) lists values of Δs, the circumferential lengths, for the various webs, as shown in Fig. 16.33. These values are divided by the web thickness in column (2). The values of q', the

TABLE 16.2

Web	Δs (1)	$\dfrac{\Delta s}{t}$ (2)	q' (3)	$q' \dfrac{\Delta s}{t}$ (4)	$2A$ (5)	$2Aq'$ (6)	q (7)
1	8	200	0	0	80	0	−0.53
2	8	200	−10	−2,000	80	−800	−10.53
3	10	156	−20	−3,120	0	0	−20.53
4	8	200	−10	−2,000	0	0	−10.53
5	8	200	0	0	0	0	−0.53
6	15.7	392	+20	7,840	239	4,780	+19.47
Total	....	1,348	...	720	399	3,980	

shear flows when web 1 is cut, are tabulated in column (3). The shear flows are considered as positive when clockwise around the outboard face of the element of Fig. 16.34. The values of $q' \Delta s/t$ are calculated in column (4). The totals of columns (2) and (4) are now substituted into Eq. 16.42.

$$1,348q_0 + 720 = 0$$
$$q_0 = -0.53 \text{ lb/in.}$$

The final shear flows in the web, q are calculated in column (7) by adding the value of q_0 to values of q' in column (3).

The position of the shear center will now be calculated from the equilibrium of torsional moments. The moment about any point can be obtained from the relation

$$T = \Sigma 2Aq$$

or, since $q = q_0 + q'$,

$$T = q_0\Sigma 2A + \Sigma 2Aq' \tag{16.43}$$

where A is the area enclosed by a web and the lines joining the end points of the web and the center of moments. The center of moments is taken as the lower right-hand corner of the box, and values of $2A$ are tabulated in column (5) and values of $2Aq'$ in column (6). The totals of columns (5) and (6) are now substituted into Eq. 16.43.

$$400\bar{x} = -0.53 \times 399 + 3,980$$
$$\bar{x} = 9.42 \text{ in.}$$

This value of $\bar{x}$ determines the horizontal location of the shear center. From symmetry, the vertical location is on the line of symmetry. For cross sections which are not symmetrical about a horizontal axis, the vertical location of the shear center may be obtained by considering a horizontal shear force to act on the section and then proceeding in the same manner as above to find the shear flows for a zero twist. The location of the resultant of these shear flows, obtained by equating torsional moments, gives the vertical position of the shear center.

16.16. Castigliano's Theorem. In some types of problems it is convenient to calculate deflections by means of Castigliano's theorem, which may be stated as follows:

The deflection of any elastic structure in the direction of a load acting on it is equal to the partial derivative of the total strain energy with respect to the load.

This theorem is used in the method of least work for the analysis of statically indeterminate structures. Manabrea stated the theory in 1858, but Castigliano published a more satisfactory treatment in 1879. Maxwell, in 1864, and Mohr, in 1868, developed the general theory for the analysis of statically indeterminate structures.[1]

Castigliano's theorem will be proved by reference to Fig. 16.35, which is assumed to represent a general case of an elastic structure loaded in any manner by gradually applied loads. The loads P_1, P_2, P_3, . . ., P_n act through the deflections δ_1, δ_2, δ_3, . . ., δ_n, measured in the directions of the loads, and produce strain energy equal to U.

$$U = \frac{P_1\delta_1}{2} + \frac{P_2\delta_2}{2} + \frac{P_3\delta_3}{2} + \cdots + \frac{P_n\delta_n}{2} \qquad (16.44)$$

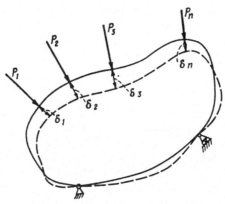

Fig. 16.35.

If any load P_n is increased an amount dP_n, the strain energy will be increased by an amount

$$\frac{\partial U}{\partial P_n} dP_n \qquad (16.45)$$

from the definition of a partial derivative. The total strain energy is independent of the order in which the loads are applied, and the load dP_n may be considered as acting when the other loads are applied. The work done by the force dP_n acting through the displacement δ_n will be $(dP_n)\delta_n$. This must be equal to the expression 16.45.

$$\left. \begin{array}{c} (dP_n)\delta_n = \dfrac{\partial U}{\partial P_n} dP_n \\[2ex] \delta_n = \dfrac{\partial U}{\partial P_n} \end{array} \right\} \qquad (16.46)$$

This equation is an algebraic statement of Castigliano's theorem, which was to be proved.

In applying this theorem to practical structures, the total strain energy may be stated in terms of a load and then differentiated with respect to the load. In most cases, the differentiation will be performed before evaluating the strain-energy terms for the various members, and the numerical work is equivalent to that for an analysis by the method of virtual work. For an elastic structure containing members loaded axially and members loaded in bending, the strain energy may be expressed as follows:

$$U = \sum \frac{S^2 L}{2AE} + \int \frac{M^2 \, dx}{2EI} \qquad (16.47)$$

The deflection of the structure in the direction of any load P_n may be obtained by differentiating Eq. 16.47 with respect to P_n.

$$\delta_n = \frac{\partial U}{\partial P_n} = \sum \frac{S \frac{\partial S}{\partial P_n} L}{AE} + \int \frac{M \frac{\partial M}{\partial P_n} \, dx}{EI} \qquad (16.48)$$

The axial loads S and the bending moments M are linear functions of P_n. The partial derivatives therefore represent the axial loads or bending moments resulting from a unit value of P_n.

$$\frac{\partial S}{\partial P_n} = u \qquad \frac{\partial M}{\partial P_n} = m \qquad (16.49)$$

The values of u and m correspond to those in Eqs. 16.11 and 16.20. Substituting these values into Eq. 16.48, the deflections are seen to correspond with those of the virtual-work equations.

$$\delta_n = \sum \frac{SuL}{AE} + \int \frac{Mm}{EI} \, dx \qquad (16.50)$$

The practical application of Castigliano's theorem is usually equivalent to the application of the virtual-work equation, since Eq. 16.48 is used in the same manner as Eq. 16.50.

PROBLEMS

16.18. Find the angle of twist of the wing shown in Fig. 16.25 if all four flanges have equal areas. Use $G = 4 \times 10^6$ psi.

16.19. Find the angle of twist of the wing shown in Fig. 16.25 if all webs have a thickness t of 0.040 in. Use $G = 4 \times 10^6$ psi.

16.20. If all webs of the beam in Fig. 16.25 have the same gage, $t = 0.040$ in., find the shear flows at the support, where warping is prevented. Find the warping angle θ between spar cross sections at a distance from the support. Use $G = 4 \times 10^6$ psi.

16.21. Calculate the location of the shear center, or elastic axis, of the wing in Fig. 16.25.

16.22. Find the location of the shear center of the beam shown in Fig. 16.33 if web 6 has a thickness of 0.064 in. and the other webs have thicknesses of 0.040 in.

16.23. Solve Example 1, Art. 16.8, by the use of Castigliano's theorem. Note that in finding the rotation it is necessary to apply a couple in the direction of the desired rotation. This couple is equated to zero after the differentiation.

REFERENCE FOR CHAPTER 16

1. WESTERGAARD, H. M.: One Hundred Fifty Years Advance in Structural Analysis, *Trans. ASCE*, 1930.

CHAPTER 17

STATICALLY INDETERMINATE STRUCTURES

17.1. Degree of Redundancy. A statically indeterminate structure is a structure for which the reactions or stresses cannot be completely determined from the conditions for static equilibrium. A stable and statically determinate structure contains only enough reactions or members for stability, and the reactions and forces in all members may be determined by the equations of statics. If one member or reaction is removed, the stable structure is changed to a mechanism or linkage which is not capable of resisting loads, or is unstable. If one member or reaction is added, the structure becomes singly statically indeterminate, and the reactions and forces must be obtained from a deflection condition in addition to the conditions of static equilibrium.

Normally, a rigid coplanar structure requires three external reactions for stability, and they may be calculated from the three equations of statics. The number of reactions, however, is not the only criterion for stability, and it is necessary to examine each particular structure in

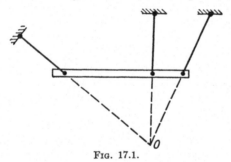

Fig. 17.1.

order to determine whether it is stable, unstable, or statically indeterminate. For example, a horizontal simple beam normally requires three reaction components. However, if the beam is supported on rollers at three points along the span, it will be unstable for resisting horizontal forces and statically indeterminate for vertical forces. Similarly, if the three reactions of a simple beam act through any common point in the plane, the moments about that point will be zero regardless of the magnitudes of the forces, and the moment equation cannot be used to obtain the reactions. Such a structure, as shown in Fig. 17.1, is a mechanism

454

which is free to rotate through a small angle about point O as an instantaneous center, and it is unstable in resisting any load which does not act through point O. If a load acts through point O, the structure is statically indeterminate.

In most structures, the number of equations of statics may be compared with the number of redundants in order to determine the conditions of stability. For such special structures as that shown in Fig. 17.1, an attempt to find the three unknown reactions from the three equations of statics will result in equations which are not independent, or one of the equations can be derived from the others. When an attempt to analyze a structure by the equations of statics results in such a condition, the structure must be examined for instability or redundancy.

The determinacy of a simple truss was discussed in Art. 1.3. It was shown that the necessary condition for stability was

$$m = 2j - 3 \qquad (1.4)$$

where m is the number of members and j the number of joints for a truss with three external reactions. Equation 1.4, like the requirement for three reaction components, is not always applicable, as shown by Fig. 1.5(b).

If a structure has one more member or reaction than is required for stability, it has single redundancy. In many cases, any one of several members or reactions may be removed without causing instability. One deflection equation must then be used in addition to the equations of statics in order to analyze the structure. If a structure has several more members or reactions than are required for stability, it has multiple redundancy. The degree of redundancy is equal to the number of redundant members, and it is equal to the number of deflection conditions which must be used in the analysis.

17.2. Trusses with Single Redundancy. A truss which is composed of elastic members and which has single redundancy will first be considered. A typical truss of this type is shown in Fig. 17.2(a). The supports are assumed to be rigid, and the members are assumed to be unstressed before the load P is applied. There are four external reactions, and only three would be required for stability. The horizontal reaction component X_a will be considered as the redundant, and the deflection equation will be obtained from the condition that the horizontal support deflection δ is zero. From Eq. 16.11,

$$\delta = \sum \frac{SuL}{AE} \qquad (16.11)$$

where S represents the force in any member of the structure of Fig. 17.2(a) and u represents the force in any member due

to a unit load applied in the direction of the desired deflection, as shown
in Fig. 17.2(c).

The force S in any member is found by superimposing the loading conditions shown in Figs. 17.2(b) and (d). If the redundant force is removed, the resulting statically determinate structure is shown in

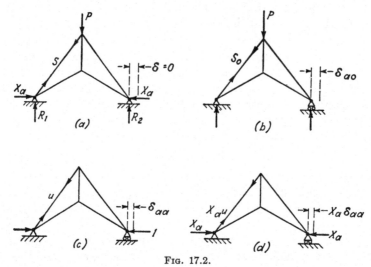

FIG. 17.2.

Fig. 17.2(b), and the applied loads produce a force S_0 in any member.
If the redundant force X_a is acting alone, it produces a force $X_a u$ in any
member, as shown in Fig. 17.2(d), since the u forces result from a unit
value of X_a. The total force S is obtained as the sum of the forces for
the two conditions.

$$S = S_0 + X_a u \qquad (17.1)$$

Substituting from Eq. 17.1 into Eq. 16.11,

$$\delta = \sum \frac{S_0 u L}{AE} + X_a \sum \frac{u^2 L}{AE} \qquad (17.2)$$

or, for $\delta = 0$,

$$X_a = -\frac{\sum \dfrac{S_0 u L}{AE}}{\sum \dfrac{u^2 L}{AE}} \qquad (17.3)$$

where all the terms on the right side of the equation may be obtained
from the loading and geometry of the structure.

Equation 17.3 is applicable to any elastic truss with single redundancy
in which the deflection in the direction of the redundant is zero. If the
deflection in the direction of the redundant δ is a known value other than

zero, this deflection may be substituted into Eq. 17.2 and the value of X_a determined for this condition.

The physical significance of the terms in Eqs. 17.2 and 17.3 will be discussed in order to visualize the action of a redundant structure. If the redundant force is removed, the structure is statically determinate, and the deflection in the direction of the redundant due to the applied loads has the following value:

$$\delta_{ao} = \sum \frac{S_0 u L}{A E} \qquad (17.4)$$

This deflection is assumed positive in the direction of the redundant. A unit value of the redundant deflects the structure a distance

$$\delta_{aa} = \sum \frac{u^2 L}{A E} \qquad (17.5)$$

which is also positive in the direction of the redundant. The value of X_a required to give a zero deflection is obtained by dividing the deflection resulting from the applied loads by the deflection resulting from the unit load, and it will be negative with the assumed sign conventions.

$$X_a = -\frac{\delta_{ao}}{\delta_{aa}} \qquad (17.6)$$

It is easier to visualize the deflection terms of Eq. 17.6 than the summation terms of Eq. 17.3.

Example 1.   Find the reactions and the stresses in the members of the structure of Fig. 17.3.   The areas of members AB, BC, and BD are 4 sq in., and the areas of members AD and DC are 3.6 sq in. Assume rigid supports and $E = 10,000$ ksi.

Solution. The numerical solution of Eq. 17.3 is made in Table 17.1, with the notation as shown in Fig. 17.2. The horizontal reaction component X_a is considered as the redundant. The forces S_0, in the truss with the redundant removed, are tabulated in column (1). The lengths L of the various members are calculated from the dimensions shown in Fig. 17.3 and are divided by the areas and modulus of elas-

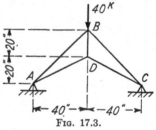

FIG. 17.3.

ticity. The values of L/AE are tabulated in column (2). The forces u, resulting from a unit value of X_a applied as shown in Fig. 17.2(c), are tabulated in column (3). Positive signs indicate tension and negative signs indicate compression. The summation terms of Eq. 17.3 are now evaluated as the sum of the values for the individual members tabulated in columns (4) and (5).

$$X_a = -\frac{\displaystyle\sum \frac{S_0 u L}{A E}}{\displaystyle\sum \frac{u^2 L}{A E}} = \frac{0.5150}{0.02008} = 25.6 \text{ kips}$$

TABLE 17.1

Members	S_0, kips (1)	$\dfrac{L}{AE}$, in./kip (2)	u (3)	$\dfrac{S_0 u L}{AE}$, in. (4)	$\dfrac{u^2 L}{AE}$, in./kip (5)	$X_a u$ (6)	S (7)
AB	−56.6	0.001414	1.414	−0.1130	0.00283	36.3	−20.3
BC	−56.6	0.001414	1.414	−0.1130	0.00283	36.3	−20.3
AD	44.8	0.001245	−2.233	−0.1245	0.00621	−57.3	−12.5
DC	44.8	0.001245	−2.233	−0.1245	0.00621	−57.3	−12.5
BD	40.0	0.000500	−2.0	−0.0400	0.00200	−51.3	−11.3
Total..................				−0.5150	0.02008		

This value is multiplied by the terms in column (3) to obtain values of $X_a u$ for all members. The final forces S in the members are obtained as the algebraic sum of terms in columns (1) and (6) and are listed in column (7).

Example 2. Find the forces in the members of the structure of Example 1, if the support at point C is deflected 0.25 in. to the right and the temperature is decreased 40°F. Assume a temperature coefficient ϵ of 10^{-5} per degree Fahrenheit.

Solution. A temperature decrease would cause the right end of the statically determinate truss of Fig. 17.2(b) to move to the left a distance of ϵLt, where L is the distance between supports, or 80 in.

$$\epsilon Lt = 10^{-5} \times 80 \times 40 = 0.032 \text{ in.}$$

The support of point C is displaced an additional 0.25 in. to the right; therefore the total displacement δ, which must be given to the right support of the truss by the strains in the members, is

$$\delta = -0.032 - 0.25 = -0.282 \text{ in.}$$

where the negative sign indicates that the deflection due to the stresses is opposite to X_a. From Eq. 17.2,

$$\delta = \sum \frac{S_0 u L}{AE} + X_a \sum \frac{u^2 L}{AE} = -0.282$$

where the summation terms are obtained in Table 17.1.

$$-0.515 + 0.02008 X_a = -0.282$$
$$X_a = 11.6 \text{ kips}$$

The values of the resulting forces in the members are found in Table 17.2, as $S = S_0 + X_a u$, where values of S_0 and u are the same as for Example 1.

Example 3. Find the forces in the members of the structure shown in Fig. 17.4. This structure is constructed by adding member AC to the structure of Example 1 and supporting the right end of the truss on frictionless rollers. Member AC has an area of 3 sq. in.

TABLE 17.2

Member	S_0	$X_a u$	S
AB	−56.6	16.4	−40.2
BC	−56.6	16.4	−40.2
AD	44.8	−25.9	18.9
DC	44.8	−25.9	18.9
BD	40.0	−23.2	16.8

Solution. The member AC is considered as a redundant, and it is assumed to resist a tension force X_a. If member AC is cut as shown, the cut ends will move apart a distance $-\delta_{ao}$, when X_a is zero, if δ_{ao} is defined by Eq. 17.4. The force X_a must be sufficient to deflect the cut ends back to their original position, a distance $X_a\delta_{aa}$, where δ_{aa} is defined by Eq. 17.5, in which the summation must include the deformation of member AC. The value of X_a will again be found from Eq. 17.3 or Eq. 17.6. The numerator will be the same as for Example 1, since the structures are the same for $X_a = 0$. The denominator will include an additional term for member AC,

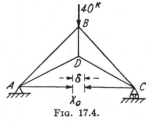

Fig. 17.4.

$$\frac{u^2 L}{AE} = \frac{1^2 \times 80}{3 \times 10,000} = 0.00267$$

where $u = 1$ for this member. From the values obtained in Example 1,

$$X_a = -\frac{\sum \dfrac{S_0 u L}{AE}}{\sum \dfrac{u^2 L}{AE}} = \frac{0.5150}{0.02008 + 0.00267} = 22.6 \text{ kips}$$

The forces in the members are found from the equation, $S = S_0 + X_a u$, where S_0 and u have values as shown in Table 17.1. For AB and BC, $S = -24.6$, for AD and DC, $S = -5.8$, and for BD, $S = -5.2$.

17.3. Other Structures with Single Redundancy. The procedure for the analysis of any type of elastic statically indeterminate structure is similar to that for a truss. The redundant reaction or member is first considered to be removed, and the deflection δ_{ao} of the remaining statically determinate structure in the direction of the redundant is calculated. The magnitude of the redundant required to produce a deflection equal to $-\delta_{ao}$ is then calculated. It is assumed that the deflection produced by the redundant force is X_a times the deflection δ_{aa} produced by a unit value of X_a, or that the structure deforms elastically under the action of the redundant.

$$X_a\delta_{aa} = -\delta_{ao}$$

This relationship is identical to that expressed in Eq. 17.6 for trusses, and it is applicable to any type of elastic structure.

$$X_a = -\frac{\delta_{ao}}{\delta_{aa}} \qquad (17.6)$$

The deflections δ_{ao} and δ_{aa} may be calculated for any type of structure by the method of virtual work. They may include axial, bending, torsional, or shearing deformations of the members of the structure. In the case of structures such as beams, arches, or rigid frames, the bending deformations are predominant, and other deformations may be neglected. The deformations of Eq. 17.6 are then obtained from Eq. 16.20.

$$X_a = -\frac{\int \frac{M_0 m}{EI} \, ds}{\int \frac{m^2}{EI} \, ds} \qquad (17.7)$$

The term M_0 represents the bending moment at any cross section in the statically determinate structure with X_a removed for the applied loading condition, and m represents the bending moment in the statically determinate structure for a load $X_a = 1$. The application of Eqs. 17.6 and 17.7 will be illustrated by several Example problems.

Example 1. Find the bending moment at any point of the semicircular arch of Fig. 17.5 if the supports do not move. The value of EI is constant for all cross sections.

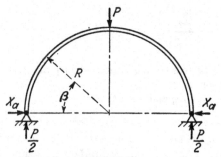

FIG. 17.5.

Solution. The structure is symmetrical about a vertical center line, and the integrals of Eq. 17.7 will be evaluated for the left half of the structure and multiplied by 2. The horizontal reaction X_a is considered as the redundant, and the value of M_0 is calculated for the statically determinate structure formed by supporting one end of the arch on frictionless rollers.

$$M_0 = \frac{PR}{2} (1 - \cos \beta)$$

The value of m is calculated for a unit load acting in the direction of X_a.

$$m = -R \sin \beta$$

A positive bending moment is assumed to produce compressive stresses on the outside of the arch. Substituting values of M_0, m, and $ds = R\, d\beta$ into Eq. 17.7, the value of X_a is obtained.

$$X_a = -\frac{2\int_0^{\frac{\pi}{2}} \frac{PR}{2EI}(1 - \cos \beta)(-R \sin \beta)R\, d\beta}{2\int_0^{\frac{\pi}{2}} \frac{(-R \sin \beta)^2}{EI} R\, d\beta} = \frac{P}{\pi}$$

The final bending moment is obtained by superimposing the values of M_0 for the applied loads and $X_a m$ for the redundant.

$$M = M_0 + X_a m$$
$$M = \frac{PR}{2}(1 - \cos \beta) - \frac{PR}{\pi} \sin \beta$$

This equation applies for $0 < \beta < \pi/2$, and the bending-moment diagram is symmetrical about a vertical center line.

Example 2. Find the reactions and bending moments for the rigid frame shown in Fig. 17.6(a). The modulus of elasticity is constant, and the supports do not move.

Solution. The horizontal component of the right-hand reaction is chosen as the redundant X_a. The bending-moment diagram for the condition $X_a = 0$ is plotted in Fig. 17.6(b). It is observed that the parabolic and triangular diagrams are superimposed, so that their areas may be easily calculated. The bending moments are considered positive when they produce compression on the outside of the frame. The values of m are plotted in Fig. 17.6(c) for a loading $X_a = 1$, and are negative at all points. The redundant X_a will now be computed from Eq. 17.7, making use of the fact that E is constant and evaluating the integrals by the method of Eq. 16.25. The values of m opposite the centroids of the various M_0 areas are shown in Fig. 17.6(c).

$$X_a = -\frac{\int \frac{M_0 m}{EI}\, ds}{\int \frac{m^2}{EI}\, ds} = -\frac{\int \frac{M_0 m}{I}\, ds}{\int \frac{m^2}{I}\, ds}$$

$$X_a = \frac{\dfrac{50 \times 2.5 \times 3.33 + 50 \times 4 \times 7}{40} + \dfrac{50 \times 6 \times 8 + 36 \times 12 \times \frac{2}{3} \times 7.5}{30}}{\dfrac{9 \times 4.5 \times 6}{40} + \dfrac{12 \times 6 \times 7.5 + 3 \times 6 \times 8}{30} + \dfrac{6 \times 3 \times 4}{20}}$$

$$= 6.08 \text{ kips}$$

The other reactions and the bending moments are now found by the equations of statics and are shown in Fig. 17.6.

Example 3.   The structure shown in Fig. 17.7(a) consists of a round tube in a horizontal plane, bent at an angle of 90°.   The free end supports a load of 2 kips,

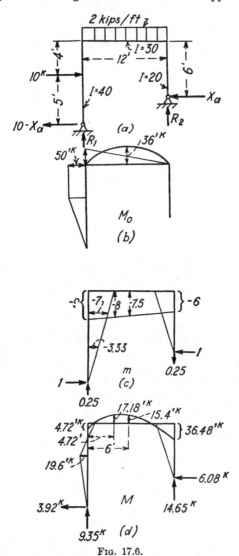

Fig. 17.6.

and it is also supported by a vertical wire.   Find the tension in the wire and the bending-moment and torsion diagrams for the tube.

Solution.   The wire will be considered as the redundant member.   The vertical deflection of the free end of the tube will be calculated, assuming the wire to be removed and the statically determinate tube to support the load of

2 kips. The bending moment and torsion in the tube under this loading are designated as M_o and T_o and are plotted in Fig. 17.7(*b*). The deflection results from bending and torsional deformation of the tube.

$$\delta_{ao} = \int \frac{M_o m}{EI}\, ds + \int \frac{T_o m_t}{JG}\, ds \qquad (17.8)$$

Fɪɢ. 17.7.

The values of m and m_t, the bending moment and torsional moment in the statically determinate structure for $X_a = 1$, are plotted in Fig. 17.7(*c*). The integrals of Eq. 17.7 will be evaluated from Eq. 16.25.

$$\delta_{ao} = -\frac{36 \times 4 + 81 \times 6}{1,000} - \frac{12 \times 6 \times 9}{800} = -1.44 \text{ in.}$$

The negative sign indicates a deflection in the opposite direction to the unit load, or a downward deflection.

The deflection δ_{aa}, resulting from a unit force X_a in the wire, consists of parts due to torsion and bending of the tube and tension in the wire.

$$\delta_{aa} = \int \frac{m^2}{EI}\, ds + \int \frac{m_t^2}{JG}\, ds + \frac{u^2 L}{AE} \tag{17.9}$$

The tension u in the wire is 1, and the values of m and m_t are shown in Fig. 17.7(c).

$$\delta_{aa} = \frac{18 \times 4 + 40.5 \times 6}{1,000} + \frac{6 \times 6 \times 9}{800} + \frac{1^2 \times 40}{200} = 0.92 \text{ in./kip}$$

This is the distance that a 1-kip tension force in the wire would move the cut ends together. The force X_a required to move the cut ends of the wire the distance $-\delta_{ao}$ is found from Eq. 17.6.

$$X_a = -\frac{\delta_{ao}}{\delta_{aa}} = \frac{1.44}{0.92} = 1.565 \text{ kips}$$

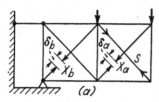

(a)

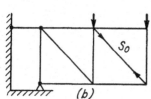

(b)

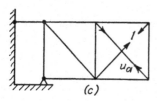

(c)

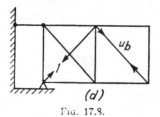

(d)

Fig. 17.8.

The bending-moment and torsion diagrams for the tube are now calculated from the equations of statics, and are shown in Fig. 17.7(d).

17.4. Trusses with Multiple Redundancy.

The procedure for analysis of structures with two or more redundants is similar to that used for a structure with one redundant member or reaction. The first step is to remove the redundant members or reactions in order to obtain a base structure which is statically determinate. The deflections of the statically determinate base structure in the directions of the redundants are then calculated in terms of the redundant forces and are equated to the known deflections, which are usually zero. The number of known deflection conditions must be equal to the number of redundants. For a structure with n redundants, the deflection conditions yield n equations which must be solved simultaneously for the values of the redundants.

The truss shown in Fig. 17.8(a) has only three reactions and is statically determinate externally, but there are two more members than are required for stability; therefore it is statically indeterminate internally. The deflection conditions which will be specified are that there are no stresses in the structure when it is not loaded, or if two members of the unloaded structure are cut, the relative deflections δ_a and δ_b of the cut ends will be zero. The deflections will now be expressed in terms of the forces X_a and X_b in the redundant members. All deflections are assumed to be elastic.

The statically determinate base structure, shown in Fig. 17.8(b), is formed by cutting or removing the redundant members. The applied loads produce forces S_0 in the members of the base structure. The forces u_a in the members are produced by a unit value of X_a applied to the base structure as shown in Fig. 17.8(c). Similarly a force $X_b = 1$ produces forces u_b in the members when applied to the base structure as shown in Fig. 17.8(d). The final force S in any member may be obtained by superimposing the forces due to the applied loads and the redundant forces.

$$S = S_0 + X_a u_a + X_b u_b \tag{17.10}$$

The deflections δ_a and δ_b of the cut ends of the members are now equated to zero.

$$\delta_a = \sum \frac{S u_a L}{AE} = 0 \tag{17.11}$$

$$\delta_b = \sum \frac{S u_b L}{AE} = 0 \tag{17.12}$$

The values of S from Eq. 17.10 will now be substituted into Eqs. 17.11 and 17.12.

$$\delta_a = \sum \frac{S_0 u_a L}{AE} + X_a \sum \frac{u_a^2 L}{AE} + X_b \sum \frac{u_a u_b L}{AE} = 0 \tag{17.13}$$

$$\delta_b = \sum \frac{S_0 u_b L}{AE} + X_a \sum \frac{u_b u_a L}{AE} + X_b \sum \frac{u_b^2 L}{AE} = 0 \tag{17.14}$$

These equations may be solved simultaneously for X_a and X_b. The final values of the forces S may be obtained from Eq. 17.10.

The preceding equations have been derived from the superposition of the stress conditions, as stated in Eq. 17.10. They may also be obtained from a superposition of the deflection conditions. The applied loads are assumed to produce deflections δ_{ao} and δ_{bo} of the redundants. A unit value of X_a produces deflections δ_{aa} at X_a and δ_{ba} at X_b. A unit value of X_b produces deflections δ_{bb} at X_b and δ_{ab} at X_a. The total deflections in the directions of the redundants may now be obtained by superimposing the effects of the various loads.

$$\delta_a = \delta_{ao} + X_a \delta_{aa} + X_b \delta_{ab} \tag{17.15}$$
$$\delta_b = \delta_{bo} + X_a \delta_{ba} + X_b \delta_{bb} \tag{17.16}$$

In most problems the deflections δ_a and δ_b are zero, but in some cases, known values of support deflections or similar deformations may be substituted. When values of $\delta_a = \delta_b = 0$ are used, and the other δ terms are obtained by the virtual-work equations, Eqs. 17.15 and 17.16 are seen to correspond with Eqs. 17.13 and 17.14.

For a structure with n redundants, a number of deflection conditions, n, must be used. These may be written in the same form as Eqs. 17.15 and 17.16.

$$\left.\begin{array}{l} \delta_a = \delta_{ao} + X_a\delta_{aa} + X_b\delta_{ab} + \cdots + X_n\delta_{an} \\ \delta_b = \delta_{bo} + X_a\delta_{ba} + X_b\delta_{bb} + \cdots + X_n\delta_{bn} \\ \qquad \cdot \qquad \cdot \qquad \cdot \qquad \cdot \qquad \cdot \qquad \cdot \qquad \cdot \\ \delta_n = \delta_{no} + X_a\delta_{na} + X_b\delta_{nb} + \cdots + X_n\delta_{nn} \end{array}\right\} \qquad (17.17)$$

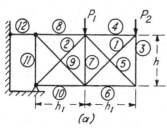
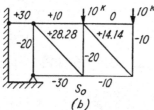
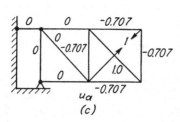
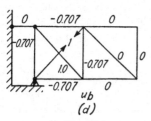

(a)

(b)

(c)

(d)

FIG. 17.9.

The terms in Eqs. 17.17 may be defined in the following manner for truss structures:

$$\delta_{no} = \sum \frac{S_0 u_n L}{AE} \qquad \delta_{mn} = \sum \frac{u_m u_n L}{AE} \qquad (17.18)$$

The applications of these equations to truss analysis will be illustrated by numerical examples.

Example 1. Find the forces in the members of the truss shown in Fig. 17.9(a) if $P_1 = P_2 = 10$ kips, $h = h_1$, and L/AE is the same for each member of the structure. The members are unstressed when $P_1 = P_2 = 0$, and stresses do not exceed the elastic limit.

Solution. The numerical values of L/AE are not required for the members, since only relative values are important. For $\delta_a = \delta_b = 0$ in Eqs. 17.13 and 17.14, the summation terms may be multiplied by any constant value. If δ_a or δ_b is not zero, it is necessary to know the numerical values. It will therefore be assumed that L/AE is unity for all members. The calculations of the summation terms are made in Table 17.3. The forces S_0 in the members of the statically determinate base structure are shown in Fig. 17.9(b) and tabulated in column (1). The values of u_a and u_b due to unit values of the redundant are shown in Figs. 17.9(c) and (d) and are tabulated in columns (3) and (4). The terms for the summations of Eqs. 17.13 and 17.14 are obtained as the totals of columns (4), (5), (6), (7), and (8). Substituting these totals into

TABLE 17.3

	S_0 (1)	u_a (2)	u_b (3)	$\dfrac{S_0 u_a L}{AE}$ (4)	$\dfrac{S_0 u_b L}{AE}$ (5)	$\dfrac{u_a^2 L}{AE}$ (6)	$\dfrac{u_b^2 L}{AE}$ (7)	$\dfrac{u_a u_b L}{AE}$ (8)	S, kips (9)
1	0	1.000	0	0	0	1.0	0	0	−8.53
2	0	0	1.000	0	0	0	1.0	0	−16.60
3	−10	−0.707	0	7.07	0	0.5	0	0	−3.97
4	0	−0.707	0	0	0	0.5	0	0	+6.03
5	14.14	1.000	0	14.14	0	1.0	0	0	+5.61
6	−10	−0.707	0	7.07	0	0.5	0	0	−3.97
7	−20	−0.707	−0.707	14.14	14.14	0.5	0.5	0.5	−2.24
8	10	0	−0.707	0	−7.07	0	0.5	0	+21.73
9	28.28	0	1.000	0	28.28	0	1.0	0	+11.68
10	−30	0	−0.707	0	21.21	0	0.5	0	−18.27
11	−20	0	−0.707	0	14.14	0	0.5	0	−8.27
12	30	0	0	0	0	0	0	0	+30.00
Total..........................				42.42	70.70	4.0	4.0	0.5	

Eqs. 17.13 and 17.14, the equations for the redundant forces X_a and X_b are obtained.

$$42.42 + 4.0X_a + 0.5X_b = 0$$
$$70.70 + 0.5X_a + 4.0X_b = 0$$

Solving these equations simultaneously, the values $X_a = -8.53$ and $X_b = -16.60$ are obtained. The final values of S are obtained from Eq. 17.10.

$$S = S_0 - 8.53u_a - 16.60u_b$$

The values of S are shown in column (9).

Example 2. Find the forces in the truss of Fig. 17.10(a). Assume that L/AE is 0.01 in./kip for all members. The right-hand support deflects 0.5 in. from the unstressed position of the truss. Assume $h = h_1$.

Solution. The redundants X_a and X_b are chosen as shown in Fig. 17.10(a), leaving the same statically determinate base structure as for Example 1. The values for S_0 and u_b are the same as for Example 1, but values of u_a must be calculated as shown in Fig. 17.10(b). The terms involving S_0 and u_b are therefore the same as those calculated in Table 17.3, except that the value of L/AE is now 0.01 for each member, whereas a value of 1.0 was used in Table 17.3. For constant values of L/AE for all members, Eqs. 17.13 and 17.14 may be written as follows:

$$\left. \begin{aligned} \frac{AE}{L} \delta_a &= \Sigma S_0 u_a + X_a \Sigma u_a^2 + X_b \Sigma u_a u_b \\ \frac{AE}{L} \delta_b &= \Sigma S_0 u_b + X_a \Sigma u_b u_a + X_b \Sigma u_b^2 \end{aligned} \right\} \tag{17.19}$$

The values of $\Sigma S_0 u_a$, $\Sigma u_a u_b$, and Σu_a^2 are calculated in Table 17.4 as the totals of columns (2), (3), and (4). Substituting $AE/L = 100$ and $\delta_a = -0.5$ in. into Eqs. 17.19, the equations for the redundant forces are obtained.

$$-100 \times 0.5 = -240 + 16X_a - 3.535X_b$$
$$0 = 70.70 - 3.535X_a + 4.0X_b$$

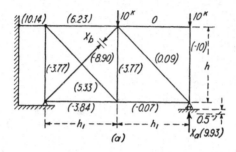

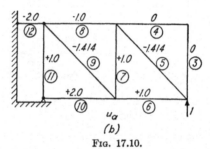

Fig. 17.10.

TABLE 17.4

	u_a (1)	$S_0 u_a$ (2)	$u_a u_b$ (3)	u_a^2 (4)	S, kips (5)
2	0	0	0		−8.90
3	0	0	0		−10.00
4	0	0	0		0
5	−1.414	−20	0	2.0	0.09
6	1.0	−10	0	1.0	−0.07
7	1.0	−20	−0.707	1.0	−3.77
8	−1.0	−10	0.707	1.0	6.23
9	−1.414	−40	−1.414	2.0	5.33
10	2.0	−60	−1.414	4.0	−3.84
11	1.0	−20	−0.707	1.0	−3.77
12	−2.0	−60	0	4.0	10.14
Total		−240	−3.535	16.0	

These equations are solved simultaneously to yield $X_a = 9.93$ and $X_b = -8.90$. The forces S are now obtained from Eq. 17.10.

$$S = S_0 + 9.93u_a - 8.90u_b$$

The values of S are tabulated in column (5) of Table 17.4, and are shown in parenthesis on the members in Fig. 17.10(a).

17.5. Other Structures with Multiple Redundancy. The procedure for analysis of any statically indeterminate structure is the same as that for trusses. The redundant members or reactions are removed, to obtain a statically determinate base structure. The deflections in the directions of the redundants resulting from the applied loads and from the redundants are then superimposed and equated to known values. The resulting equations will then be the same as Eqs. 17.17, regardless of whether the members are stressed axially, in bending, or in shear.

For a structure with three redundants in which the deflections δ_a, δ_b, and δ_c are zero, the solution of Eqs. 17.17 for X_a is as follows:

$$X_a = -\frac{\begin{vmatrix} \delta_{ao} & \delta_{ab} & \delta_{ac} \\ \delta_{bo} & \delta_{bb} & \delta_{bc} \\ \delta_{co} & \delta_{cb} & \delta_{cc} \end{vmatrix}}{\begin{vmatrix} \delta_{aa} & \delta_{ab} & \delta_{ac} \\ \delta_{ba} & \delta_{bb} & \delta_{bc} \\ \delta_{ca} & \delta_{cb} & \delta_{cc} \end{vmatrix}} \qquad (17.20)$$

and the solution for X_b and X_c will involve similar determinants. It is obvious that the numerical work of computing the deflections δ and of solving the simultaneous equations increases considerably as the number of redundants increases.

For structures such as beams, frames, and arches in which the deflections result from bending, the bending moment at any point may be obtained by superposition of the bending moments M_o resulting from the applied loads and the bending moments resulting from the redundants.

$$M = M_o + X_a m_a + X_b m_b + X_c m_c \qquad (17.21)$$

The bending moments m_a, m_b, and m_c result from unit values of X_a, X_b, and X_c acting on the base structure. The deflection terms which are used in Eq. 17.20 are defined as follows:

$$\delta_{no} = \int \frac{M_o m_n}{EI}\, ds \qquad \delta_{mn} = \int \frac{m_m m_n}{EI}\, ds \qquad (17.22)$$

where the first subscript of δ indicates the point of the deflection and the second subscript indicates the load producing the deflection. The deflection δ_{mn} is the deflection in the direction of a redundant X_m caused by a unit value of a redundant X_n. From Maxwell's theorem of reciprocal displacements, $\delta_{mn} = \delta_{nm}$. The work involved in evaluating the various displacements is reduced by making use of Maxwell's theorem.

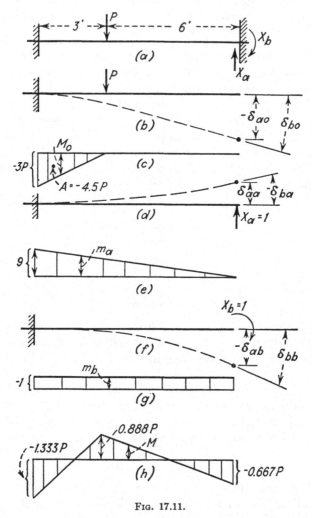

Fig. 17.11.

For structures resisting torsional moments or shearing stresses, the deflection terms of Eqs. 17.17 may be evaluated by the methods of virtual work previously used.

Example 1. Find the reactions and bending moments for the beam of constant cross section shown in Fig. 17.11(a). The ends are fixed against rotation, and there is no bending moment in the beam before the load P is applied.

Solution. The redundants are selected as the shear and bending moment at the right-hand support. The statically determinate base structure is the cantilever beam shown in Fig. 17.11(*b*). The deflections of this beam resulting from the load P are δ_{ao} in the direction of X_a and the rotation δ_{bo} in the direction of X_b. These deflections are obtained from the diagrams for M_o, m_a, and m_b shown in Fig. 17.11(*c*), (*e*), and (*g*).

$$EI\delta_{ao} = \int M_o m_a \, dx = -4.5P \times 8 = -36P$$

$$EI\delta_{bo} = \int M_o m_b \, dx = -4.5P \times -1 = 4.5P$$

The terms δ_{aa} and δ_{ba} represent the deflection and rotation of the free end of the beam due to a unit value of X_a, as shown in Fig. 17.11(*d*).

$$EI\delta_{aa} = \int m_a^2 \, dx = 40.5 \times 6 = 243$$

$$EI\delta_{ba} = \int m_a m_b \, dx = 40.5 \times -1 = -40.5$$

The terms δ_{ab} and δ_{bb} represent the rotation and deflection of the free end due to a unit value of X_b, as shown in Fig. 17.11(*f*). It is not necessary to evaluate δ_{ab} again, because it is equal to δ_{ba}, from Maxwell's theorem.

$$EI\delta_{bb} = -9 \times -1 = 9$$

The equations for obtaining the redundants may be set up as follows:

$$\delta_{ao} + X_a\delta_{aa} + X_b\delta_{ab} = 0$$
$$\delta_{bo} + X_a\delta_{ba} + X_b\delta_{bb} = 0$$

where each equation represents a superposition of the deflections due to the three loading conditions. These equations may be multiplied by EI, and the numerical values of the deflections may be inserted.

$$-36P + 243X_a - 40.5X_b = 0$$
$$4.5P - 40.5X_a + 9X_b = 0$$

The simultaneous solution of these equations yields $X_a = 0.259P$ and $X_b = 0.667P$. The final bending moment M may now be obtained from the equation

$$M = M_o + X_a m_a + X_b m_b$$

or

$$M = M_o + 0.259Pm_a + 0.667Pm_b$$

where the diagrams for M_o, m_a, and m_b are shown in Fig. 17.11. The final diagram for M is shown in Fig. 17.11(*h*).

Example 2. Find the bending-moment diagram for the frame shown in Fig. 17.12(*a*). The value of EI is constant, and the members are fixed against rotation at the supports.

Solution. The structure has three redundant reactions. If the frame is cut at the left support, the remaining structure is stable and statically determinate. The two force components and the couple, X_a, X_b, and X_c are assumed to be the redundant reactions. The bending-moment diagram for the base structure

under the action of the applied load is shown in Fig. 17.12(b). All bending
moments are plotted on the compression side of the members and will not be
designated as positive or negative. The product of two bending moments is

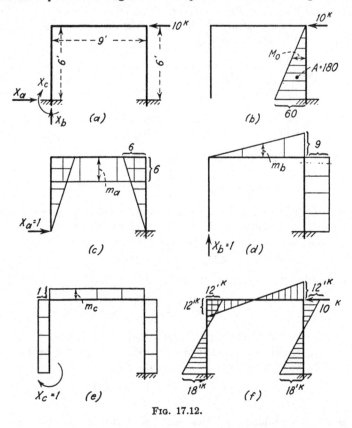

Fig. 17.12.

positive if they are both plotted on the same side of the member. The bending
moments m_a, m_b, and m_c, due to unit values of the redundants, are plotted in
Figs. 17.12(c), (d), and (e). The various deflection terms will be evaluated
semigraphically, by reference to the moment diagrams.

$$EI\delta_{ao} = \int M_o m_a \, dx = 180 \times 2 = 360$$

$$EI\delta_{bo} = \int M_o m_b \, dx = -180 \times 9 = -1{,}620$$

$$EI\delta_{co} = \int M_o m_c \, dx = -180 \times 1 = -180$$

$$EI\delta_{aa} = \int m_a^2 \, dx = 2 \times 18 \times 4 + 6 \times 9 \times 6 = 468$$

$$EI\delta_{bb} = \int m_b^2 \, dx = 40.5 \times 6 + 54 \times 9 = 729$$

$$EI\delta_{cc} = \int m_c^2\, dx = 1 \times 1 \times 21 = 21$$

$$EI\delta_{ab} = \int m_a m_b\, dx = -6 \times 40.5 - 18 \times 9 = -405$$

$$EI\delta_{ac} = \int m_a m_c\, dx = -1 \times 18 \times 2 - 1 \times 54 = -90$$

$$EI\delta_{bc} = \int m_b m_c\, dx = 1 \times 40.5 + 1 \times 54 = 94.5$$

The equations for the redundants are expressed as follows:

$$\delta_{ao} + X_a\delta_{aa} + X_b\delta_{ab} + X_c\delta_{ac} = 0$$
$$\delta_{bo} + X_a\delta_{ba} + X_b\delta_{bb} + X_c\delta_{bc} = 0$$
$$\delta_{co} + X_a\delta_{ca} + X_b\delta_{cb} + X_c\delta_{cc} = 0$$

These equations may be multiplied by EI, and the numerical values of the deflections substituted.

$$360 + 468X_a - 405X_b - 90X_c = 0$$
$$-1,620 - 405X_a + 729X_b + 94.5X_c = 0$$
$$-180 - 90X_a + 94.5X_b + 21X_c = 0$$

The values $X_a = 5$, $X_b = 2.667$, and $X_c = 18$ are obtained from the simultaneous solution of these equations. The final bending-moment diagram is now obtained by statics; it is shown in Fig. 17.12(f).

PROBLEMS

17.1. Analyze the truss of Fig. 17.3, assuming L/AE constant for all members.

17.2. Analyze the truss of Fig. 17.3 for a horizontal load of 20 kips at B, in addition to the vertical load shown. Assume AE constant for all members.

17.3. Repeat Example 3, Art. 17.2, assuming member BD as the redundant.

17.4. Repeat Example 3, Art. 17.2, assuming member BC as the redundant.

17.5. Analyze the truss of Fig. 17.4, assuming that member BD is 0.1 in. too long because of manufacturing tolerances. Assume no external loads on the structure. The areas are the same as used in the examples of Art. 17.2; $AC = 3$ sq in., $AB = BC = BD = 4$ sq in., $AD = DC = 3.6$ sq in., and $E = 10,000$ ksi.

17.6. Analyze the structure of Fig. 17.5 if the load P is acting horizontally at the same point. Assume a constant value of EI.

17.7. Analyze the structure of Fig. 17.5 for the loading shown, assuming the supports to be spread horizontally a distance of 0.5 in. Use the numerical values $R = 50$ in., $P = 2$ kips, $I = 1.0$ in.⁴, and $E = 10,000$ ksi.

17.8. Analyze the structure of Fig. 17.6, considering the horizontal load of 10 kips to be moved upward 4 ft and to act at the corner of the structure.

17.9. Determine the bending moments in the structure shown in Fig. 17.6 if the right support is displaced outward 0.5 in. Assume that no external loads are acting. The values of I are in units of in.⁴, and $E = 28,000$ ksi.

17.10. Analyze the structure of Fig. 17.6 for a temperature drop of 100°F, if $\epsilon = 0.0000067$ per degree Fahrenheit and $E = 28,000$ ksi.

17.11. Analyze the structure of Fig. 17.7, assuming the 2-kip load to be applied at the point where the tube is bent.

17.12. Repeat Example 1, Art. 17.4, assuming $h = 40$ in., $h_1 = 30$ in., and $AE = 10,000$ kips for each member.

17.13. Repeat Example 2, Art. 17.4, if $h = 40$ in., $h_1 = 30$ in., and $AE = 10,000$ kips for each member. Member 12 has a length of 10 in.

17.14. Repeat Prob. 17.13 assuming member 12 as a redundant in place of the reaction X_a.

17.15. Repeat Example 1, Art. 17.5, assuming the moments at the supports to be the redundants.

17.16. Repeat Example 1, Art. 17.5, assuming the shear and moment at the left support to be the redundants.

17.17. Analyze the frame of Fig. 17.12(a), assuming the left support to be pin-connected and the right support to be fixed.

17.18. Analyze the frame of Fig. 17.12(a), assuming an additional load of 20 kips to act down at the center of the structure.

17.6. Selection of Redundants. In most statically indeterminate structures there are several methods by which members or reactions may be removed in order to obtain a stable and statically determinate base structure. In the truss of Fig. 17.4, for example, any one of the six members might be removed to obtain the base structure. The choice of members or reactions to be considered as redundants depends on the individual structure, and no general rules are applicable. The choice of redundants should be such as to make the numerical work of analysis as simple and as accurate as possible. In many cases the analysis may be simplified by choosing the base structure in such a way that many of the members of the base structure are not stressed by the applied loads, or by some of the redundant forces.

In an exact analysis of an elastic structure, the same final stresses are obtained regardless of which member is selected as the redundant. When slide-rule calculations are used in an analysis, it is desirable to avoid calculations involving the subtraction of numbers of approximately the same magnitudes. This situation exists when the forces or bending moments in the base structure are much larger than the final forces or bending moments. It is therefore desirable to choose the base structure in such a way that it has approximately the same stresses as the final structure. In the truss of Fig. 17.4, for example, the members AD, BC, and BD resist much smaller loads than the other three members. This truss was analyzed in Example 3, Art. 17.2, by choosing member AC as the redundant. The final force in member BD was found as $S = S_0 + X_a u = 40.0 - 45.2 = -5.2$ by slide-rule calculations. Thus, a slide-rule error of 0.5 per cent in S_0, X_a, or u would yield a 4 per cent error in S. A better choice of redundant would be one of the members AD, BD, or DC. A solution of this problem using member BD as redundant yields a value of $S = -5.09$ for member BD and $S = 20 + 5.09 \times 0.5 = 22.55$ for member AC. This solution is more accurate because the values of S_0 are closer to the final values of S, and the slide-rule errors are not magnified by subtraction.

The process of solving simultaneous equations usually involves the subtraction of numbers, and consequently introduces large slide-rule errors. The solution of simultaneous equations is also very tedious, particularly when more than three equations must be solved. It has been shown by Muller-Breslau[1] and by Krivoshein[2] that it is possible to select the redundants for any structure in such a manner that it is not necessary to solve simultaneous equations. The method of selecting the redundants will be discussed for a few types of structures.

The general equations for the analysis of any statically indeterminate structure are given in Eqs. 17.17.

$$\left.\begin{array}{l} \delta_a = \delta_{a0} + X_a\delta_{aa} + X_b\delta_{ab} + \cdots + X_n\delta_{an} \\ \delta_b = \delta_{b0} + X_a\delta_{ba} + X_b\delta_{bb} + \cdots + X_n\delta_{bn} \\ \cdot \quad \cdot \quad \cdot \quad \cdot \quad \cdot \quad \cdot \quad \cdot \quad \cdot \quad \cdot \\ \delta_n = \delta_{n0} + X_a\delta_{na} + X_b\delta_{nb} + \cdots + X_n\delta_{nn} \end{array}\right\} \qquad (17.17)$$

If the redundants are selected in such a manner that each redundant force does not deflect the base structure in the direction of another redundant, the following terms vanish:

$$\delta_{ab} = \delta_{ac} = \delta_{bc} = \cdots \delta_{mn} = 0 \qquad (17.23)$$

Equations 17.17 may now be solved as follows for zero values of δ_a, $\delta_b, \ldots, \delta_n$.

$$X_a = -\frac{\delta_{a0}}{\delta_{aa}} \qquad X_b = -\frac{\delta_{b0}}{\delta_{bb}} \qquad \cdots \qquad X_n = -\frac{\delta_{n0}}{\delta_{nn}} \qquad (17.24)$$

Equations 17.24 are obviously much more convenient to use than Eqs. 17.17. The number of conditions expressed by Eqs. 17.23 is $(n^2 - n)/2$, or one condition for a structure with two redundants and three conditions for a structure with three redundants. The method of selecting the redundants so that Eqs. 17.23 are satisfied will be considered for a few typical structures.

For a structure with two redundants X_a and X_b, Eqs. 17.23 and 17.24 are obtained if $\delta_{ab} = 0$. This condition is easy to obtain in a structure which is symmetrical about a center line. One redundant is chosen as symmetrical so that the stresses and deflections due to the redundant have the same values at corresponding points on opposite sides of the center line. The other redundant is made antisymmetrical, or the stresses and deflections on opposite sides of the center line will have equal numerical values but opposite signs.

The beam shown in Fig. 17.13 is similar to that analyzed in Example 1, Art. 17.5. The structure is symmetrical with respect to a vertical center line and is twofold statically indeterminate. It is desired to select

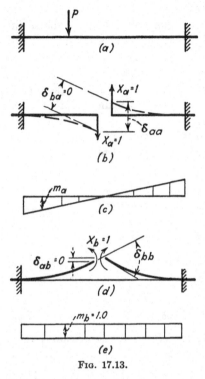

Fig. 17.13.

the redundants X_a and X_b so that $\delta_{ab} = 0$, in order to avoid the solution of simultaneous equations. If the base structure is formed by cutting the structure at the center line, the two resulting cantilever beams will be stable and statically determinate. The shear X_a at the cross section will be antisymmetrical, or will produce bending moments and deflections which are numerically equal but opposite in sign for corresponding points on opposite sides of the center line. The deflection δ_{aa} will be the relative vertical displacement of the cut ends due to the loads $X_a = 1$, as shown in Fig. 17.13(b). The deflection δ_{ba} will be the relative rotation of the tangents at the cut ends, and it will be zero because the anti-symmetrical loading will produce the same clockwise rotation at each cut end. The bending moment X_b produces a symmetrical loading and deflection, as shown in Fig. 17.13(d). The deflection δ_{bb} is the relative rotation of the cut ends. The deflection δ_{ab} is the relative vertical displacement of the cut ends, and it is zero because of the symmetrical loading.

The values of the deflection terms for beams are obtained from the bending-moment diagrams. The m_a diagram is antisymmetrical, as shown in Fig. 17.13(c), and the m_b diagram is symmetrical as shown in Fig. 17.13(e). The deflection term

$$\delta_{ab} = \int \frac{m_a m_b}{EI}\, ds = 0$$

when evaluated for the entire span is zero because the negative value of the integral for the left half of the structure must be equal numerically to the positive value for the right half of the structure.

The two redundants for a symmetrical truss, such as that shown in Fig. 17.14, may also be selected so that one of the redundants is symmetrical with respect to the center line and the other is antisymmetrical. A unit value of the redundant shear X_a produces forces u_a in the truss members which will be antisymmetrical.   A unit value of the redundant

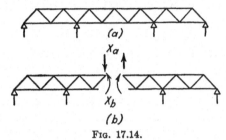

FIG. 17.14.

couple X_b produces forces u_b in the members which are symmetrical with respect to the center line.   The condition

$$\delta_{ab} = \sum \frac{u_a u_b L}{AE} = 0$$

must be satisfied because the values of the summation term for the two sides of the structure will be equal numerically but opposite in sign.

In the case of unsymmetrical structures, it is also possible to select the redundants so that Eq. 17.23 is satisfied, but it may not be possible to select the redundants by inspection.   The rigid frame shown in Fig. 17.15 is fixed at the right support and hinged at the left support.

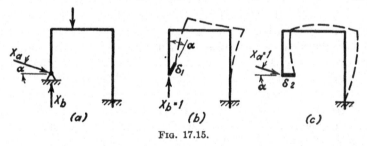

FIG. 17.15.

The two reaction components at the left support may be used as the redundants X_a and X_b.   The condition $\delta_{ab} = 0$ may be satisfied by assuming the direction of one of the redundant forces arbitrarily and then calculating the direction of the other redundant to satisfy the given condition.   If the direction of the redundant force X_b is arbitrarily assumed to be vertical, as shown in Fig. 17.15(b), the force will deflect the base structure an amount δ_1, at an angle α to its direction.   If the redundant force X_a acts perpendicular to the deflection δ_1, the deflection δ_{ab} is zero because it is the component of δ_1 in the direction of X_a.   The redundant force X_a then acts at an angle α with the horizontal, and the deflection δ_2 caused by this force will be horizontal, since from Maxwell's theorem $\delta_{ba} = \delta_{ab}$. The term δ_{ba} is the deflection in the direction of X_b due to a unit value of X_a, or it is the vertical component of the displacement δ_2.

In the unsymmetrical structure it was necessary to calculate the angle of one of the redundants in order to eliminate the solution of simultaneous equations. The angle α of Fig. 17.15 would be calculated from the components of the deflection δ_1. In some cases the extra calculations involved in obtaining the locations of the redundant forces might require as much effort as solving the simultaneous equations. If more than one loading condition of a structure is investigated, it will usually be easier to calculate the positions of the redundants than to solve the simultaneous equations.

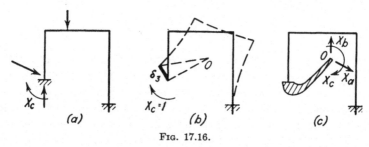

FIG. 17.16.

In the case of a threefold statically indeterminate structure, it is necessary to satisfy three conditions, $\delta_{ab} = \delta_{ac} = \delta_{bc} = 0$, in order to determine the redundant forces from Eqs. 17.24 instead of from the simultaneous equations, Eqs. 17.17. The structure shown in Fig. 17.16(a) has three redundant reactions, which could be assumed as the two force components and the couple at the left support, as shown in Fig. 17.16(a). These redundants cannot satisfy the desired conditions, however, because a couple X_c, applied as shown in Fig. 17.16(b), produces a deflection δ_3, as shown. One of the force components could be perpendicular to δ_3, but both of them could not be, and consequently the couple would produce a displacement in the direction of one of the redundants. However, during the displacement produced by the couple, the free end of the structure must rotate about some point O as an instantaneous center. If the redundants X_a and X_b could be applied at point O, the couple X_c would not displace X_a or X_b, and the two conditions $\delta_{ac} = \delta_{bc} = 0$ would be satisfied. It is therefore assumed that the free end of the structure is extended to point O, and the redundant forces are applied at that point. The bracket which extends to O is assumed to have infinite rigidity, so that it transmits the forces but does not affect the elastic properties of the structure. The point O is called the elastic center. The direction of the redundant X_b is assumed arbitrarily, and the direction of X_a is calculated as in Fig. 17.15 so that $\delta_{ab} = 0$.

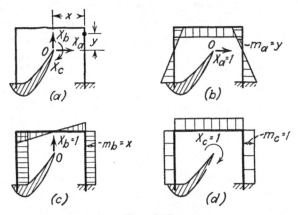

Fig. 17.17.

The elastic center for a symmetrical structure is obtained readily. The structure shown in Fig. 17.17 is symmetrical with respect to a vertical center line. The bending-moment diagram m_c due to a unit value of the redundant couple X_c is symmetrical with respect to the center line. The elastic center O must be on the center line of symmetry in order for the bending-moment diagram m_b for a unit value of the vertical redundant force X_b to be antisymmetrical. The force X_a must be horizontal in order to produce a symmetrical diagram for m_a. Thus, by inspection,

$$\delta_{bc} = \delta_{cb} = \int \frac{m_b m_c}{EI} ds = 0 \qquad (17.25)$$

and

$$\delta_{ba} = \delta_{ab} = \int \frac{m_a m_b}{EI} ds = 0 \qquad (17.26)$$

The vertical location of the elastic center depends on the elastic properties of the structure. Substituting $m_a = y$ and $m_c = 1$ from Figs. 17.17(b) and (d),

$$\delta_{ac} = \delta_{ca} = \int \frac{m_a m_c}{EI} ds = \int \frac{y\,ds}{EI} = 0 \qquad (17.27)$$

where the last integral states that the elastic center is at the centroid of the ds/EI values for the structure.

The use of the elastic center as the point of application of the redundant forces does not completely determine the stable and statically determinate base structure. In Fig. 17.17, the base structure was assumed to be fixed at the right support and free at the left support. The base structure could have been obtained by cutting the frame at the center line and considering both supports to be fixed as shown in Fig. 17.18(a). The redundant forces X_a, X_b, and X_c are again applied at the elastic center, and the bending-moment diagrams for m_a, m_b, and

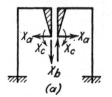
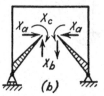

(a)   (b)

FIG. 17.18.

m_c are identical with those shown in Fig. 17.17; therefore Eqs. 17.25 to 17.27 are obviously satisfied. The base structure could also be selected as shown in Fig. 17.18(b), in which the fixity is removed at both supports and the horizontal restraint is removed at the left support. The redundants are again applied at the elastic center, and the diagrams for m_a, m_b, and m_c are identical with those of Fig. 17.17.

The choice of base structure may depend on the applied loading. The final bending moments M at any point are obtained from the equation

$$M = M_o + X_a m_a + X_b m_b + X_c m_c \qquad (17.28)$$

and it is desirable for the value of M_o to approximate the final value of M in order to minimize slide-rule errors. For vertical loads on the top member, the base structure of Fig. 17.18(b) would be preferable, whereas for the loading of Fig. 17.12 the base structure of Fig. 17.17 would be preferable.

Example 1. Find the bending moments for the beam of Fig. 17.11 by the method indicated in Fig. 17.13.

Solution. The bending-moment diagram M_o for the base structure will be the same as that shown in Fig. 17.11(c). The bending-moment diagram for m_a is as shown in Fig. 17.13(c) and has maximum values of 4.5 at the supports. The bending-moment diagram for m_b is shown in Fig. 17.13(e). The redundant forces may now be found without solving simultaneous equations, since $\delta_{ab} = 0$.

$$X_a = -\frac{\delta_{ao}}{\delta_{aa}} = -\frac{EI\delta_{ao}}{EI\delta_{aa}} = -\frac{\displaystyle\int M_o m_a \, dx}{\displaystyle\int m_a^2 \, dx}$$

$$= -\frac{4.5P \times 3.5}{20.25 \times 3} = -0.259P$$

$$X_b = -\frac{\delta_{bo}}{\delta_{bb}} = -\frac{\int M_o m_b \, dx}{\int m_b^2 \, dx} = \frac{4.5P}{1^2 \times 9} = 0.5P$$

The bending-moment diagram may now be obtained as $M = M_o + X_a m_a + X_b m_b$, shown in Fig. 17.11.

Example 2. Solve Example 2, Art. 17.5, by the elastic-center method.

Solution. The base structure will be chosen as shown in Fig. 17.17. The structure is symmetrical, and the elastic center will be on the center line. The vertical location of the elastic center will be at the centroid of the ds/EI values. Taking moments of the ds/EI terms about a horizontal line through the supports, the distance of the elastic center from this line is obtained.

$$\bar{y} = \frac{\dfrac{6 \times 2 \times 3}{EI} + \dfrac{9 \times 6}{EI}}{\dfrac{6 \times 2}{EI} + \dfrac{9}{EI}} = 4.29 \text{ ft}$$

The redundants are applied as shown in Fig. 17.19(a). The bending moments M_o due to the applied load acting on the base structure are shown in Fig. 17.19(b).

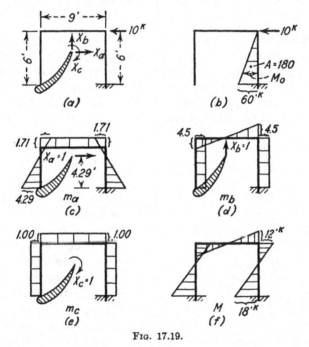

Fig. 17.19.

The bending moments due to unit values of the redundants are shown in Figs. 17.19(c) to (e). The values of the redundants are now calculated.

$$X_a = -\frac{\delta_{ao}}{\delta_{aa}} = -\frac{\int M_o m_a \, ds}{\int m_a^2 \, ds}$$

$$= \frac{180 \times 2.29}{\dfrac{2 \times (4.29)^3}{3} + \dfrac{2 \times (1.71)^3}{3} + (1.71)^2 \times 9} = \frac{412}{82.4} = 5 \text{ kips}$$

$$X_b = -\frac{\delta_{bo}}{\delta_{bb}} = -\frac{\int M_o m_b \, ds}{\int m_b^2 \, ds} =$$

$$= \frac{180 \times 4.5}{2(4.5)^2 \times 6 + \dfrac{2(4.5)^3}{3}} = \frac{810}{303.7} = 2.66 \text{ kips}$$

$$X_c = -\frac{\delta_{co}}{\delta_{cc}} = -\frac{\int M_o m_c \, ds}{\int m_c^2 \, ds} = \frac{180 \times 1}{1^2 \times 21} = 8.57 \text{ ft-kips}$$

The final bending moments are calculated from the equation $M = M_o + X_a m_a + X_b m_b + X_c m_c$, and are plotted in Fig. 17.19(f).

This solution may be compared with the solution of Example 2, Art. 17.5. The simplification of the analysis by the use of the elastic center is obvious.

17.7. Circular Fuselage Rings. The concentrated loads applied to semimonocoque structures must be distributed to the skin and stringer shell structure by means of bulkheads. The wing bulkheads are termed ribs and are frequently constructed as solid webs, although webs with access holes, or trusses, may be used. The fuselage bulkheads are termed rings, or frames, and must usually be constructed so that the interior of the fuselage is unobstructed. The fuselage rings usually extend only a few inches inside the skin and must resist deformation by means of bending stresses.

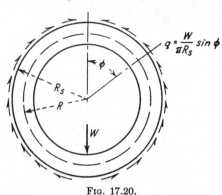

$q = \dfrac{W}{\pi R_s} \sin \phi$

Fig. 17.20.

Many fuselage cross sections are circular or approximately circular. In such cases, the bending moment in a fuselage ring may usually be obtained with sufficient accuracy by assuming the ring to be circular and to have a constant moment of inertia. Such a ring will be analyzed for a concentrated load acting radially.

The ring shown in Fig. 17.20 transfers the load W to the shell structure.

The method of calculating the distribution of the shear flow reactions on the ring was discussed in Art. 8.2. It will be assumed that the skin and stringers are distributed uniformly around the circumference of the shell and that they may be replaced by an effective skin of thickness t_e. The moment of inertia of the shell structure may be found from the equation

$$I = \pi R_s^3 t_e \tag{17.29}$$

where t_e is small compared with the radius R_s. The shear flow is now found from the equation

$$q = \frac{W}{I} \int y \, dA = \frac{W}{I} \int_0^\phi R_s \cos \phi R_s t_e \, d\phi \tag{17.30}$$

where $W/2$ is assumed to be resisted by the skin on the right side of the center line because of symmetry. The shear flow at any point is obtained by substituting I from Eq. 17.29 into Eq. 17.30 and integrating.

$$q = \frac{W}{\pi R_s} \sin \phi \tag{17.31}$$

The direction of the shear flow is shown in Fig. 17.20.

A stable and statically determinate base structure may be obtained by cutting the ring at any point. Three redundants, the bending moment, shearing force, and axial force, must be obtained at the cut section. Because of symmetry, the shearing force is zero at the upper center line, and the solution is simplified by cutting the ring at this cross section. The base structure is then as shown in Fig. 17.21(a). All the stresses are symmetrical about the center line; therefore only one-half of the ring need be considered, as shown in Fig. 17.21(b). It is noticed that the structure of Fig. 17.21(b) does not satisfy the condition that the vertical displacement δ_b of the redundant X_b is zero, whereas the structure of Fig. 17.21(a) satisfies this condition; therefore the structure of Fig. 17.21(a) must be used in cases where the loading is unsymmetrical and X_b must be found from the condition $\delta_b = 0$.

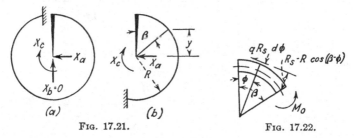

FIG. 17.21.   FIG. 17.22.

The bending moment M_0 in the base structure due to the applied loads is obtained by considering the portion of the ring shown in Fig. 17.22

as a free body. The shear flow on an element of skin of length $R_s\, d\phi$ has a moment arm of $R_s - R \cos (\beta - \phi)$, as shown in Fig. 17.22. The bending moment M_0 is found by integrating the expression for the bending moment due to the load on the element.

$$M_o = -\int_0^\beta qR_s[R_s - R \cos (\beta - \phi)]\, d\phi \tag{17.32}$$

The value of q may be substituted from Eq. 17.31 into Eq. 17.32.

$$M_o = -\frac{W}{\pi} \int_0^\beta [R_s \sin \phi - R \sin \phi \cos (\beta - \phi)]\, d\phi$$

Integrating, and substituting the limits of integration, the value of M_o at any angle β is obtained.

$$M_o = -\frac{WR_s}{\pi} \left(1 - \cos \beta - \frac{R}{2R_s} \beta \sin \beta \right) \tag{17.33}$$

The bending moments m_a and m_c resulting from unit values of the redundants X_a and X_c are found from the following equations:

$$m_a = y = R \cos \beta$$
$$m_c = 1 \tag{17.34}$$

where positive bending moments produce compression on the outside of the ring. The redundants are now obtained from the following equations:

$$X_a = -\frac{\delta_{ao}}{\delta_{aa}} = -\frac{\displaystyle\int \frac{M_o m_a}{EI}\, ds}{\displaystyle\int \frac{m_a^2}{EI}\, ds} = -\frac{\displaystyle\int M_o m_a\, ds}{\displaystyle\int m_a^2\, ds} \tag{17.35}$$

$$X_c = -\frac{\delta_{co}}{\delta_{cc}} = -\frac{\displaystyle\int \frac{M_o m_c}{EI}\, ds}{\displaystyle\int \frac{m_c^2}{EI}\, ds} = -\frac{\displaystyle\int M_o m_c\, ds}{\displaystyle\int m_c^2\, ds} \tag{17.36}$$

The redundants are now evaluated by substituting values from Eqs. 17.33 and 17.34 into Eqs. 17.35 and 17.36 and integrating.

$$X_a = -\frac{\displaystyle\int_0^\pi -\frac{WR_s}{\pi} \left(1 - \cos \beta - \frac{R}{2R_s} \beta \sin \beta \right) R^2 \cos \beta\, d\beta}{\displaystyle\int_0^\pi (R \cos \beta)^2 R\, d\beta}$$

$$X_a = -\frac{W}{\pi} \frac{R_s}{R} \left(1 - \frac{R}{4R_s}\right) \tag{17.37}$$

$$X_c = -\frac{\int_0^\pi -\frac{R_sW}{\pi}\left(1 - \cos\beta - \frac{R}{2R_s}\beta\sin\beta\right)R\,d\beta}{\int_0^\pi R\,d\beta}$$

$$X_c = \frac{WR_s}{\pi}\left(1 - \frac{R}{2R_s}\right)$$

(17.38)

The final bending moments are now obtained by superimposing the moments from the applied loads and the moments from the redundants.

$$M = M_o + X_a m_a + X_c m_c$$

$$M = -\frac{WR}{2\pi}\left(1 - \frac{\cos\beta}{2} - \beta\sin\beta\right)$$

(17.39)

This equation for the bending moment does not contain the term R_s, or the bending moment depends only on the radius of the ring neutral axis and not on the depth of the ring cross section. The values of the bending moment are calculated at 10° intervals from Eq. 17.39 and are tabulated in Table 17.5.

TABLE 17.5

β, deg	M	β, deg	M	β, deg	M	β, deg	M
0	$-0.0796WR$	50	$-0.0016WR$	100	$0.1006WR$	150	$-0.0197WR$
10	$-0.0760WR$	60	$0.0250WR$	110	$0.1008WR$	160	$-0.0820WR$
20	$-0.0654WR$	70	$0.0508WR$	120	$0.0897WR$	170	$-0.1555WR$
30	$-0.0486WR$	80	$0.0735WR$	130	$0.0663WR$	180	$-0.2387WR$
40	$-0.0268WR$	90	$0.0908WR$	140	$0.0298WR$		

A circular ring may be analyzed for a couple loading or a tangential load at any point in the same manner as for a radial load. Bending-moment curves have been published for such loadings.[3] Any possible ring loading may then be obtained as a superposition of radial, tangential, and couple loads, which are reacted by the shear flows in the skin. Bending-moment curves for fuselage rings of elliptical shape have also been published.[4] The bending-moment diagrams for most practical fuselage rings may be approximated from these published curves with sufficient accuracy.

17.8. Irregular Fuselage Rings. In the previous analyses of rings and frames, the moment of inertia was assumed constant, and the bending moment could be expressed by a simple equation. The deflection terms could therefore be evaluated by integration or by semigraphic integration. In many problems the load distribution varies in such a manner that the bending moment cannot be expressed as a simple equation, or the moment of inertia varies. In such cases it is necessary to consider numerous small increments of length over which the bending moment and moment of inertia may be assumed constant and to evaluate the deflection terms by a summation process. The following equations apply:

$$X_a = -\frac{\delta_{ao}}{\delta_{aa}} = -\frac{\sum \frac{M_o m_a}{EI} \Delta s}{\sum \frac{m_a^2}{EI} \Delta s} \tag{17.40}$$

$$X_b = -\frac{\delta_{bo}}{\delta_{bb}} = -\frac{\sum \frac{M_o m_b}{EI} \Delta s}{\sum \frac{m_b^2}{EI} \Delta s} \tag{17.41}$$

$$X_c = -\frac{\delta_{co}}{\delta_{cc}} = -\frac{\sum \frac{M_o m_c}{EI} \Delta s}{\sum \frac{m_c^2}{EI} \Delta s} \tag{17.42}$$

The summation process can best be studied by means of a numerical example.

Example. Calculate the bending moments for the ring shown in Fig. 17.23. The fuselage stringers have equal areas and are equally spaced around the cir-

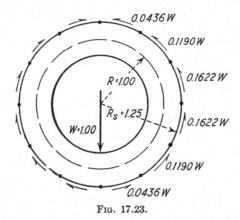

Fig. 17.23.

cumference. The load W and the radius R are assumed to have unit values, and the products of the shear flows and web lengths are shown as the ring reactions. The value of EI is assumed constant.

Solution. The base structure and the redundants are selected as shown in Fig. 17.21(*b*). The increments of length Δs are selected as shown in Fig. 17.24 as lengths of 30° arcs. The bending moments at the centers of these increments are calculated and are assumed constant over the increment. The increments

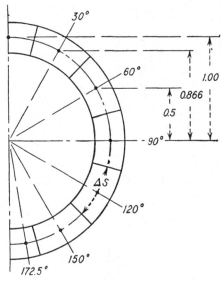

Fig. 17.24.

at the top and bottom of the ring have only one-half the length of other increments and are assumed to have relative values of $\Delta s/EI = 0.5$, as compared with values of $\Delta s/EI = 1.0$ for the other increments. From symmetry, $X_b = 0$. The values $m_a = y$ and $m_c = 1$ are obtained for unit values of the redundants. Substituting these values in Eqs. 17.40 and 17.42,

$$X_a = -\frac{\sum \dfrac{M_0 y \, \Delta s}{EI}}{\sum \dfrac{y^2 \, \Delta s}{EI}} \tag{17.43}$$

$$X_c = -\frac{\sum \dfrac{M_0 \Delta s}{EI}}{\sum \dfrac{\Delta s}{EI}} \tag{17.44}$$

for which the terms are evaluated in Table 17.6. The values of M_0 listed in column (2) are obtained taking moments of the shear flow forces of Fig. 17.23 about the ring neutral axis, assuming the ring to be cut as in Fig. 17.21. Since only relative values of $\Delta s/EI$ are required, most of them are assumed unity in column (3), and the summation of this column yields the denominator of Eq. 17.44.

TABLE 17.6

β, deg (1)	M_0 (2)	$\dfrac{\Delta s}{EI}$ (3)	$\dfrac{M_0 \, \Delta s}{EI}$ (4)	$m_a = y$ (5)	$\dfrac{M_0 y \, \Delta s}{EI}$ (6)	$\dfrac{y^2 \, \Delta s}{EI}$ (7)	M (8)
0	0	0.5	0	1.000	0	0.500	−0.0766
30	−0.0138	1.0	−0.0138	0.866	−0.0120	0.750	−0.0483
60	−0.0563	1.0	−0.0563	0.500	−0.0261	0.250	0.0245
90	−0.1486	1.0	−0.1486	0	0	0	0.0889
120	−0.3080	1.0	−0.3080	−0.500	0.1540	0.250	0.0865
150	−0.5335	1.0	−0.5335	−0.866	0.6420	0.750	−0.0240
172.5	−0.7290	0.5	−0.3645	−0.991	0.3621	0.491	−0.1797
180	−0.7970						−0.2454
Total		6.0	−1.4247	. . . .	0.9400	2.991	

The values in column (4) are obtained as the product of terms in columns (2) and (3), and the summation of the column yields the numerator of Eq. 17.44. The summations of columns (6) and (7) yield the numerator and denominator of Eq. 17.43. The numerical values may now be substituted in Eqs. 17.43 and 17.44.

$$X_a = -\frac{0.9400}{2.991} = -0.3141$$

$$X_c = \frac{1.4247}{6} = 0.2375$$

The final bending moments are obtained by superposition.

$$M = M' - 0.3141y + 0.2375$$

These values are listed in column (8), and are within a few per cent ot the values in Table 17.5. There is a difference of only $2\frac{1}{2}$ per cent in the maximum bending moment. This discrepancy is primarily due to the error involved in taking increments Δs, which are comparatively large. The difference in bending moments due to the different distribution of shear flow is less than 1 per cent of the maximum bending moment.

PROBLEMS

17.19. Obtain the bending-moment diagram for the frame shown if $P = 0$ and $w = 4.0$ kips/ft. Solve by the following methods:

a. Use a base structure as shown in Fig. 17.17(a).

b. Use a base structure as shown in Fig. 17.18(a).

c. Use a base structure as shown in Fig. 17.18(b).

17.20. Obtain the bending-moment diagram for the frame shown if $w = 0$, $P = 10$ kips, and $a = b = 4.5$ ft.

17.21. Obtain the bending-moment diagram for the frame shown if $w = 0$, $P = 10$ kips, $a = 3$ ft, and $b = 6$ ft.

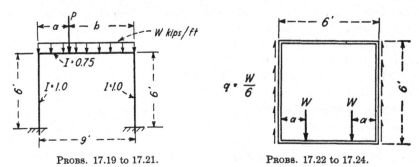

PROBS. 17.19 to 17.21.   PROBS. 17.22 to 17.24.

17.22. The frame for a fuselage of rectangular cross section is shown. Calculate the bending moments if $W = 3$ kips and $a = 1.5$ ft. Assume EI constant.

17.23. Repeat Prob. 17.22 if EI for the bottom frame member is four times as large as the value of EI for the other members.

17.24. Repeat Prob. 17.22 if $2W = 6$ kips and $a = 3.0$ ft.

17.25. Repeat the Example problem of Art. 17.8 if the moment of inertia of the ring cross section is doubled for the sections from $\beta = 105°$ to $\beta = 180°$. Note that it is necessary to calculate the position of the elastic center or else to solve simultaneous equations for the redundants.

17.9. Redundancy of Box Beams. The only type of box beam which is stable and statically determinate consists of three flange areas and three webs, as shown in Fig. 17.25. In this beam the three unknown

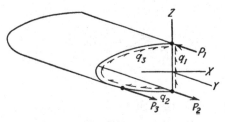

FIG. 17.25.

flange forces, P_1, P_2, and P_3, and the three unknown shear flows, q_1, q_2, and q_3, may be obtained from the six equations of static equilibrium, $\Sigma F_x = 0$, $\Sigma F_y = 0$, $\Sigma F_z = 0$, $\Sigma M_x = 0$, $\Sigma M_y = 0$, and $\Sigma M_z = 0$. As in any other statically determinate structure, the internal forces are independent of the areas or stiffness properties of the members. In any statically indeterminate structure, the areas and elastic properties affect the distribution of the internal forces in the members.

The internal bending stress distribution in all common beams is statically indeterminate, and the deformations are considered in deriving the flexure formula $f = My/I$. This equation is so common that it is not customary to think of such beams as statically indeterminate. The bending stress distribution in a box beam containing more than three flanges, such as that shown in Fig. 17.26, depends on the area and elastic properties of the flanges. The shear flow distribution, which is obtained from the bending stress distribution, also depends on the areas of the flanges, and it is therefore statically indeterminate.

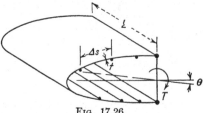

Fig. 17.26.

In this and the following article, it will be assumed that the bending stress is obtained by the simple formula and that a pure torsion load such as shown in Fig. 17.26 produces no axial stresses in the flanges. These assumptions have been used in previous shear flow analyses and have been shown in Art. 16.12 to be accurate in most cases. The shear flow distribution in a single cell box may then be obtained from the conditions of statics, and such a box will be considered as statically determinate for shear flow calculations. With the assumption that a torsional moment produces no axial stresses in the flanges, the equation $q = T/2A$ for the shear flows is obtained from the conditions of statics.

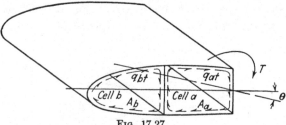

Fig. 17.27.

A two-cell box such as that shown in Fig. 17.27 cannot be analyzed by the equations of statics. It is assumed that the wing ribs have sufficient rigidity so that the two cells deflect through the same angle θ. This deflection condition and the equations of statics are then sufficient for the shear flow analysis, and the structure has a single redundancy. It is of course assumed that the torsion produces no axial load in the flanges, and the flanges are therefore not shown in the sketch. A box structure with several cells has one less redundant than the number of cells, since webs in all but one cell may be cut to leave a single cell as a statically determinate base structure.

The angle of twist of a box beam was found by Eq. 16.37,

$$\theta = \sum \frac{q \,\Delta s \, L}{2AtG} \qquad (16.37)$$

where the terms are indicated in Fig. 17.26. This equation may be used for the angle of twist of a multicell structure, if the summation is evaluated around any closed path, and the area A is enclosed by this closed path. Thus, for a three-cell structure, the summation may be evaluated around the entire perimeter enclosing the three cells, it may be evaluated around the entire perimeter enclosing any one cell, or it may be evaluated around the area enclosing two cells. This procedure is sometimes defined as a line integral, as follows:

$$\theta = \oint \frac{qL}{2AtG} \, ds \qquad (17.45)$$

where the integral represents an evaluation along a closed path, returning to the starting point. The values of the summation or integral are considered as positive when going clockwise around the enclosed area.

17.10. Torsion of Multicell Box Beams. For the two-cell box of Fig. 17.27, the angle of twist θ_a for cell a must equal the angle θ_b for cell b. A unit length L may be considered, since L is always the same for the two cells.

$$\sum_a \frac{q \,\Delta s}{2A_a tG} = \sum_b \frac{q \,\Delta s}{2A_b tG} \qquad (17.46)$$

The first summation must be evaluated around the total perimeter of cell a, including the interior web, and the second summation must include all webs of cell b, also including the interior web.

The value of q for any exterior web of cell a is q_{at}, and for the interior web it is $q_{at} - q_{bt}$. Similarly, the value of q for an exterior web of cell b is q_{bt}, and for the interior web it is $q_{bt} - q_{at}$. Equation 17.46 may now be rewritten, by making these substitutions, moving constant terms outside the summation signs, and assuming G to be constant in all webs.

$$\frac{q_{at}}{A_a} \sum_a \frac{\Delta s}{t} - \frac{q_{bt}}{A_a}\left(\frac{\Delta s}{t}\right)_{a-b} = \frac{q_{bt}}{A_b} \sum_b \frac{\Delta s}{t} - \frac{q_{at}}{A_b}\left(\frac{\Delta s}{t}\right)_{a-b} \qquad (17.47)$$

The following abbreviations will be used for the terms in Eq. 17.47:

$$\delta_{aa} = \sum_a \frac{\Delta s}{t} \qquad \delta_{bb} = \sum_b \frac{\Delta s}{t} \qquad \delta_{ab} = \left(\frac{\Delta s}{t}\right)_{a-b} \qquad (17.48)$$

The term δ_{aa} represents a summation around the entire perimeter of cell a, δ_{bb} a summation around the entire perimeter of cell b, and δ_{ab} the value for the interior web. The terms δ_{aa} and δ_{bb} both include the term δ_{ab} for the interior web. The δ terms do not have quite the same significance as the similar terms used in previous structures, because the constants are eliminated for simplicity and the redundants are taken as shear flows. The abbreviations of Eqs. 17.48 may now be substituted into Eq. 17.47.

$$\frac{1}{A_a}(q_{at}\delta_{aa} - q_{bt}\delta_{ab}) = \frac{1}{A_b}(q_{bt}\delta_{bb} - q_{at}\delta_{ab}) \tag{17.49}$$

The equation for equilibrium of moments about a torsional axis may now be obtained by reference to Fig. 17.27.

$$T = 2A_a q_{at} + 2A_b q_{bt} \tag{17.50}$$

Equations 17.49 and 17.50 may now be solved simultaneously.

$$\left.\begin{aligned}
q_{at} &= \frac{T}{2}\frac{A_a\delta_{bb} + A_b\delta_{ab}}{A_a^2\delta_{bb} + 2A_aA_b\delta_{ab} + A_b^2\delta_{aa}}\\
q_{bt} &= \frac{T}{2}\frac{A_b\delta_{aa} + A_a\delta_{ab}}{A_a^2\delta_{bb} + 2A_aA_b\delta_{ab} + A_b^2\delta_{aa}}
\end{aligned}\right\} \tag{17.51}$$

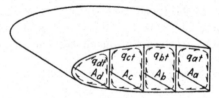

Fig. 17.28.

The shear flows resulting from pure torsion may be obtained in a similar manner for a box beam with any number of cells. For the four-cell structure shown in Fig. 17.28, three equations may be obtained by equating the angles of twist for all cells, $\theta_a = \theta_b = \theta_c, = \theta$.

$$\left.\begin{aligned}
&\frac{1}{A_a}(q_{at}\delta_{aa} - q_{bt}\delta_{ab})\\
&= \frac{1}{A_b}(q_{bt}\delta_{bb} - q_{at}\delta_{ab} - q_{ct}\delta_{bc})\\
&= \frac{1}{A_c}(q_{ct}\delta_{cc} - q_{bt}\delta_{bc} - q_{dt}\delta_{cd})\\
&= \frac{1}{A_d}(q_{dt}\delta_{dd} - q_{ct}\delta_{cd})
\end{aligned}\right\} \tag{17.52}$$

The δ terms are defined by Eqs. 17.48 and by the conditions

$$\delta_{nn} = \sum_n \frac{\Delta s}{t} \qquad \delta_{mn} = \delta_{nm} = \left(\frac{\Delta s}{t}\right)_{m-n} \tag{17.53}$$

where the summation includes all webs around the circumference of the cell and the term δ_{mn} applies to the interior web between cells m and n. The equation for equilibrium of torsional moments is written as follows:

$$T = 2A_a q_{at} + 2A_b q_{bt} + 2A_c q_{ct} + 2A_d q_{dt} \qquad (17.54)$$

The four unknowns may now be obtained by a simultaneous solution of Eqs. 17.52 and 17.54.

17.11. Beam Shear in Multicell Structures. Box beams usually resist transverse shearing forces in addition to the torsional moments which have been considered. It is usually convenient to consider the two effects separately, as a shearing force applied at the shear center and as a torsional moment about the shear center. It was found necessary to use one deflection equation in Art. 16.15 in locating the shear center of a single cell box. For a multicell box it is necessary to use as many deflection equations as there are cells in order to obtain the redundant shear flows and the shear center location.

In a multicell box such as that shown in Fig. 17.29, the increments of flange loads, ΔP, may be calculated from the bending stresses at two cross sections, or from the shear equations as used in Chap. 6 for a single cell box. If one web is cut in each cell, the shear flows q' may be obtained from the equilibrium of spanwise forces on the stringers. The structure shown in Fig. 17.29 is unstable for torsional moments, but the system of shear flows q' will be in equilibrium with external shear forces acting at the shear center of the open section.

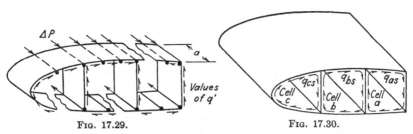

FIG. 17.29.   FIG. 17.30.

The shear flows q_{as}, q_{bs}, and q_{cs} in the cut webs may be obtained and superimposed on the shear flows q' to give a system of shear flows which have a resultant equal to the external shearing force acting at the shear center of the closed multicell box. A superposition of the conditions shown in Figs. 17.29 and 17.30 yields the shear flows in a closed multicell box with no twist. The values of q_{as}, q_{bs}, and q_{cs} are obtained from the condition that the angles of twist θ_a, θ_b, and θ_c for each cell are zero. After q_{as}, q_{bs}, and q_{cs} are obtained, the equation of torsional moments yields the position of the shear center of the closed box.

The external torque about the shear center may then be computed, and the shear flows resulting from this torque computed by the methods of Art. 17.10.

The conditions that the angles of twist for each cell are zero yield the following equation for each cell:

$$\sum \frac{q\,\Delta s}{2AGt} = 0$$

Since G is usually constant and $2A$ is always constant, these terms may be canceled from the equation. Then

$$\sum \frac{q\,\Delta s}{t} = 0 \tag{17.55}$$

For cell a, Eq. 17.55 yields the equation

$$\sum_a \frac{q'\,\Delta s}{t} + q_{as}\sum_a \frac{\Delta s}{t} - q_{bs}\left(\frac{\Delta s}{t}\right)_{a-b} = 0 \tag{17.56}$$

in which summations are evaluated around the entire perimeter of the cell including the interior web and the last term applies only to the interior web. The terms in Eq. 17.56 may be abbreviated, and similar equations may be written for the other cells.

$$\left.\begin{array}{c} \delta_{ao} + q_{as}\delta_{aa} - q_{bs}\delta_{ab} = 0 \\ \delta_{bo} + q_{bs}\delta_{bb} - q_{as}\delta_{ab} - q_{cs}\delta_{bc} = 0 \\ \delta_{co} + q_{cs}\delta_{cc} - q_{bs}\delta_{bc} = 0 \end{array}\right\} \tag{17.57}$$

The following abbreviations are used:

$$\delta_{ao} = \sum_a \frac{q'\,\Delta s}{t} \qquad \delta_{bo} = \sum_b \frac{q'\,\Delta s}{t} \qquad \delta_{co} = \sum_c \frac{q'\,\Delta s}{t} \tag{17.58}$$

in addition to the abbreviations given in Eqs. 17.48 and 17.53.

Equations 17.57 may now be solved simultaneously for q_{as}, q_{bs}, and q_{cs}. These equations are applicable to a two-cell structure if all terms containing the subscript c are dropped. Similar equations may also be written for any number of cells.

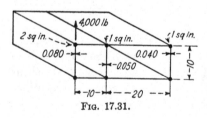

FIG. 17.31.

Example. Find the shear flows in the two-cell box of Fig. 17.31. The horizontal webs have gages of $t = 0.040$ in. Assume G constant for all webs. The cross section is symmetrical about a horizontal center line.

Solution 1. The shear flows may be obtained by superposition of the values of q' for the structure shown in Fig. 17.32(a) and the values of q_a and q_b shown in Fig. 17.32(b). The shear flows q_a and q_b will be computed as the sum of values for a load at the shear center and for a pure torsion loading.

$$\left.\begin{aligned} q_a &= q_{as} + q_{at} \\ q_b &= q_{bs} + q_{bt} \end{aligned}\right\} \quad (17.59)$$

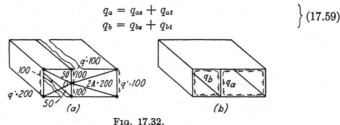

FIG. 17.32.

First considering the shear flows q_{as} and q_{bs} required to produce no twist of the structure, the following equation is obtained for cell a:

$$\sum_a \frac{q\,\Delta s}{t} = q_{as}\left(\frac{20}{0.040}\right) + (q_{as} - 100)\left(\frac{10}{0.040}\right) + q_{as}\left(\frac{20}{0.040}\right)$$

$$+ (q_{as} - q_{bs} + 100)\left(\frac{10}{0.050}\right) = 0$$

$$1{,}450q_{as} - 200q_{bs} - 5{,}000 = 0 \qquad (17.60)$$

A similar equation is written for cell b.

$$\sum_b \frac{q\,\Delta s}{t} = q_{bs}\left(\frac{10}{0.040}\right) + (q_{bs} + 200)\left(\frac{10}{0.080}\right) + q_{bs}\left(\frac{10}{0.040}\right)$$

$$+ (q_{bs} - q_{as} - 100)\left(\frac{10}{0.050}\right) = 0$$

$$825q_{bs} - 200q_{as} + 5{,}000 = 0 \qquad (17.61)$$

Equations 17.60 and 17.61 are now solved simultaneously, yielding $q_{as} = 2.7$ lb/in., and $q_{bs} = -5.4$ lb/in. The negative sign indicates that the shear flow q_{bs} is opposite to the assumed direction, or counterclockwise around the box. These values represent the shear flows for a load applied at the shear center. The torsional moment about the reference point O of the shear flows is found as

$$\Sigma 2Aq' + 2A_a q_{as} + 2A_b q_{bs}$$

where A_a and A_b represent the enclosed areas of the cells.

$$100 \times 200 - 100 \times 200 + 2 \times 200 \times 2.7 - 2 \times 100 \times 5.4 = 0$$

The external shearing force of 4,000 lb acting at the shear center produces no torsional moment about point O, or the shear center of the cross section is at point O.

The actual load of 4,000 lb acting at the left-hand web has a moment arm of 10.0 in. about the shear center. The shear flows q_{at} and q_{bt} must now be obtained for a pure torque of $T = 4{,}000 \times 10 = 40{,}000$ in-lb. From Eq. 17.50,

$$40{,}000 = 400q_{at} + 200q_{bt}$$

and from Eq. 17.47,

$$\frac{q_{at}}{200}\left(\frac{2 \times 20}{0.040} + \frac{10}{0.040} + \frac{10}{0.050}\right) - \frac{q_{bt}}{200}\left(\frac{10}{0.050}\right)$$

$$= \frac{q_{bt}}{100}\left(\frac{2 \times 10}{0.040} + \frac{10}{0.050} + \frac{10}{0.080}\right) - \frac{q_{at}}{100}\left(\frac{10}{0.050}\right)$$

These two equations are solved simultaneously and yield $q_{at} = 66.7$ lb/in. and $q_{bt} = 66.7$ lb/in. The final shear flows in the cut webs are obtained from Eqs. 17.59, as $q_a = 69.4$ lb/in. and $q_b = 61.3$ lb/in. These values are now superimposed on the values of q', and the final shear flows are shown in Fig. 17.33.

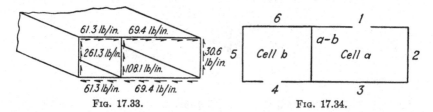

FIG. 17.33.　　　　　　　　FIG. 17.34.

Solution 2. It is usually more convenient to arrange the calculations in tabular form. The calculations are therefore performed in Table 17.7. Column (1)

TABLE 17.7

Web (1)	Δq (2)	q' (3)	$\dfrac{\Delta s}{t}$ (4)	$q'\dfrac{\Delta s}{t}$ (5)	$2A$ (6)	$2Aq'$ (7)	q (8)
		$\Sigma(2)$		$(3) \times (4)$		$(3) \times (6)$	
1	−100	0	500	0	100	0	69.4
2	100	−100	250	−25,000	200	−20,000	−30.6
3	100	0	500	0	100	0	69.4
ab		100	200	+20,000	0	0	108.1
Σ_a			1,450	−5,000	400		
4	200	0	250	0	50	0	61.3
5	−200	200	125	25,000	100	20,000	261.3
6	−100	0	250	0	50	0	61.3
ab		−100	200	−20,000	0	0	−108.1
Σ_b			825	5,000	200	0	

lists the web numbers as shown in Fig. 17.34. The webs are numbered clockwise around the perimeter, with webs for each cell listed separately, and the redundant webs listed first for each cell. In column (2) the change in shear flow at the flange adjacent to each web is tabulated, as in the analysis of single cell boxes in Chap. 6. The shear flows q' are then obtained in column (3) as a summation of the terms in column (2), considering each cell separately. Clockwise shear flows on the faces shown in Fig. 17.33 are positive. The shear flow in the intermediate web ab is clockwise around cell a and counterclockwise around cell b. The values of $\Delta s/t$ for the webs are shown in column (4), and values of $q' \Delta s/t$ are calculated in column (5). The summations of these columns for each cell yield values of the δ terms defined by Eqs. 17.48 and 17.58. The values of $2A$, as shown in Fig. 17.32, are listed in column (6), and values of $2Aq'$, the moments of the shear flows q', are listed in column (7). From the summations given in Table 17.7 and the definitions of Eqs. 17.48 and 17.58,

$$\delta_{aa} = 1,450 \qquad \delta_{ao} = -5,000 \qquad 2A_b = 200$$
$$\delta_{bb} = 825 \qquad \delta_{bo} = 5,000 \qquad \Sigma 2Aq' = 0$$
$$\delta_{ab} = 200 \qquad 2A_a = 400$$

From Eqs. 17.57, if terms with the subscript c are omitted,

$$-5,000 + 1,450q_{as} - 200q_{bs} = 0$$
$$5,000 + 825q_{bs} - 200q_{as} = 0$$

from which $q_{as} = 2.7$ and $q_{bs} = -5.4$.

The external torque T about the shear center may now be found from the torque T_0 about point O.

$$T = T_0 - \Sigma 2Aq' - 2A_a q_{as} - 2A_b q_{bs}$$
$$= 40,000 - 0 - 400 \times 2.7 + 200 \times 5.4 = 40,000$$

In this problem, the assumed center of moments coincided with the shear center. The shear flows resulting from the pure torque are now obtained from Eqs. 17.51, as $q_{at} = 66.7$ lb/in. and $q_{bt} = 66.7$ lb/in. The shear flow in web 1 is $q_a = q_{as} + q_{at} = 69.4$ lb/in., and the shear flow in web 4 is $q_b = q_{bs} + q_{bt} = 61.3$ lb/in. The final web shear flows are listed in column (8) of Table 17.7 and correspond with values shown in Fig. 17.33. The final shear flows for webs 1, 2, and 3 are $q' + q_a$ and for webs 4, 5, and 6 are $q' + q_b$. For web ab the final shear flow is $q' + q_a - q_b$, if it is considered as part of cell a.

17.12. Analysis of Practical Multicell Structures. In the two previous articles, several simplifying assumptions have been made. The cross section was assumed to remain constant along the span, and the webs were assumed to deform elastically in shear, so that $f_s = G\gamma$. These conditions seldom exist in practical structures, but the analysis may be readily modified to account for these items.

The thin webs of aircraft structures frequently wrinkle as tension field webs. In such cases the total shearing deformation γ is defined by Eq. 15.21 for flat webs. An effective shearing modulus may then be assumed as $G_e = f_s/\gamma$, where γ depends on the diagonal tension strain in the web and on the strain in the web stiffeners. The values of G_e will usually vary from $0.25E$ for a highly developed tension field to $0.40E$ for a shear resistant web of a material with Poisson's ratio $\mu = 0.25$. Thus, for a flat web, G_e may vary from $0.625G$ to G. Since G was eliminated from most of the deflection equations, it is usually more convenient to use an effective web thickness t_e, which may vary from $0.625t$ to t, depending on the extent to which the tension field has developed. Curved tension field webs have even less shearing rigidity than flat tension field webs. In practical problems, the designer must usually estimate the effective modulus G_e or the effective thickness t_e, as there are no simple equations for calculating these values.

The methods of considering the taper of a multicell beam are the same as for a single-cell beam. Perhaps the most convenient method is the unit method proposed by Shanley and Cozzone. The bending stresses at two cross sections are obtained either from the general bending equation, Eq. 7.12, or from the bending moments about the principal axes of the cross sections. The loads on the flange areas at the two cross sections are then obtained. The spanwise equilibrium of each stringer element is then considered in finding the shear flows q' in the box with the redundant webs cut. As shown in Chap. 7, this procedure automatically accounts for the components of the flange loads in the plane of the cross section. The center of torsional moments is then carefully selected so that the in-plane components of the flange forces have no appreciable moment about this point. It was shown in Chap. 7 that the in-plane components usually have a negligible torsional moment about a spanwise axis joining the centroids of the cross sections.

Example. The wing cross section shown in Fig. 17.35 is constructed with the 14 stringers, as shown, and with all webs 0.025 in. thick. The wing has taper in both planform and depth. The bending stresses have been calculated at

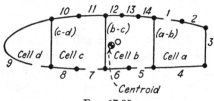

Fig. 17.35.

cross sections 155 in. and 135 in. from the center of the airplane. These bending stresses have been multiplied by the stringer areas to obtain the loads shown in columns (3) and (4) of Table 17.8. The numbers correspond to stringers which are clockwise from webs of the corresponding number. The double areas enclosed by the webs and lines joining the web ends to the point O at station 145 are tabulated in column (2). The torque about an axis O, joining the centroids of the cross-sectional areas, is 13,000 in-lb at station 145. It is desired to find the shear flows at station 145.

Solution. The axial stringer loads P at the ends of each stringer are subtracted and divided by 20. These values of $\Delta P/20$ are tabulated in column (5) of Table 17.8 and represent the change in shear flow at each flange area. The values of the shear flows q' in the structure with redundant webs cut are obtained in column (6) by summing the increments in column (5). It is noticed that the value of q' for web 10 includes values from webs 9 and cd, as well as the stringer 9. Similarly, for web 12 the value of q' is the sum of values from stringer 11, web 11, and web bc. The torsional moment of the shear flows q' about point O is obtained in column (7).

The increments of peripheral length, Δs, of the skin elements are tabulated in column (9). The values of the effective skin thickness t_e are tabulated in column (10) and are obtained by estimating the amount of tension field wrinkling in the webs. The terms $\Delta s/t_e$ are tabulated in column (11), in such a manner that they can be totaled for all the webs in each cell, including the interior webs. Column (11) therefore lists the values for the interior webs twice, since these webs ab, bc, and cd form walls of each of the adjacent cells. The values of q' are again tabulated in column (12). The values of q' in exterior webs are always positive when clockwise around the perimeter, but values of q' for interior webs

TABLE 17.8

Web No. (1)	$2A$ (2)	P Sta. 155 (3)	P Sta. 135 (4)	$\dfrac{\Delta P}{20}$ (5)	q' (6)	$2Aq'$ (7)
				$\dfrac{(4)-(3)}{20}$	$\Sigma(5)$	$(2) \times (6)$
1(a)	21.26	−1,100	−1,260	−8.0	0	0
2	14.56	−1,540	−1,690	−7.5	−8.0	−116
3	70.42	560	970	20.5	−15.5	−1,092
4	29.75	1,290	2,290	50.0	5.0	149
ab	−42.19				55.0	−2,320
5(b)	18.83	1,880	2,910	51.5	0	0
6	15.03	2,170	3,550	69.0	51.5	774
bc	16.33				120.5	1,968
7(c)	15.71	0	3,370	168.5	0	0
8	18.21	3,330	3,270	−3.0	168.5	3,068
cd	73.63				165.5	12,186
9(d)	131.01	−227	−1,420	−59.5	0	0
10	23.06	−1,200	−2,620	−71.0	106.0	2,444
11	20.56	−2,360	−3,050	−34.5	35.0	719
12	8.89	0	−2,870	−143.5	121.0	1,076
13	10.58	−2,310	−2,830	−26.0	−22.5	−238
14	18.88	−520	−600	−4.0	−48.5	−916
Σ	......	......	......	......	......	17,702

TABLE 17.8 (Continued)

Web No. (8)	Δs (9)	$t_e = 0.025 \frac{G_e}{G}$ (10)	$\frac{\Delta s}{t_e}$ (11)	q' (12)	$q' \frac{\Delta s}{t_e}$ (13)	q (14)
		Est.	(9)/(10)	(6)	(11) × (12)	

Cell a

Web No. (8)	Δs (9)	$t_e = 0.025 \frac{G_e}{G}$ (10)	$\frac{\Delta s}{t_e}$ (11)	q' (12)	$q' \frac{\Delta s}{t_e}$ (13)	q (14)
1(a)	5.46	0.025	218	0	0	−3.2
2	4.78	0.025	191	−8.0	−1,530	−11.2
3	4.07	0.025	163	−15.5	−2,530	−18.7
4	10.20	0.025	408	5.0	2,040	1.8
ab	5.65	0.020	283	55.0	15,570	71.7
Σ	.....	.....	1,260	......	13,550	

Cell b

Web No. (8)	Δs (9)	$t_e = 0.025 \frac{G_e}{G}$ (10)	$\frac{\Delta s}{t_e}$ (11)	q' (12)	$q' \frac{\Delta s}{t_e}$ (13)	q (14)
5(b)	5.19	0.025	208	0	0	−19.9
6	5.52	0.025	221	51.5	11,380	31.6
bc	6.60	0.0175	377	120.5	45,430	135.1
12	2.98	0.025	119	121.0	14,400	101.1
13	2.70	0.025	108	−22.5	−2,430	−42.4
14	5.08	0.0175	291	−48.5	−14,110	−68.4
ab			283	−55.0	−15,570	
Σ	.....	.....	1,607	.....	39,100	

Cell c

Web No. (8)	Δs (9)	$t_e = 0.025 \frac{G_e}{G}$ (10)	$\frac{\Delta s}{t_e}$ (11)	q' (12)	$q' \frac{\Delta s}{t_e}$ (13)	q (14)
7(c)	4.65	0.0213	218	0	0	−34.5
8	5.40	0.0187	289	168.5	48,700	134.0
cd	5.85	0.020	293	165.5	48,490	86.1
10	5.33	0.0238	224	106.0	23,740	71.5
11	4.70	0.025	188	35.0	6,580	0.5
bc			377	−120.5	−45,430	
Σ	.....	.....	1,589	......	82,080	

Cell d

Web No. (8)	Δs (9)	$t_e = 0.025 \frac{G_e}{G}$ (10)	$\frac{\Delta s}{t_e}$ (11)	q' (12)	$q' \frac{\Delta s}{t_e}$ (13)	q (14)
9(d)	15.30	0.020	765	0	0	44.9
cd	5.85		293	−165.5	−48,490	
Σ	.....	.....	1,060	......	−48,490	

will have opposite signs for adjacent cells. Thus the value of q' for web ab is clockwise for cell a but is counterclockwise or negative for cell b. In column (6) the shear flows for interior webs were considered positive up, and the values of $2A$ in column (2) were negative if a positive shear flow caused a counterclockwise moment. The values of $q' \Delta s/t_e$ are given in column (13), and are summed separately for each cell.

The following summations are therefore obtained from Table 17.8.

$$
\begin{array}{lll}
\delta_{ao} = 13{,}550 & \delta_{aa} = 1{,}260 & \delta_{ab} = 283 \\
\delta_{bo} = 39{,}100 & \delta_{bb} = 1{,}607 & \delta_{bc} = 377 \\
\delta_{co} = 82{,}080 & \delta_{cc} = 1{,}589 & \delta_{cd} = 293 \\
\delta_{do} = -48{,}490 & \delta_{dd} = 1{,}060 &
\end{array}
$$

The equations for the redundant shear flows for a resultant load at the shear center are obtained by extending Eqs. 17.57 as follows:

$$
\left.
\begin{array}{l}
\delta_{ao} + \delta_{aa}q_{as} - \delta_{ab}q_{bs} + 0 + 0 = 0 \\
\delta_{bo} - \delta_{ab}q_{as} + \delta_{bb}q_{bs} - \delta_{bc}q_{cs} + 0 = 0 \\
\delta_{co} + 0 - \delta_{bc}q_{bs} + \delta_{cc}q_{cs} - \delta_{cd}q_{ds} = 0 \\
\delta_{do} + 0 + 0 - \delta_{cd}q_{cs} + \delta_{dd}q_{ds} = 0
\end{array}
\right\} \tag{17.62}
$$

where the zero terms are included to indicate the nature of the equations. Substituting numerical values obtained above and solving the equations simultaneously, the values $q_{as} = -19.9$, $q_{bs} = -40.9$, $q_{cs} = -55.8$, and $q_{ds} = 30.3$ are obtained.

The enclosed areas of the cells may be obtained by summing terms in column (2) for each cell, using negative values for interior webs.

$$
2A_a = 93.80 \qquad 2A_b = 130.73 \qquad 2A_c = 134.83 \qquad 2A_d = 57.38
$$

The torque about the shear center is obtained by taking moments about point O.

$$
T = T_0 - \Sigma 2Aq' - 2A_a q_{as} - 2A_b q_{bs} - 2A_c q_{cs} - 2A_d q_{ds}
$$
$$
T = 8{,}010 \text{ in-lb}
$$

The shear flows resulting from this torque are now obtained by the method of Art. 17.10.

$$
q_{at} = 16.7 \qquad q_{bt} = 21.0 \qquad q_{ct} = 21.3 \qquad q_{dt} = 14.6
$$

The redundant shear flows are now obtained by superimposing the shear flows from beam shear and torsion.

$$
\begin{array}{l}
q_a = q_{as} + q_{at} = -3.2 \\
q_b = q_{bs} + q_{bt} = -19.9 \\
q_c = q_{cs} + q_{ct} = -34.5 \\
q_d = q_{ds} + q_{dt} = 44.9
\end{array}
$$

The final shear flows are given in column (14) of Table 17.8, and are obtained by superimposing the redundant shear flows q_a, q_b, q_c, and q_d on the shear flows q'.

17.13. Shear Lag. It was pointed out in Art. 16.12 that many of the assumptions made in deriving the simple beam flexure theory are somewhat in error. The assumptions that plane sections remain plane after bending and that bending stresses are proportional to the distance from the neutral axis are less accurate for semimonocoque structures than they are for heavy structures, because the shearing deformations in thin webs are not always negligible.

The effect of shearing deformations in redistributing the bending stresses in a box beam is commonly known as *shear lag*. The effect may be illustrated by considering the cantilever box beam shown in Fig. 17.36. For simplicity, it is assumed that the beam cross section is symmetrical about a vertical center line and that the load is applied along this center line, so that there is no torsional deformation. An analysis by the simple beam theory shows that all the stringers on the upper surface of the beam resist equal bending stresses and that the shearing stresses are the same for all cross sections. As a result of these shearing stresses, an originally plane cross section will deform to the position indicated by line $a'b'c'$.

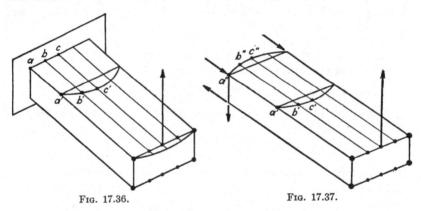

FIG. 17.36. FIG. 17.37.

At the support, however, the cross section is restrained from warping out of its original plane, and the line abc of Fig. 17.36 remains straight. Since the distance cc' is greater than the distance aa', the stringer at c resists a smaller compressive stress than the stringer at a. The bending stress at a must therefore be greater than that calculated by the simple flexure theory and the bending stress at c must be less than indicated by the simple theory. In this case, all the cross sections at some distance from the support warp the same amount, and all the stringers therefore have approximately the same bending stress and strain. The shear-lag effect is greatest at the support and is something of a local effect.

Many wing structures are spliced only at the spars, so that the stringers resist no bending stress at the splice. The box beam shown in Fig. 17.37 is spliced in this manner, so that only the corner flanges resist axial loads at the left-hand support. In this case the cross section at the support deforms as indicated by line $a''b''c''$, in an opposite direction to the deformation $a'b'c'$ of cross sections some distance from the support. The middle stringer has a final length $c'c''$ which is considerably greater than the final length $a'a''$ of the corner stringer. The shear-lag effect is greater in this beam than in the beam with the entire cross section restrained. The effect is also localized near the support, as the stringers at some distance from the support resist bending stresses which are approximately as calculated by the simple flexure theory.

The effect of shear lag may be desirable, since it permits a structure to resist higher ultimate bending moments than are calculated from the simple flexure theory. The allowable bending stresses for the stringers between the spars is smaller than the allowable stresses for the corner flanges, or spar caps. The stringers tend to fail as columns with lengths equal to the rib spacing. The spar caps are supported vertically by the spar web and horizontally by the skin, and they will usually resist high compressive stresses.

When a rectangular box beam is subjected to torsion, a cross section tends to warp from its original plane, as shown in Fig. 16.28. When one end is restrained against warping, axial loads are induced in the flanges, and the shear flows are redistributed near the fixed end. This also is an effect of the shear deformation and is sometimes referred to as a shear-lag effect.

17.14. Spanwise Variation of Warping Deformation. It has been stated that the effects of restraining a cross section from warping are localized to a comparatively short length along the span. The extent of this effect can be studied by considering the stresses resulting from a few simple conditions of loading and then superimposing them with other stress conditions. The structure with two webs and three stringers, shown in Fig. 17.38(a), is assumed to extend for an indefinite length in the x direction and to be loaded as shown. The distribution of the loads and deformations in the x direction will be investigated.

The force P in the center stringer is a function of the distance x. In a length dx the force changes an amount dP, and the web shearing deformation changes an amount $d\gamma$, as shown in Fig. 17.38(b). From the spanwise equilibrium of a stringer, the load increase dP results from the shearing stress f_s in the web.

$$dP = -2f_s t\, dx \qquad (17.63)$$

The deformation γ results from the web shearing stress.

$$f_s = G\gamma \tag{17.64}$$

The change in the angle γ results from the axial elongation of the stringers.

$$b\,d\gamma = -\left(\frac{P}{AE} + \frac{P}{2A_1E}\right)dx \tag{17.65}$$

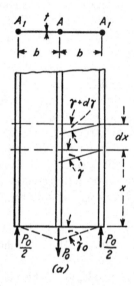

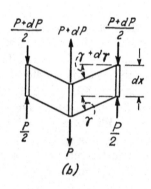

(b)

(a)

FIG. 17.38.

The variables f_s and γ may be eliminated from these three equations in order to obtain a differential equation for P as a function of x. Differentiating Eq. 17.63 and substituting from Eq. 17.64,

$$\frac{d^2P}{dx^2} = -2tG\frac{d\gamma}{dx} \tag{17.66}$$

The value of $\frac{d\gamma}{dx}$ may be substituted from Eq. 17.65.

$$\frac{d^2P}{dx^2} = k^2P \tag{17.67}$$

where

$$k^2 = \frac{2tG}{bE}\left(\frac{1}{A} + \frac{1}{2A_1}\right) \tag{17.68}$$

Equation 17.67 may be integrated as follows:

$$P = C_1e^{kx} + C_2e^{-kx} \tag{17.69}$$

where C_1 and C_2 are constants of integration. The load P approaches zero at a large value of x: thus for $x = \infty$, $P = 0$ and $C_1 = 0$. At the loaded end, $x = 0$ and $P = P_0$, or $C_2 = P_0$. Equation 17.69 therefore has the following value:

$$P = P_0 e^{-kx} \tag{17.70}$$

Equation 17.70 may be differentiated and equated to Eq. 17.63 in order to obtain an expression for f_s.

$$f_s = \frac{P_0 k}{2t} e^{-kx} \tag{17.71}$$

The displacement δ of the force P_0 is equal to γb, or $f_s b / G$ for $x = 0$.

$$\delta = \frac{P_0 k b}{2tG} \tag{17.72}$$

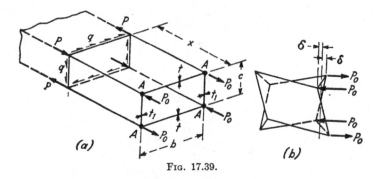

FIG. 17.39.

The structure shown in Fig. 17.39 may be analyzed in a similar manner to the structure of Fig. 17.38. The flange forces P at any cross section are defined by the equation

$$P = P_0 e^{-kx} \tag{17.73}$$

where

$$k^2 = \frac{4G}{AE} \frac{(1/b) + (1/c)}{(1/t) + (1/t_1)} \tag{17.74}$$

The shear flows q may be similarly defined by the expression

$$q = \frac{P_0 k}{2} e^{-kx} = q_0 e^{-kx} \tag{17.75}$$

where q_0 is the shear flow where $x = 0$. The shear flow must be the same in all four webs to satisfy the equilibrium condition for torsional moments. The warping displacement δ of each of the forces P_0 from the plane of the original cross section is measured as shown in Fig. 17.39(b).

$$\delta = \frac{q_0[(1/t) + (1/t_1)]}{2G[(1/b) + (1/c)]} = \frac{P_0 k}{4G} \frac{[(1/t) + (1/t_1)]}{[(1/b) + (1/c)]} \tag{17.76}$$

17.15. Numerical Example of Shear-Lag Calculations. The application of the equations of the previous article to shear-lag calculations will be illustrated by a simple numerical example. The box beam of Fig. 17.40(*a*) is assumed to have a constant cross section for the length

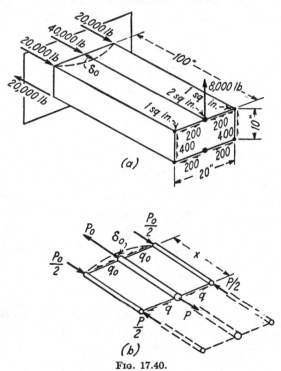

FIG. 17.40.

of the beam. All the webs have a thickness $t = 0.020$ in., and the material has the properties $E = 10^7$ psi and $G = 0.4E$. The simple beam theory yields shear flows which are constant for the length of the span, with the values shown, and values of the axial stringer loads as shown. For this theory to apply, however, the cross section at the support must warp so that the middle stringer is displaced the distance δ_0 from the original plane.

$$\delta_0 = \frac{f_s}{G} b = \frac{200 \times 10}{0.020 \times 4,000,000} = 0.025 \text{ in.}$$

If the cross section at the support is restrained from warping, the center stringer resists a compression force smaller than 40,000 lb and the corner stringers resist compression forces larger than 20,000 lb. The force P_0 acting as shown in Fig. 17.40(*b*), which is required to displace the structure a distance δ_0, will be calculated from Eq. 17.72. The system of forces shown in Fig. 17.40(b) will then be superimposed on those obtained by the simple flexure theory in Fig. 17.40(a).

The structure of Fig. 17.40(b) is equivalent to that of Fig. 17.38. From Eq. 17.68,

$$k^2 = \frac{2 \times 0.020 \times 0.4}{10}\left(\frac{1}{2} + \frac{1}{2}\right) = 0.0016$$

or

$$k = 0.04$$

Substituting $\delta = 0.025$ in Eq. 17.72 and solving for P_0,

$$P_0 = \frac{2tG\delta}{kb} = \frac{2 \times 0.020 \times 4 \times 10^6 \times 0.025}{0.04 \times 10} = 10,000 \text{ lb}$$

From Eq. 17.70,

$$P = 10,000e^{-0.04x}$$

and from Eq. 17.71

$$f_s = 10,000e^{-0.04x}$$

or $q = f_s t = 200e^{-0.04x}$. Thus, at the support, the corner stringers each resist compression forces of 25,000 lb, and the center stringer a compression force of 30,000 lb. The shear flow is zero at this cross section, which is obviously necessary for the assumed condition of no shearing deformation.

The values of P and q at various distances x from the fixed support are calculated in Table 17.9. These values tabulated must be superim-

TABLE 17.9

x	$e^{-0.04x}$	$P = 10,000e^{-0.04x}$	$q = 200e^{-0.04x}$
0	1	10,000	200
5	0.817	8,170	163
10	0.670	6,720	134
20	0.450	4,500	90
40	0.202	2,020	40
100	0.019	190	4

posed on the values shown in Fig. 17.40(a). It is observed that the correction forces at a cross section 20 in. from the support are less than one-half the values at the support.

The loading condition shown in Fig. 17.40 requires larger corrections for the effects of shear lag than are required in the normal airplane wing. The shear loads in an airplane wing are usually resisted at the side of the fuselage, but the cross section at the center of the fuselage is prevented from warping. The cross section at the side of the fuselage, which has the maximum shear flows, is therefore permitted to warp and distribute the shear flows in almost the same manner as predicted by the simple flexure theory.

The effects of shearing deformation on the distribution of torsional stresses may be calculated for simple structures in a similar manner to the previous calculations. The shear flows are first calculated from the simple theory and the warping of the cross section obtained as in Art. 16.13. The axial flange loads and correcting shear flows required to produce the given restraint of the cross section are then calculated and superimposed.

17.16. Ultimate Strength of Indeterminate Structures. The equations for the analysis of statically indeterminate structures were based on the assumption that the stresses were below the elastic limit. In the design of an aircraft structure, however, it is necessary to predict the ultimate strength. The stress distribution obtained for stresses below the elastic limit will usually provide a conservative method of estimating the ultimate strength, but the results may not be very accurate.[5]

To compare the stress distribution at failure with the stress distribution below the elastic limit, the simple truss of Fig. 17.42 will be studied. Any member of this truss may be considered as redundant, and an analysis by means of the elastic theory shows tension forces of $0.661P$ in member 1 and $0.424P$ in member 2 and a compression force of $-0.254P$ in member 3. Member 3 is considered as rigid, and the variation of stress with load for members 1 and 2 will be studied. An assumed stress-strain curve for the material is shown in Fig. 17.41. The ultimate tensile strength is 100,000 psi, and the stress-strain curve is assumed horizontal from elongations of 0.02 to 0.05. Members 1 and 2 have areas of 1 sq in. and therefore have ultimate tensile strengths of 100,000 lb each. If member 1 resisted

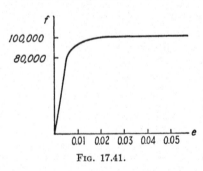

Fig. 17.41.

$0.661P$ at failure, the ultimate load which could be resisted by the truss would be $P = 100,000/0.661 = 151,000$ lb. This is not the correct load, however, because member 2, which has 64 per cent as much stress as member 1 below the elastic limit, has 64 per cent as much unit strain above the elastic limit. If member 1 fails at a strain of 0.05, member 2 has a strain of 0.032 and consequently also has a stress of 100,000 psi at failure. This redistribution of stress occurs at much smaller strains, for member 2 will reach this stress at a 0.02 strain, and member 1 then has a strain of 0.02/0.64 = 0.031, as shown in Fig. 17.41.

It might first appear that the design of an indeterminate structure for ultimate loads is more complicated than the elastic analysis. Most materials have stress-strain curves which are almost horizontal near the ultimate strength, however, and many structures can be designed from a consideration of statics only. The ultimate load resisted by the truss of Fig. 17.42 is determined by assuming that members 1 and 2 are both stressed to their ultimate strengths and then obtaining the load $P =$ 180,000 lb from the statics equations.

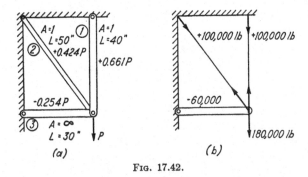

Fig. 17.42.

Aircraft structures must not be stressed beyond the yield point in the limit load condition. For limit loads, the elastic stress distribution must be used. When member 1 of the truss of Fig. 17.42 resists a yield stress of 80,000 lb, the allowable limit load P is 80,000/0.661 = 121,000 lb. The ultimate strength of 180,000 lb is therefore critical for this structure, since the allowable ultimate load is less than 1.5 times the limit load.

The bending-moment diagrams for beams with statically indeterminate reactions depend on the elastic properties of the beams. These bending-moment diagrams are different for stresses which are above the elastic limit than for stresses which are below the elastic limit. The beam shown in Fig. 17.43(a) is assumed to have a constant section and to be loaded by a gradually increasing load. The bending-moment diagram is shown in Fig. 17.43(b) for stresses below the elastic limit. The maximum bending moments at the supports are twice the maximum moments at mid-span, and the stresses at the supports may reach the yield point when the stresses at mid-span are only one-half the yield stresses. As the beam yields in bending at the support, the end-fixity is reduced. An increasing load then causes the bending moment to increase at mid-span but to remain almost constant at the support. In order to satisfy conditions of statics, the bending-moment diagram will remain parabolic, and the sum of the moments at supports and at mid-span will be $wL^2/8$. At the time failure occurs, the bending moment at mid-span will be equal to the bending moment at the supports, or both will be equal to $wL^2/16$, as shown in Fig. 17.43(c).

If the load is removed from the beam shortly before failure, the structure will have a permanent set as shown in Fig. 17.43(d). The beam will have a residual bending moment which will be approximately $wL^2/48$ at all points, if w is the final load. This residual mo-

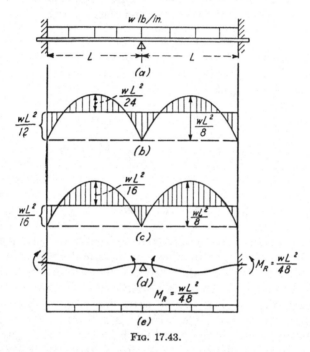

Fig. 17.43.

ment represents the difference between the bending moment existing in the beam after yielding and that which would exist if yielding had not occurred. The permanent set in bending of the beam at the supports is equivalent to a rotation of the supports and produces a couple loading at the supports. If the beam of Fig. 17.43(d) is again loaded with the final value of w, it will behave elastically in regard to the additional deflections and bending moments. The final bending moments of Fig. 17.43(c) are obtained by superimposing the values for an elastic beam shown in Fig. 17.43(b) and the residual moments of Fig. 17.43(e).

17.17. Method of Least Work. The theorem of least work provides a useful method for obtaining the redundant forces or the deformed configuration for many types of statically indeterminate structures. For the case of a structure in which the deformations are elastic and the external reactions do no work during the deformations, the theorem of least work may be stated as follows:

The redundant forces and the deformed configuration of a structure are such as to make the total strain energy a minimum.

The theorem of least work was proposed by Manabrea in 1858 and developed by Castigliano in 1879. The proof follows directly from the method of obtaining deflections which was discussed in Art. 16.16. For a structure with redundant forces X_a, X_b, . . ., X_n, the displacements in the directions of these redundants were stated in Eq. 16.46.

$$\delta_a = \frac{\partial U}{\partial X_a} \qquad \delta_o = \frac{\partial U}{\partial X_b} \qquad \cdots \qquad \delta_n = \frac{\partial U}{\partial X_n} \qquad (17.77)$$

The term U is the total strain energy of the structure. For the type of elastic structure under consideration, the deflections δ_a, δ_b, . . ., δ_n are zero. Equations 17.77 may therefore be written in the following forms.

$$\frac{\partial U}{\partial X_a} = 0 \qquad \frac{\partial U}{\partial X_b} = 0 \qquad \cdots \qquad \frac{\partial U}{\partial X_n} = 0 \qquad (17.78)$$

Equations 17.78 represent the conditions that the strain energy is a minimum.

The practical application of the theorem of least work to the analysis of a conventional statically indeterminate structure is similar to the application of the method of virtual work, as explained in Art. 16.16. Thus, for structures of the type considered in this chapter, there is little advantage in applying the least work method.

The theorem of least work is very useful in obtaining the deformed configuration of many structures. For example, an elastic column might have any one of an infinite number of deflection curves which would satisfy the conditions of statics. The true deflection curve is the one corresponding to the condition of minimum strain energy. Similarly, the deflection surface to which an elastic plate buckles corresponds to a condition of minimum strain energy for the plate.

PROBLEMS

17.26. Find the shear flows in the webs of the structure shown in Fig. 17.31. The horizontal webs have gages of 0.064 in. and G is constant.

17.27. Find the shear flows in the webs of the structure shown in Fig. 17.31 if all flange areas are 1 sq in. The horizontal webs have gages of 0.064 in. and G is constant.

17.28. Find the shear flows in the webs of the structure shown if all flanges have areas of 1 sq in. and all webs have gages of 0.040 in.   Assume $V = 3,000$ lb. $e = 8$ in., and G constant for all webs.

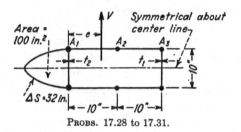

PROBS. 17.28 to 17.31.

17.29. Repeat Prob. 17.28 if $e = 0$.

17.30. Find the shear flows in the webs of the structure shown if $A_1 = A_3 = 1$ sq in., $A_2 = 2$ sq in., $t_1 = t_2 = 0.064$ in., $V = 4,000$ lb, and $e = 10$ in.   The other webs have gages of 0.040 in., and G is constant for all webs.

17.31. Repeat Prob. 17.30, assuming an additional vertical web of 0.064 in. gage at the flange A_2.

17.32. Assume the box beam shown in Fig. 17.40(a) to be loaded by a torsional couple of 160,000 in-lb at the free end instead of the vertical load shown. Assume the cross section to be symmetrical about a horizontal center line and all web gages to be $t = 0.020$ in.   Calculate the warping displacements of a cross section which is free to warp, and calculate the axial flange loads and web shear flows at the wall and at cross sections at 10-in. intervals along the span. Note that the stringer of 2 sq in. area resists no load and does not affect the analysis.   Assume $E = 10^7$ psi and $G = 0.4E$.

REFERENCES FOR CHAPTER 17

1. MULLER-BRESLAU, H. F. B.: "Die graphische Statik der Baukonstruktionen," 3 Vols., Alfred Kroener, Leipzig, 1920–1927.
2. KRIVOSHEIN, G. G.: "Simplified Calculation of Statically Indeterminate Bridges," published by the author, Prague, 1930.
3. WISE, J. A.: Analysis of Circular Rings for Monocoque Fuselages, *J. Aeronaut. Sci.*, September, 1939.
4. BURKE, W. F.: Working Charts for the Stress Analysis of Elliptic Rings, *NACA TN* 444, 1933.
5. VAN DEN BROEK, J. A.: "Theory of Limit Design," John Wiley & Sons, Inc., New York, 1948.

CHAPTER 18

SPECIAL METHODS OF ANALYSIS

18.1. Area Moments. The method of virtual work may be used in obtaining the deflections or the redundant forces for any type of structure. However, other methods are frequently more convenient to use in the analysis of certain special types of structures. Several of these methods will be discussed in this chapter.

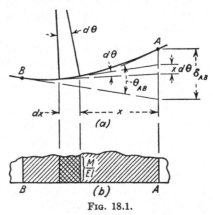

The method of area moments is very convenient to use in finding beam deflections or in analyzing statically indeterminate beams. The line AB in Fig. 18.1(a) represents the elastic curve of an initially straight elastic beam. The deflections are assumed small, as in previous problems. The angular deformation $d\theta$ in a length dx is found from Eq. 14.3.

$$d\theta = \frac{M}{EI} dx \qquad (14.3)$$

The total angle θ_{AB} between tangents to the elastic curve at points A and B is obtained as a sum of all of the increments $d\theta$.

$$\theta_{AB} = \int_A^B \frac{M}{EI} dx \qquad (18.1)$$

The integrand is equal to the area under the M/EI diagram between the points, which is shown as the shaded area in Fig. 18.1(b). Equation 18.1 may be stated as follows:

First Area Moment Principle. The change in slope of the elastic curve between any two points of a beam is equal to the area under the M/EI diagram between the points.

The distance δ_{AB} of Fig. 18.1(a) may be obtained as the sum of the increments $x\ d\theta$.

$$\delta_{AB} = \int x\ d\theta \qquad (18.2)$$

513

Substituting the value of $d\theta$ from Eq. 14.3,

$$\delta_{AB} = \int_A^B \frac{Mx}{EI}\,dx \qquad (18.3)$$

where the integrand represents the moment of the shaded area under the M/EI curve about point A. This procedure may be stated as follows:

Second Area Moment Principle. *The deflection of point A on the elastic curve of a beam from a tangent to the elastic curve at point B is equal to the moment of the area under the M/EI diagram between the points, taken about point A.*

The area moment principles were published by Professor Greene of the University of Michigan, in 1873. They have been used extensively in structural analysis and design.

The method of virtual work is sometimes used in the derivation of Eqs. 18.1 and 18.3. In finding the rotation θ_{AB}, a unit virtual couple would be applied at A and resisted at B, producing a virtual bending moment of $m = 1$. Thus Eq. 16.20 corresponds with Eq. 18.1 when $m = 1$. In obtaining the deflection δ_{AB}, a unit virtual load is applied at point A and the beam is assumed to be fixed at point B, as in Fig. 16.20. The virtual bending moment which is substituted into Eq. 16.20 is now $m = x$, and Eq. 18.3 is obtained.

Example. Find the reactions for the fixed-end beam shown in Fig. 18.2(a).

Solution. The structure is twofold statically indeterminate. The redundants are selected as the reactions at the right-hand support, and the bending-moment

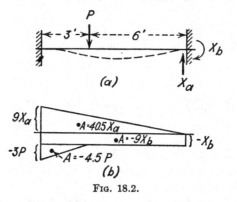

(a)

(b)

Fig. 18.2.

diagram is plotted as shown in Fig. 18.2(b). The two deflection equations are obtained from the condition that there is no relative rotation of the tangents at the two ends of the beam and from the condition that there is no deflection of one end of the beam from a tangent at the other end. The two area moment principles are now applied by equating the total moment area to zero and by equating the moment of the moment area about the left end to zero.

$$40.5X_a - 9X_b - 4.5P = 0$$
$$3 \times 40.5X_a - 4.5 \times 9X_b - 1 \times 4.5P = 0$$

These equations are solved simultaneously to yield $X_a = 0.259P$ and $X_b = 0.667P$. The remaining reactions and bending moments may be obtained from the conditions of statics. The final bending-moment diagram is shown in Fig. 17.11(h).

18.2. Conjugate Beam Method. The method of area moments is very convenient to use if the beam is fixed at a support or if the direction of the tangent to the elastic curve is known at any point. In the case of an unsymmetrical simple beam, however, the direction of the tangent to the elastic curve cannot be determined by inspection for any point on the beam. In order to obtain the deflection curve for such a beam by the method of area moments, it is necessary to consider the geometric relationships for the deflected beam and to obtain first the slope at one support by dividing the deflection of the other support from the tangent at the first support by the distance between supports. Additional applications of the moment area principles are then necessary in order to find the deflections. The conjugate beam method is more convenient to use for such problems. The procedure for calculating beam rotations and deflections is reduced to the more familiar process of calculating shear and bending-moment diagrams.

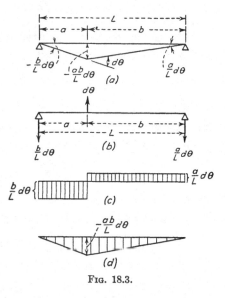

Fig. 18.3.

It will first be assumed that only one increment of length, dx, of a simple beam is elastic. This increment bends through an angle $d\theta$, as shown in Fig. 18.3(a). This single bend produces constant slopes of $-b\,d\theta/L$ to the left of the increment and $a\,d\theta/L$ to the right of the increment. The maximum deflection is $-ab\,d\theta/L$, as shown, if deflections are assumed positive upward.

A fictitious beam called a conjugate beam, which has the same length as the real beam and which resists a concentrated load $d\theta$, is shown in

Fig. 18.3(*b*). The shear diagram for the conjugate beam is shown in Fig. 18.3(*c*), and the values of the shear are observed to coincide with the slopes of the real beam. Similarly, the bending moments in the conjugate beam, shown in Fig. 18.3(*d*), are identical with the deflections of the real beam, shown in Fig. 18.3(*a*).

If the angular displacements of all elements of the beam are considered, the conjugate beam will be loaded with a distributed load of $d\theta/dx$ per unit length for the entire span, rather than the single concentrated load of $d\theta$. The slope of the deflection curve of the real beam is equal to the shear in the conjugate beam, and the deflection of the real beam is equal to the bending moment in the conjugate beam. The distributed load per unit length on the conjugate beam may be expressed in several forms by using various relationships from Eqs. 14.1 to 14.6.

$$\frac{d\theta}{dx} = \frac{1}{R} = \frac{d^2y}{dx^2} = \frac{M}{EI} = \frac{f}{Ec} \tag{18.4}$$

The notation is similar to that previously used, with R the radius of curvature of the deflection curve, y the upward deflection of a point, and f the bending stress at a distance c from the neutral axis.

It was shown in Chap. 5 that the process of plotting shear and bending-moment diagrams was equivalent to integration of the equation

$$w = \frac{dV}{dx} = \frac{d^2M}{dx^2} \tag{18.5}$$

where the loading w per unit length is known, and curves for V and M are determined. Similarly, the process of obtaining deflection curves is equivalent to integration of the equations

$$\frac{M}{EI} = \frac{d\theta}{dx} = \frac{d^2y}{dx^2} \tag{18.6}$$

where M/EI is known and θ and y are determined. From the nature of Eqs. 18.5 and 18.6, it is obvious that they may be solved in the same manner, or that Eq. 18.6 may be solved by loading a conjugate beam with M/EI per unit length and calculating shear and bending-moment diagrams for this loading. In Figs. 18.4(*a*) to (*c*), the load, shear, and bending-moment diagrams for a beam are shown. In Figs. 18.4(*d*) to (*f*), a conjugate beam is loaded with the M/EI diagram to obtain values of the slope θ and deflection y. The procedures are obviously identical. The conjugate beam loading shown corresponds to a down load on the real beam rather than to the loading shown in Fig. 18.4(*a*).

In obtaining shear and bending-moment diagrams by integration, the constants of integration must be determined from the conditions at the supports of a beam. In the case of a simple beam, the slope has maximum values and the deflection zero values at the supports. The conjugate beam also has simple supports at the end to provide maximum shear and zero bending-moment values. The cantilever beam shown

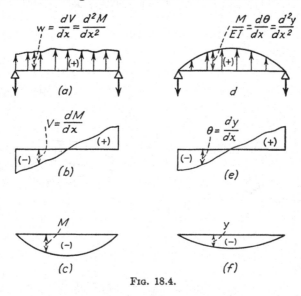

$$w = \frac{dV}{dx} = \frac{d^2M}{dx^2}$$

(a)

$$\frac{M}{EI} = \frac{d\theta}{dx} = \frac{d^2y}{dx^2}$$

d

$$V = \frac{dM}{dx}$$

(b)

$$\theta = \frac{dy}{dx}$$

(e)

M

(c)

y

(f)

Fɪɢ. 18.4.

in Fig. 18.5(a) has zero values of the slope and deflection at the left end. The conjugate beam is shown in Fig. 18.5(b), with the left end free to provide zero values of the shear and bending moment.

The relationship between the support conditions for a real beam and its conjugate beam is shown by Fig. 18.6, in which each beam is the

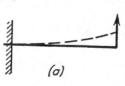

(a)

conjugate of the other. An interior support in a real beam, for example, permits no deflection, but permits equal rotations for the beam on each side of the support. The conjugate beam replaces such supports by an unsupported hinge, which

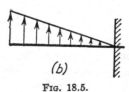

(b)

Fɪɢ. 18.5.

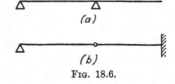

(a)

(b)

Fɪɢ. 18.6.

permits no bending moment, but requires equal shears on each side of the hinge. Similarly an unsupported hinge in the real beam would require an interior support in its conjugate beam.

Example. For the airplane wing shown in Fig. 18.7(a), the bending stress at station 0 is 40,000 psi at a distance c of 8 in. from the neutral axis. At station 200 the bending stress is 20,000 psi for $c = 2$ in. Calculate the deflection curve if $E = 10,000,000$ psi and f/c has a linear variation between the two stations as shown in Fig. 18.7(b).

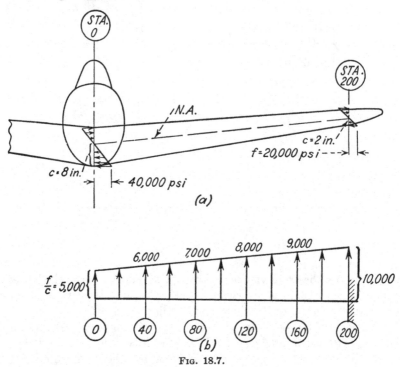

FIG. 18.7.

Solution. The conjugate beam is shown in Fig. 18.7(b), with the left end free and the right end clamped. The shear and bending-moment diagrams for this conjugate beam are calculated in Table 18.1. The conjugate beam may be loaded either with the diagram for M/EI or the diagram for f/Ec, since $f/c = M/I$. The values of f/Ec are given in column (2) for stations at 40-in. intervals. The area under this diagram between any two stations represents the change in slope between the stations, and it is tabulated in column (3). The values in column (3) are obtained by multiplying the average of two values in column (2) by 40, the distance between stations, and represent the trapezoidal areas shown in Fig. 18.7(b). The slope θ of the tangent to the elastic curve at any station is obtained in column (4) by a summation of the values in column (3). The slope is equal to the shear in the conjugate beam, which is equal to the area under the f/Ec load curve to the left of the point. The areas under the shear, or θ, curve are obtained in column (5) by multiplying the average of two values in column (4) by the distance, 40, between stations. The deflections y are

TABLE 18.1

Station (1)	$\dfrac{f}{Ec}$ (2)	$\Delta\theta$ (3)	θ, rad. (4)	Δy (5)	y, in. (6)
0	0.0005		0		0
		0.022		0.44	
40	0.0006		0.022		0.44
		0.026		1.40	
80	0.0007		0.048		1.84
		0.030		2.52	
120	0.0008		0.078		4.36
		0.034		3.80	
160	0.0009		0.112		8.16
		0.038		5.24	
200	0.0010		0.150		13.40

obtained in column (6) by a summation of values in column (5). The deflections are equal to bending moments in the conjugate beam, which are obtained as the area under the shear diagram to the left of the point.

The deflections calculated in column (6) are slightly in error because the increments of area, Δy, were assumed trapezoidal. The true shear curve for the conjugate beam of Fig. 18.7 is concave up, and the true area is slightly less than that obtained in column (6). An exact value of the deflection at station 200 may be obtained by taking moments of the area shown in Fig. 18.7(b).

$$y = \frac{5{,}000 \times 200 \times 100 + 5{,}000 \times (200/2) \times (200/3)}{10{,}000{,}000} = 13.333 \text{ in.}$$

The value shown in Table 18.1 is therefore only 0.5 per cent too large.

18.3. Truss Deflections by Elastic Loads. The deflection curves for beams were calculated in the previous article by obtaining the bending-moment diagrams for conjugate beams loaded with the angle changes. This method permits the simultaneous calculation of deflections at several points along the span, and it has several advantages over other methods which require a separate operation for each deflection. The deflection curve for a truss may also be obtained by a similar method, which is variously termed *elastic loads, elastic weights, angle loads,* or *angle weights.* These terms are sometimes applied also to the method of calculating beam deflection curves.

The diagrams of Fig. 18.3 show that the deflection curve of a beam with a single angle change $d\theta$ is identical with the bending-moment curve of a conjugate beam loaded with a single load $d\theta$. These diagrams would apply equally well to a truss which has a single change in angle, ϕ. For the truss of Fig. 18.8, a change in angle, ϕ, at any point on the line *ABCDEFG*, produces deflections of the line which are identical with the bending moments in a conjugate beam resisting a load equal to ϕ. The methods of supporting the conjugate beam correspond with those

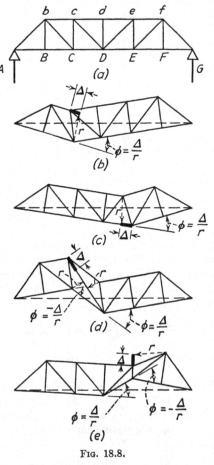

used in obtaining beam deflections. The angle changes for a beam were distributed along the span, but the angle changes for the truss of Fig. 18.8 are concentrated at points *B*, *C*, *D*, *E*, and *F*.

The angle changes produced by the deformations of individual truss members may be calculated from geometric considerations. If the member *bc* of Fig. 18.8(*a*) is shortened a distance Δ, the two parts of the truss have a relative rotation about point *C* of $\phi = \Delta/r$, as shown in Fig. 18.8(*b*). The distance r is the perpendicular distance from the member *bc* to the center of relative rotation *C*. Similarly, if member *DE* is increased in length a distance Δ, the two parts of the structure have a relative rotation about point *e* of Δ/r, as shown in Fig. 18.8(*c*). The deflection curve resulting from these angle changes is the same as the bending-moment curve for a conjugate beam resisting concentrated loads, ϕ, equal to the angle

Fig. 18.8.

changes. The loads act at panel points which correspond to the centers of rotation for the deformations.

Elongations of the vertical or diagonal members of the truss of Fig. 18.8 produce equal and opposite angle changes at two adjacent panel points. An increase Δ in the length of diagonal member *cD* causes rotations of Δ/r about point *d* and $-\Delta/r$ about point *C*, as shown in Fig. 18.8(*d*). The distances r are measured perpendicular to the deformed member from the center of rotation. A decrease Δ in the length of the vertical member eE causes rotations of Δ/r about point D and $-\Delta/r$ about point f, as shown in Fig. 18.8(e). The distance r is the horizontal panel length, as shown. The positive angle changes are represented by upward loads on the conjugate beam, and the negative angle changes are represented by downward loads. The final truss-deflection curve will be obtained by superimposing the deflections resulting from the elongations of all members. The bending moment in a conjugate beam which resists the loads ϕ for all members will be identical with the deflection curve for the truss.

Some trusses may not have horizontal upper chord members or vertical web members. The angle changes for such trusses have different values but may be obtained by a method similar to that shown in Fig. 18.8. The values of ϕ for such trusses will not be given here, as this information is available in many reference books and the values may be obtained from a consideration of the geometry of any particular structure.[1,2]

Example. Calculate the deflection curve for the truss shown in Fig. 18.9. The member dD has no stress but the other members all resist stresses of 29,000 psi and have values of $E = 29,000,000$ psi. The directions of the stresses correspond to those resulting from equal downward loads at all of the lower panel points.

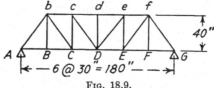

Fig. 18.9.

Solution. The elastic loads, or angle changes, are calculated in Table 18.2. The structure is symmetrical about the center line; therefore only members for the left half are listed in the table. The deformations Δ are negative when compression and positive when tension and are calculated in column (2) as fL/E for each member. The distances r are tabulated in column (3), and the signs are determined by inspection so that the values of Δ/r will have the proper sign. A study of Fig. 18.8 indicates that the values of Δ/r are positive for the upper and lower chord members. The web members each have one positive and one negative value of Δ/r, and the signs are determined from a study of the geometry of the deflected structure, as shown in Fig. 18.8. The values of Δ/r are calculated in column (4), and the point of application of each value is listed in column (5). Member bB is not considered in the table, since the elongation of this member affects only the location of point B and can be considered separately.

The values of the elastic loads, $\phi = \Delta/r$, are totaled for panel points B, C, and D, and are listed in column (2) of Table 18.3. These loads act on the con-

TABLE 18.2

Member (1)	Δ (2)	r (3)	$\dfrac{\Delta}{r}$ (4)	Point (5)
AB	0.03	40	0.00075	B
BC	0.03	40	0.00075	B
CD	0.03	40	0.00075	C
Ab	−0.05	−24	0.00208	B
bc	−0.03	−40	0.00075	C
cd	−0.03	−40	0.00075	D
bC	0.05	$\begin{cases} -24 \\ +24 \end{cases}$	−0.00208 0.00208	B C
cC	−0.04	$\begin{cases} +30 \\ -30 \end{cases}$	−0.00133 0.00133	C D
cD	0.05	$\begin{cases} -24 \\ +24 \end{cases}$	−0.00208 0.00208	C D

jugate beam as shown in Fig. 18.10(a). The reactions at A and G must be in equilibrium with the elastic loads, and they are found to be −0.00583 rad, as

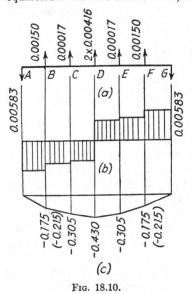

shown. The shear in the conjugate beam is calculated in column (3) as a summation of terms in column (2) of Table 18.3. The shear diagram, shown in Fig. 18.10(b), represents the slope of the loaded truss in radians. The area under each section of the shear diagram is calculated in column (4) by multiplying values in column (3) by the panel length of 30. The deflections y of the panel points are equal to the bending moments in the conjugate beam and are obtained in column (5) by the summation of terms in column (4). The value of −0.175 in. for point B is for a zero elongation of member Bb and is now corrected to −0.215 in. because member Bb increases in length by 0.040 in. The deflection curve is plotted in Fig. 18.10(c).

FIG. 18.10.

PROBLEMS

18.1. A uniform cantilever beam of length L supports a load P at the free end. Using the method of area moments, find the vertical and angular deflections (a) at the load and (b) at mid-span.

Table 18.3

Point (1)	Angle loads (2)	Shear, Σ(2) (3)	Shear area, 30 × (3) (4)	y, in. Σ(4) (5)
A	−0.00583			0
		−0.00583	−0.175	
B	0.00150			−0.175(−0.215)
		−0.00433	−0.130	
C	0.00017			−0.305
		−0.00416	−0.125	
D(×½)	0.00416			−0.430

18.2. A uniform simple beam of length L resists a uniformly distributed load of w lb/in. over the entire span and a concentrated load P at mid-span. Find the deflection at mid-span and the rotation at the supports by the method of area moments.

Hint: Use the tangent at the center line as the reference line.

18.3. A uniform beam of span L is fixed at the left end and is simply supported at the right end. A load of w lb/in. is distributed over the left end of the beam for a length of $L/2$. Find the reactions, using the method of area moments.

18.4. Repeat Prob. 18.1, using the conjugate beam method. Also obtain the equation of the elastic curve by finding the deflection at a point a distance x from the support.

18.5. Repeat Prob. 18.2, using the conjugate beam method. Find the equation of the deflection curve for the left half of the span.

18.6. Repeat Prob. 18.3, using the conjugate beam method. Find the position and the magnitude of the maximum deflection.

Hint: The shear in the conjugate beam is zero at the point of maximum bending moment in the conjugate beam.

18.7. A uniform beam of span L is fixed at both ends and resists a concentrated load W at mid-span. Determine the bending moments and the maximum deflection (a) by area moments and (b) by the conjugate beam method. Note that the conjugate beam has both ends free and that the bending-moment area must produce no reactions on the conjugate beam.

18.8. Solve Example 1, Art. 16.5 (Fig. 16.9), by the conjugate beam method.

18.9. Solve Example 2, Art. 16.5 (Fig. 16.10), by the conjugate beam method. Also find the deflection of the right end of the beam.

18.10. Find the deflection curve for the truss of Fig. 16.4(a) by the method of elastic loads. Assume that five vertical members are added at the panel points and that these vertical members have no elongation. The elongations of other members are given in Example 1, Art. 16.3.

18.4. Beam Columns. In the structures which have been previously considered, it was assumed that the deflections were proportional to the loads and that the deflections were small enough so that they did not appreciably affect the moment arms of the forces. These assumptions are not sufficiently accurate in the case of slender beams which resist axial compressive forces. In these members, called *beam columns*, the lateral deflections appreciably change the moment arms of the compression forces. The deflections are not proportional to the compression loads but have a relation similar to that shown in Fig. 14.5, in which the deflections are small for small compressive loads but increase to large values as the compression load approaches the critical or Euler load for the member.

The beam column shown in Fig. 18.11 resists lateral loads and end moments. The bending moment in a simple beam with no axial load but with the same lateral loads and end moments will be termed the

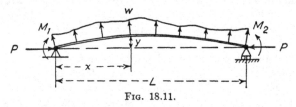

FIG. 18.11.

primary bending moment M'. The axial load P produces an additional or secondary bending moment of $-Py$, because of the deflection y. As the deflection y results from both the primary and secondary bending moments, it cannot be computed directly by the beam-deflection methods which have been previously used. It is, of course, possible to calculate y for the primary bending moment first and then make further calculations of secondary effects by successive approximations. These approximations converge rapidly if the axial loads are small, but converge very slowly if the axial loads approach the Euler load.

The bending moment at any cross section of the beam column of Fig. 18.11 is found from the equation

$$M = M' - Py \tag{18.7}$$

where M' is the bending moment when $P = 0$. Differentiating Eq. 18.7 twice with respect to x, the following expression is obtained:

$$\frac{d^2M}{dx^2} = \frac{d^2M'}{dx^2} - P\frac{d^2y}{dx^2} \tag{18.8}$$

It was shown in Chap. 5 that the second derivative of M' was equal to the load intensity w. The following relation will also be used:

$$\frac{d^2y}{dx^2} = \frac{M}{EI} \tag{18.8a}$$

Substituting these values into Eq. 18.8, the following differential equation for the bending moment in any beam column is obtained:

$$\frac{d^2M}{dx^2} + \frac{PM}{EI} = w \tag{18.9}$$

The terms M and w are functions of x. It will be assumed that w is a linear function of x so that its second derivative is zero. The general solution of Eq. 18.9 is as follows:

$$M = C_1 \sin \frac{x}{j} + C_2 \cos \frac{x}{j} + wj^2 \tag{18.10}$$

The term j is defined as follows:

$$j^2 = \frac{EI}{P} \tag{18.11}$$

The terms C_1 and C_2 are undetermined constants which must be evaluated from the boundary conditions. They may have different values for different sections of the beam. For the beam shown in Fig. 18.12, for example, the constants C_1 and C_2 would have one set of values from 0 to a, another set from a to b, and a third set from b to L.

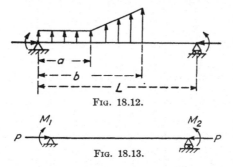

Fig. 18.12.

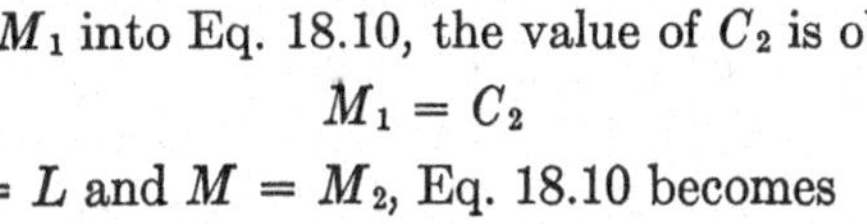

Fig. 18.13.

The constants C_1 and C_2 of Eq. 18.10 will be evaluated for some simple conditions of loading. For the beam of Fig. 18.13, the value of w is zero, and only the end moments M_1 and M_2 need be considered. Substituting $x = 0$ and $M = M_1$ into Eq. 18.10, the value of C_2 is obtained.

$$M_1 = C_2 \tag{18.12a}$$

Similarly, for $x = L$ and $M = M_2$, Eq. 18.10 becomes

$$M_2 = C_1 \sin \frac{L}{j} + M_1 \cos \frac{L}{j}$$

or

$$C_1 = \frac{M_2 - M_1 \cos \dfrac{L}{j}}{\sin \dfrac{L}{j}} \tag{18.12b}$$

For a beam column resisting a uniformly distributed transverse load, as shown in Fig. 18.14, the terms C_1 and C_2 may be evaluated by considering the conditions at the beam supports. From Eq. 18.10, if $x = 0$

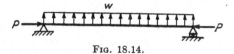

FIG. 18.14.

and $M = 0$, the value $C_2 = -wj^2$ is obtained. Similarly, for $x = L$ and $M = 0$, the value

$$C_1 = \frac{wj^2[\cos (L/j) - 1]}{\sin (L/j)} \tag{18.13}$$

$M = C_1 \sin \dfrac{x}{j} + C_2 \cos \dfrac{x}{j} + wj^2 \; ; \; j^2 = \dfrac{EI}{P}$		
LOADING	C_1	C_2
END MOMENTS, $w = 0$ 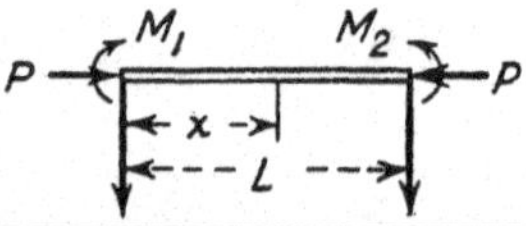	$\dfrac{M_2 - M_1 \cos \dfrac{L}{j}}{\sin \dfrac{L}{j}}$	M_1
UNIFORM LOAD, $w = w_0$ 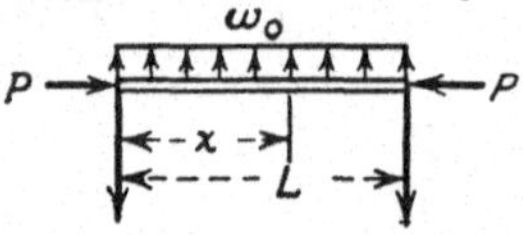	$\dfrac{w_0 j^2 \left(\cos \dfrac{L}{j} - 1\right)}{\sin \dfrac{L}{j}}$	$-w_0 j^2$
TRIANGULAR LOADING, $w = \dfrac{x}{L} w_0$ 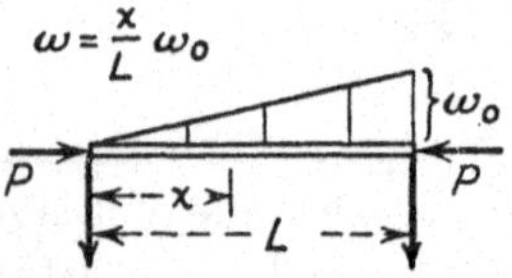	$\dfrac{-w_0 j^2}{\sin \dfrac{L}{j}}$	0
CONCENTRATED LOAD, W, $w = 0$ 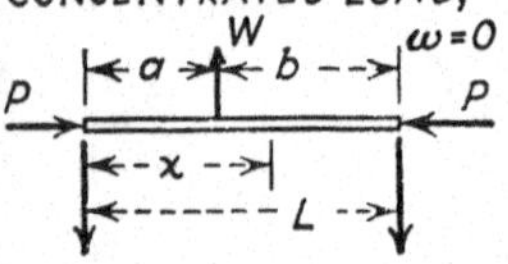 $x < a$	$-\dfrac{Wj \sin \dfrac{b}{j}}{\sin \dfrac{L}{j}}$	0
$x > a$	$\dfrac{Wj \sin \dfrac{a}{j}}{\tan \dfrac{L}{j}}$	$-Wj \sin \dfrac{a}{j}$
COUPLE, $w = 0$ 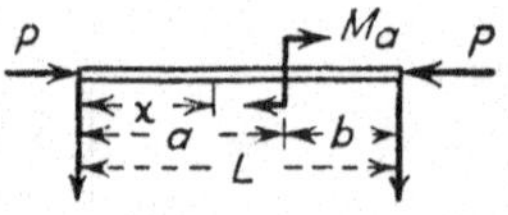 $x < a$	$-\dfrac{M_a \cos \dfrac{b}{j}}{\sin \dfrac{L}{j}}$	0
$x > a$	$-\dfrac{M_a \cos \dfrac{a}{j}}{\tan \dfrac{L}{j}}$	$M_a \cos \dfrac{a}{j}$

FIG. 18.15.

is obtained. Other loading conditions may be considered in the same manner. The values of the constants C_1 and C_2 are tabulated for common conditions of loading in Fig. 18.15. It will be observed that for the last two loading conditions shown in Fig. 18.15 the equation for bending moment must change at the point $x = a$. Two sets of values of C_1 and C_2 must be obtained, or a total of four conditions must be used. Two of these conditions are that the bending moments must be zero at the supports. The other two conditions are based on the conditions of shear and bending moment at the point of application of the load. For the concentrated load condition, the bending moment just to the left of the load must be equal to the bending moment to the right of the load, and the shear to the left of the load must be less than the shear to the right of the load by the amount of the load W. Similarly, for the couple loading, the shear does not change at the point of application of M_a, but the bending moment changes an amount M_a.

In obtaining the shear at any cross section of a beam column, it is necessary to consider the slope of the elastic curve. From the diagrams of Fig. 18.15, it is observed that the vertical reactions at the supports may always be obtained from the equations of statics and that they will be the same as for a simple beam with no axial compression, since the force P has no moment about the supports. Similarly, the shear on a

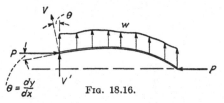

Fig. 18.16.

vertical cross section of the beam may be obtained by statics and will have a value V' which is the same as the shear in a simple beam with $P = 0$. In order to make use of the common beam relationship, $V = dM/dx$, it is necessary to consider the shear V to be normal to the elastic curve, as shown in Fig. 18.16. The components of P and V' normal to the beam may be obtained from Fig. 18.16.

$$V = V' \cos \theta - P \sin \theta \qquad (18.14)$$

From the geometry of small angles, $\cos \theta = 1$, and $\sin \theta = \theta = \tan \theta$. Equation 18.14 becomes

$$V = V' - P\theta \qquad (18.15)$$

where the second term is not negligible. If Eq. 18.7 is differentiated, it will yield an expression for V.

$$V = \frac{dM}{dx} = \frac{dM'}{dx} - P\frac{dy}{dx} \qquad (18.16)$$

The term dM'/dx is the derivative of the simple beam bending moment, and it is equal to the simple beam shear V'. The term dy/dx is equal to tan θ, or for small angles is equal to the angle in radians, or θ. Equation 18.15 is therefore obtained from Eq. 18.16 by making these substitutions. It is now apparent that the last term of Eq. 18.15 for the beam shear corresponds to the last term of Eq. 18.7 for the beam bending moment.

18.5. Superposition of Beam-column Loadings. The deflections and bending moments in beam columns are not proportional to the axial loads; therefore the combined effects of two or more axial loads cannot be obtained by superimposing the effects of the loads acting separately. If the axial load remains constant, however, the deflections and bending

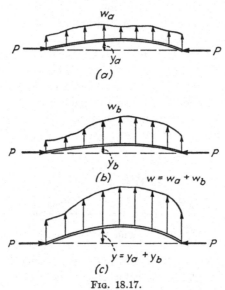

Fig. 18.17.

moments resulting from any system of lateral loads are proportional to the loads. The deflections or bending moments for two or more systems of lateral loads may therefore be superimposed, if the axial load is considered to act with each system. The procedure for superposition of loading conditions is shown in Fig. 18.17, in which the deflections and bending moments for the beam of Fig. 18.17(c) are obtained as the sum of those in Figs. 18.17(a) and (b). The axial load P must act for each condition.

The principle of superposition for beam columns may be proved by considering the differential equations for the beam deflections. For the beam of Fig. 18.17(a),

$$EI \frac{d^2y_a}{dx^2} = M_{a'} - Py_a \qquad (18.17)$$

where $M_{a'}$ represents the simple beam bending moment, or the bending moment for $P = 0$. Similarly, for the beam of Fig. 18.17(b),

$$EI \frac{d^2y_b}{dx^2} = M_{b'} - Py_b \qquad (18.18)$$

where $M_{b'}$ is the simple beam bending moment. Equations 18.17 and 18.18 are now added.

$$EI \frac{d^2}{dx^2} (y_a + y_b) = M_{a'} + M_{b'} - P(y_a + y_b) \qquad (18.19)$$

The sum of the simple beam moments $M_{a'}$ and $M_{b'}$ is equal to the simple beam moment M' for the total of the loads w_a and w_b. Equation 18.19 must therefore correspond to the differential equation for the beam of Fig. 18.17(c).

$$EI \frac{d^2y}{dx^2} = M' - Py \qquad (18.20)$$

The following terms are defined:

$$M' = M_{a'} + M_{b'} \qquad (18.21)$$
$$y = y_a + y_b \qquad (18.22)$$

Equation 18.22 was to be proved. This equation must obviously be satisfied if Eqs. 18.19 and 18.21 are to yield Eq. 18.20.

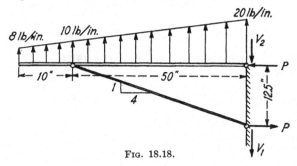

FIG. 18.18.

Example. The horizontal beam shown in Fig. 18.18 is constructed of a 1⅜- by 0.058-in. steel tube, which is braced by a tension member. Plot the bending-moment diagram, and determine the margin of safety for the tube.

Solution. The reactions are first calculated by the equations of statics.

$$\Sigma M_0 = 8 \times 60 \times 30 + 12 \times 60 \times \tfrac{1}{2} \times 20 - 12.5P = 0$$
$$P = 1,728 \text{ lb}$$
$$V_1 = \frac{P}{4} = 432 \text{ lb}$$
$$V_2 = 480 + 360 - 432 = 408 \text{ lb}$$

The bending moment at the wire attachment is

$$M_1 = 80 \times 5 + 10 \times 3.3 = 433 \text{ in-lb}$$

The section of the tube between supports is now considered as a beam column with a compression load of $P = 1,728$ lb, an end moment of $M_1 = 433$ in-lb, and a trapezoidal distributed load varying from 10 lb/in. at the left support to

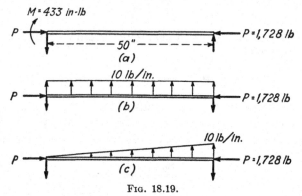

$M = 433$ in-lb

P ⟵ $P = 1,728$ lb

50"

(a)

10 lb/in.

P ⟵ $P = 1,728$ lb

(b)

10 lb/in.

P ⟵ $P = 1,728$ lb

(c)

FIG. 18.19.

20 lb/in. at the right support. The three cases shown in Fig. 18.19 may be analyzed by the equations shown for the first three cases of Fig. 18.15, and the results superimposed.

For the 1⅜- by 0.058-in. steel tube, $I = 0.0521$ in.⁴, $E = 29,000,000$ psi, and $P = 1,728$ lb.

$$j = \sqrt{\frac{EI}{P}} = 29.6 \text{ in.} \qquad \frac{L}{j} = \frac{50}{29.6} = 1.69 \text{ rad}$$

$$\sin\frac{L}{j} = 0.993 \qquad \cos\frac{L}{j} = -0.1184$$

The constants C_1 and C_2 are obtained from Fig. 18.15. For the end moment, $M_1 = 433$ and $M_2 = 0$:

$$C_1 = \frac{433 \times 0.1184}{0.993} = 51.7 \qquad C_2 = 433 \qquad wj^2 = 0$$

For the uniform load of $w_0 = 10$ lb/in.,

$$C_1 = \frac{10 \times (29.6)^2 \times -1.1184}{0.993} = -9,860 \qquad C_2 = -8,760 \qquad wj^2 = 8,760$$

For the triangular load of $w = 0.2x$,

$$C_1 = \frac{-8,760}{0.993} = -8,830 \qquad C_2 = 0 \qquad wj^2 = 175.2x$$

The final bending moment is obtained from Eq. 18.10 by superimposing the constants obtained above.

$$M = C_1 \sin\frac{x}{j} + C_2 \cos\frac{x}{j} + wj^2$$

$$M = -18,640 \sin\frac{x}{29.6} - 8,330 \cos\frac{x}{29.6} + 8,760 + 175.2x \qquad (18.23)$$

TABLE 18.4

1	x	0	10	20	27.0	30	40	50
2	M, in-lb (from Eq. 18.23)	430	−3,530	−5,910	−6,350	−6,190	−4,240	0
3	M', in-lb (for simple beam)	430	−2,450	−4,140	−4,505	−4,430	−3,114	0

Numerical values of the bending moment are calculated from Eq. 18.23 and are tabulated in line 2 of Table 18.4. For comparison, values of the bending moment M' in a simple beam with no compression load are tabulated in line 3. The curves for M and M' are plotted in Fig. 18.20.

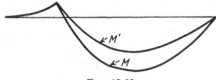

FIG. 18.20.

In designing a tube to resist the combined bending and compression, the method of stress ratios may be used if secondary bending stresses are considered. The following equation is given in *ANC*–5.

$$\text{MS} = \frac{1}{R_b + R_c} - 1 \tag{18.24}$$

where R_b and R_c are the stress ratios in bending and compression. For the $1\frac{3}{8}$- by 0.058-in. tube, $I/y = 0.0758$ in.^{3}, $A = 0.2400$ in.^{2}, $\rho = 0.4661$ in., and $D/t = 23.70$. The unit stresses are as follows:

$$f_b = \frac{My}{I} = \frac{6,350}{0.0758} = 83,800 \text{ psi}$$

$$f_c = \frac{P}{A} = \frac{1,728}{0.24} = 7,200 \text{ psi}$$

$$f = f_b + f_c = 91,000 \text{ psi}$$

For steel with an ultimate tensile strength of 100,000 psi, the values $F_b = 112,000$ psi and $F_c = 25,000$ psi are obtained from *ANC*–5. The yield stress is $F_{cy} = 85,000$ psi. The stress ratios are $R_b = f_b/F_b = 0.748$ and $R_c = f_c/F_c = 0.288$. From Eq. 18.24,

$$\text{MS} = \frac{1}{0.748 + 0.288} - 1 = -0.035$$

This tube is unsatisfactory. Even if the analysis indicated a small positive margin of safety, the tube would be questionable because the combined stress f exceeds the yield stress, and the true deflections and secondary bending moments are therefore larger than the values calculated for an elastic tube. The tube will be heat-treated to an ultimate tensile strength $F_{tu} = 125{,}000$ psi, with $F_{cy} = 100{,}000$ psi. The stress ratio in compression remains the same, since the member is a long column. From ANC-5, $F_b = 141{,}000$ psi, and $R_b = f_b/F_b = 0.594$. From Eq. 18.24,

$$MS = \frac{1}{0.594 + 0.288} - 1 = \underline{0.13}$$

The end fittings may be welded to the tube after heat-treatment, since the strength of the tube is questionable only at the cross sections of high bending moment.

18.6. Approximate Method for Beam Columns. In the design of a beam column it is necessary to assume a value of EI before the values of L/j can be calculated. If the bending-moment calculations show the assumed cross section to be unsatisfactory, a new section must be assumed, and the bending moments must be recalculated. It is obviously desirable to estimate the cross section required rather accurately before starting the bending-moment calculations in order to avoid numerous trials. The following approximate method of analysis enables the designer to select a cross section before applying the equations of Fig. 18.15.

If the bending moment in a beam column with a compressive load P is compared to the bending moment in a simple beam with no axial load, the following approximate relationship is obtained:

$$M = \frac{M'}{1 - (P/P_{cr})} \tag{18.25}$$

where M is the bending moment in the beam column, M' is the bending moment in a simple beam resisting the same loading, and P_{cr} is the Euler load, $\pi^2 EI/L^2$. The relationship is obtained from the first term in the Fourier series, $y = a \sin (\pi x/L)$, and is exact for beams which deflect to this curve.[3] The deflection curves for most beams and beam columns with no end moments approximate this sine curve; therefore the relationship is accurate for such cases.

For the beam column analyzed in the Example of Art. 18.5, the value of P/P_{cr} was 0.288. From Eq. 18.25,

$$M = \frac{M'}{1 - 0.288} = 1.404M'$$

The bending moments M' in line 3 of Table 18.4 will be used to obtain the approximate values of M in Table 18.5. The resulting values of M

TABLE 18.5

x	10	20	27	30	40
1	M'	$-2,450$	$-4,140$	$-4,505$	$-4,430$	$-3,114$
2	$M = 1.440M'$	$-3,440$	$-5,810$	$-6,330$	$-6,210$	$-4,370$
3	Error, per cent	-2.5	-1.7	-0.3	0.3	3.0

are very close to the exact values in line 2 of Table 18.4. Slide-rule calculations have been used for both values, and the slide-rule errors may be as large as the differences in the bending moments at the cross section of maximum bending moment.

18.7. Beams in Tension. An axial tension load on a beam has an effect which is similar to that of an axial compression load. The tension load reduces the deflections and bending moments from the values calculated for a simple beam, whereas the compression load increased them. The equations for beams in tension may be developed in a similar manner to those for beams in compression. The ultimate strength of a member in tension and bending is usually greater than the yield strength, and formulas obtained by assuming elastic deflections are usually not applicable in obtaining the ultimate strength. Beam columns usually fail at stresses which do not greatly exceed the yield stress; therefore the equations which were derived in the preceding articles are more useful than similar equations for members in tension.

The exact equations for elastic members in tension and bending may be obtained by changing the sign of P in Eqs. 18.7 and 18.9. The value of $1/j$ will be replaced by the term

$$\frac{1}{j'} = \sqrt{\frac{-P}{EI}} = \sqrt{-1}\sqrt{\frac{P}{EI}} = \frac{i}{j}$$

where i is the imaginary term $\sqrt{-1}$.

The trigonometric functions will be replaced by hyperbolic functions, from the relationships,

$$\sin \frac{ix}{j} = i \sinh \frac{x}{j} \quad \text{and} \quad \cos \frac{ix}{j} = \cosh \frac{x}{j}$$

The following equation will replace Eq. 18.10:

$$M = C_3 \sinh \frac{x}{j} + C_4 \cosh \frac{x}{j} - wj^2 \tag{18.26}$$

where C_3 and C_4 are constants which must be determined from the boundary conditions. The constants will be similar in form to those in

Fig. 18.15, but hyperbolic functions will replace the trigonometric functions and some of the signs will be changed.

An approximate equation for the bending moment in an elastic beam in tension is obtained by changing the sign of P in Eq. 18.25.

$$M = \frac{M'}{1 + (P/P_{cr})} \qquad (18.27)$$

This equation is accurate enough for most practical design problems in which the end moments are zero.

PROBLEMS

18.11. Check the values of C_1 and C_2 shown for the triangular loading in Fig. 18.15.

18.12. Check the values of C_1 and C_2 shown for the concentrated load in Fig. 18.15.

18.13. A beam column resists a uniformly distributed load of 20 lb/in. The values $P = 3,000$ lb, $L = 50$ in., $E = 29,000,000$ psi, and $I = 0.10$ in.[4] are assumed. Find the bending moments at $x = 12.5$ and $x = 25$, and compare with values obtained from Eq. 18.25.

18.14. Repeat Prob. 18.13 for a triangular load varying from 0 to 50 lb/in.

18.15. Design a steel-tube beam column to resist a uniformly distributed load of 15 lb/in. over a span of 50 in. and a compression load of 2,000 lb. First determine a trial size from Eq. 18.25, and then check the stresses by the equations of Fig. 18.15.

18.16. Design a steel-tube beam column to resist a load of 500 lb at the midpoint, with a span of 40 in. and a compression load of 2,000 lb.

18.8. Moment Distribution. In the classical methods of analysis for statically indeterminate structures, a base structure is first obtained by removing the redundant members or reactions. One deflection equation is then written for each redundant and the equations solved simultaneously. When structures have more than three redundants, the process of solving simultaneous equations becomes very tedious. The method of moment distribution permits the analysis of several types of highly redundant structures without the necessity of solving simultaneous equations.

The method of moment distribution was introduced in 1932 by Prof. Hardy Cross.[4, 5] It is applicable for the analysis of continuous beams and rigid frames. The structure is first considered as a series of single-span fixed-end beams. The moment restraints at the supports are successively removed one by one, and the bending moments are corrected by successive approximations until the conditions for the actual structure are obtained. Numerical values are used in the successive approximations, and it is not necessary to set up algebraic equations for the redundants.

The bending moments in single-span beams with fixed ends must first be calculated for the loading conditions of the structure. A single-span beam with clamped ends has two redundant reactions, and it may be analyzed by the methods of area moments, conjugate beam, or virtual work. The fixed-end moments for beams of uniform cross section are shown in Fig. 18.21 for various loadings. Similar values for other loadings are tabulated in most engineering handbooks.

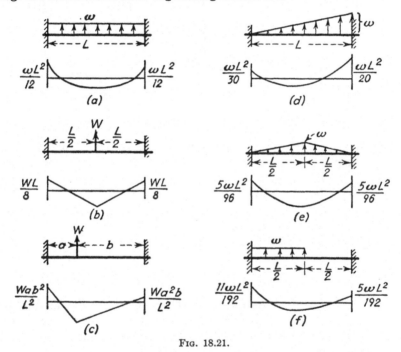

Fig. 18.21.

It is necessary to calculate the angle θ and the bending moment M_B for a known bending moment M_A acting on the beam shown in Fig. 18.22(a). The end B is fixed against rotation, and the end A is prevented from deflecting vertically from the tangent at B. From the second area-moment theorem, the moment of the moment area about point A must be zero. The moment area is shown in Fig. 18.22(b), but is more conveniently represented as shown in Fig. 18.22(c)

$$\frac{M_B L}{2} \cdot \frac{2L}{3EI} - \frac{M_A L}{2} \cdot \frac{L}{3EI} = 0$$

$$M_B = 0.5 M_A \qquad (18.28)$$

The ratio of M_B to M_A is termed the *carry-over factor* and is always 0.5 for a beam of uniform cross section, but it has other values if EI varies in the span.

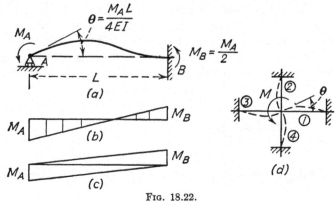

FIG. 18.22.

The angle θ of Fig. 18.22 represents the rotation of the tangent at point A from the tangent at point B and is obtained by the first theorem of area moments. The area of the M/EI diagram is equal to the angle θ.

$$\theta = \frac{M_A L}{2EI} - \frac{M_B L}{2EI} = \frac{M_A L}{4EI} \tag{18.29}$$

The bending moment M_A required to produce a unit angle θ is called the *stiffness factor* K for the member.

$$K = \frac{4EI}{L} \tag{18.30}$$

The stiffness factors are used in order to determine the distribution of a bending moment M to several members at a joint, when the joint is rotated as shown in Fig. 18.22(d). All members meeting at the joint are rotated through the same angle θ and resist bending moments in proportion to their stiffness factors K. The total moment M is resisted by all members.

$$M = K_1\theta + K_2\theta + K_3\theta + K_4\theta = \theta\Sigma K$$

where K_1, K_2, K_3, and K_4 are the stiffness factors for the members and have a sum ΣK. The bending moments M_1, M_2, M_3, and M_4 in the members are therefore found by substituting the value of $\theta = M/\Sigma K$ into Eqs. 18.29 and 18.30.

$$M_1 = \frac{K_1 M}{\Sigma K} \qquad M_2 = \frac{K_2 M}{\Sigma K} \qquad M_3 = \frac{K_3 M}{\Sigma K} \qquad M_4 = \frac{K_4 M}{\Sigma K}$$

The *distribution factor* for a member represents the proportion of the moment M at a joint which is resisted by the member and is defined by the following relationship:

$$\text{Distribution factor} = \frac{K}{\Sigma K}$$

The physical concepts involved in the procedure of moment distribution will be considered for the structure of Fig. 18.23(a). Each span is

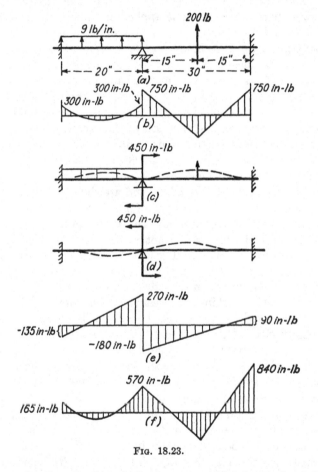

FIG. 18.23.

first assumed to have both ends fixed, or an artificial restraint is imposed at the center support which prevents the rotation of the beam. The fixed-end moments are obtained from Figs. 18.21(a) and (b) as 300 in-lb for the left-hand span and 750 in-lb for the right-hand span, as shown in Fig. 18.23(b). The unbalanced moment at the center support is the difference between the fixed-end moments in the adjacent spans, or 450 in-lb. Thus, an external clockwise couple of 450 in-lb must be acting as shown in Fig. 18.23(c), in addition to the real support restraints in order to produce the bending-moment diagram shown in Fig. 18.23(b). The true conditions of support restraints may now be obtained by superimposing the couple loading of Fig. 18.23(d) on the loading of Fig. 18.23(c).

The couple loading of Fig. 18.23(d) is equivalent to the loading of Fig. 18.22(a) in which a couple is applied at the simply supported end of each span and each span is fixed at the opposite end. The bending-moment diagram for the couple loading is equivalent to that of Fig. 18.22(b). The stiffness factor K for the right-hand member is only two-thirds of the stiffness factor for the left-hand member; therefore 40 per cent of the couple, or 180 in-lb, is resisted by the right-hand member and 60 per cent of the applied couple, or 270 in-lb, is resisted by the left-hand member. The bending moments at the fixed supports are one-half the bending moments at the center support. The bending-moment diagram for the loading of Fig. 18.23(d) is therefore as shown in Fig. 18.23(e). The final bending moments are obtained by superimposing those of Figs. 18.23(b) and (e) and are shown in Fig. 18.23(f).

The superposition of the loadings of Figs. 18.23(c) and (d) satisfies all the conditions of equilibrium and continuity in the structure. The tangents at the end supports do not rotate, and there is no external moment at the center support. The loads resisted correspond with those in the original beam. Most problems in moment distribution require that artificial restraints be provided at more than one support. The restraints consist of locking the beam against rotation. When each support is unlocked, an unbalanced moment is carried over to the adjacent supports. If these adjacent supports are not fixed ends, the carry-over moment results in an additional unbalance. The balancing process must then be repeated in successive cycles. The unbalanced moments become smaller with each cycle, and the balancing process is repeated until they are negligible.

The three-span beam shown in Fig. 18.24(a) is fixed at the end supports and simply supported at the two intermediate points. The structure is symmetrical about the center line, and the two left-hand spans are similar to those of Fig. 18.23(a). The value of EI is constant at all cross sections. The first step in an analysis by moment distribution is to lock the beam at the interior supports and compute the fixed-end moments. The fixed-end moments are 750 in-lb and 300 in-lb, as shown in Fig. 18.24(b). The artificial restraints necessary to prevent rotation at the interior supports are represented by couples of 450 in-lb.

The beam is now unlocked at joint *B* by applying the couple of 450 in-lb, as shown in Fig. 18.24(*c*), assuming joints *A* and *C* to remain fixed. Because of the relative stiffnesses, 60 per cent of this moment, or

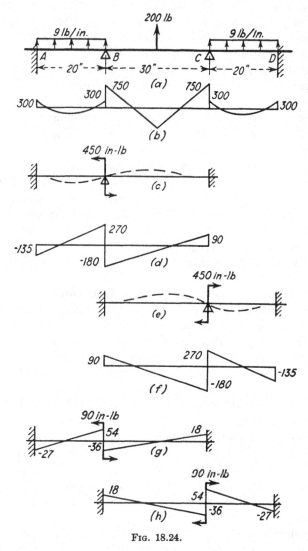

Fig. 18.24.

270 in-lb, goes to member *AB*, and 40 per cent, or −180 in-lb, goes to member *BC*. One-half of these moments at joint *B* is carried over to the other ends of the members, yielding bending moments of −135 in-lb at *A* and 90 in-lb at *C*, as shown in Fig. 18.24(*d*). Joint *C* is then un-

locked in the same manner as joint B, by applying the 450 in-lb couple as shown in Fig. 18.24(e), which produces the bending moments shown in Fig. 18.24(f).

The process of unlocking joint B introduced a carry-over moment of 90 in-lb at joint C, and the process of unlocking joint C introduced a carry-over moment of 90 in-lb at joint B. Thus, instead of completely removing the moment restraints at these joints, the beam was rotated through an angle θ and relocked. The angle θ was not sufficient to remove the total restraint. Additional cycles of balancing are necessary.

The second cycle of balancing consists of applying a 90 in-lb couple at point B, as shown in Fig. 18.24(g). The bending moments for this loading are one-fifth of those shown in Fig. 18.24(d). A similar couple is applied at joint C, yielding the bending moments shown in Fig. 18.24(h). This second cycle leaves unbalanced moments of 18 in-lb at B and C. A third cycle of distribution would yield bending moments of one-fifth the values shown in Figs. 18.24(g) and (h) and would leave unbalanced moments of 3.6 in-lb. These would be negligible, and a fourth cycle would not be required. The final bending-moment diagram may now be obtained by superimposing the diagrams of Figs. 18.24(b), (d), (f), (g), and (h).

18.9. Practical Procedure for Moment Distribution. The numerical work represented in Fig. 18.24 may be tabulated in a more compact form. The bending-moment diagrams and deflection curves are shown only for the purpose of representing the physical action involved. After some practice in the use of the method, a designer may visualize the deflections and may calculate the numerical values of the bending moments without drawing the individual deflection curves and moment diagrams. A systematic procedure of tabular computations is desirable.

The bending moments for the beam of Fig. 18.24 are calculated by the customary tabular method in Fig. 18.25. The sign convention used here is customary for horizontal beams, in which positive moments produce compression on the upper side of the beam. This sign convention was used by Professor Cross in his original paper on moment distribution. A line sketch is first made for the structure, showing the nature of the supports. The relative stiffness factors of the members, 0.60 and 0.40, are shown in boxes at the supports. The fixed-end moments are now shown in line 1, below the corresponding points on the members. The differences between these fixed-end moments represent unbalanced couples at the supports which must be distributed to the members in proportion to their stiffness factors. The balancing moments, shown in line 2, are computed as 60 per cent and 40 per cent of the unbalanced

450 in-lb couple. A horizontal line is drawn below line 2 to indicate that the total moments above the line are balanced at all joints.

The carry-over moments are obtained in line 3 of Fig. 18.25, and represent one-half of the balancing moments at the opposite ends of the members, as indicated by the arrows. The sum of the moments in the first three lines of Fig. 18.25 represents the superposition of the bending-moment diagrams of Figs. 18.24(*b*), (*d*), and (*f*). The second cycle of

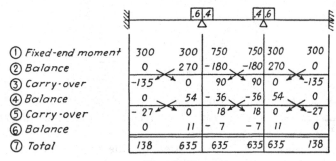

	.6	.4	.4	.6		
① *Fixed-end moment*	300	300	750	750	300	300
② *Balance*	0	270	-180	-180	270	0
③ *Carry-over*	-135	0	90	90	0	-135
④ *Balance*	0	54	-36	-36	54	0
⑤ *Carry-over*	-27	0	18	18	0	-27
⑥ *Balance*	0	11	-7	-7	11	0
⑦ *Total*	138	635	635	635	635	138

FIG. 18.25.

balancing is carried out in line 4 of Fig. 18.25. The unbalanced moments are multiplied by 0.60 and 0.40 and are recorded below the proper points of the structure. A horizontal line is again drawn to indicate that the moments above the line are in balance at all joints. One-half of these moments is carried over to opposite ends of the members in line 5, as indicated by the arrows. The moments in lines 4 and 5 of Fig. 18.25 correspond to the moments in Figs. 18.24(*g*) and (*h*).

If the first five lines of Fig. 18.25 are totaled at each point in the beam, the resulting moments will correspond with those obtained by superimposing the diagrams of Fig. 18.24. The bending moments at the interior supports will have a small unbalance. Since it is desirable for the approximate bending moments to be balanced, it is customary to terminate the distribution process at the completion of a balancing operation rather than at the completion of a carry-over operation. In line 6 of Fig. 18.25, the moments are balanced, but the resulting values are not carried over, because they are negligible. The algebraic sums of the moment terms in each column are shown in line 7.

The sign convention used in Figs. 18.23 to 18.25 permits the plotting of beam bending moments in the customary manner. It is unsatisfactory, however, for the analysis of rigid frames containing vertical members. The application of the method of moment distribution is simplified if certain rules can be set up and followed automatically. A sign convention which is applicable for all types of structures will therefore be adopted, and it will be used for horizontal beams as well as rigid frames. In moment distribution problems, the end moments will be considered positive when they represent a clockwise couple acting on the joint, or a counterclockwise couple when acting on the member, as shown in Fig. 18.26. This sign convention applies only to the end moments

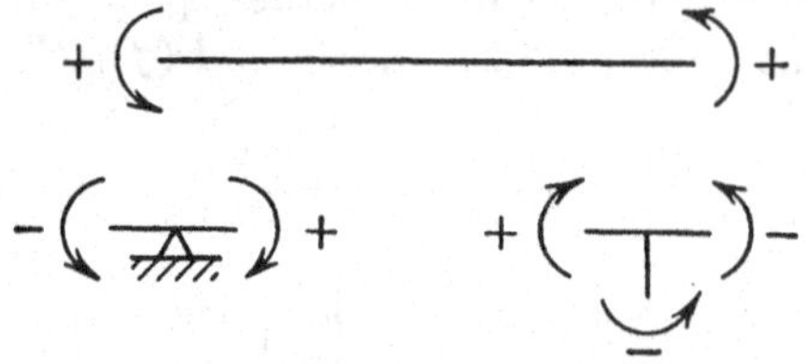

FIG. 18.26.

and not to the bending moments along the member, or it applies to moments as tabulated in Fig. 18.25 but not to moment diagrams as plotted in Figs. 18.23 and 18.24.

The following rules will be stated for the process of moment distribution.

1. *Compute the fixed-end moments for each loaded span, and record them with the correct signs at their proper locations on a diagram of the structure. Positive moments tend to rotate the member counterclockwise.*

2. *Compute the distribution factors as $K/(\Sigma K)$ for the members at each joint, and record them on the members at each joint.*

3. *Balance the moments at each joint by multiplying the unbalanced moment by the distribution factor for each member, changing the sign, and recording the balancing moment below each fixed-end moment. The unbalanced moment is the algebraic sum of the fixed-end moments at a joint. At a support fixed against rotation, the unbalanced moment is resisted by the support and the balancing moment is recorded as zero, which means that the joint is not unlocked in the balancing process.*

4. *Draw a horizontal line below the balancing moments. The algebraic sum at any joint of all moments above the horizontal line must be zero.*

5. *Record the carry-over moments at the opposite ends of the members, indicating the carry-over process by diagonal arrows. The carry-over moments have the same sign as the corresponding balancing moments and have one-half of their magnitudes.*

6. *Repeat the process for the balance and carry-over of moments for as many cycles as desired. The unbalanced moment for each cycle will be the algebraic sum of the moments at the joint recorded below the last horizontal line.*

7. *Obtain the final moment at the end of each member as the algebraic sum*

of all moments tabulated at the point. The total of the final moments for all members at any joint must be zero.

The calculations of Fig. 18.25 are repeated in Fig. 18.27, using the adopted sign convention. The structure is that of Fig. 18.24(*a*). After

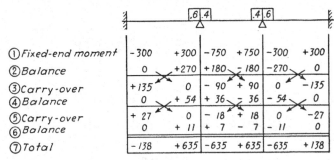

		.6	.4		.4	.6	
① Fixed-end moment	-300	+300	-750	+750	-300	+300
② Balance	0	+270	+180	- 180	-270	0
③ Carry-over	+135	0	- 90	+ 90	0	-135
④ Balance	0	+ 54	+ 36	- 36	- 54	0
⑤ Carry-over	+ 27	0	- 18	+ 18	0	-27
⑥ Balance	0	+ 11	+ 7	- 7	- 11	0
⑦ Total	- 138	+635	-635	+ 635	-635	+ 138

Fig. 18.27.

the signs of the fixed-end moments are determined by inspection, the signs for the remaining terms are recorded automatically from rules 3 and 5 above.

The structure shown in Fig. 18.28 will be analyzed by the method of moment distribution. The line diagram of Fig. 18.29 shows the nature of the supports and also shows the distribution factors recorded on the proper members. Joints A and B are successively locked and unlocked, and the unbalanced moments will be distributed at these joints. If the unloaded structure is rotated at joint A, the cantilever end will resist no

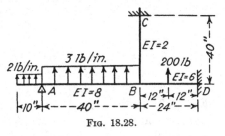

Fig. 18.28.

moment, and the member AB will resist the entire moment applied at A; therefore the distribution factors are 0.0 and 1.0 at joint A, as shown. The stiffness factors K will be computed by Eq. 18.30 for the members meeting at joint B.

Member AB: $\quad K = \dfrac{4EI}{L} = \dfrac{4 \times 8}{40} = 0.8$

Member BC: $\quad K = \dfrac{4EI}{L} = \dfrac{4 \times 2}{40} = 0.2$

Member BD: $\quad K = \dfrac{4EI}{L} = \dfrac{4 \times 6}{24} = \underline{1.0}$

$$\Sigma K = 2.00$$

If the unloaded structure is rotated at joint B with joint A locked, the unbalanced moment will be distributed to the members in proportion to the stiffness factors K. The total unbalanced moment is therefore multiplied by the value of $K/(\Sigma K)$ for each member, or by the distribution factors of 0.4, 0.1, and 0.5 for members AB, BC, and BD, respectively.

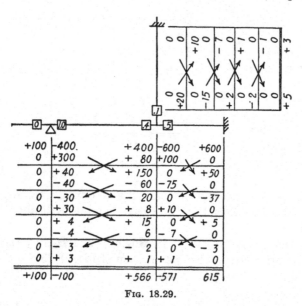

Fɪɢ. 18.29.

The fixed-end moment for the overhanging end is 100 in-lb, and it represents the true bending moment at A. The fixed-end moments are -400 in-lb and 400 in-lb for member AB and are obtained from Fig. 18.21, with the sign convention shown in Fig. 18.26. The fixed-end moments of -600 in-lb and 600 in-lb for member BD are obtained in a similar manner. Member BC is unloaded and has no fixed-end moment.

The fixed-end moments are now distributed by the procedure previously outlined. The unbalanced moment at each joint is obtained as the algebraic sum of moments in all members meeting at the joint. This moment is balanced by multiplying by the distribution factor for each member and changing the sign. The moments are carried over as indicated by the arrows and the carry-over factors are plus one-half in all cases. The procedure is repeated until the corrections are negligible. The calculations are always terminated after a balancing operation in order to facilitate checking. The total moments in all members at a joint must add up to zero.

18.10. Joint Displacement. In the preceding discussion of moment distribution, it was assumed that the joints were free to rotate but that they had no motion in translation. It is frequently necessary to determine the effects of a support deflection, or to analyze a rigid frame in which the joints are free to deflect in translation. The effects of the joint translation may be considered in obtaining the fixed-end moments.

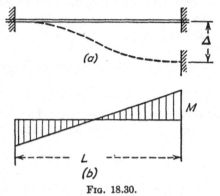

If one end of a uniform member is displaced a distance Δ, as shown in Fig. 18.30(a), and neither end is permitted to rotate, the bending-moment diagram will be as shown in Fig. 18.30(b). The displacement may be obtained from the second theorem of area moments by taking moments of the moment area about one end of the span.

Fig. 18.30.

$$\Delta = \frac{ML}{4EI} \times \frac{5L}{6} - \frac{ML}{4EI} \times \frac{L}{6} = \frac{ML^2}{6EI} \qquad (18.31)$$

Thus, for known values of Δ, L, and EI, a fixed-end moment may be calculated from Eq. 18.31. This moment may be distributed in the same manner as the fixed-end moments resulting from other types of loading.

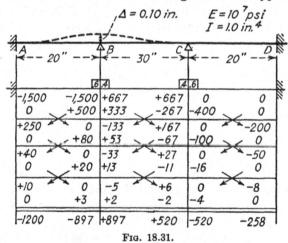

Fig. 18.31.

The method of obtaining the bending moments in a horizontal beam in which a support is displaced will be illustrated by a numerical example. The beam shown in Fig. 18.31 has a constant cross section with

$E = 10^7$ psi and $I = 1.0$ in.[4] The support B is displaced upward a distance of $\Delta = 0.01$ in. The fixed-end moments in members AB and BC will first be calculated for a condition in which joints B and C are locked against rotation but joint B is displaced relative to the tangents at A and C. From Eq. 18.31, for member AB,

$$M = \frac{6EI\Delta}{L^2} = \frac{6 \times 10^7 \times 0.01}{(20)^2} = 1{,}500 \text{ in-lb}$$

This fixed-end moment is clockwise on both ends of the member and is recorded as negative at A and B. For member BC the fixed-end moment is positive at both ends.

$$M = \frac{6EI\Delta}{L^2} = \frac{6 \times 10^7 \times 0.01}{(30)^2} = 667 \text{ in-lb}$$

Both joints B and C rotate as a result of the support deflection. The rotation is now considered by a conventional moment distribution process, and the final end moments are obtained in Fig. 18.31.

For loaded beams in which the supports are displaced, the effects of the applied loads and the support displacement may be superimposed in either of two ways. The combined fixed-end moments due to applied loads and support displacement may be first obtained, and these combined moments may be balanced by one process of moment distribution. In some cases, it is desirable to separate the two effects, and the support displacement moments may be distributed as in Fig. 18.31 and the applied load moments distributed as in Figs. 18.24 and 18.27 and then the results of the two distributions superimposed.

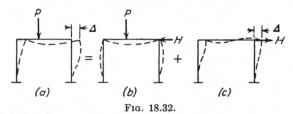

Fig. 18.32.

The analysis of rigid frames in which the joints are displaced requires two separate processes of moment distribution. For the rigid frame shown in Fig. 18.32, the two upper joints move sidewise for an unsymmetrical loading. If the sidesway is prevented by a horizontal force H, as shown in Fig. 18.32(b), the bending moments resulting from the loads P and H may be obtained by the conventional moment distribution analysis. The force H required to prevent the sidesway is then determined from the equations of statics. An equal and opposite force H is then applied as shown in Fig. 18.32(c), and the structure is analyzed for this force. The final bending moments for the structure of Fig. 18.32(a) are now obtained by superimposing the bending moments for Figs. 18.32(b) and (c).

The analysis of a rigid frame for the load H, shown in Fig. 18.32(c), will be explained by means of a numerical example. If the structure is displaced a known distance Δ, the fixed-end moments could be obtained from Eq. 18.31. Usually, however, the force H is known but Δ is unknown. It is necessary to assume an arbitrary value of Δ and to balance the fixed-end moments. The value of H is then determined for the resulting moments. The moments for any other value of H are then found by direct proportion.

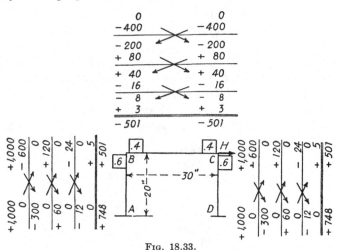

FIG. 18.33.

For the frame of Fig. 18.33 in which EI is constant, joints B and C are assumed to be displaced to the right with no rotation. The fixed-end moments in member BC are therefore zero, and the displacement Δ is assumed to be sufficient to produce fixed-end moments of 1,000 in-lb in members AB and CD, as shown. The numerical value of Δ could now be computed from Eq. 18.31, but is not required. The fixed-end moments of 1,000 in-lb are balanced in Fig. 18.33 to yield final moments of 748 in-lb at A and D and 501 in-lb at B and C. Each member of the frame is now considered as a free body, as shown in Fig. 18.34. The shearing forces at the ends of the members are found from the equilibrium of moments for the members as free bodies. The axial loads for the individual members are not shown, but they can be obtained from the shearing forces of the adjacent members. The total external forces at the supports A and D are shown for the entire structure as a free body. The value of H is found to be 125 lb from the equilibrium of horizontal forces on the entire structure. The bending moments for any other value of H may now be obtained by direct proportion.

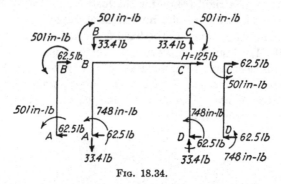

Fig. 18.34.

The structure will be analyzed for an unsymmetrical vertical load P by the steps shown in Fig. 18.32. A load of 2,000 lb, acting 10 in. from B, will be first considered as acting on a structure which is restrained from horizontal motion at point C, as shown in Fig. 18.35. The fixed-end

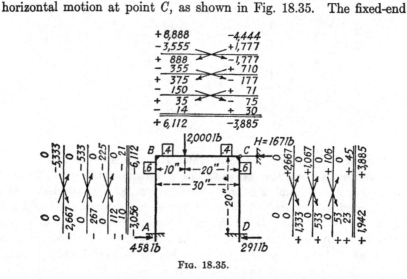

Fig. 18.35.

moments in members AB and CD are zero, since joints B and C are restrained against both translation and rotation. The fixed-end moments in member BC are obtained from Fig. 18.21(c) as 8,888 in-lb at B, and $-4,444$ in-lb at C. These moments are distributed in the customary manner in Fig. 18.35. After the end moments are obtained, other forces may be found by applying the equations of statics to each member as a free body, as was done in Fig. 18.34.

For the structure restrained against sidesway, the horizontal reactions are 458 lb to the right at A and 291 lb to the left at D. The force at C required to prevent sidesway is obtained from the horizontal equilibrium for the entire structure, as $H = 458 - 291 = 167$ lb, acting to the left. These forces are shown in Fig. 18.35.

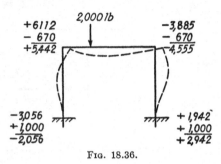

FIG. 18.36.

The bending moments for a force $H = 167$ lb, applied as shown in Fig. 18.32(c), are now obtained by multiplying the moments shown in Figs. 18.33 or 18.34 by the ratio of $^{167}/_{125}$. These moments are now superimposed on the moments found in Fig. 18.35. The superposition is shown in Fig. 18.36. The total moments represent the end moments in a structure which is not restrained against sidesway.

18.11. Moment Distribution for Axially Loaded Members. Many continuous beams and frames contain slender members in tension or compression. The beam-column formulas which were previously discussed applied only to single spans, although the case of a single span resisting end moments was considered. Consequently, if the end moments for any span are obtained by moment distribution, the bending moment for any point in the span may be obtained from the equations of Fig. 18.15.

The process of moment distribution consists of the superposition of the bending moments for fixed-ended spans and the bending moments due to successive rotation of the joints. It has been shown in Art. 18.5 that the principle of superposition applies for beam columns, provided that the axial load remains the same for all the transverse loading effects. The bending moments and deflections due to end moments or end rotations may also be superimposed, if the axial load is acting when all of the moments are applied.

The fixed-end moments for a beam in compression are greater than those for a similar beam with no axial load. For a member in tension, the fixed-end moments are smaller than for a beam with no axial loads. The fixed-end moments for axially loaded beams may be obtained from the differential equations of the deflection curves by equating the slopes to zero at the ends of the beam. Such calculations have been made by B. W. James, in NACA Technical Note 534,[6] for various loadings.

Figures 18.37 and 18.38 are taken from this paper, and similar information is available for various other loadings.   In Fig. 18.37, the fixed-end moments for a uniformly loaded beam are given in the form

$$M = \frac{wL^2}{C_1}$$

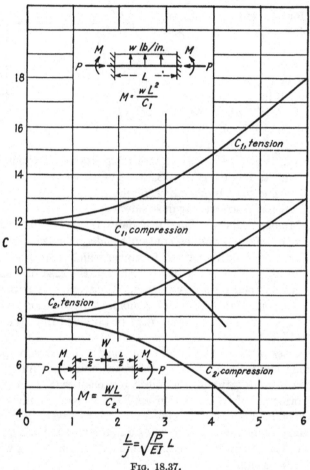

Fig. 18.37.

where C_1 is a function of the axial load and is plotted against the term L/j which was used in Art. 18.4.

$$\frac{L}{j} = \sqrt{\frac{P}{EI}}\, L$$

For no axial load, $C_1 = 12$, and for increasing tension loads the value of C_1 increases, or the moment decreases.   For increasing compression loads, the value of C_1 decreases, which means that the fixed-end moment increases.

The fixed-end moments for an axially loaded member resisting a concentrated transverse load W are also shown in Fig. 18.37, in the following form:

$$M = \frac{WL}{C_2}$$

The coefficient C_2 is also plotted as a function of L/j.

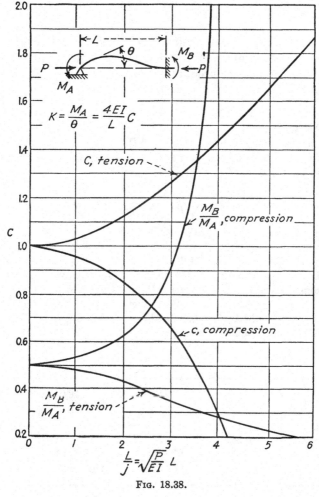

Fig. 18.38.

The carry-over factors and stiffness factors for axially loaded members are shown in Fig. 18.38. They are also obtained from the differential equations of the deflection curve for a beam column loaded with counterclockwise end moments by equating the slope at the right end to zero. The ratio of the end moments, M_B/M_A, is 0.5 for no axial load, but increases with an increase in axial compression and decreases with an increase in axial tension. This ratio is, of course, the carry-over factor for the span.

The stiffness factor for an axially loaded beam is determined as C times the stiffness factor for a beam with no axial load.

$$K = \frac{M_A}{\theta} = \frac{4EI}{L} C$$

The values of C are plotted against L/j in Fig. 18.38. After obtaining the fixed-end moments, stiffness factors, and carry-over factors for axially loaded beams, the procedure for moment distribution is identical with that for beams with no axial loads.

Example. The beam of Fig. 18.24(a) resists an axial compression load of $P = 10,000$ lb and has a constant cross section with $EI = 10^6$ lb-in.^{2} Find the bending moments at points A, B, C, and D.

Solution. The fixed-end moments, carry-over factors, and stiffness factors depend on the value of L/j. From Eq. 18.11,

$$j = \sqrt{\frac{EI}{P}} = \sqrt{\frac{10^6}{10^4}} = 10$$

For spans AB and CD, $L/j = 2.0$, and for span BC, $L/j = 3.0$. The fixed-end moments are obtained from Fig. 18.37. For spans AB and CD, $C_1 = 11.2$ and $M = wL^2/C_1 = 9 \times (20)^2/11.2 = 321$ in-lb. For span BC, $C_2 = 6.9$ and $M = WL/C_2 = 200 \times 30/6.9 = 875$ in-lb. The carry-over factors are obtained from Fig. 18.38. For spans AB and CD, $M_B/M_A = 0.62$, and for span BC, $M_B/M_A = 0.91$. The stiffness factors are obtained from Fig. 18.38. For members AB and CD, $C = 0.86$, and for member BC, $C = 0.66$. The distribution factors are tabulated below.

TABLE 18.6

Member	C	$\dfrac{4EI}{L}$	$K = \dfrac{4EI}{L} C$	$\dfrac{K}{\Sigma K}$
AB or CD	0.86	2×10^5	1.72×10^5	0.66
BC	0.66	1.33×10^5	0.88×10^5	0.34
ΣK			2.60×10^5	

The numerical work of the moment distribution is carried out in Fig. 16.39. The carry-over factors are shown in circles at the centers of the members. In previous problems these were not shown, since they were one-half for all members. The final bending moments are obtained as the totals for the columns in Fig. 18.39, and may be compared with those shown in Fig. 18.27 for the same beam with no axial load. The effect of the axial load is to increase the deflections in the center span, with a corresponding increase in the bending moments at points B and C. The larger deflection of the center span reduces the deflections

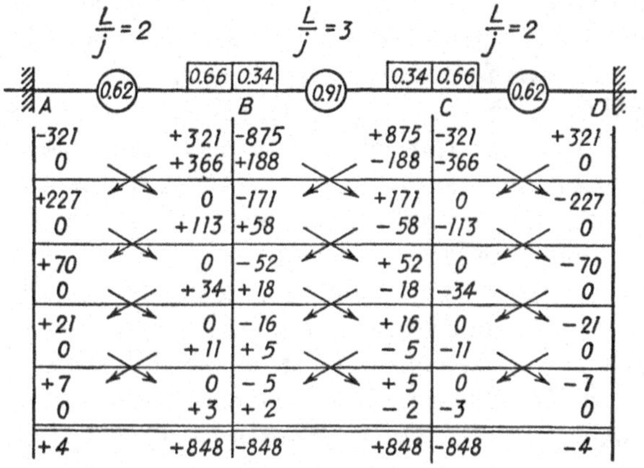

$$\frac{L}{j}=2 \qquad \frac{L}{j}=3 \qquad \frac{L}{j}=2$$

	(0.62)	0.66	0.34	(0.91)	0.34	0.66	(0.62)	
A		B			C			D
-321		+321	-875		+875	-321		+321
0		+366	+188		-188	-366		0
+227		0	-171		+171	0		-227
0		+113	+58		-58	-113		0
+70		0	-52		+52	0		-70
0		+34	+18		-18	-34		0
+21		0	-16		+16	0		-21
0		+11	+5		-5	-11		0
+7		0	-5		+5	0		-7
0		+3	+2		-2	-3		0
+4		+848	-848		+848	-848		-4

Fig. 18.39.

of the end spans and reduces the bending moments at points A and D. If the concentrated lateral load in the center span were larger, the bending moments at A and D would change signs. If the bending-moment curves in the spans are desired, they may be obtained from the equations of Fig. 18.15 by considering each span separately and by substituting the end moments obtained in Fig. 18.39 for M_1 and M_2 of the beam-column equations.

18.12. Other Applications of Moment Distribution. Since the introduction of the method of moment distribution by Professor Cross in 1932, numerous applications and techniques have been developed. In addition to the applications in the analysis of beams and frames, the method of successive approximations for numerical solutions has been used for shear lag analysis, for finding deflections of plates, for calculating vibration frequencies, and for the flow of water in networks of pipes. The general procedure represents one of the most powerful methods of engineering analysis. The discussion here will be confined to problems of beam and frame analysis, however.

Only the fundamental idea of the moment distribution method has been presented. A designer who has frequent use for the method should consult other books which have a more extensive treatment of the subject.[1, 2] One of the most useful short cuts is that for treating a simply supported or overhanging end of a beam. Instead of first assuming the support fixed, as was done at support A of Fig. 18.28, it may be assumed simply supported at the start. Only joint B would be unlocked, and the correct moments would be obtained after the first cycle of distribution.

Beams in which the cross sections vary in each span may also be analyzed readily by moment distribution. Numerous curves have been published for fixed-end moments, stiffness factors, and carry-over factors for spans with variable moment of inertia and various loadings.[7] The procedure for using these curves is similar to the procedure for using Figs. 18.37 and 18.38 for axially loaded beams.

In the rigid frame shown in Fig. 18.32, only one restraint was required to fix all joints against translation. In many rigid frame structures, several artificial restraints are required. Such structures are more difficult to analyze, and several special methods of analysis have been developed and published.[8, 9]

The method of moment distribution is frequently used in the calculation of secondary bending stresses for members of a simple truss.[1, 9] In the analysis of trusses, the members are assumed to be pin-connected at the joints although in most actual structures the members are rigidly riveted or welded at the joints. Thus a moment applied to a truss member by a transverse load or by an eccentricity at a joint is distributed to all the other truss members because of the rotation of the joints. The bending moments in the members may be obtained by the conventional procedure of moment distribution. The axial deformations of the members of a rigidly welded truss also cause some bending of the members. The analysis for such bending moments requires a special procedure for obtaining the fixed-end moments.

The procedure of moment distribution may be used to determine the condition of elastic stability of a structure.[10, 11] For example, the moment distribution for the continuous beam in compression of Fig. 18.39 converged more slowly than the distribution of Fig. 18.27 for the same beam with no axial load. If a somewhat larger compressive load were applied to the beam, the moment distribution calculations would not converge, but the unbalanced moments would become larger after each cycle of the distribution. This corresponds to the case where the compression load is greater than the load at which elastic buckling occurs. For example, a continuous beam with several equal spans, and with the ends simply supported, would buckle when the compression load reached the Euler load, $\pi^2 EI/L^2$, or when L/j reached the corresponding value of π. If such a beam were analyzed by moment distribution, the carry-over factor would be 1.00 for each span, and the distribution would not converge. In trusses which have members in both tension and compression, the general condition of stability may be determined by the convergence of the moment distribution process.

PROBLEMS

18.17. Analyze the structure of Fig. 18.24(*a*) if the distributed loads are changed to 18 lb/in. in each side span.

18.18. Analyze the structure of Fig. 18.28 if the stiffness properties of member *BC* are changed so that *EI* = 10.

18.19. Analyze the rigid frame shown by the method of moment distribution. Use $P_2 = 0$, $P_1 = 100$ lb, and $w = 2$ lb/in. Note that there is no sidesway because of symmetry.

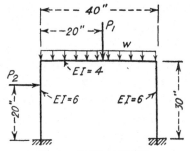

18.20. Analyze the rigid frame shown in Fig. 17.12(*a*) by the method of moment distribution. The value of *EI* is constant.

18.21. Analyze the rigid frame shown by the method of moment distribution. Use $P_1 = 100$ lb, $P_2 = 200$ lb, and $w = 0$.

PROBS. 18.19 and 18.21.

18.22. Analyze the rigid frame shown in Fig. 17.6(*a*) by the method of moment distribution.

18.23. Repeat the Example problem of Art. 18.11 for a tension load *P* of 10,000 lb and for compression loads *P* of 4,000 lb, 8,000 lb, and 12,000 lb.

REFERENCES FOR CHAPTER 18

1. BRUHN, E. F.: "Analysis and Design of Airplane Structures," Tri-State Offset Co., Cincinnati, 1943.
2. GRINTER, L. E.: "Theory of Modern Steel Structures," Vol. 2, The Macmillan Company, New York, 1937.
3. TIMOSHENKO, S.: "Strength of Materials," Part II, D. Van Nostrand Company, Inc., New York, 1930.
4. CROSS, H.: Analysis of Continuous Frames by Distributing Fixed-end Moments, *Trans. ASCE*, 1932.
5. CROSS, H., and N. D. MORGAN: "Continuous Frames of Reinforced Concrete," John Wiley & Sons, Inc., New York, 1932.
6. JAMES, B. W.: Principal Effects of Axial Load on Moment Distribution Analysis of Rigid Structures, *NACA TN 534*, 1935.
7. EVANS, L. T.: "Rigid Frames," Edwards Bros., Inc., Ann Arbor, Mich., 1936.
8. MAUGH, L. C.: "Statically Indeterminate Structures," John Wiley & Sons, Inc., New York, 1946.
9. NILES, A. S., and J. S. NEWELL: "Airplane Structures," Vol. II, John Wiley & Sons, Inc., New York, 1943.
10. LUNDQUIST, E. E.: A Method for Estimating the Critical Buckling Load for Structural Members, *NACA TN 717*, 1939.
11. HOFF, N. J.: Stable and Unstable Equilibrium of Plane Frameworks, *J. Aeronaut. Sci.*, January, 1941.

APPENDIX

TABLE 1. PROPERTIES OF ROUND TUBING

Diam.	Gage	A	ρ	I	I/Y	D/t	Weight, lb/100 in.	
							Steel	Dural
¼	0.022	0.01576	0.0810	0.000103	0.000825	11.38	0.45	0.16
	0.028	0.01953	0.0791	0.000122	0.000978	8.93	0.55	0.20
⅜	0.028	0.03053	0.1231	0.000462	0.002466	13.39	0.86*	0.31
	0.035	0.03739	0.1208	0.000546	0.002912	10.72	1.06	0.38
	0.049	0.05018	0.1166	0.000682	0.003636	7.65	1.43	0.51
½	0.028	0.04152	0.1672	0.001160	0.004641	17.85	1.17	0.42
	0.035	0.05113	0.1649	0.001390	0.005559	14.28	1.45*	0.52*
	0.049	0.06943	0.1604	0.001786	0.007144	10.20	1.96	0.70
⅝	0.028	0.05252	0.2113	0.002345	0.007503	22.30	1.49	0.54
	0.035	0.06487	0.2090	0.002833	0.009065	17.85	1.84*	0.66*
	0.049	0.08867	0.2044	0.003704	0.011852	12.77	2.51	0.90
	0.058	0.10331	0.2016	0.004195	0.013425	10.79	2.93	1.05
¾	0.028	0.06351	0.2555	0.004145	0.011052	26.80	1.80	0.65
	0.035	0.07862	0.2531	0.005036	0.013429	21.42	2.23*	0.80*
	0.049	0.10791	0.2485	0.006661	0.017762	15.30	3.06	1.09
	0.058	0.12609	0.2455	0.007601	0.02027	12.94	3.57	1.28
	0.065	0.13988	0.2433	0.008278	0.02208	11.53	3.96	1.42
⅞	0.028	0.07451	0.2996	0.006689	0.015289	31.23	2.11	0.76
	0.035	0.09236	0.2973	0.008161	0.018653	25.00	2.62*	0.94*
	0.049	0.12715	0.2925	0.010882	0.02487	17.85	3.60	1.29
	0.058	0.14887	0.2896	0.012484	0.02853	15.10	4.22	1.51
	0.065	0.16541	0.2865	0.013653	0.03121	13.47	4.66	1.68
1	0.035	0.10611	0.3414	0.012368	0.02474	28.56	3.01*	1.07*
	0.049	0.14640	0.3367	0.016594	0.03319	20.40	4.15	1.48
	0.058	0.17164	0.3337	0.019111	0.03822	17.25	4.86	1.74
	0.065	0.19093	0.3314	0.020970	0.04193	15.38	5.41	1.93
1⅛	0.035	0.11985	0.3856	0.01782	0.03168	32.10	3.40*	1.21
	0.049	0.16564	0.3808	0.02402	0.04270	22.95	4.68*	1.68
	0.058	0.19442	0.3780	0.02775	0.04933	19.40	5.51	1.97
	0.065	0.21650	0.3755	0.03052	0.05425	17.30	6.14	2.20
1¼	0.035	0.13360	0.4297	0.02467	0.03948	35.70	3.78*	1.35*
	0.049	0.18488	0.4250	0.03339	0.05342	25.50	5.23*	1.87*
	0.058	0.2172	0.4219	0.03867	0.06187	21.55	6.15	2.20
	0.065	0.2420	0.4196	0.04260	0.06816	19.22	6.86	2.45
1⅜	0.035	0.1473	0.4739	0.03309	0.04814	39.25	4.17	1.49
	0.049	0.2041	0.4691	0.04492	0.06534	28.05	5.78*	2.07
	0.058	0.2400	0.4661	0.05213	0.07583	23.70	6.80	2.43
	0.065	0.2675	0.4638	0.05753	0.08367	21.15	7.58	2.70
1½	0.035	0.1611	0.5181	0.04324	0.05765	42.80	4.56	1.63
	0.049	0.2234	0.5132	0.05885	0.07847	30.60	6.32*	2.20*
	0.058	0.2628	0.5102	0.06841	0.09121	25.85	7.45	2.66
	0.065	0.2930	0.5079	0.07558	0.10079	23.05	8.30	2.97
	0.083	0.3695	0.5018	0.09305	0.12407	18.08	10.47	3.74
1⅝	0.035	0.1748	0.5622	0.05528	0.06803	46.40	4.95	1.77
	0.049	0.2426	0.5575	0.07540	0.09279	33.15	6.87*	2.46
	0.058	0.2855	0.5544	0.08776	0.10801	28.00	8.09	2.89
	0.065	0.3186	0.5520	0.09707	0.11948	25.00	9.05	3.23
	0.083	0.4021	0.5459	0.11985	0.14751	19.58	11.40	4.06
1¾	0.035	0.1885	0.6065	0.06936	0.07927	50.00	5.32	1.91
	0.049	0.2618	0.6017	0.09478	0.10832	35.70	7.42*	2.65*
	0.058	0.3083	0.5986	0.11046	0.12624	30.20	8.73*	3.12
	0.065	0.3441	0.5962	0.12230	0.13977	26.90	9.75	3.48
	0.083	0.4347	0.5901	0.15136	0.17299	21.10	12.32	4.40

* AN Standard Tubing

TABLE 1. PROPERTIES OF ROUND TUBING (*Continued*)

Diam.	Gage	A	ρ	I	I/Y	D/t	Weight, lb/100 in.	
							Steel	Dural
1⅞	0.035	0.2023	0.6507	0.08565	0.09136	53.60	5.73	2.04
	0.049	0.2811	0.6458	0.11720	0.12500	38.25	7.95	2.84
	0.058	0.3311	0.6427	0.13677	0.14589	32.30	9.38	3.35
	0.065	0.3696	0.6404	0.15156	0.16166	28.80	10.47	3.74
	0.083	0.4673	0.6342	0.18797	0.20050	22.60	13.25	4.73
2	0.049	0.3003	0.6900	0.14299	0.14299	40.80	8.50	3.04
	0.058	0.3539	0.6869	0.16696	0.16696	34.45	10.03*	3.58*
	0.065	0.3951	0.6845	0.18514	0.18514	30.75	11.19	4.00
	0.083	0.4999	0.6783	0.2300	0.2301	24.10	14.16	5.06
	0.095	0.5685	0.6744	0.2586	0.2586	21.05	16.11	5.76
2¼	0.049	0.3388	0.7783	0.2052	0.1824	45.90	9.59	3.43
	0.058	0.3994	0.7753	0.2401	0.2134	38.80	11.30*	4.05*
	0.065	0.4462	0.7728	0.2665	0.2369	34.60	12.64	4.52
	0.083	0.5651	0.7667	0.3322	0.2953	27.15	16.01	5.72
	0.095	0.6432	0.7626	0.3741	0.3325	23.70	18.22	6.51
2½	0.049	0.3773	0.8667	0.2834	0.2267	51.00	10.68	3.82
	0.058	0.4450	0.8635	0.3318	0.2655	43.10	12.60	4.50
	0.065	0.4972	0.8613	0.3688	0.2950	38.45	14.09*	5.03*
	0.083	0.6302	0.8550	0.4607	0.3686	30.10	17.85	6.38
	0.095	0.7178	0.8509	0.5197	0.4158	26.30	20.34	7.27
2¾	0.049	0.4158	0.9551	0.3793	0.2759	56.10	11.78	4.20
	0.058	0.4905	0.9521	0.4446	0.3233	47.40	13.90	4.96
	0.065	0.5483	0.9496	0.4944	0.3596	42.30	15.50*	5.55*
	0.083	0.6954	0.9434	0.6189	0.4501	33.15	19.70	7.04
	0.095	0.7924	0.9393	0.6991	0.5084	28.95	22.48	8.03
3	0.058	0.5361	1.0403	0.5802	0.3868	51.70	15.18	5.42
	0.065	0.5993	1.0380	0.6457	0.4305	46.20	16.95	6.06
	0.083	0.7606	1.0318	0.8097	0.5398	36.15	21.55*	7.70*
	0.095	0.8670	1.0276	0.9156	0.6104	31.58	24.56	8.78
	0.120	1.0857	1.0191	1.1276	0.7518	25.00	30.76	11.00
3¼	0.058	0.5816	1.1287	0.7410	0.4560	56.10	16.47	5.89
	0.065	0.6504	1.1263	0.8251	0.5077	50.00	18.40	6.58
	0.083	0.8258	1.1201	1.0361	0.6376	39.15	23.38*	8.35*
	0.095	0.9416	1.1160	1.1727	0.7217	34.20	26.66	9.52
	0.120	1.1800	1.1074	1.4472	0.8906	27.10	33.43	11.95
3½	0.065	0.7014	1.2147	1.0349	0.5914	53.80	19.85	7.09
	0.083	0.8910	1.2085	1.3012	0.7435	42.20	25.20	9.01
	0.095	1.0162	1.2043	1.4739	0.8422	36.85	28.70*	10.25*
	0.120	1.2742	1.1958	1.8220	1.0411	29.15	36.00	12.89
3¾	0.065	0.7525	1.3031	1.2777	0.6814	57.60	21.30	7.60
	0.083	0.9562	1.2968	1.6080	0.8576	45.20	27.06	9.67
	0.095	1.0908	1.2927	1.8228	0.9722	39.50	30.84*	11.04*
	0.120	1.3685	1.2841	2.2565	1.2035	31.25	38.70	13.82
4	0.065	0.8035	1.3915	1.5557	0.7779	61.50	22.75	8.12
	0.083	1.0214	1.3852	1.9597	0.9799	48.20	28.95	10.32
	0.095	1.1655	1.3810	2.2228	1.1114	42.10	32.95	11.78
	0.120	1.4627	1.3725	2.7552	1.3776	33.33	41.40*	14.80*
4¼	0.134	1.7327	1.4557	3.6732	1.7408	31.75	49.10*	17.55*
4½	0.156	2.1289	1.5369	5.0282	2.2347	28.80	60.40*	21.55*
4¾	0.188	2.6944	1.6143	7.0213	2.9563	25.25	76.25*	27.20*

* AN Standard Tubing

TABLE 2. ALLOWABLE STRENGTH OF PROTRUDING HEAD RIVETS IN SINGLE SHEAR, LB/RIVET

Sheet alloy		Clad 24S–T					
Rivet alloy		A17S–T				24S–T	
Rivet diameter		$\frac{3}{32}$	$\frac{1}{8}$	$\frac{5}{32}$	$\frac{3}{16}$	$\frac{3}{16}$	$\frac{1}{4}$
Hole diameter (d)		0.096	0.1285	0.159	0.191	0.191	0.257
Sheet thickness	0.020	202					
	0.025	210	357				
	0.032	217	374	552			
	0.040	217	386	574	803	870*	
	0.051	217	388	593	838	1,110*	1,490*
	0.064	...	388	596	862	1,180	1,970*
	0.072	...	388	596	862	1,180	2,080
	0.081	...	...	596	862	1,180	2,110
	0.091	...	...	596	862	1,180	2,120

* Bearing stress is critical for these values, and the allowable load is 114,000 td. All other values in table will apply for any sheet material stronger than the clad 24S–T, or with F_{br} greater than 114,000 psi.

Allowable shear strength obtained as F_{su} times the area of hole, when d/t is less than 3.0. When d/t exceeds 3.0, a factor of $[1 - 0.04(d/t - 3)]$ is used.

When d/t exceeds 5.5, the allowable loads must be obtained from tests.

TABLE 3. ALLOWABLE STRENGTH OF FLUSH RIVETED JOINTS, 100° HEAD ANGLE, LB/RIVET

Dimple countersunk rivets					
Sheet alloy	Clad 24S–T				
Rivet alloy	A17S–T				24S–T
Rivet diameter, in.	$\frac{3}{32}$	$\frac{1}{8}$	$\frac{5}{32}$	$\frac{3}{16}$	$\frac{3}{16}$
Sheet thickness 0.020	209	299			
0.025	235	360	474		
0.032	257	413	568	722	744
0.040	273	451	635	839	941
0.051		484	693	940	1,110
0.064			736	1,012	1,236
0.072			755	1,045	1,291

Machine Countersunk Rivets					
Sheet thickness 0.025	156				
0.032	178	272			
0.040	193	309	418		
0.051	206	340	479	628	758
0.064	216	363	523	705	886
0.072		373	542	739	942
0.081			560	769	992
0.091			575	795	1,035
0.102				818	1,073
0.125				853	1,131

INDEX